THE EVOLUTION OF THE MILKY WAY

ASTROPHYSICS AND SPACE SCIENCE LIBRARY

VOLUME 255

THE EVOLUTION OF THE MILKY WAY

edited by

FRANCESCA MATTEUCCI

Dipartimento di Astronomia,
University of Trieste, Italy

and

FRANCO GIOVANNELLI

Instituto di Astrofisica Spaziale,
Fuescoti, Italy

KLUWER ACADEMIC PUBLISHERS

DORDRECHT / BOSTON / LONDON

Library of Congress Cataloging-in-Publication Data

The evolution of the Milky Way : stars versus clusters / edited by Francesca Matteucci and Franco Giovannelli.
p. cm. -- (Astrophysics and space science library ; v. 255)
Includes index.
ISBN 0-7923-6679-4 (alk. paper)
1. Milky Way--Congresses. 2. Galaxies--Evolution--Congresses. I. Matteucci, F. II. Giovannelli, Franco. III. Series.

QB857.7 .E96 2000
523.1'13--dc21

00-049311

ISBN 0-7923-6679-4

Published by Kluwer Academic Publishers,
P.O. Box 17, 3300 AA Dordrecht, The Netherlands.

Sold and distributed in North, Central and South America
by Kluwer Academic Publishers,
101 Philip Drive, Norwell, MA 02061, U.S.A.

In all other countries, sold and distributed
by Kluwer Academic Publishers,
P.O. Box 322, 3300 AH Dordrecht, The Netherlands.

Printed on acid-free paper

Printed in the Netherlands.

TABLE OF CONTENTS

IV-HALO, BULGE AND DISK

V-STELLAR EVOLUTION AND NUCLEOSYNTHESIS

VI-FORMATION AND CHEMICAL EVOLUTION OF THE MILKY WAY

CONCLUSIONS

PREFACE

This book contains the proceedings of the meeting held in the lovely island of Vulcano, close to the noth coast of Sicily (Italy) on September 20-24, 1999. Vulcano is one of the group called Eolian islands. The name derives from the fact that the island is a volcano, and legend says that it was inhabited by the greek god Volcano.

The subject of the conference was "The Chemical Evolution of the Milky Way : Stars versus Clusters". This colloquium was a project devised by a group of italian astronomers (Franco Giovannelli, Francesca Matteucci and Sergio Ortolani) together with Beatriz Barbuy (Brazil) and R.M. Rich (U.S.A.), on the basis of the fact that all of them actively work in the field and that the last meeting, held in Italy on this subject, was the Elba meeting in 1989 on "Chemical and Dynamical Evolution of Galaxies". The reason for having the meeting in Vulcano was that the Istituto di Astrofisica Spaziale (Tor Vergata, Roma) has a long tradition of organizing astronomical meetings in Vulcano at the pleasant Eolian Hotel, which provides a secluded, beautiful and relaxing environment.

The main purpose of this workshop was to gather a representative community of astrophysicists involved in the study of the chemical evolution of the Milky Way both from a theoretical and an observational point of view. Concerning the observational point of view we wanted to give emphasis to the comparison between the information obtained for field stars on one hand and clusters on the other hand.

Star clusters can be used as tracers of Galactic chemical and dynamical evolution; it is possible to obtain their ages through Colour-Magnitude-Diagrams (CMDs), and therefore abundance derivation is made easier, since a star's location in the CMD constrains its stellar parameters (Teff, log g). The limited number of clusters in the Galaxy forces one however to rely more on field stars, which cover a wider range of metallicities and ages, and can be found in the solar neighbourhood. With the availability of Hipparcos data, it is possible now to date samples of nearby field stars, and, with the improvement of detectors and telescopes, it is possible to carry out detailed analyses of individual stars in a larger number of clusters and also a larger number of stars per cluster, as well as larger and more distant samples of field stars (e.g. abundances in B stars to trace gradients along the Galaxy radius). It appeared important to define, in this meeting, the best sets of abundances for cluster and field stars, to identify disagreements between cluster and field, and for which elements the data available provide the necessary requirements modelling of chemical evolution.

On the other hand, we were expecting that modellers of chemical evolution, would compare the results of their models for the Galaxy with the data and try to impose constraints on the mechanism of formation of the Milky Way, the final purpose of the conference.

In fact, what was discussed during the meeting, by means of many excellent reviews and contributed papers, was the most up- to- date information on abundances in field stars, HII regions, planetary nebulae, supernovae and globular clusters. Then we discussed the IMF in globular clusters as well as their relative ages, all important constraints for understanding the formation and evolution of the Galactic stellar halo. A large number of interesting talks were devoted to stellar evolution and nucleosynthesis in both single and binary stars, fundamental ingredients for computing the chemical evolution of galaxies. The histories of the star formation and the stellar mass function in the Milky Way and its satellites were also discussed. Finally, it was the turn of the various modellers who elaborated their models for the chemical evolution of the Milky Way. The encouraging conclusion was that there is a general agreement on the interpretation of the observational constraints in terms of the mechanism of formation and evolution of the Galaxy. An important characteristic of this conference was the large number of contributions presented by young researchers indicating that galaxy evolution is indeed an active and interesting astronomical field of research.

We think that the blue sky, the green sea and the late summer sicilian Sun have helped the participants to enjoy a fruitful and interesting conference, evidenced by the many enthusiastic comments we received after the conference.

A very important component fot the success of the conference was the excellent organization provided by the roman group; special thanks go to Lidia Barbanera and Massimo Frutti, without whom the conference could not have been held. Thanks also to the Trieste group, Dario Chiummariello (Department of Astronomy, Trieste University) and Simonetta Fabrizio (Astronomical Observatory of Trieste), who provided help during the conference. One of us (Francesca Matteucci) expressese her gratitude to her students Simone Recchi and Donatella Romano, as well as to Simonetta Fabrizio for help in editing this book. We also thank:

the Director of the Astronomical Observatory of Trieste, Fabio Mardirossian, together with the Council of the Observatory and,

the University of Trieste,

for financial support;

the director of the Istituto di Astrofisica Spaziale, Andrea Preite-Martinez, for providing personnel and computing facilities.

Franco Giovannelli and Francesca Matteucci

THE CHEMICAL EVOLUTION OF THE MILKY WAY:
STARS VERSUS CLUSTERS

Scientific Organizing Committee
Beatriz Barbuy (IAGUSP, Sao Paulo, Brazil)
Franco Giovannelli (IAS, Tor Vergata, Italy)
Francesca Matteucci (Trieste University, Italy)
Sergio Ortolani (Padova University, Italy)
Michael Rich (UCLA, U.S.A.)

Local Organizing Committee
Lidia Barbanera (IAS, Tor Vergata, Italy)
Franco Giovannelli (IAS, Tor Vergata, Italy)
Francesca Matteucci (University of Trieste, Italy)

LIST OF PARTICIPANTS

Carlos Abia	cabia@goliat.ugr.es
Massimo Badiali	badiali@ias.rm.cnr.it
Beatriz Barbuy	barbuy@orion.iagusp.usp.br
Michele Bellazzini	bellazzini@astbo3.bo.astro.it
Giuseppe Bono	bono@oat.ts.astro.it
Angela Bragaglia	angela@astbo3.bo.astro.it
Roberto Buonanno	buonanno@coma.mporzio.astro.it
W. Butler Burton	burton@strw.leidenuniv.nl
Giovanni Carraro	carraro@pd.astro.it
Eugenio Carretta	carretta@pd.astro.it
Santi Cassisi	cassisi@astrte.te.astro.it
Cristina Chiappini	chiappin@astro.columbia.edu
Alessandro Chieffi	achieffi@ias.rm.cnr.it
Cesare Chiosi	chiosi@pd.astro.it
John Danziger	danziger@oat.ts.astro.it
Massimo della Valle	massimo@arcetri.astro.it
Erwin De Donder	ededonde@vub.ac.be
Inma Dominguez	pk@huascaran.ugr.es
Michael Feast	MWF@artemisia.ast.uct.ac.za
Francesco R. Ferraro	ferraro@astbo3.bo.astro.it
Roberto Gallino	gallino@ph.unito.it
Gerard Gilmore	gil@ast.cam.ac.uk
Franco Giovannelli	franco@ias.rm.cnr.it
Léo Girardi	lgirardi@pleiadi.pd.astro.it
Raffaele G. Gratton	gratton@pd.astro.it
Michel Grenon	Michel.Grenon@obs.unige.ch
Garik Israelian	gil@iac.es
Chiaki Kobayashi	chiaki@galaxy.astron.s.u-tokyo.ac.jp
Ginluca Imbriani	imbriani@na.infn.it
Marco Limongi	marco@nemo.mporzio.astro.it
Walter J. Maciel	maciel@iagusp.usp.br
Andre Maeder	Andre.Maeder@obs.unige.ch
Maria Lucia Malagnini	malagnini@ts.astro.it
Claudia Maraston	maraston@usm.uni-muenchen.de
Marcella Marconi	marcella@na.astro.it
Paola Marigo	marigo@pd.astro.it
Francesca Matteucci	matteucci@ts.astro.it
Carlo Morossi	morossi@ts.astro.it
Emma Nasi	nasi@pd.astro.it

Livia Origlia	origlia@astbo3.bo.astro.it
Sergio Ortolani	ortolani@pd.astro.it
Bernard E.J. Pagel	bejp@star.cpes.susx.ac.uk
Nino Panagia	panagia@stsci.edu
Francesco Paresce	fparesce@eso.org
Randy L. Phelps	phelps@saclink.csus.edu
Giampaolo Piotto	piotto@pd.astro.it
Laura Portinari	portinari@pd.astro.it
Nikos Prantzos	prantzos@iap.fr
Luigi Pulone	pulone@coma.mporzio.astro.it
Vincenzo Ripepi	ripepi@cerere.na.astro.it
Donatella Romano	romano@sissa.it
Robert T. Rood	rtr@ninkasi.astro.virginia.edu
Sean G. Ryan	sgr@ast.cam.ac.uk
Stephen J. Smartt	sjst@ing.iac.es
Jesper Sommer-Larsen	jslarsen@tac.dk
Letizia Stanghellini	lstanghe@stsci.edu
Oscar Straniero	straniero@astrte.te.astro.it
Friedrich-Karl Thielemann	fkt@quasar.physik.unibas.ch
Daniel Thomas	daniel@usm.uni-muenchen.de
Monica Tosi	tosi@bo.astro.it
Takuji Tsujimoto	tsuji@misty.mtk.nao.ac.jp
Massimo Turatto	turatto@pd.astro.it
Dany Vanbeveren	dvbevere@vub.ac.be
Karel A. Van der Hucht	k.a.van.der.hucht@sron.nl
Paolo Ventura	paolo@coma.mporzio.astro.it
Patricia Whitelock	paw@saao.ac.za
Rosemary F.G Wyse	wyse@tardis.pha.jhu.edu
Manuela Zoccali	mzoccali@eso.org

I-ABUNDANCES IN STARS

During the lunch break ……..

THE CHEMICAL EVOLUTION OF THE MILKY WAY: SOME IMPORTANT FACTS

F. MATTEUCCI
Dipartimento di Astronomia, Università di Trieste
Via G.B. Tiepolo, 11, I-34100, Trieste
and SISSA/ISAS, Via Beirut 2-4, I-34100 Trieste

1. Abstract

I review the important observational constraints relevant to the chemical evolution of the Milky Way and compare the various theoretical interpretations given for these constraints in the past years. In particular, I stress the important role played by supernovae of different type in understanding the element- to- element ratios versus metallicity as well as the meaning of the G-dwarf metallicity distribution and the abundance gradients along the disk relative to the formation and the evolution of the Galaxy. Some open questions are addressed.

2. Introduction

What we call *the chemical evolution of the Galaxy* is the evolution in space and time of the abundances of the most common chemical species in the interstellar gas. Stars form out of the gas, evolve and synthesi He and heavier elements in their interiors. These elements are then restored into the interstellar medium (ISM) either through stellar winds or supernova (SN) explosions with the net effect of increasing the He and metal content of the gas. From this gas new stars will form and their chemical composition will be different from that of the previous generations. In principle, the process will continue until the exhaustion of the gas itself. Fundamental factors influencing the chemical evolution of galaxies in general are the stellar evolution and nucleosynthesis, the birthrate function, including the star formation rate (SFR) and the initial mass function (IMF) and possible gas flows (infall, outflows and radial flows).

The basic observational facts that allow us to impose constraints on the formation and the evolution of the Milky Way are:

F. Matteucci and F. Giovannelli (eds.), The Evolution of the Milky Way, 3–13.

- The relative abundances in stars in the solar neighbourhood and in the bulge ([X/Fe] versus [Fe/H]). From the measured trends we can infer the timescales for the formation and evolution of different components of the Galaxy (halo, disk, bulge) by using the abundance ratios as cosmic clocks, and at the same time we can impose constraints on the early phases of the Galactic evolution.
- The G-dwarf metallicity distribution, namely the distribution of the existing stars, imposes constraints on the mechanism of the formation of the disk in the solar vicinity. In particular, in the framework of infall models for the formation of the Galactic disk, the fit to the G-dwarf metallicity distribution can indicate the timescale of disk formation in the solar vicinity.
- Abundance gradients (abundances from HII regions, PNe, B stars, open clusters) along the Galactic disk constrain the mechanism of formation of the whole disk. In fact, it can be shown that models with a constant (relative to the galactocentric distance) timescale for disk formation do not reproduce abundance gradients, whereas an *inside-out* formation of the disk produces results in good agreement with observations.
- The relative ages of globular clusters give precious information about the timescale for the formation of the Galactic halo.

3. Abundances in stars

Overabundances of α-elements relative to Fe in metal poor stars in the solar neighbourhood are confirmed by all the most recent studies (e. g. Ryan this conference). In Figure 1 we report as an example a very large compilation of data relative to α-elements (O, Mg and Ca). Superimposed on the data are the predictions of Chiappini et al. (1999) relative to two different sets of yields for massive stars (Woosley and Weaver, 1995; Thielemann et al. 1996). The common interpretation of the overabundance of α-elements relative to Fe is the so-called *time-delay model*, based on the different timescales of production of the different elements (SNe Ia produce the bulk of Fe whereas SNe II produce the bulk of the α-elements plus some Fe (Tinsley, 1980, Greggio and Renzini, 1983 and Matteucci and Greggio, 1986). This interpretation is the most widely adopted now and it is based on the assumption of an IMF constant in space and time and that O and Fe are primary elements from the nucleosynthetic point of view, in other words that they are produced directly from H and He. In the time-delay model the plateau observed in the [α/Fe] ratios at low metallicities is interpreted to be due to the influence of the type II SNe alone, since at those early times the type Ia SNe have not yet started to occur, since they are thought to originate from white dwarfs in binary systems thus having stellar progenitors with masses below $8M_\odot$. Therefore, the abundance ratios should reflect the production ratios in massive stars. The decline of the abundance ratios in disk stars is then due to the occurrence of type Ia SNe restoring the bulk of Fe ($\sim$ 70 % of the total Fe production). It should be noted that at the moment there is no general agreement about the existence of a plateau in the [O/Fe] ratio at low metallicities, as indicated by the results of Israelian et al. (1998) who found a steep increase of [O/Fe] with decreasing metallicity. Personally, I find it strange that this effect is seen only in the [O/Fe] ratio and not in the other α-elements, since they are produced by the same massive stars, and therefore I would be cautious in drawing conclusions on this point. The data shown in Figure 1 do not contain the Israelian et al. data but nevertheless show a more marked increase of the [O/Fe] ratio with decreasing metallicity relative to the other α-elements. The theoretical models also show such an increase which is more marked for the lowest metallicities where the observations are still lacking, This is due to the adopted stellar yields, in particular those of massive stars which predict different ratios (particularly for O/Fe) as functions of the initial stellar mass. The comparison in Figures 1-2 shows that the time-delay model coupled with the most recent stellar yields produces good agreement with most of the observational data.

We should mention that Edmunds et al. (1991) proposed an alternative interpretation suggesting that the trend [α/Fe] versus [Fe/H] could be explained by suppressing the type Ia SNe and simply assuming a metallicity dependent yield of Fe. This interpretation is known as the *secondary-like behaviour of Fe* in SNe II. Edmunds et al. (1991) also proposed a test to discriminate between the time-delay model and the secondary-like behaviour of Fe, since the former predicts that the [α/Fe] ratios in different environments with different histories of star formation should be different (Matteucci, 1992), whereas the latter predicts the same behaviour irrespective of the star formation history. Therefore, the Galactic bulge represents a good place to test the two interpretations, since the star formation there should have been quite different from that in the solar vicinity. In particular, the secondary-like behaviour of Fe would predict for the bulge the same behaviour of the [α/Fe] versus [Fe/H] as in the solar neighbourhood, namely overabundance for the metallicities below [Fe/H] < -1.0 and steady decline for higher metallicities, irrespective of the assumed star formation rate.

On the other hand, the time-delay model for the bulge, where the star formation is assumed to have been faster and stronger than in the solar neighbourhood, predicts that most of the bulge stars should show overabundances of α-elements relative to Fe, since the pollution due to type Ia

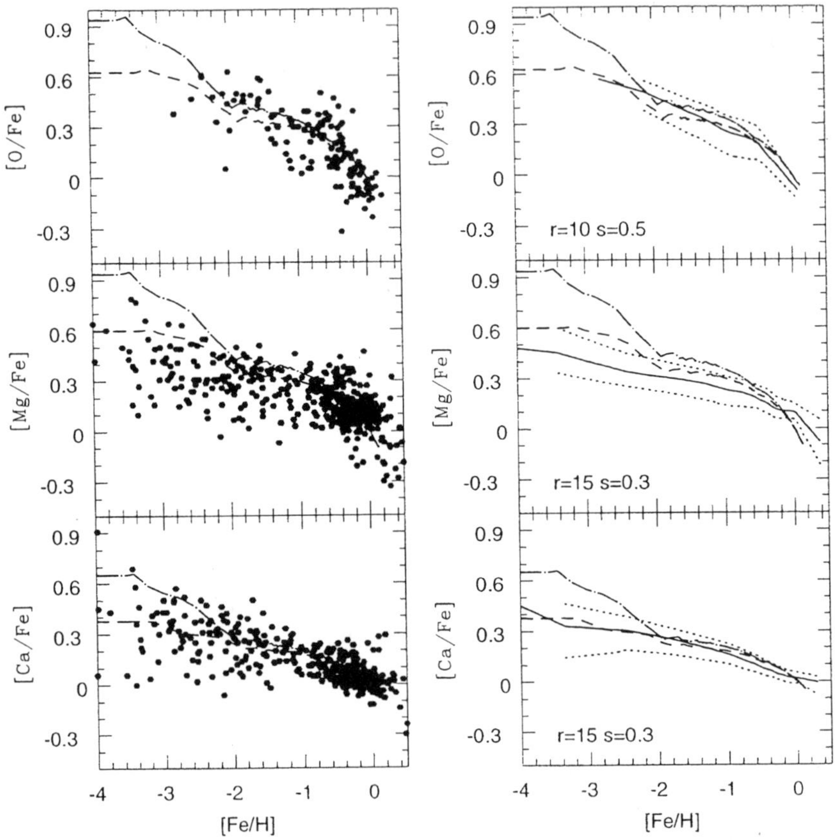

Figure 1. Observed [X/Fe] versus [Fe/H] (X=O, Mg and Ca) in solar vicinity stars. The data sample is homogeneous for what concerns the adopted solar abundances (Anders and Grevesse 1989), for the data source see Chiappini et al. (1999). The curves represent the predictions of the time-delay model when adopting different stellar yields in the domain of massive stars. In particular the dashed lines refer to the yields of Thielemann et al. (1996) whereas the dashed-dotted lines refer to the yields of Woosley and Weaver (1995). The left panels show a comparison between the model predictions and the bestfit to the data (solid line). The dotted lines indicate the area inside which the models should fall. For details see Chiappini et al. (1999).

SNe should arise when already most of the gas has been turned into stars (Matteucci and Brocato 1990; Matteucci et al. 1999). The assumed fast evolution of the Galactic bulge can be justified by the following argument: the deep bulge potential well could not allow self-regulation or blowout by pressures from young stars as in the disks or dwarf galaxies, and therefore the bulge formation should have been at the maximum rate, namely most

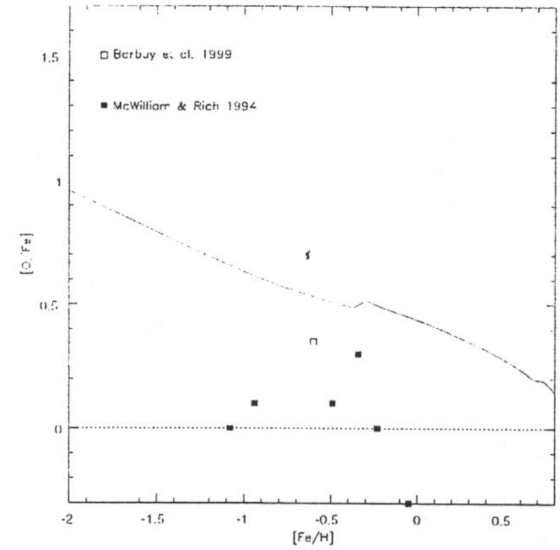

Figure 2. Chemical evolution of the Galactic bulge: predicted [O/Fe] versus [Fe/H] for the bulge stars by the time-delay models of Matteucci et al. (1999). The data points are from Barbuy (this conference) and McWilliam and Rich (1994).

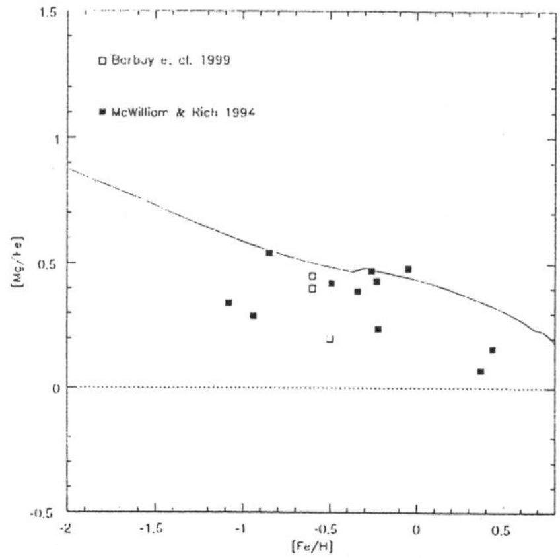

Figure 3. Predicted [Mg/Fe] versus [Fe/H] for the bulge stars by the time-delay models of Matteucci et al. (1999). The data points are from Barbuy (this conference) and McWilliam and Rich (1994).

of the gas must have been converted into stars in only a few dynamical timescales ($\sim 10^8$ years) (Elmegreen, 1999). In Figures 2,3,4,5 we show the predicted and observed [α/Fe] ratios for bulge stars. With the exception of oxygen for which the data are particularly uncertain and scarce, the predictions and the observations for Mg, Si and Ca seem to be in agreement.

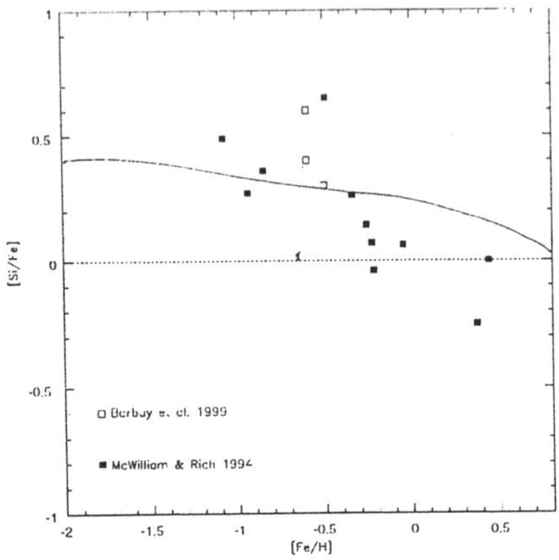

Figure 4. Predicted [Si/Fe] versus [Fe/H] for the bulge stars by the time-delay models of Matteucci et al. (1999). The data points are from Barbuy (this conference) and McWilliam and Rich (1994).

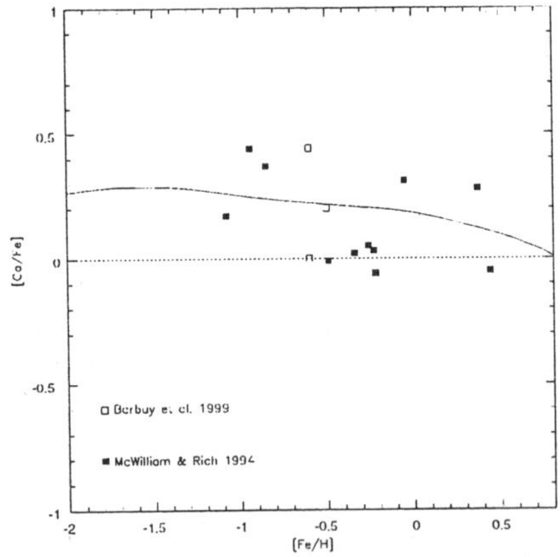

Figure 5. Predicted [Ca/Fe] versus [Fe/H] for the bulge stars by the time-delay models of Matteucci et al. (1999). The data points are from Barbuy (this conference) and McWilliam and Rich (1994).

4. Abundance gradients along the disk

It is reassuring that most of the more recent studies on HII regions, PNe and B stars (see Maciel, Smartt, this conference) agree on the value of the oxygen gradient along the Galactic disk of $\sim 0.07 dex/kpc$.

From the theoretical point of view there is mainly one way of producing the observed abundance gradients such as to assume a SFR strongly varying with galactocentric distance together with an *inside-out* formation of the disk (longer timescale of formation at larger radii) (Matteucci and François 1989; Boissier and Prantzos 1999). However, in general it is necessary to do a fine tuning of the various parameters to reproduce at the same time the gas distribution and the abundance gradients (see Portinari and Chiosi 1999). In particular, the gas distribution along the disk is not only the product of simple chemical evolution but mainly of the dynamical evolution of the disk, including the presence of a bar in the bulge and therefore a correct representation of such distribution awaits more complex chemo-dynamical approaches.

An alternative way of producing negative abundance gradients is to vary the IMF. However, plausibly variable (in space and time) IMFs do not seem to work (see Chiappini this conference), since any reasonable assumption (more massive stars formed in the lower metallicity regions) leads to the destruction of negative abundance gradients, as it is intuitive. Finally, radial flows can improve the gradients but cannot be entirely responsible for their formation. In particular, radial flows coupled with a radial variation of the star formation rate can in principle reproduce the observed gradients even with a constant timescale of disk formation (see Portinari, this conference), although from a physical point of view the *inside-out formation* of the galactic disk seems more physical and in agreement with dynamical models.

5. Other constraints

Other important constraints on the chemical evolution of the Milky Way are:

- the recent finding of Nissen and Schuster (1997) that some halo stars, probably born in the outer halo, have lower [α/Fe] ratios than the stars with the same metallicities in the solar neighbourhood. This is a possible indication of inside-out formation of the halo, namely that the outer halo formed more slowly than the inner one leaving open the possibility that it could have formed by slow accretion from satellites. The difference in ages among the globular clusters in the inner and outer halo goes in the same direction ($\sim 2-4$ Gyrs, Lee et a. 1994, Piotto this conference).
- Edvardsson et al. (1993) found lower [α/Fe] ratios in stars born in the outer disk and this could be interpreted to be due to the inside-out formation of the disk, since in this case the outermost Galactic disk should have had the slowest evolution, thus attaining lower [α/Fe] ratios than the innermost regions for the same [Fe/H]. This is a direct consequence of the time-delay model coupled with the differential formation of the disk.

– Wyse and Gilmore (1992) and Ibata and Gilmore (1995) showed that the distribution of the angular momentum per unit mass in the halo stars is quite different from that of the stars in the disk (thick+thin) whereas it is very similar to that of the stars in the bulge. This implies that halo and bulge formed out of the same gas whereas the disk formed out mostly of extragalactic gas! This finding rules out all of those models for Galaxy formation which imply that the disk formed out of gas shed by the halo.

6. Different Approaches to the study of Chemical Evolution

Most of the existing models of the chemical evolution of the Galaxy refer only to the disk evolution and the reason is that it is extremely difficult to model the halo. This is because the halo and disk must have evolved almost independently but the observed abundances and abundance ratios do not show clear discontinuities. However, in the plot [Fe/O] versus [O/H] distribution, as suggested by Gratton et al. (1996), one could envisage a discontinuity occurring at [O/H] $\sim$ -0.3 dex, indicating that the O abundance stayed constant for a certain period of time while the Fe abundance continued to increase (this is shown by an increase in the [Fe/O] ratio at a fixed [O/H]). This same behaviour has been noticed more recently also in the plot [Fe/Mg] versus [Mg/H] by Furhmann (1998).This is a clear sign that a hiatus in the star formation rate occurred probably during the transition from the halo to the disk.

In the past, attempts to model both the halo and disk evolution have been made using an approach which I will define as a *serial approach*, with the Galaxy forming as a continuous process and the halo representing just the early phases of a unique infall event (e.g. Matteucci and François, 1989).

Later on, Pardi et al. (1995) developed what I will refer to as the *parallel approach*, where the various Galactic components start forming at the same time and from the same gas but then evolve at different rates according to their different star formation histories.

These two approaches seem now to be ruled out by the finding that the gas of the halo did not form the disk and therefore a different approach has been devised, the *two-infall approach*, where the evolution of the halo and disk are totally disentangled since they formed during independent infall episodes (Chiappini et al. 1997; 1999).

Another recent interesting approach is the *stochastic approach* where it is assumed that the gas mixes only on small scales ($\sim 10^5 M_\odot$) and allows different Galactic regions of small size (compared to previous models) to evolve at slightly different rates (Copi 1997). This approach is mainly devised to predict some spread in the observed abundances. Other models assuming an inhomogeneous halo evolution have been recently suggested, such as the one of Tsujimoto et al. (this conference).

Still a lot of work needs to be done to clarify the mechanism of Galaxy formation and in the future this cannot be done without some dynamical treatment coupled with the chemical one. However, I stress that in all the previously discussed approaches there is a *necessary* assumption which is the time-delay between the production of Fe and α-elements; this implies that the stellar yields and SN progenitors are of fundamental importance, and unfortunately these quantities are still known with large uncertainties (see Chieffi, Thielemann, this conference) and perhaps chemical evolution models could be used to impose constraints on the yields. For example, the percentage of Fe contributed by type Ia SNe (one of the most uncertain quantities) to the solar Fe abundance should be not less than $60-70\%$, in order to explain the majority of the [X/Fe] vs. [Fe/H] independently of the particular approach adopted to model the chemical evolution of the Milky Way, as shown by most of the previously mentioned models. Alternatively, we could use stellar yields of Fe empirically derived from SNe (see Danziger, this conference).

7. Open questions

I would like to conclude by mentioning some of the still open questions relevant to the chemical evolution of the Galaxy, hoping that some answers will emerge from this conference. The questions are:

- 1 Do the cluster abundances tell us the same story as the star abundances?
- 2 Is the "plateau" in the [α/Fe] ratios in halo stars real? What globular clusters tell us about it?
- 3 Do we really need an inhomogeneous halo evolution to explain the observed spread in some abundance ratios at low metallicities?
  4 Can we infer a consistent picture about the formation and the evolution of the Milky Way by means of information on abundances and kinematics of stars?
- 5 Do we understand the progenitors of type Ia SNe? Do we understand the evolution of binary systems?
- 6 Do we have a *unique model* to explain the chemical evolution of the Galaxy?

I would like to thank Cristina Chiappini and Donatella Romano for their scientific contribution to this paper and their patience in assisting me during the organization of this conference.

References

1. Anders, E., Grevesse, N. 1989, *Geochimica and Cosmochimica Acta*, 53, 197
2. Boissier, S. Prantzos, N. 1999, *MNRAS*, 307, 857
3. Chiappini, C., Matteucci, F., & Gratton, R. G. 1997, *ApJ*, 477, 765
4. Chiappini, C., Matteucci, F., Beers, T. C., & Nomoto, K. 1999, *ApJ*, 515, 226
5. Copi, C. J. 1997, *ApJ*, 487, 704
6. Edmunds, M. G., Greenhow, R. M., Johnson, D., Klueckers, V., & Vila, M. B. 1991, *MNRAS*, 251, 33
7. Elmegreen, B. G. 1999, *ApJ*, 517, 103
8. Edvardsson, B., Andersen, J., Gustasfsson, B., Lambert, D. L., Nissen, P. E., & Tomkin, J. 1993, *A&A*, 275, 101
9. Furhmann, K. 1998, *A&A*, 338, 161
10. Gratton, R. G., Carretta, E., Matteucci, F., & Sneden, C. 1996, in *Formation of the Galactic Halo... Inside and out*, eds. Morrison, H. Sarajedini, A., ASP Conf. Series 92, 307
11. Greggio, L., Renzini, A. 1983, in *The First Stellar Generations*, Mem. Soc. Astron. It., Vol. 54, p.311
12. Ibata, R. A., Gilmore, G. 1995, *MNRAS*, 275, 605
13. Israelian, G., Garcia-Lopez, R. J., & Rebolo, R. 1998, *ApJ*, 507, 805
14. Lee, Y.W., Demarque, P., & Zinn, R. 1994, *ApJ*, 423, 248
15. Matteucci, F. 1992,in *Morphological and Physical Classification of Galaxies*, ed. G. Longo & al., Kluwer Academic Publ., p.245
16. Matteucci, F., Brocato, E. 1990, *ApJ*, 365, 539
17. Matteucci, F., François, P. 1989, *MNRAS*, 239, 885
18. Matteucci, F., Greggio, L. 1986, *A&A*, 154, 279
19. Matteucci, F., Romano, D., & Molaro, P. 1991, *A&A*, 341, 458
20. McWilliam, A., Rich, R. M. 1994, *ApJ Suppl.*, 91, 749
21. Nissen, P. E., Schuster, W. J. 1997, *A&A*, 326, 751
22. Pardi, C., Ferrini, F., & Matteucci, F. 1995, *ApJ*, 444, 207
23. Portinari, L., Chiosi, C. 1999, *astro-ph/9908326*
24. Tinsley, B.M. 1980, *Fund. Cosmic Physics*, 5, 287
25. Thielemann, F. K., Nomoto, K., & Hashimoto, M. 1996, *ApJ*, 460, 480
26. Woosley, S. E., Weaver, T. A. 1995, *ApJ*, 101, 181
27. Wyse, R. F. G., Gilmore, G. 1992, *ApJ*, 104, 144

Speaker: F. MATTEUCCI

A. MAEDER:
You have shown this very interesting starburst model for the bulge. What do you consider as corresponding evidences (and predictions) for highly redshifted galaxies?

F. MATTEUCCI:
I do not think that the high redshift objects observed insofar (e.g. Damped Lyman-α Systems) can represent the early stages of bulge evolution but rather the outer regions of galactic disks. From the theoretical point of view I expect that high redshift bulges should show a very high overabundance of α-elements relative to Fe in a very narrow redshift range.

B. PAGEL: Before interpreting O/Fe as a function of metallicity it is important to establish the actual behaviour, since abundances from forbidden lines are quite different from what you showed.

F. MATTEUCCI: Yes, I agree, although the abundances from forbidden lines show a behaviour of oxygen quite different from all the other α-elements and this is a bit strange.

F.K. THIELEMANN's Comment: I want to support fully what Alessandro Chieffi said. We do not have any self-consistent explosion model of type II SNe, yet. Any Ni/Fe ejecta, which are the products of explosive burning closest to the neutron star mass cut, can only be based on "reasonable" estimates on the position of the mass cut (our prescription) or positioning a "piston" (Woosley and Weaver). Where possible, we based our choice on light curve observations (^{56}Ni) of SNeII with known progenitor mass.

R. GALLINO: There has been a recent theoretical suggestion that all AGB stars would produce a substantial contribution of oxygen, up to $\sim 20\%$ of the solar oxygen (Herwig et al. 1999). Would this have an influence on the analysis of the galactic oxygen trend?

F. MATTEUCCI: I do not know the paper and to answer precisely I should try the predicted yields in my chemical evolution code. What I can say now is that a substantial oxygen production from AGB stars would lower the predicted [O/Fe] at low [Fe/H].

ABUNDANCE RATIOS IN GALACTIC METAL-POOR STARS

S.G. RYAN
Physics Department, The Open University, Walton Hall, Milton Keynes MK7 6AA, UK *s.g.ryan@open.ac.uk*

Abstract. Metal-poor stars provide information on the characteristics and chemical evolution of the halo population of the Galaxy, the first epoch of star formation and Galaxy formation (not just locally but with relevance to high-redshift objects), and big bang nucleosynthesis. This paper looks at recent developments in this subject. A more comprehensive review can be found elsewhere (Ryan 2000).

1. Introduction

Halo stars constitute the oldest and most extended stellar population *known* in the Galaxy, and have the lowest metallicities. They provide the first data in the sequence of Galactic chemical evolution (GCE).

The halo field and globular cluster metallicity distributions (Laird et al. 1988a, Laird, Carney & Latham 1988b; Ryan & Norris 1991) are *lower* in metallicity than the damped Lyman-α systems (DLAs) having redshifts $z \sim 1$–3 (Pettini et al. 1997). That is, very metal poor stars are amongst the lowest metallicity objects in the known universe. There is debate about what the DLAs are: spiral disks/protodisks/thick disks (e.g. Wolfe et al. 1986; Lu et al. 1996), dwarf galaxies (Pettini, Boksenberg & Hunstead 1990; Pettini et al. 1999a), or ejecta from dwarf galaxies (Nulsen, Barcons & Fabian 1998). Galactic stars, on the other hand, have well understood histories and physical states, have unblended spectra, and their abundances are measurable for many elements, unaffected by depletion onto dust grains.

The value of halo stars for studying early epochs of the universe may be further illustrated by considering metallicity-redshift space. Figure 1 shows a number of Galactic and high-redshift objects, along with three GCE models assuming Galaxy formation begins at redshift $z = 5$ (Edmunds & Phillipps 1997). ($H_0 = 50$ km s^{-1} Mpc^{-1} and a flat cosmology

F. Matteucci and F. Giovannelli (eds.), The Evolution of the Milky Way, 15–26.

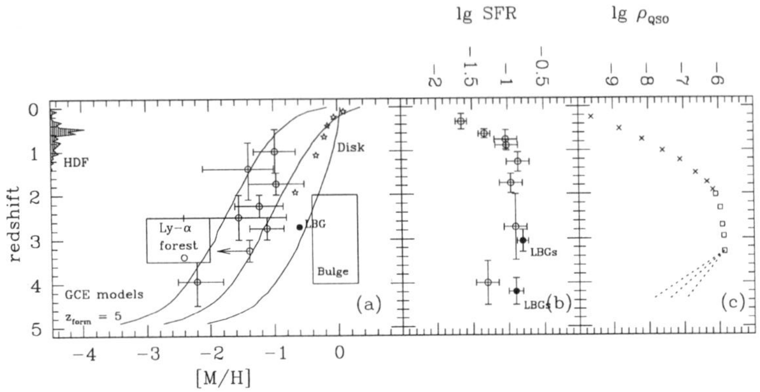

Figure 1. Metallicity distributions of systems. **(a)** Galactic and high-redshift objects in the redshift–metallicity plane, the former located at their redshift of formation (assuming $H_0 = 50$ km s^{-1} Mpc^{-1}). *solid curves* = GCE models; *open circles* = DLAs; *stars* = disk star sequence; *solid circle* = LBG. **(b)** Star formation history of extragalactic objects. **(c)** Quasar space density. (See text for discussion)

were used for the age-redshift relation.) The Galactic disk and bulge data are from the observations of Edvardsson et al. (1993) and Sadler, Rich & Terndrup (1996). The high-redshift objects are the DLAs from Pettini et al. (1997) binned in redshift, Molaro et al. (1996) and Lu et al. (1996), a region corresponding to the Lyman-α forest (Hellsten et al. 1997), and one Lyman break galaxy (LBG) from Pettini et al. (1999b). The redshift distribution in the Hubble Deep Field (Bouwens, Broadhurst & Silk 1998) is shown as an inset. In adjoining panels are the extinction-corrected star formation rate (Steidel et al. 1999) and quasar space density (Warren, Hewitt & Osmer 1994) whose steep fall at $z > 3.5$ indicates that these objects were still forming prior to this epoch, presumably along with galaxies. Several points can be noted.

- The overlapping of high redshift and Galactic objects in this epoch-metallicity plane (redshift translates to epoch), e.g. the LBG and the Galactic bulge, and the Lyman-α forest and metal poor stars, emphasises that these objects provide complementary views on the formation and evolution of galaxies. No one class should be considered in isolation from the others.
- Galactic stars with [Fe/H] < -3, corresponding to $z \gtrsim 4$–5, uniquely probe the earliest star formation. Many elements can be measured in these well understood objects. Furthermore, they owe their existence to very few previous generations of stars, possibly only one (Ryan, Norris & Bessell 1991). They mark the beginning of GCE, and as such will be the main topic of this review.

2. Enrichment by a Single Supernova

Searle's (1977) stochastic GCE model, fit to the globular cluster metallicity distribution, implies a mean of ten enrichments per fragment (Ryan & Norris 1991), i.e. that a single enrichment could produce [Fe/H] $\simeq -2.5$.

Star-to-star variations in neutron-capture abundances in stars having [Fe/H] < -2.5 (e.g. Gilroy et al. 1988) suggested that prior nucleosynthesis involved small numbers of supernovae (SN). Ryan et al. (1991) calculated the metallicity produced by a single supernova as [Fe/H] $= -3.8$, based on a SN II progenitor mass of 25 $M_\odot$ and a primordial cloud mass of 10^6 $M_\odot$.[1] This coincided with the huge abundance variations they observed for strontium, and was comparable to the lowest stellar metallicities.

The star-to-star differences also showed that chemical inhomogeneities existed when these stars formed. Audouze & Silk (1995) showed that mixing timescales in the halo were long, explaining why inhomogeneities were not erased, and supporting the proposition that the progeny of the first supernovae would be found around [Fe/H] $= -4$. Mounting examples of neutron-capture element variations from star to star (e.g. Norris, Peterson & Beers 1993; McWilliam et al. 1995; Ryan 1996; Ryan et al. 1996) strengthened the view that the most metal-poor stars exhibit the ejecta of very few supernova.

Adopting the supernova remnant (SNR) model of Cioffi, McKee & Bertschinger (1988), Ryan, Norris & Beers (1996) calculated the cloud mass with which SN ejecta would mix as

$$m_{\rm ISM} = 3.4 \times 10^4 E_{51}^{0.95} n_0^{-0.10} \zeta_m^{-0.15} (\beta C_{06})^{-1.29} \quad \mathrm{M}_\odot,$$

where E, n, ζ, and βC_{06} refer to the explosion energy, cloud density, cloud metallicity, and shock speed. The main features of this result were:

- the mass of gas enriched is almost independent of the cloud characteristics (n, ζ) and depends strongly (almost linearly) on the energy of the SN;
- the typical enriched mass of the cloud would be 7×10^4 $M_\odot$;
- the typical metallicity following this first enrichment would be [Fe/H] $= -2.7$, matching changes in the behaviour of iron-peak and neutron-capture elements and the lowest metallicity globular clusters.

Recent GCE models have attempted to include SNR physics and inhomogeneous mixing (e.g. Shigeyama & Tsujimoto 1998; Tsujimoto & Shigeyama 1998; Ishimaru & Wanajo 1999a; Argast & Samland 1999). It will be possible to use observed abundances to constrain the model inputs, such

[1]The SN mass was a compromise between higher mass stars being rarer and lower mass stars having lower yields. The cloud mass was based on large globular clusters, giant molecular clouds, and the collapse of metal poor gas.

as the shape of the IMF of their progenitors (Population III stars!), mass limits for SN of Pop III stars, and their yields.

3. The Lightest Elements

In the context of globular clusters versus field stars, it is useful to compare M92 subgiants (Boesgaard et al. 1998) and field halo turnoff stars (Ryan, Norris & Beers 1999); see Figure 2. The latter show no intrinsic spread ($\sigma_{\rm int} < 0.02$ dex) once a small metallicity dependence (believed to be due to GCE) is taken into account, whereas M92 shows a range of Li. This suggests a difference between field and globular cluster populations, presumably related to their very different environments, specifically the stellar density. Although single stars seldom exchange angular momentum, protostellar disks may have done so in nascent globular clusters but not in lower density clusters destined ultimately to produce field stars (Kraft 1998, private communication).

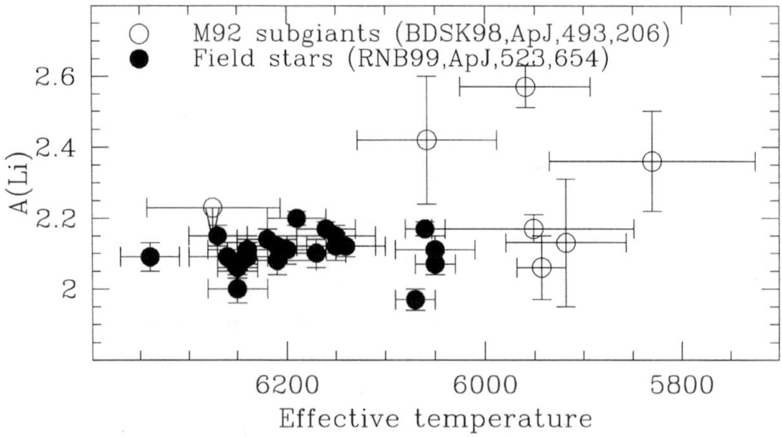

Figure 2. Lithium abundances in very metal-poor halo field stars near the turnoff , compared with globular cluster subgiants in M92.

4. Intermediate-Mass Elements

Figure 3 shows O abundances derived for giants and dwarfs from the forbidden lines [O I], near-infrared triplet O I, and ultraviolet OH lines.

Results for the infra-red triplet have been diverse, prompting arguments concerning inconsistencies between different lines (e.g. Magain 1988), $T_{\rm eff}$

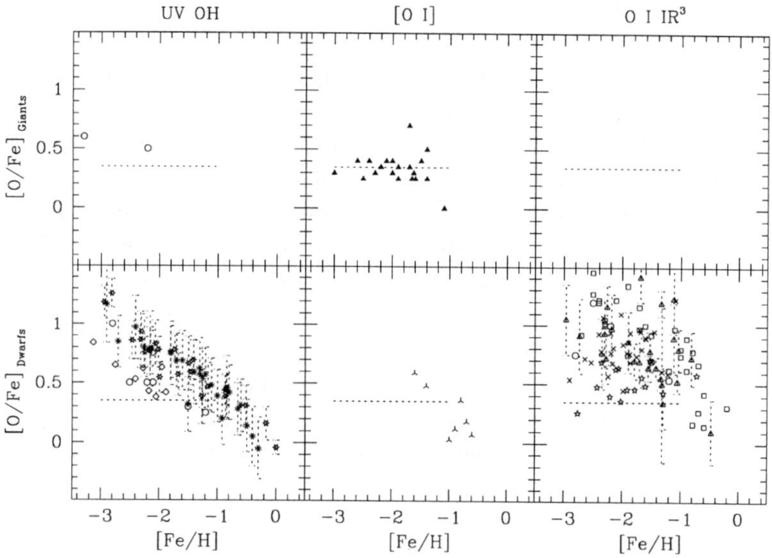

Figure 3. Oxygen in metal poor field stars. *circles:* Bessell, Sutherland & Ruan 1991; *filled triangles:* Barbuy 1988; *diamonds:* Nissen et al. 1994; *six-pointed stars:* Boesgaard et al. 1999; *asterisks:* Israelian, García Lopéz & Rebolo 1998; *inverted 'Y's:* Spite & Spite 1991; *squares:* Abia & Rebolo 1989; *crosses:* Tomkin et al. 1992; *five-pointed stars:* King 1993; *open triangles:* Cavallo, Pilachowski & Rebolo 1997

dependences, NLTE effects, and errors in the $T_{\rm eff}$ scale and equivalent widths. The lack of agreement attests to the difficulties in using these lines. Two recent studies using the UV OH lines (Israelian et al. 1998; Boesgaard et al. 1999) found trends towards high O overabundances in the most metal-poor stars. Bessell, Sutherland & Ruan (1991) concluded that their OH line measurements were consistent with Barbuy's (1988) giants. However, Bessell et al's dwarfs, and Nissen et al's (1994), lie parallel to the recent OH trend, albeit displaced by $\simeq -0.3$ dex. Gratton (1999) warns of the difficulty of fitting the continuum near the UV OH lines for the more metal-rich stars, and the need for accurate molecular line parameters.

Oxygen and the α elements are formed separately, so do not *have* to exhibit the same dependence on metallicity, but current models predict that they do (e.g. Timmes, Woosley & Weaver 1995). While the Fe yields in the models could be wrong, the OH observations require that $[O/\alpha]$ is also strongly dependent upon metallicity. So, if the theoretical yields are to fit the OH line data, significantly higher O production will be required, irrespective of what changes are made to Fe.

Carbon and nitrogen measurements in very metal-poor stars (Israelian

& Rebolo 1999) give approximately solar ratios. How the primary nitrogen is produced is unclear. Significant numbers of C-rich stars have been found at the lowest iron abundances (Beers, Preston & Shectman 1992); while CS 22892-052 has r-process element overabundances (Sneden et al. 1994, 1996), the majority exhibit s-process patterns (Norris, Ryan & Beers 1997a; Barbuy et al. 1997; Norris et al. 2000). Note, however, that not all C-rich stars exhibit heavy element anomalies (Norris, Ryan & Beers 1997b).

Recent studies of the α-elements (Mg, Si, Ca) have concentrated on star-to-star variations, with King (1997), Carney et al. (1997), and Nissen & Schuster (1997) identifying stars with low $[\alpha/\mathrm{Fe}]$ ratios, predominantly retrograde kinematics, and young ages. Low $[\alpha/\mathrm{Fe}]$ ratios resemble those proposed by Matteucci & Brocatto (1990) and Gilmore & Wyse (1991) for when star formation rates are low, and SN Ia appear, leading to low $[\alpha/\mathrm{Fe}]$ ratios, before metallicities typical of Galactic disk stars are reached.[2] The dual halo models of Zinn (1993) and Norris (1994) fit the characteristics of the stars now observed. Clearly there is much to learn about the Galactic halo from relative abundances.

Hanson et al. (1998) have found differences for [Na/Fe] and [Mg/Fe] in globular cluster and field stars of the same evolutionary state. They find the two elements to be positively correlated in the field, but in clusters they find an overall anti-correlation suggestive of the proton-capture chain. Seeing these changes at the stellar surface suggests that deep mixing has occurred in the globular cluster giants but not in the field giants. Why this should be so has yet to be resolved; the Li results (§3) may be related.

5. The Iron-Peak Elements

The iron-peak elements usually observed in very metal-poor stars are Sc, Ti, Cr, Mn, Fe, Co, and Ni. Ti is believed to be produced in the same region of stars as the other iron-peak elements, but its abundance appears more like the α-elements[3]. Mn and Cr become very underabundant in stars with [Fe/H < -2.5 (McWilliam et al. 1995; Ryan et al. 1996), while Co becomes overabundant. These changes were not predicted by supernova computations.

The iron-peak species are synthesised close to the mass-cut. Efforts to understand the abundance trends have focussed on the possible α-rich freeze-out (Woosley & Hoffman 1992; McWilliam et al. 1995), the location of the mass-cut (Nakamura et al. 1999), and the dependence of yields on

[2]There is possible conflict with the work of Kobayashi et al. (1998) who propose that the SN Ia mechanism has lower efficiency at [Fe/H] < -1 due to weaker winds.

[3]Truran (1999, private communication) reports that the experimental $^{44}\mathrm{Ti}(\alpha,\mathrm{p})^{47}\mathrm{V}$ rate is five times higher than the theoretical one used in previous nucleosynthesis calculations, so the theoretical yields of some iron-peak elements will be subject to revision.

stellar mass, metallicity, and neutron excess (Nakamura et al. 1999; Hix & Thielemann 1996). The explanation must not distort the [X/Fe] ratios of species formed well outside the mass-cut, such as the α-elements; this is particularly challenging for models which alter the Fe yield! 3D hydrodynamical models may provide more realistic treatments of the explosion (e.g. Müller, Fryxell & Arnett 1991); observational evidence for mixing from the mass-cut to the photosphere (Stathakis 1996; Fassia et al. 1998) provides additional clues and constraints.

6. Neutron-Capture Elements

Spite & Spite (1978) showed that the Ba in halo stars owed its origin primarily to the r-process. Truran (1981) argued that the s-process should be viewed as a secondary process while the r-process was primary, and thus that first generation — Population III — stars would not execute the s-process. Gilroy et al. (1988) showed that the neutron-capture elements exhibited abundance patterns more consistent with the r-process than the s-process. McWilliam et al. (1995, 1998) extended the Ba and Eu comparison to the lowest metallicities ([Fe/H] $= -4$), confirming the existence of r-process (rather than s-process) ratios.

Spectral line profiles can in rare cases provide constraints on isotope ratios. Magain (1995) inferred an s-process rather than r-process pattern for Ba in HD 140283, at [Fe/H] $= -2.6$. In contrast, Gacquer & Francois (1998, private communication) find an r-process signature for this star.

The reliance on Ba and Eu as diagnostics of the s- vs r-process fractions has diminished with *HST* UV spectra for other elements including Ag, Pt, Os, and Pb (Crawford et al. 1998; Cowan et al. 1996; Sneden et al. 1998).

Gilroy et al. (1988) also observed significant star-to-star variations in the most metal poor objects ([Fe/H] $\lesssim -2.5$), building upon earlier cases. Variations are greatest for Sr, where ranges of a factor of more than 100 were found amongst dwarfs (Ryan et al. 1991) and giants (Norris et al. 1993). Moreover, as Figure 4 shows, such extreme variations are *not* shared by Ba, which exhibits a fairly steady trend towards lower [Ba/Fe] at lower [Fe/H].

Reconciling Sr and Ba is a challenge. If the r-process alone is responsible, then Figure 4 tells us that the r-process cannot be universal. Whilst there is no theoretical reason why it *must* be universal, halo star abundance patterns generally resemble the r-process contribution in the sun (e.g. Gilroy et al. 1988; McWilliam et al. 1995; Cowan et al. 1995, Sneden et al. 1996). We seek a source of additional Sr in the lowest metallicity objects involving low neutron exposures that synthesise only species around the atomic number of Sr (Ryan et al. 1996, 2000; Ishimaru & Wanajo 1999b).

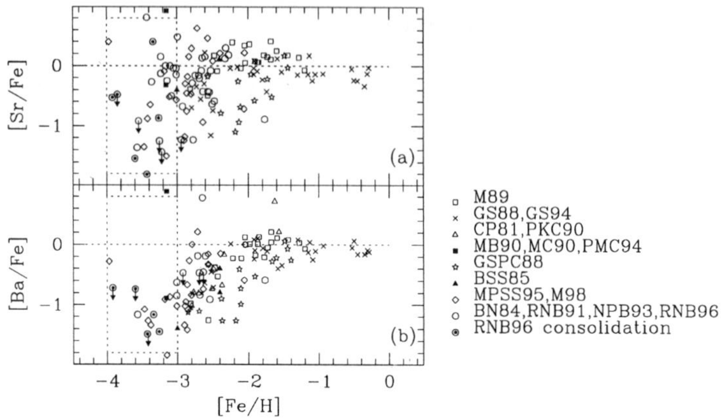

Figure 4. Abundances of Sr and Ba in halo stars. The dashed box outlines the same region of [X/Fe] and [Fe/H] in each plot. Whereas [Sr/Fe] exhibits a range by a factor of more than 100, [Ba/Fe] has a well-behaved trend towards lower values at the lowest [Fe/H].

The decrease in [Ba/Fe] and [Eu/Fe] at [Fe/H] < -2 indicates that these elements are not produced abundantly in the most massive supernovae. Mathews & Cowan (1990) associated the production with 10–11 $M_\odot$ stars. Although the evolution in their models is now seen to be too steep, the concept of low-mass stars being responsible has survived, and a more gradual emergence is seen in the models of Travaglio et al. (1999) which use a 8–10 $M_\odot$ range for the process. Travaglio et al. also treat the enrichment of the cloud inhomogeneously, along the lines described in §2 of this review, as do Ishimaru & Wanajo (1999a).

Assuming that the neutron-capture elements in the r-process-rich star, CS 22892-052, have the same proportions as in the sun, Cowan et al. (1997) derive as its age 17±4 Gyr, based on radioactive Th. This is consistent with globular cluster ages from isochrone fitting of $12.8^{+4.2}_{-2.8}$ Gyr (Chaboyer 1999). Improved data and GCE models will hopefully provide a more accurate Th chronometer.

References

1. Abia, C. & Rebolo, R. 1989, ApJ, 347, 186
2. Argast, D. & Samland, M. 1999, The Galactic Halo: from Globular Clusters to Field Stars, 35th Liège Int. Astroph. Coll., in press
3. Audouze, J. & Silk, J. 1995, ApJ, 451, L49
4. Barbuy, B. 1988, A&A, 191, 121
5. Barbuy, B. Cayrel, R., Spite, M., Beers, T. C., Spite, F., Nordstrom, B. & Nissen, P. E. 1997, A&A, 317, L63

6. Barbuy, B., Spite, F., & Spite, M. 1985, A&A, 144, 343
7. Beers, T. C., Preston, G. W. & Shectman, S. A. 1992, AJ, 103, 1987
8. Bessell, M. S. & Norris, J. E. 1984, ApJ, 285, 622
9. Bessell, M. S., Sutherland, R. S. & Ruan, K. 1991, ApJ, 383, L71
10. Boesgaard, A. M., Deliyannis, C. P., Stephens, A. & King, J. R. 1998, ApJ, 493, 206
11. Boesgaard, A. M., King, J. R., Deliyannis, C. P. & Vogt, S. S. 1999, AJ, 117, 492
12. Bouwens, R., Broadhurst, T., & Silk, J. 1998, ApJ, 506, 557
13. Carney, B. W. & Peterson, R. C. 1981, ApJ, 245, 238
14. Carney, B. W., Wright, J. S., Sneden, C., Laird, J. B., Aguilar, L. A. & Latham, D. W. 1997, AJ, 114, 363
15. Cavallo, R. M., Pilachowski, C. A. & Rebolo, R. 1997, PASP, 109, 226
16. Chaboyer, B. 1999, The Galactic Halo: from Globular Clusters to Field Stars, 35th Liège Int. Astroph. Coll., in press
17. Cioffi, D. F., McKee, C. F. & Bertschinger, E. 1988, ApJ, 334, 252
18. Cowan, J. J., Burris, D. L., Sneden, C., McWilliam, A. & Preston, G. W. 1995, ApJ, 439, L51
19. Cowan, J. J., McWilliam, A., Sneden, C., & Burris, D. L. 1997, ApJ, 480, 246
20. Cowan, J. J., Sneden, C., Truran, J. W. & Burris, D. L. 1996, ApJ, 460, L115
21. Crawford, J. L., Sneden, C., King, J. R., Boesgaard, A. M. & Deliyannis, C. P. 1998, AJ, 116, 2489
22. Edmunds, M. G. & Phillipps, S. 1997, MNRAS, 292, 733
23. Edvardsson, B., Andersen, J., Gustafsson, B., Lambert, D. L., Nissen, P. E. & Tomkin, J. 1993, A&A, 275, 101
24. Fassia, A., Meikle, W. P. S., Geballe, T. R., Walton, N. A., Pollaco, D. L., Rutten, R. G. M. & Tinney, C. 1998, MNRAS, 299, 150
25. Gilmore, G. & Wyse, R. F. G. 1991, ApJ, 367, L55
26. Gilroy, K. K., Sneden, C., Pilachowski, C. A. & Cowan, J. J. 1988, ApJ, 327, 298
27. Gratton, R. 1999, The Galactic Halo: from Globular Clusters to Field Stars, 35th Liège Int. Astroph. Coll., in press
28. Gratton, R. G. & Sneden, C. 1988, A&A, 204, 193
29. Gratton, R. & Sneden, C. 1994, A&A, 287, 927
30. Hanson, R. B., Sneden, C., Kraft, R. P. & Fulbright, J. 1998, AJ, 116, 1286
31. Hellsten, U., Dave, R., Hernquist, L., Weinberg, D. H. & Katz, N. 1997, ApJ, 487, 482
32. Hix, W. R. & Thielemann, K.-F. , 1996, ApJ, 460, 869
33. Ishimaru, Y. & Wanajo, S. 1999a, ApJ, 511, L33
34. Ishimaru, Y. & Wanajo, S. 1999b, MPA/ESO Workshop on the First Stars, Garching 1999, in press.
35. Israelian, G., García López, R. & Rebolo, R. 1998, ApJ, 507, 805
36. Israelian, G. & Rebolo, R. 1999, The First Stars, MPA/ESO Workshop, Garching August 1999, in press
37. King, J. R. 1993, AJ, 106, 1206
38. King, J. R. 1997, AJ, 113, 2302
39. Kobayashi, C., Tsuijmoto, T., Nomoto, K. Hachisu, I. & Kato, M. 1998, ApJ, 503, L155
40. Laird, J. B., Carney, B. W. & Latham, D. W. 1988b, AJ, 95, 1843
41. Laird, J. B., Rupen, M. P., Carney, B. W. & Latham, D. W. 1988a, AJ, 96, 1908
42. Lu, L., Sargent, W. L. W., Barlow, T. A., Churchill, C. W. & Vogt, S. S. 1996, ApJS, 107, 475
43. Magain, P. 1988, The Impact of Very High S/N Spectroscopy on Stellar Physics, ed. G. Cayrel de Strobel & M. Spite (Dordrecht, Kluwer), 485
44. Magain, P. 1989, A&A, 209, 211
45. Magain, P. 1995, A&A, 297, 686

46. Mathews, G. J. & Cowan, J. J. 1990, Nature, 345, 491
47. Matteucci, F. & Brocato, E. 1990, ApJ, 365, 539
48. McWilliam, A., 1998, AJ, 115, 1640
49. McWilliam, A., Preston, G. W., Sneden, C. & Searle, L. 1995, AJ, 109, 2757
50. Molaro, P. & Bonifacio, P. 1990, A&A, 236, L5
51. Molaro, P. & Castelli, F. 1990, A&A, 228, 426
52. Molaro, P., D'Odorico, S., Fontana, A., Savaglio, S. & Vladilo, G. 1996, A&A, 308, 1
53. Müller, E., Fryxell, B. & Arnett, D. 1991, A&A, 251, 505
54. Nakamura, T., Umeda, Nomoto, K., Thielemann, F.-K. & Burrows, I. 1999, ApJ, in press (astro-ph/9809307)
55. Nissen, P. E., Gustafsson, B., Edvardsson, B., & Gilmore, G. 1994, A&A, 285, 440
56. Nissen, P. E. & Schuster, W. M. 1997, A&A, 326, 751
57. Norris, J. E. 1994, ApJ, 431, 645
58. Norris, J. E., Peterson, R. C. & Beers, T. C. 1993, ApJ, 415, 797
59. Norris, J. E., Ryan, S. G. & Beers, T. C. 1997a, ApJ, 488, 350
60. Norris, J. E., Ryan, S. G. & Beers, T. C. 1997b, ApJ, 489, L169
61. Norris, J. E., Ryan, S. G., Beers, T. C. & Aoki, W. 2000, in prep.
62. Nulsen, P. E. J., Barcons, X. & Fabian, A. C. 1998, MNRAS, 301, 168
63. Peterson, R., Kurucz, R. L., & Carney, B. W. 1990, ApJ, 350, 173
64. Pettini, M., Boksenberg, A. & Hunstead, R. W. 1990, ApJ, 348, 48
65. Pettini, M., Ellison, S. L., Steidel, C. C. & Bowen, D. V. 1999a, ApJ, 510, 576
66. Pettini, M., Smith, L. J., King, D. L. & Hunstead, R. W. 1997, ApJ, 486, 665
67. Pettini, M., Steidel, C. C., Adelberger, K. L., Dickinson, M. & Giavalisco, M. 1999b, preprint
68. Primas, F., Molaro, P., & Castelli, F. 1994, A&A, 290, 885
69. Ryan, S. G. 1996, ASP Conf.Ser. 92, 113
70. Ryan, S. G. 2000, The Galactic Halo: from Globular Clusters to Field Stars, 35$^{\rm th}$ Liège Int. Astroph. Coll., in press
71. Ryan, S. G., Blake, L., Norris, J. E. & Beers, T. C. 2000, in preparation
72. Ryan, S. G. & Norris, J. E. 1991, AJ, 101, 1865
73. Ryan, S. G., Norris, J. E. & Bessell, M. S. 1991, AJ, 102, 303
74. Ryan, S. G., Norris, J. E., & Beers, T. C. 1996, ApJ, 471, 254
75. Ryan, S. G., Norris, J. E. & Beers, T. C. 1999, ApJ, 523, 654
76. Sadler, E. M., Rich, R. M. & Terndrup, D. M. 1996, AJ 112, 171
77. Searle, L. 1977 In: Tinsley, B. M. & Larson, R. B. (eds) The Evolution of Galaxies and Stellar Populations, Yale Uni. Obs., New Haven, p.219
78. Shigeyama, T. & Tsujimoto, T. 1998, ApJ, 507, L135
79. Sneden, C., Cowan, J. J., Burris, D. L. & Truran, J. W. 1998, ApJ, 496, 235
80. Sneden, C., Preston, G. W., McWilliam, A. & Searle, L. 1994, ApJ, 431, L27
81. Sneden, C., McWilliam, A., Preston, G. P., Cowan, J. J., Burris, D. L. & Armosky, B. J., 1996, ApJ, 467, 819
82. Spite, M. & Spite, F. 1978, A&A, 67, 23
83. Spite, M. & Spite, F. 1991, A&A, 252, 689
84. Stathakis, R. A. 1996, PhD Thesis, Sydney Uni.
85. Steidel, C. C., Adelberger, K. L., Giavalisco, M., Dickinson, M. & Pettini, M., 1999, astro-ph/9811399
86. Timmes, F. X., Woosley, S. E., & Weaver, T. A. 1995, ApJS, 98, 617
87. Tomkin, J., Lemke, M., Lambert, D. L. & Sneden, C. 1992, AJ, 104, 1568
88. Travaglio, C., Galli, D., Gallino, R., Busso, M., Ferrini, F. & Straniero, O. 1999, ApJ, 521, 691
89. Truran, J. W. 1981, A&A, 97, 391
90. Tsujimoto, T. & Shigeyama, T. 1998, ApJ, 508, L151
91. Warren, S. J., Hewett, P. C. & Osmer, P. S. 1994, ApJ, 421, 412

92. Wolfe, A. M., Turnshek, D. A., Smith, H. E. & Cohen, R. D. 1986, ApJS, 61, 249
93. Woosley, S. E. & Hoffman, R. D. 1992, ApJ, 395, 202
94. Zinn, R. 1993, ASP.Conf.Ser, 48, 38

Speaker: S. RYAN

F. MATTEUCCI: Are you sure that the spreads observed at low metallicity for s- and r-process elements are real?

S. RYAN: For Sr there is absolutely no question of a large spread of a factor of 100 or more in [Sr/Fe], but for the heavier species Ba and Eu the spread is much less, and almost certainly for Eu part of it is due to observational errors. Eu has a very weak spectral line, so the errors are much larger. In fact, Ba definitely has a much smaller spread than Sr, so [Sr/Ba] is very non-uniform. This rules out a unique r-process pattern as being responsible for these elements in very metal-poor stars.

B. PAGEL's Comment: An added complication is the work by Wasserburg et al. on radioactive ^{129}I and ^{182}Hf in the Solar System, implying two different r-process progenitors for the respective abundance peaks going off with different frequencies in the solar neighbourhood. Yet the relative abundances in the famous CS r-process star are not very different from the solar r-process.

R. GALLINO's Comment: The observed large spread in [Ba/Sr] in low [Fe/H] stars is one major indication that the r-process mechanism is by far more complex that a current unique r-process. As to Sr-Y-Zr in metal poor stars I would prefer to speak of a neutron capture mechanism", not in the classical sense, but certainly not s-weak, which would be a "secondary-like" process and not a grossly primary one.

S/α/FE ABUNDANCE RATIOS IN HALO FIELD STARS: IS THERE A GLOBULAR CLUSTER CONNECTION?

B.E.J. PAGEL
Astronomy Centre
CPES, Sussex University, Brighton BN1 9QJ, UK

AND

G. TAUTVAIŠIENĖ
Institute of Theoretical Physics and Astronomy
Goštauto 12, Vilnius 2600, Lithuania

Abstract.

We try to understand the s- and r-process elements vs Ti/Fe plots derived by Jehin et al. (1999) for mildly metal-poor stars within the framework of the analytical semi-empirical models for these elements by Pagel & Tautvaišienė (1995, 1997). Jehin et al. distinguished two Pop II subgroups: IIa with α/Fe and s-elements/Fe increasing together, which they attribute to pure SNII activity, and IIb with constant α/Fe and a range in s/Fe which they attribute to a prolonged accretion phase in parent globular clusters. However, their sample consists mainly of thick-disk stars with only 4 clear halo members, of which two are 'anomalous' in the sense defined by Nissen & Schuster (1997). Only the remaining two halo stars (and one in Nissen & Schuster's sample) depart significantly from Y/Ti (or s/α) ratios predicted by our model.

1. Introduction

The distribution of s-process abundances in stellar populations is one of the more mysterious features of Galactic chemical evolution (GCE). The problems are well illustrated in the review by McWilliam (1997) where his Figs 9 and 10 show typical s-element (Sr and Ba) to iron ratios and Ba and La to Eu ratios as functions of metallicity [Fe/H], Eu being representative

F. Matteucci and F. Giovannelli (eds.), The Evolution of the Milky Way, 27–33.

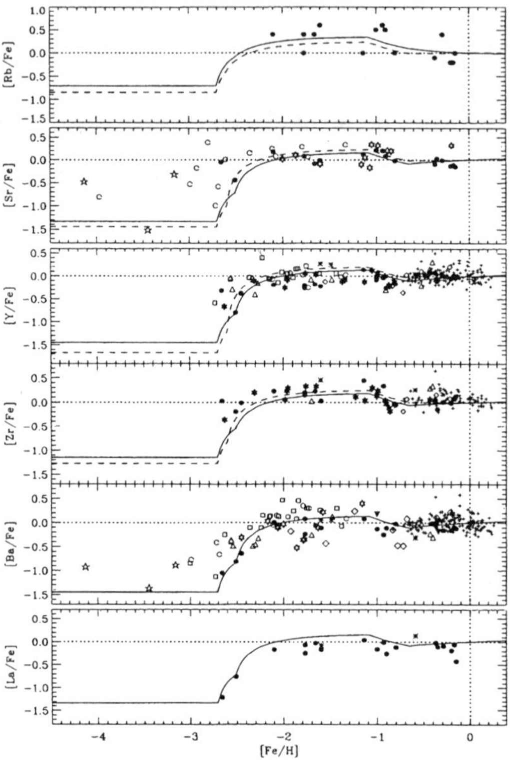

Figure 1. s-process element-to-iron ratios as a function of [Fe/H] according to the model by Pagel & Tautvaišienė (1997).

of a nearly pure r-process. For [Fe/H] ≤ -2.5, there is a large scatter above a lower limit [Ba/Eu] $\simeq -0.8$ representing a pure r-process, with higher values presumably due to internal mixing or contamination by a companion; but between [Fe/H] $= -2$ and about -1 there is a constant plateau with [Ba/Eu] $\simeq -0.3$, which Pagel & Tautvaišienė (1997) attributed to a general contribution to GCE of a primary s-process not readily understandable in terms of the expected age and metallicity dependence. Figs. 1 and 2 show element-to-iron ratios resulting from our *ad hoc* model, in which the s-process was treated as primary with a superposition of different time delays, noting that any secondary or other dependence of the yields on chemical composition (e.g. Travaglio et al. 1999) could be obscured by scatter in the metallicities at any given time.

In our model we supposed that the first batch of s-process synthesis came from rather massive progenitors with a typical time delay of 40 Myr corresponding to about $8.5M_\odot$ to get the plateau in Ba/Eu for $-2 \leq$ [Fe/H] < -1, long before the onset of the bulk of SNIa, and a second batch more

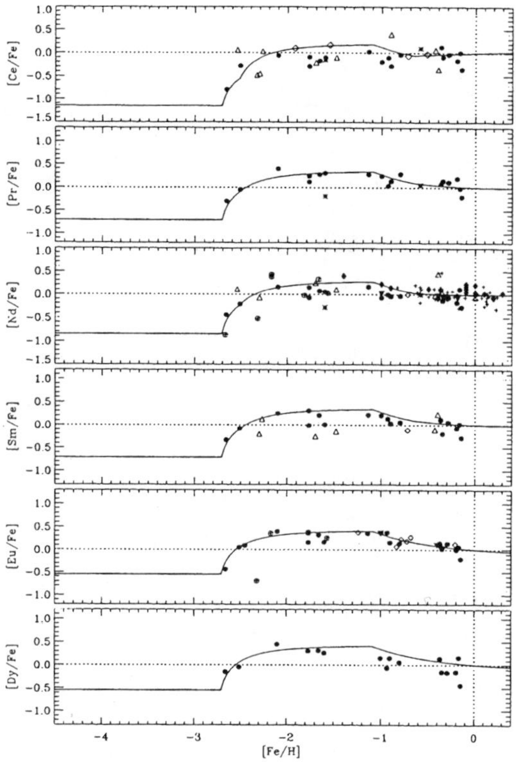

Figure 2. s- and r-process element-to-iron ratios as a function of [Fe/H] according to the model by Pagel & Tautvaišienė (1997).

like the conventional model for the s-process with a time delay of the order of 3 Gyr corresponding to $1.5M_\odot$ and longer than for typical SNIa leading to the decline followed by a rise in Ba/Fe that appears near solar metallicity in Fig 1. The overall fit to the data in Figs 1 and 2 is quite good, although at the lowest metallicites one should take into account the scatter in Eu/Fe that has been discussed by Tsujimoto, Shigeyama & Yoshii (1999).

In the last few years there have been substantial developments, of which we should like to mention two here:

- The work of Nissen & Schuster (1997) who investigated disk and halo stars with overlapping metallicity and found 'anomalous' halo stars which have too much iron for their content in O, α- and s-process elements represented by Y and Ba, and which might represent a slower chemical evolution such as may have occurred in the Magellanic Clouds (Pagel & Tautvaišienė 1998).
- A very interesting paper by Jehin et al. (1999), where they select a group of stars in the restricted metallicity range $-1.2 \leq [Fe/H] \leq -0.6$,

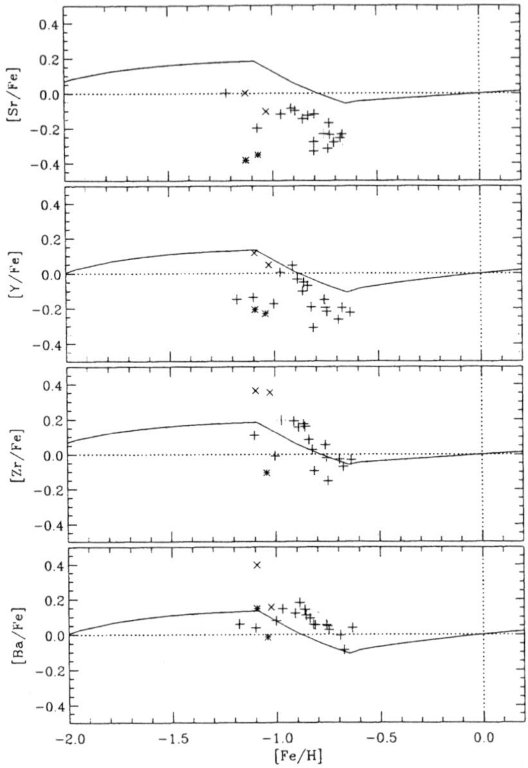

Figure 3. Data by Jehin et al. (1999) compared to the model by Pagel & Tautvaišienė (1997) for s/Fe ratios in mildly metal-deficient stars. Symbols as in Fig 4.

which is again the region of metallicity overlap between the halo and thick disk and also just the range where SNIa are believed to kick in.

2. Results of Jehin et al.

Jehin et al. determined very precise abundances for a number of metals: Fe, Mg, Ca, Ti, Y, Sr, Ba and Eu, among others. However, in presenting their results they ignore metallicity as such (except in the case of Ti/Fe itself) and plot correlation diagrams [X/Fe] vs [Ti/Fe], the latter being the most accurate representative of $[\alpha/\mathrm{Fe}]$. Thus α-elements and Eu are found to track [Ti/Fe] quite precisely, except in the case of the two 'anomalous' halo stars (in the sense of Nissen & Schuster) in their sample, which have excess europium and other r-process elements – an intriguing result not yet explained, although it may indicate the role of an r-process with a significant time delay. However, the behaviour of s-process elements was found to be different: instead of running more or less parallel to [Ti/Fe], there appear to be two sequences, of which one (which they call Pop IIa

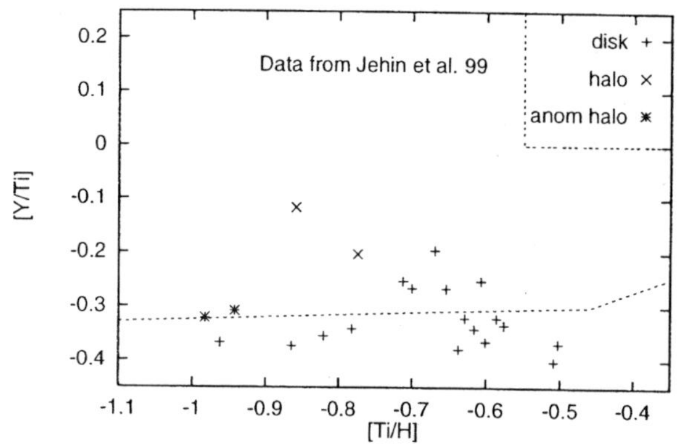

Figure 4. Data by Jehin et al. (1999) compared to the model by Pagel & Tautvaišienė (1997) for [Y/Ti] vs. [Ti/H] in mildly metal-deficient stars.

and includes the 'anomalous' stars) does run more or less parallel, while the other (which they call Pop IIb) starts at the end of the previous sequence and then runs up vertically at [Ti/Fe] = 0.24. This inspired the authors to put forward what they call the EAS scenario — Evaporation, Accretion, Self-enrichment — in which all halo and thick-disk stars are assumed to form in globular clusters or proto-clusters undergoing chemical evolution. Some clusters were disrupted at an early stage leading to Pop IIa with pure SNII ejecta, whereas others lasted longer enabling dwarf stars to accrete s-process material from nearby AGB stars before the clusters evaporated, leading to Pop IIb.

3. Relation with metallicity

While the hypothesis of Jehin et al. is interesting and possibly even right, we do not think their results can be understood without looking at the data as a function of metallicity; this is done in Fig 3 which distinguishes stellar population types and also presents the model by Pagel & Tautvaišienė (1997). The upper 7 or 8 stars in each panel are Type IIb and the 'anomalous' stars appear near bottom left.

According to our model, there is an overall downward trend due to the impact of Type Ia supernovae contributing extra iron in just this range of metallicity, and that is nicely confirmed by the mean trend of the new data, although the absolute fit is better for some elements than for others, and the results of Jehin et al. appear to be a scatter around this trend, possibly related to their scenario, whereas Ti/Fe reaches a sharply defined plateau

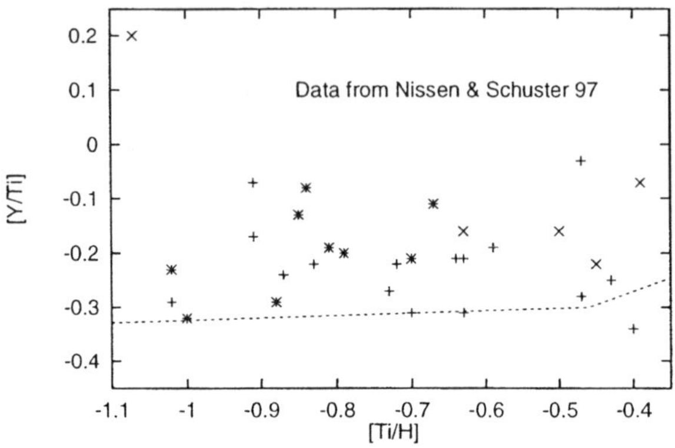

Figure 5. Data by Nissen & Schuster (1997) compared to the model by Pagel & Tautvaišienė (1997) for [Y/Ti] vs. [Ti/H] in mildly metal-deficient stars. Symbols as in Fig 4.

representing pure SNII production on the low-metallicity side, where the s-process points spread out forming a wedge-shaped distribution.

How much of this represents really significant deviations from conventional GCE? To investigate this point, we have plotted in Fig 4 what seems to be the best determined s-process result, [Y/Ti], against [Ti/H], which is chosen as the best available 'clock' (cf. Wheeler, Sneden & Truran 1989). Fig 5 shows corresponding data from the paper by Nissen & Schuster, where scatter in the determinations is somewhat greater, but one star appears as a still more extreme case of Jehin et al.'s Pop IIb. The dotted line in each case shows the prediction of our model. We feel that the departure of any thick-disk or anomalous-halo stars in these samples from our model predictions in this plane are at most marginal, but the effect discovered by Jehin et al. is certainly there among at least some of the non-anomalous halo stars. Clearly more and better statistics would be useful.

References

1. Jehin, E., Magain, P., Neuforge, C., Noels, A., Parmentier, G. & Thoul, A.A. 1999, *A & A*, **341**, 241
2. McWilliam, A. 1997, *Ann. Rev. Astr. Ap.*, **35**, 503
3. Nissen, P.E. & Schuster, W.J. 1997, *A & A*, **326**, 751
4. Pagel, B.E.J. & Tautvaišienė, G. 1995, *MNRAS*, **276**, 505
5. Pagel, B.E.J. & Tautvaišienė, G. 1997, *MNRAS*, **288**, 108
6. Pagel, B.E.J. & Tautvaišienė, G. 1998, *MNRAS*, **299**, 535
7. Travaglio, C., Galli, D., Gallino, R., Busso, M., Ferrini, F. & Straniero, O. 1999, *Ap.J.*, **521**, 691
8. Tsujimoto, T., Shigeyama, T, & Yoshii, Y. 1999, *Ap. J. Lett.*, **519**, L63

Speaker: B.E.J. PAGEL

R. GALLINO: Among the existing globular clusters, there are no indications of self s- process accretion, except for the atypical, massive, ωCen. I wonder if, in the idea of Jehin et al., a major fraction of disrupted globular clusters feeding the anomalous Pop II field stars in [Fe/H]~ -1.5, would be objects similar to ωCen. The Jehin et al. observations are very interesting, while the interpretation may be needs a major effort, as you indeed stressed.

B. PAGEL: Since the clusters are hypothetical, I suppose they could be hypothetical ωCens as well as any other type, but I basically agree with you.

M. RYAN's Comments: Comment 1 (following Danziger): Just because [Ti/Fe] resembles [α/Fe] does not mean Ti is produced in the same way. As an iron-peak element, it is produced near the mass cut and could have a metallicity-dependence in its synthesis as Mn appears to.

Comment 2 (following Matteucci): Some data show well defined trends (e.g. Jehan et al.) which seem to rule out the scatter. On the other hand, some data compilations that Bernard showed include data from many sources, some 10 years old, and which inevitably involve different systematics which broaden the trends.

EARLY GALACTIC EVOLUTION OF CARBON, NITROGEN AND OXYGEN

G. ISRAELIAN[1], R. J. GARCIA LOPEZ[1,2] AND R. REBOLO[1]
1. Instituto de Astrofísica de Canarias
E-38200 La Laguna, Tenerife, Spain
2. Departamento de Astrofísica, Universidad de La Laguna
E-38071 La Laguna, Tenerife, Spain

Abstract.

We present results on carbon, nitrogen, and oxygen abundances for a sample of unevolved metal-poor stars with metallicities in the range $-0.3<$ [Fe/H]< -3. Oxygen abundances derived from different indicators are compared showing consistently that in the range 0.3 >[Fe/H]> -3.0, the [O/Fe] ratio increases from approximately 0 to 1. We find a good agreement between abundances based on the forbidden line, the OH and IR triplet lines when gravities based on *Hipparcos* parallaxes are considered for the sample stars. Gravities derived from LTE ionization balance in metal-poor stars with [Fe/H]< -1 are likely too low, and could be responsible for an underestimation of the oxygen abundances derived using the [O I] line. [C/Fe] and [N/Fe] ratios appear to be constant, independently of metallicity, in the same range. However, they show larger scatter than oxygen at a given metallicity, which could reflect the larger variety of stellar production sites for these other elements.

1. Introduction

CNO abundances in metal-poor stars can tell us about nucleosynthesis and mixing processes in stars, and about several fundamental parameters of Galactic chemical evolution (initial mass function, star formation rate, etc.). Unevolved stars play a key role in this respect because the original surface abundances of CNO in evolved giant stars can be altered by internal nucleosynthesis and mixing between the core and outer layers. Mixing is expected to occur when a star becomes a red giant following the exhaustion

F. Matteucci and F. Giovannelli (eds.), The Evolution of the Milky Way, 35–46.

of hydrogen in the core (so called "the first dredge-up"). Important results have been obtained from the analysis of CNO molecular bands and atomic lines in metal-poor stars (see Wheeler et al. 1989 and references therein). However, such analyses have been either based on high resolution spectra of small samples (less than 10) or low/intermediate resolution spectra of large samples ($\sim$ 100) of stars. We are studying CNO abundances from high resolution spectra of more than 40 halo dwarfs. In this paper we report preliminary results based on 24 halo dwarfs discussed by Israelian, García López and Rebolo (1998) and briefly discuss the current understanding of the galactic evolution of these elements.

2. Observations and Analysis

The observations were carried out in different runs using the UES ($R = \lambda/\Delta\lambda \sim 50000$) of the 4.2-m WHT at the Observatorio del Roque de los Muchachos (La Palma), and the UCLES ($R \sim 60000$) of the 3.9-m AAT. The final signal-to-noise ratio (S/N) varies for the different echelle orders, being in the range 30–100 for most of the stars.

A grid of LTE, plane-parallel, constant flux, and blanketed model atmospheres provided by Kurucz (1992), computed with ATLAS9 without overshooting, and interpolated for given values of $T_{\rm eff}$, $\log g$, and [Fe/H] was used. Details on the analysis of the OH lines have been presented in Israelian et al. (1998). We have derived carbon abundances from the CH band at 3145 Å using the list of Kurucz (1992). Nitrogen abundances have been derived from the NH band at 3360 Å using the atomic and molecular data from Norris (1999, private communication). Synthetic spectra were computed with the WITA3 code by Pavlenko (1991). Effective temperatures ($T_{\rm eff}$) for our stars were estimated using the Alonso et al. (1996) calibrations versus $V-K$ and $b-y$ colors, which were derived applying the infrared flux method, and cover a wide range of spectral types and metal content. Metallicities were adopted from literature values obtained from high resolution spectra. Gravities were derived using the accurate parallaxes measured by *Hipparcos* (ESA 1997). These gravity values are larger by 0.28 dex in average than the values adopted in our previous analysis of OH lines (Israelian et al. 1998). This implies a mean small reduction of 0.09 dex in the oxygen abundances inferred in the latter paper, which does not affect the previously observed linear relationship between [O/Fe] and [Fe/H].

3. Abundances of CNO in metal poor dwarfs

3.1. OXYGEN

Type II SNe are expected to produce significant amounts of oxygen. Iron is produced in both, Type II and in Type I SNe. Since the latter come from longer lifetime progenitors, it has been argued for a long time that oxygen must be overabundant in very old stars. Evidence for high [O/Fe] ratios in many metal-poor stars has been reported during the last decades. A so called "traditional" view is based on the study of [O I] lines at 6300 and 6363 Å in giants (though the second line at 6363 Å is not visible in very metal-poor stars and the analysis is based *only on one line*) by Barbuy (1988), Gratton & Ortolani (1992), Sneden et al. (1991) and Kraft et al. (1992). These authors found that [O/Fe]= $0.3-0.4$ dex at [Fe/H]< -1 and is constant with decreasing metallicity. In contrast, oxygen abundances derived in dwarfs using the O I IR triplet at 7774 Å by Abia & Rebolo (1989), Tomkin et al. (1992), King & Boesgaard (1995), and Cavallo, Pilachowski, & Rebolo (1997) point towards increasing [O/Fe] values with decreasing [Fe/H], reaching a ratio ~ 1 for stars with [Fe/H]~ -3, suggesting a higher production of oxygen during the early Galaxy.

New oxygen abundances derived from near-UV OH lines (which form in the same layers of the atmosphere as [O I]) for 24 metal-poor stars have been presented by Israelian, García López, & Rebolo (1998). It is shown how the [O/Fe] ratio of metal-poor stars increases from 0.6 to 1 between [Fe/H]=-1.5 and -3 with a slope of -0.31 ± 0.11 (Fig 1). Contrary to the previously accepted picture (Bessell, Sutherland, & Ruan 1991, who used older model atmospheres with a coarser treatment of the opacities in the UV), these new oxygen abundances derived from low-excitation OH lines, agreed well with those derived from high-excitation lines of the O I IR triplet at 7774 Å. The comparison with oxygen abundances derived using O I data from Tomkin et al. (1992) showed a mean difference of 0.00 ± 0.11 dex for the stars in common. On the other hand, Boesgaard et al. (1999) have obtained high quality Keck spectra of many metal-poor stars in the near UV, and recently concluded their analysis of a different set of OH lines. They find a very good agreement with the results obtained by Israelian et al. (1998), and basically the same dependence of [O/Fe] versus metallicity. The mean difference in oxygen abundance for ten stars in common is 0.00 ± 0.06 dex when the differences in stellar parameters are taken into account. In Fig. 1, upper panel, we plot these oxygen abundances based on OH lines as a function of metallicity.

Balachandran & Bell (1998) have pointed out that the continuous opacity of the Sun is not fully accounted for in the spectral syntheses performed in the near UV region. Although this is still a matter of debate, it would

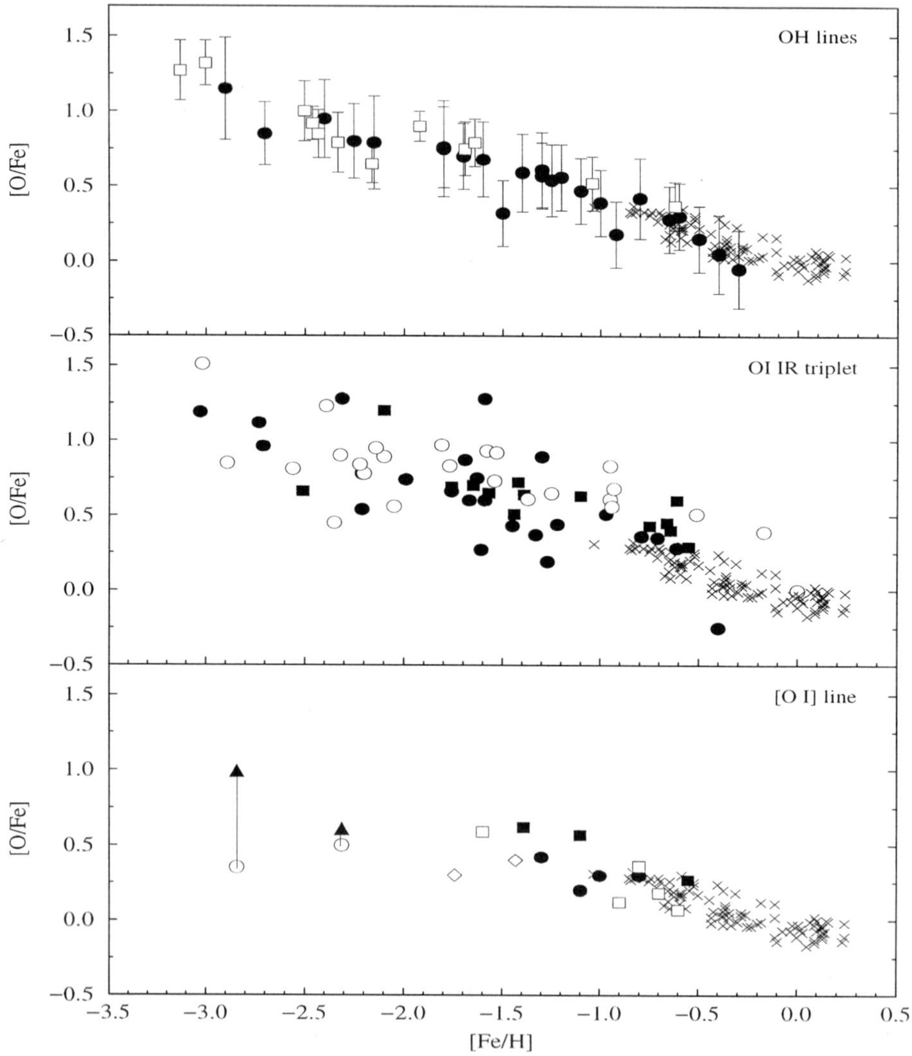

Figure 1. [O/Fe] vs. [Fe/H] for unevolved stars. Abundances from OH lines were derived by Israelian et al. (1998; filled circles) and Boesgaard et al. (1999; open squares, corrected to the scale of stellar parameters adopted by Israelian et al.). Abundances from the IR triplet were derived in NLTE by Mishenina et al. (2000; filled circles), Cavallo et al. (1997; filled squares, corrected for NLTE effects by Mishenina et al.), and in LTE by Boesgaard et al. (1999; open circles). Finally, abundances from the [O I] line come from Spiesman & Wallerstein (1991; open diamonds), Spite & Spite (1991; open squares), Israelian et al. (1998; filled circles), Mishenina et al. (2000; filled squares), and Fulbright & Kraft (1999; open circles). Filled triangles indicate the change in abundances associated with the change in gravity according to by the *Hipparcos* parallaxes for the two stars studied by Fulbright & Kraft. The abundances derived by Edvardsson et al. (1993; crosses) are shown in the three plots to indicate the trend in metal-rich stars.

have a minor effect on the recent OH results since most of the stars in the samples of Israelian et al. and Boesgaard et al. are hotter than the Sun (and very metal-poor), and the corrections to oxygen abundances for individual stars due to this effect would be lower than 0.15 dex. This would not affect significantly the [O/Fe] vs. [Fe/H] trend.

Recently, Mishenina et al. (2000) performed a non-LTE analysis of the O I IR triplet to re-derive oxygen abundances for a sample of 38 metal-poor stars from the literature. They confirmed earlier results (Abia & Rebolo 1989; Tomkin et al. 1992; Kiselman 1993) indicating that the mean value of the non-LTE correction in unevolved metal-poor stars is typically 0.1 dex and never exceeds 0.2 dex. Mishenina et al. found the same linear trend as Israelian et al. (1998) and Boesgaard et al. (1999) from the OH lines, and confirmed that oxygen abundances do not show any trend with $T_{\rm eff}$ or $\log g$ (Boesgaard et al. 1999). Furthermore, Asplund et al. (1999) showed that the O I IR triplet is not affected by 3D effects, convection and small-scale inhomogeneities in the stellar atmosphere. In addition, oxygen abundances derived form this triplet are not significantly affected by chromospheric activity either, and we can conclude that the O I IR triplet provides reliable oxygen abundances in metal-poor dwarfs. This conclusion should also apply to the oxygen abundances derived from the UV OH lines given the good agreement shown between both indicators. In Fig. 1, mid-panel, we plot oxygen abundances based on the oxygen triplet. The larger scatter observed, as compared with the measurements based on OH lines, can be associated with the different scales of stellar parameters ($T_{\rm eff}$, gravities, and metallicities) adopted by the authors of each set of stars, and to the fact that some measurements have not been corrected for NLTE effects.

Israelian et al. (1998) found four dwarfs (HD 22879, HD 76932, HD 103095 and HD 134169) in their sample for which oxygen abundances had been previously derived using [O I]. They synthesized the forbidden oxygen line for these stars adopting the same set of stellar parameters than for the OH analysis, the gf value given by Lambert (1978), and the equivalent widths provided in the literature for the λ 6300 Å line. The estimated abundances were in reasonable agreement with those derived from OH but still slightly lower. The abundances found in that work using *Hipparcos* gravities when analyzing both indicators are in better agreement, which strongly suggests that a reliable gravity scale may indeed be key to explain the discrepancies on oxygen abundances from forbidden and permitted lines in unevolved metal-poor stars.

In the lower panel of Fig.1 we compile oxygen measurements for unevolved stars based on the [O I] λ 6300 Å line. The presence of a linear trend of [O/Fe] versus metallicity strongly depends on the only two measurements available at [Fe/H]≤ -2. These two measurements have been recently reported by Fulbright & Kraft (1999) for the stars BD +37 1458 and BD +23 3130, which were also considered by Israelian et al.(1998) and Boesgaard et al. (1999; only BD +37 1458 in this case). There is an apparent discrepancy between the results obtained from the forbidden and the OH lines. However, we argue here that this discrepancy cannot be sustained when a critical analysis of the uncertainties involved in the determination from the forbidden line is performed. The analysis carried out by Fulbright & Kraft is based on gravities derived from LTE iron ionization balance of these subgiants where it is well known that NLTE effects are strong. In a recent paper, Allende Prieto et al. (1999) have shown that gravities derived using this technique in metal-poor stars do not agree with the gravities inferred from accurate *Hipparcos* parallaxes, which casts a shadow upon oxygen abundance analyses of very metal-poor stars based on gravities derived from the ionization balance. They find that gravities are systematically underestimated when derived from ionization balances and that upward corrections of 0.5 dex or even higher can be required at metallicities similar to those of our stars. We remark here that any underestimation of gravities will also strongly underestimate the abundances inferred from the forbidden line. For the two stars under discussion our *Hipparcos* based gravities are 0.45 and 1.05 dex (for BD +37 1458 and BD +23 3130, respectively) higher than derived by Fulbright & Kraft, and would imply the corrections in the oxygen abundances indicated in Fig. 1 (the details of the analysis are out of the scope of these proceedings and will be presented in a forthcoming paper). Our conclusion is that the uncertainties in the gravities of these subgiants allow the abundances inferred from the forbidden line to be consistent with those estimated from the OH lines or the triplet. Actually, consistency with the other oxygen indicators is achieved for the high gravities inferred from *Hipparcos*, and this could be taken as an indication that the high gravities are indeed the correct ones.

Chemical evolution models of the early Galaxy where stellar lifetimes are taken into account and assuming that Type Ia SN appear at a Galactic age of 30 million years can also explain the evolution of oxygen delineated in Fig. 1. (Chiappini et al. 1999.). The evolution of oxygen proposed in this paper also helps to understand the evolution of ^{6}Li versus [Fe/H] and the ^{6}Li/Be ratio at low metallicities in the framework of standard Galactic Cosmic Ray Nucleosynthesis (Fields & Olive 1999). In addition, Ramaty et al. (1999) have proposed that a delay between the effective deposition times into the ISM of Fe and O (only a fraction of which condensed in oxide grains) can explain a linear trend of [O/Fe].

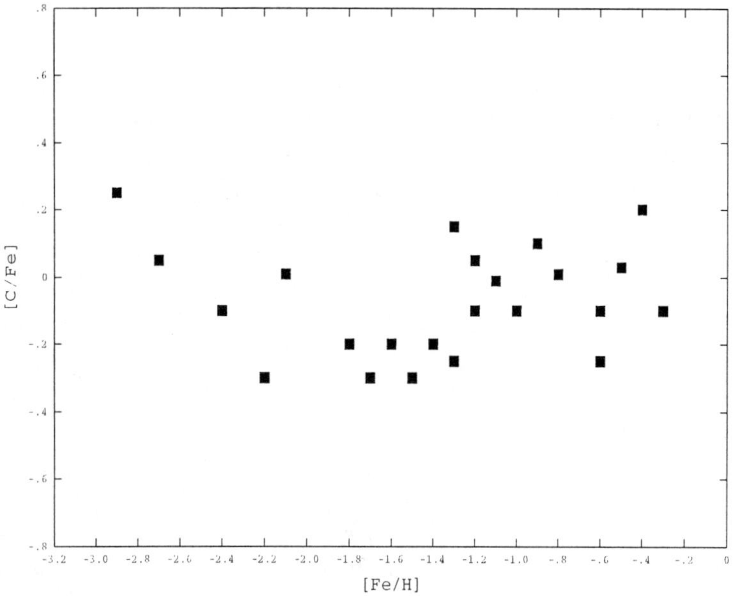

Figure 2. Carbon abundances in halo dwarfs. Typical error bar is 0.3 dex.

3.2. CARBON

Until very recently, it has been commonly accepted that intermediate mass stars ($M_{\odot} < M < 10M_{\odot}$) are the main source of Galactic C (Wheeler et al. 1989). It has been shown (Laird 1985, Carbon et al. 1987, Tomkin et al. 1992) that [C/Fe] is approximately zero independently of [Fe/H]. Note that all these studies were based on the 4300 Å feature of CH. Tomkin et al. (1992) have demonstrated that C I at $\sim$ 9100 Å provides an average [C/Fe]=+0.3±0.2, whereas the CH band provides [C/Fe]=−0.1±0.2. Recently, Gustafsson et al. (1999) performed an abundance analysis of carbon in a sample of 80 unevolved disk stars. They found that [C/Fe] increases with decreasing [Fe/H] with a slope of −0.17±0.03. This result was explained by carbon enrichment from superwinds of metal-rich massive stars. Our preliminary analysis for more metal-poor stars confirms earlier results that [C/Fe]$\sim$0 in a wide range of [Fe/H] (Fig. 2). Studies of very low metallicity halo stars ([Fe/H]< -3) have revealed a significant number of stars with very high overabundance of carbon, up to 1-2 dex (Beers et al. 1992). Given these new results, it appears necessary to review our knowledge about carbon production sites in the Galaxy.

The isotope ^{14}N is synthesized from ^{12}C and ^{16}O through the CNO cycles in the H-burning layer. Observations of the NH band at 3360 Å have allowed to delineate the Galactic evolution of N down to [Fe/H]~ -2.8. Tomkin & Lambert (1984) used high resolution spectra of 8 disk and 6 halo stars ($-0.3 <$ [Fe/H]< -2.3) and found [N/Fe]$\simeq -0.25$. Laird (1985) and Carbon et al. (1987) obtained [N/Fe]$= -0.67 \pm 0.14$ (intermediate resolution spectra of 116 stars) and [N/Fe]$= -0.11 \pm 0.06$ (low resolution spectra of 76 stars), respectively. However, we should stress that this band is blended with Ti and Sc lines and it is preferable to use high resolution spectra in order to avoid any systematics due to the overabundance of Ti (α-element) and Sc in metal-poor stars. Our preliminary analysis confirms previous results that [N/Fe] is constant in a wide range of [Fe/H]. It would be extremely interesting to check how [N/O] behaves in ultra metal-poor stars with [Fe/H]< -3. The trend of [N/Fe] at very low metallicities is not yet investigated but the existence of N rich halo subdwarfs has been already demonstrated. It is very important to check the [N/O] trend at [Fe/H]< -3 as this may help to understand the possible sites for a production of nitrogen as primary element. Recently, Maeder & Meynet (2000) have shown that the average rotation of massive stars can possibly be faster at lower metallicities. They have also found a primary N production in rapidly rotating stars in the mass range 10-15 $M_{\odot}$. To our knowledge these are the only models showing a primary nitrogen production in normal stars (see also Maeder, A., this conference).

4. Concluding remarks

We have derived carbon, nitrogen and oxygen abundances for a sample of metal-poor unevolved stars. [C/Fe] and [N/Fe] ratios appear to be constant, independently of metallicity, in the range $0.3 >$[Fe/H]> -3.0, while the [O/Fe] ratio increases from approximately 0 to 1, with consistent oxygen abundances derived from different indicators. Carbon and nitrogen abundances show larger scatter than oxygen at a given metallicity, which could reflect the larger variety of stellar production sites for these elements. Work is in progress to derive CNO abundances for a larger sample of unevolved metal-poor stars using other features in the optical and IR spectral regions, which should allow a fully reliable estimate of the Galactic evolution of these elements. Consistent abundances from different spectral features of C, N and O will be used in order to delineate [N/O], [C/O] and [(C+N+O)/Fe] behaviour as a function of [O/H] and [Fe/H].

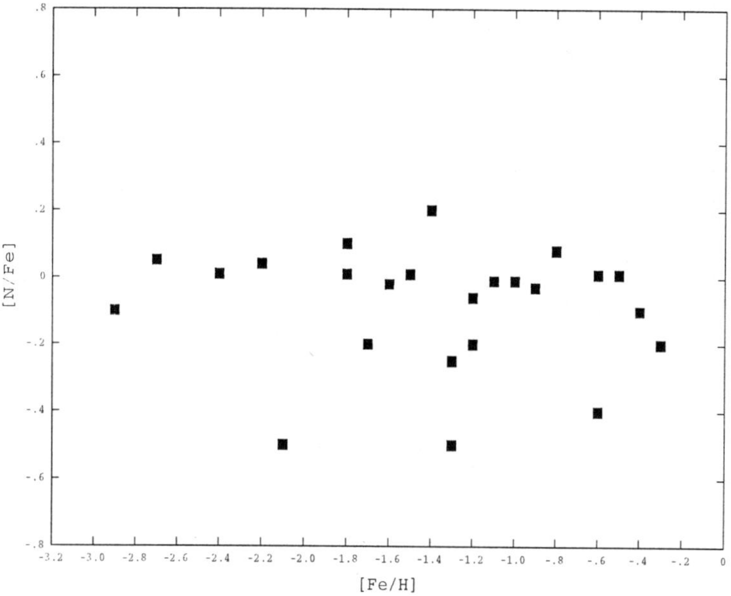

Figure 3. Nitrogen abundances in halo dwarfs. Typical error bar is 0.2 dex.

References

1. Abia, C., & Rebolo, R. 1989 *ApJ*, **347**, 186
2. Allende Prieto, C., García López, R. J., Lambert, D. L., & Gustafsson, B. 1999, *ApJ*, **527**, 879
3. Alonso, A., Arribas, S., & Martíinez-Roger, C. 1996, *A&AS*, **313**, 873
4. Asplund, M., Nordlund, Å, Trampedach, R., & Stein, R. 1999, *A&A*, **346**, L17
5. Balachandran, S., & Bell, R. 1998, *Nature*, **392**, 791
6. Barbuy, B. 1988, *A&A*, **191**, 121
7. Beers, T., Preston, G. & Shectman, S. 1992, *Astron. J*, **103**, 1987
8. Bessell, M. S., Sutherland, R. S., & Ruan, K. 1991, *ApJ*, **383**, L71
9. Boesgaard, A.M., King, J.R., Deliyannis, C. P., & Vogt, S.S. 1999, *Astron. J*, **117**, 492
10. Carbon, D., Barbuy, B., Kraft, R., Friel, E., & Suntzeff, N. 1987, *PASP*, **99**, 335
11. Cavallo, R., Pilachowski, C., & Rebolo, R. 1997, *PASP*, **109**, 226
12. Chiappini, C., Matteucci, F., Beers, T.C., & Nomoto, K. 1999, *ApJ*, **515**, 226
13. Edvardsson, B., Andersen, J., Gustafsson, B., Lambert, D. L., Nissen, P. E., & Tomkin, J. 1993, *A&A*, **275**, 101
14. ESA 1997, The Hipparcos and Tycho Catalogues, ESA SP-1200
15. Fields, B.D., & Olive, K.A. 1999, *ApJ*, **516**, 797
16. Fulbright, J., & Kraft, R. 1999, *Astron. J*, **118**, 527
17. Gratton, R., & Ortolani, S. 1999, *A&A*, **169**, 201
18. Gustafsson, B., Karlsson, T., Olsson, E., Edvardsson, B. & Ryde, N. 1999, *A&A*, **342**, 426
19. Israelian, G., García López, R. J., & Rebolo, R. 1998, *ApJ*, **507**, 805

20. King, J.R. & Boesgaard, A.M. 1995, *Astron. J*, **109**, 383
21. Kiselman, D. 1993, *A&A*, **275**, 269
22. Kraft, R., Sneden, C., Langer, G., & Prosser, C. 1992, *Astron. J*, **104**, 645
23. Kurucz, R. 1992, CD ROMs, Atlas9 Stellar Atmospheres Programs, SAO, Cambridge
24. Laird, J. 1985, *ApJ*, **289**, 556
25. Maeder, A. & Meynet, G. 2000, *Ann. Rev. Astron.& Astrophys.*, in press
26. Mishenina, T., Korotin, S., Klochkova, V., & Panchuk, V. 2000, *A&A*, in press
27. Pavlenko, Ya. V. 1991, *Soviet Ast.*, **35**, 212
28. Ramaty, R., Vangioni-Flam, E., Casse, M., & Olive, K. 1999, *PASP*, **111**, 651
29. Sneden, C., Kraft, R., Prosser, C., & Langer, G. 1991, *Astron. J*, **102**, 2001
30. Spiesman, W. & Wallerstein, G. 1991, *Astron. J*, **102**, 1790
31. Tomkin, J., Lemke, M., Lambert, D. L., & Sneden, C. 1992, *Astron. J*, **104**, 1568
32. Tomkin, J., & Lambert, D. 1984, *ApJ*, **279**, 220
33. Wheeler, J., Sneden, C., & Lambert, D. 1989, *Ann. Rev. Astron.& Astrophys.*, **27**, 279

Speaker: G. ISRAELIAN

B. BARBUY: The problem with the use of UV OH lines is the poor knowledge of the molecular constants. You have fitted the solar spectrum for each OH line, but the solar spectrum shows many blends and unknown lines in the UV. Also, one should not fit line strenghts of molecular lines in a line-by-line basis.

G. ISRAELIAN: We did our best in selecting unblended OH lines. We were guided by Moor et al. identification line list. The lines selected are perfectly symmetric and do not show any signatures of blends. However, we cannot exclude very weak blends on the top of the OH lines. But in that case, those blends are so weak, that they do not significantly change our semiempirical "gf"-values.

B. PAGEL: I think it is relevant to mention the work on solar beryllium by Balachandran and Bell, who allowed for extra opacity not included in Kurucz models and increased the estimate of Be/H by about 0.2 dex bringing it into agreement with meteorites. You "fudged" theoretical f-values of a similar amount and I expect your [O/Fe] values are also about 0.2 dex too high, but not more, just from this effect.

G. ISRAELIAN: I know this problem. However, Bob Kurucz and others claim that ATLAS 9 models do not have problems with opacities and we should not modify them. Instead, one can modify "gf"-values of many weak blended lines to make what is called "the best fit" to the profile.

R. GRATTON: Leaving aside the issue of the zero point of your [Fe/O], a very interesting feature of yours and Boesgaard et al. O abundances is the small scatter, which seems to suggest either a good mixing of the ISM, or that the [Fe/O] ratio does not depend much on initial mass that by the way would contradict the theoretical framework used to explain the trend of [Fe/O] with [Fe/H]. So, I look forward with much interest to your more extended sample.

G. ISRAELIAN: I think there are three reasons for the small scatter:
1) we use many OH lines which give us very accurate abundance,
2) we use unevolved stars and
3) there was a good mixing of the ISM started [Fe/H] ≥ -3.

A. MAEDER: I would like to mention that there is no discontinuity to expect in yields from Z=0. models and low-Z models. While the MS and He-burning phase are very different, at the end of the He-phase, the differences are quite small.

Now a question, why are you not plotting the [C/O] ratios versus [Fe/H] to estimate the primary and secondary production of C?

G. ISRAELIAN: I did not show the plot [C/O] versus [Fe/H] because I consider [C/Fe] plot as not final. We still must check if other indicators of C abundance (CI IR lines, CH band at 430 nm) show the same trend. The work is in progress and soon we will present the final results of this analysis.

THE ORIGIN AND KINEMATICAL EVOLUTION OF SMR STARS

M. GRENON
Geneva Observatory
CH-1290 Sauverny

1. Introduction

Thirty years ago, in September 1969, Spinrad and Taylor (1969) published the first detailed investigation of supermetalllicity in late type stars, from spectrum scanner data on field and cluster stars. Since most of their conclusions are still valid, let us recall part of them :

- Evolved K stars exist with $[M/H] \geq [M/H]_{\text{Hyades}}$, up to +0.6 dex. The strong spectral features are not due to gravity or microturbulence, nor to self-enrichment, but to greater metal abundances.
- Tight correlations exist between N, Na and Fe abundances, with some departures for Ca and Mg.
- Metal-rich K giants and subgiants have orbital eccentricities up to 0.5 and maximum heights above the galactic plane of 1.5 kpc.
- An abrupt cutoff in the production of metal-rich stars took place after the formation of M67. In the field, SMRs represent 10% of stars later than F8 and 0% of earlier type stars.
- SMRs have a variety of birthplaces.

The Spinrad and Taylor metal abundance for cool stars in M67 and NGC 188 were estimated as three times solar, an incorrect result as both clusters have nearly solar abundances. This paper opened a 30 year controversy whether the "supermetallicity" is true and primordial, apparent and due to peculiar thermal structures of the atmospheres, or due to evolutionnary processes as star coalescence.

The debate was non-converging, mainly because evolved stars were s-elected for analysis, with non-initial surface chemical compositions due to dredge-up effects. The cooling effect by CO and CN molecular bands was

F. Matteucci and F. Giovannelli (eds.), The Evolution of the Milky Way, 47–57.

suspected to play the dominant role. Prior to Hipparcos, distances were too poor to assign a precise evolutionnary phase for SMR giants.

Since SMRs represent the metal-rich tail of the metallicity distribution, the number of stars accepted as SMRs dramatically changes with the adopted threshold. Spinrad & Taylor (1969) defined the SMRs by [M/H] $\geq +0.15$ dex, Grenon(1972), the MR by [M/H] = [+0.08 to +0.20] and the SMR by [M/H] $> +0.20$, Grenon(1989) redefined the genuine SMRs by [M/H] = [+0.30 to +0.50], Taylor(1990) adopted a threshold at +0.20 dex and Chavez et al.(1998) at $> +0.10$. The age, origin and kinematical properties of MR and SMR are very hetrogeneous, thus a highly confusing situation occurs when samples are defined with different limits in [M/H].

The debate on the SMRs origin was partially biased by anthropocentric or at least heliocentric considerations, with some reluctance to admit that stars could be more metal-rich than our Sun.

2. The atmospherical origin

In K giants, the CN molecule cools the upper photosphere by few hundred Kelvin. SMR giants were supected to be N-enriched instead of metal-rich. Peterson(1976) concluded that μ Leo is not a genuine SMR and the strong lined spectrum results from an anomalous ratio [N/Fe] = +0.3 dex. Taylor(1990) reobserved μ Leo, obtained [Fe/H] = +0.25 $\pm$0.04 and nevertheless concluded that "Neither μ Leo nor any other giant is known to be SMR ($>$ 0.2 dex) ... among nearby giants or in Baades's window". In the same paper he listed 7 luminosity class IV,V stars with [Fe/H] +0.30 to +0.45 dex, all confirmed as genuine SMRs with Geneva photometry. Mishenina (1997) found no C,N,O anomalies, nor r-process enrichment and concluded that evolved SMRs are not the progenitors of R type stars. The isotopic ratios $^{16}O/^{17}O$ and $^{16}O/^{18}O$ measured by Campbell et al. (1990) ruled out the star coalescence as a possible origin of the SMR phenomenon. In case of coalescence, with an input of stellar core material, namely He and ^{17}O, a decrease of H abundance and an increase of [M/H] are expected. The signature would be a very low $^{16}O/^{17}O$ ratio. As the abundances of C and O isotopes are identical in the atmospheres of SMRs and normal K giants, in the same evolutionnary phase, their firm conclusion was that SMR giants are normal giants formed from metal-rich ISM. Later Luck & Challenner (1995) expressed no opinion about the origin of SMRs when coining the label VSL (Very Strong-Lined). From tiny overabundances in Fe, N and C, they conclude that VSL giants represent the normal evolution of dwarfs with somewhat metal-rich and C-rich initial composition.

3. The photometric [M/H]

Photometrists derive the metallicity from the measurement of the break at 430nm, due to a strong absorption increase towards shorter wavelengths by atomic and molecular lines. The relative importance of CH and CN bands contribution to the total absorption depends on the bands widths and locations. The contamination by CN is maximum with the DDO system, so when Taylor(1990) uses relations as [Fe/H] = 4.42 δCN -0.20, the metallicity estimator provides [M+αCN/H], with $\alpha < 1$, rather than [Fe/H] or even [M/H].

A similar effect exists in Strmgren and Geneva photometry, with a reduced amplitude. With these two systems, the [M/H] overestimation in the case of N-enriched stars could reach 0.2 dex at maximum. When Luck et al. [Fe/H] are compared to Geneva [M/H], the VSL stars follow the same relation as the normal giants, without noticeable effect of CN anomalies.

Since C,N,O abundances vary in lockstep with that of Fe, prior to the first dredge-up, photometric metallicities provide reliable estimates of the initial [Fe/H] for main sequence and subgiant stars. The spread in mass among old disc giants is small enough to allow reliable photometric [M/H] determinations also for GB and RHB giants.

4. Dredge-up effects

A cosmic scatter in CN abundance must exist among giant stars in GB and RHB phases because of the spread in ages, hence in masses. On the main-sequence, the $^{14}N/^{12}C$ ratio is 0.21; after the first dredge-up this ratio increases to 0.39, 0.55, 0.75 and 0.96 for 1.0, 1.5, 2.0, 2.5 $M_\odot$ giants, according to Schaller et al.(1992) models. Field giants are old in majority with 1.3 $M_\odot$ as mean mass. The N-enrichment for the Hyades generation is typically twice that in old disc giants. Anomalous N-enrichment may also result from meridional circulation in binaries. We have so a possible explanation for the apparent supermetallicity of some of the Hyades group giants showing [M/H] as high as +0.25 to +0.30, whereas dwarfs metallicities peak around [M/H] +0.15.

5. The evolved field SMRs

The proportion of MRs and SMRs depends strongly on the star samples constitution. As most SMRs are old, the turn-off for genuine SMRs is around G7IV according to Grenon (1989), genuine SMRs are never found among A and F stars samples. The fraction of SMRs formed during the galactic disc life may be estimated only from unevolved K dwarfs. It was estimated to be 4% for stars presently seen in the solar neighbourhood.

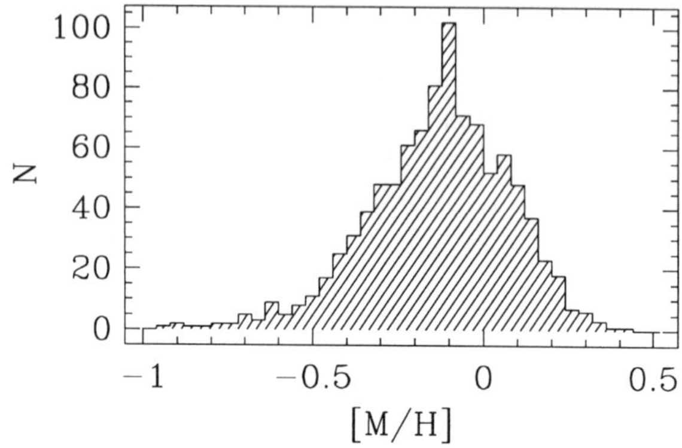

Figure 1. The distribution of metallicities from Geneva photometry for 965 G and K giants from Hipparcos Survey, with Mv = [0.0,+3.5] and d < 200pc. The typical standard error se[M/H] is 0.03 to 0.05 dex.

The distribution of [M/H] for nearby red giants provides a lower limit for the initial proportion of SMRs because of the age mixing and the presence of giants with A to F type progenitors. In Figure 5, the [M/H] distribution shows that only $\leq 3.8\%$ of giants have [M/H] > +0.20 dex and $\leq 1.9\%$ [M/H] > +0.20 dex.

The HR-diagram of field metal-rich giants shows an age dichotomy, with a group of Hyades generation forming a clump of stars centered around T_{eff} = 4000 K and Mv from 0.0 to +1.0, and a older population with a well defined giant branch (GB) and an associated RHB extending from 4750 to 4450 K, and Mv +1.0. Genuine SMR giants are very uncommon among bright stars. They are mostly high-velocity objects with small pericentric radius Rp. Typical cases are :

Star	U	V	W	[M/H]
HD 4088	+61	-73	-14	+0.34
HD 15555	+72	-55	+ 6	+0.32
HD 51266	+64	-33	-10	+0.38
HD 221148	+11	-62	-28	+0.54:

The star μ Leo was considered for many years as the prototype of field SMRs. Its space velocities, (U,V,W)=(+38,-19,-16), mimic a membership to Hyades' supercluster (Hyades and Praesepe velocities are respectively (+45,-18,-6) and (+34,-16,-5) km/s). Half a dozen stars share the μ Leo space velocities and metallicity. The W dispersion, σW indicates a several Gyr old population, confirmed by the HR-diagram morphology. With [M/H] = +0.20 to +0.26, its origin is expected to be 3 - 4 kpc inside the present solar orbit, according to the galactic metallicity gradient. μ Leo itself appears to belong the RHB, near its red edge.

From the scarcity of field SMR giants, we could question the existence of SMRs. At least one open cluster is known to be SMR, NGC 6791 with [Fe/H] = +0.27 to +0.44 according to Tripicco et al.(1995). From counts in its HR-diagram, the number of giants with Mv -1.5 to 2.5 represents about 4.7 % that of main sequence stars with Mv $\leq$ 6.5. Since we identified about 250 southern dwarf SMRs within 125pc, we expect only half a dozen genuine SMRs among the red giants listed in Catalogue of Bright Stars.

6. The dwarf MRs and SMRs

The scarcity of halo, thick-disc and very old disc stars among stars nearer than 25pc, led us to start a large observational programme on proper-motions stars. Among the 7900 NLTT stars observed by Hipparcos, 5500 have now physical parameters determined from Geneva photometry and 2900 both [M/H] and space velocities. A total of 250 SMRs was found within 125pc.

Figure 6 shows the space velocities plane (U,V) for MR stars with [M/H] centered on the Hyades metallicity +0.15. A clumpy distribution is found with, as expected, a large number of Hyades supercluster members. Mild metal-rich stars are found with V component as negative as -120 km/s and U components between +130 and -100 km/s. The excess of stars with positive U is striking, see section 9.

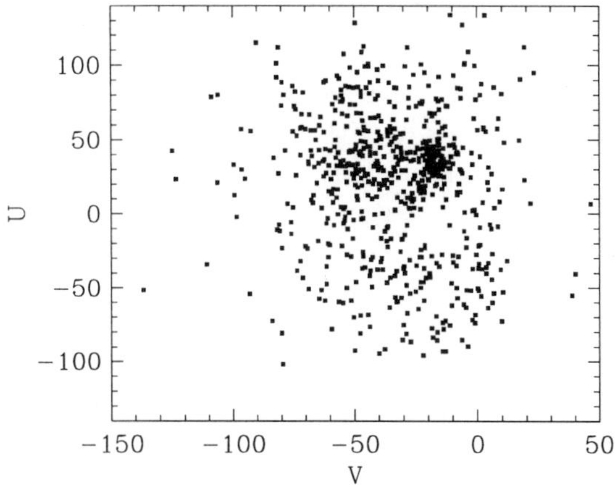

Figure 2. The (U,V) plane for 671 mild metal-rich dwarfs within 125pc, with [M/H] = [+0.05,+0.25].

Figure 6 shows the same diagram but for SMR dwarfs with [M/H] > +0.35. Stars with the smallest V show little scatter in U, an indication that these star have their apocenter at the solar galactocentric distance. In the V component range -50 to -120 km/s, an overlap exists between the SMR and the thick disc populations, the main differences are the mean [M/H], +0.38 for SMRs and -0.70 for thick disc stars, and the scale height, shown by σW = 18 and 50 km/s respectively.

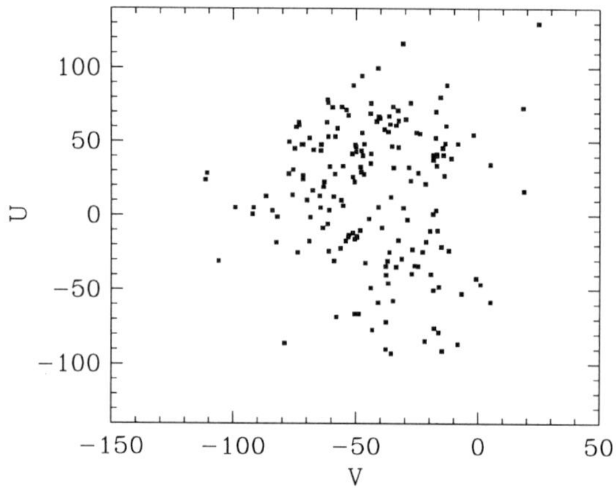

Figure 3. The same as in Figure 2, but for 180 SMR dwarfs with [M/H] = [+0.35,+0.55].

7. The ages of SMRs

The HR-diagram of mild metal-rich stars show a continuum of ages, from 0.7 Gyr to about 10 Gyr. SMRs appear older. Minimum ages of about 6 Gyr are observed for SMRs with pericentric radius Rp $<$ 7.5 kpc. When the sample is restricted to stars with Rp $<$ 5.5 kpc, the HR-diagram shows a group about 9 Gyr old, see Figure 7. The isochrones are too close and the observational scatter too large to decide wether we have a single burst of SMR formation or an age continuum, which seems more probable.

8. The origin of SMRs

The origin of MRs and SMRs is clearly multiple : The Hyades-like and μ Leo-like stars were formed inside the thin disc at galactocentric distances R smaller than $R_\odot$, 0.7 to few Gyr ago, from an ISM more metal-rich than the present ISM in the solar vicinity. Their higher metallicity results from the radial metallicity gradient.

The metallicity distribution of the flat component of the oldest disc is identical to that of the bulge as observed in situ, with a solar mean metallicity and a bell shaped distribution extending from -.55 dex to +0.55

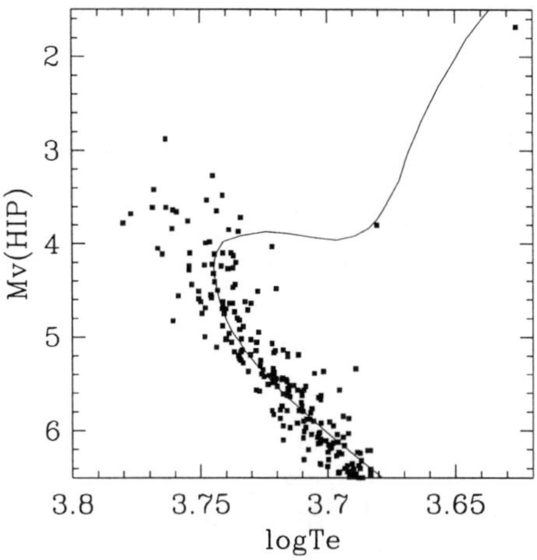

Figure 4. The HR-diagram of NLTT stars with Rp $\leq$ 5.5 and [M/H] = [+0.15,+0.55]. The isochrone for Z = 0.04 corresponds to a = 9 Gyr .

dex. A common origin for bulge stars and the extended flat old disc is then a plausible hypothesis.

The high velocity SMRs may be either :

- the metal-rich tail of the inner disc, with stars scattered inside the bulge or outside in the thin disc and crossing the solar orbit when travelling on eccentric orbits;
- the metal-rich tail of the bulge where the maximum [M/H] corresponds to the metallicity reached at the end of the prompt enrichment phase.

If the high-metallicity results from a prompt enrichment by SN II ejecta, the signature is be an overabundance of O relative to Fe. A first search for [O/Fe] anomaly was made by Barbuy & Grenon (1990) in dwarf SMRs of various galactic origin. The result was an O overabundance by +0.2 dex at [M/H] = +0.4 instead of -0.2 dex in the case of a slow enrichment in the disc. This early evidence for a prompt enrichment was partially infirmed by Castro et al. (1997) who observed an high eccentricity sample of dwarf SMRs selected by the author. The high photometric [M/H] was confirmed up to +0.5 dex. The [O/Fe] excess seemed not to exceed +0.2 dex, although a systematic offset in O abundance may exist for K dwarfs, as noticed in the Feltzing et al.(1997) analysis. The Feltzing's program contains mainly intermediate age SMRs and only 8 stars whith very small Rp.

If the evidence for a prompt enrichment is weak from the [O/Fe] excesses, a possible r-process anomaly was found by Castro et al. The [Eu/Fe] ratio show excesses by +0.3 dex and up to +0.5 dex for the high velocity star G 161-29 with U,V,W = -133, -84, +39 and [Fe/H]=+0.1.

In the studied sample, the [Ba/Fe] ratio is always subsolar, -0.3 to -0.5 dex. If true, this anomaly would indicate that the time scale for the metal-enrichment up to the SMRs metallicity is shorter than the time delay for the onset of s-process elements production in AGBs. We have here a possible chronometer for the early disc chemical evolution. This anomaly should disappear for intermediate age SMRs.

Even if the oldest SMRs were formed early in the inner Galaxy, their abundances pattern indicate a heavy elements production by both SN II and Sn Ia supernovae.

9. The SMRs migration

The SMRs show apparently conflicting properties. They are among the oldest disc stars, younger than the inner halo by 1 - 2 Gyr at most, but showing the smallest scale height when seen in the solar vicinity, not following the classical σW versus age relation. From 219 K type SMRs with eccentricity $e > 0.25$, the mean maximum distance to the galactic plane is only 0.16 kpc.

In addition SMRs had to be born inside or close to the bulge according to to their metallicity and the observed slope of abundance gradient (-0.05 dex/kpc inside the solar orbit) but their smallest Rp are as large as 3.5 to 5.0 kpc for those observable now in the solar vicinity. Orbital diffusion mechanisms have to be invoked to explain their present kinematical properties.

Mergers seem unefficient since the induced additional σU, σV, σW are small, typically 8-10 km/s and little angular momentum change is produced in case of merging. Stochastic encounters with molecular clouds seem also unsufficient because the acceleration saturates with time, i.e. σU grows as $t^{1/2}$, and little global outwards migration is produced.

On the contrary the galactic bar action is a mechanism very efficient to modify the the inner disc and bulge structures. Models by Fux et al.(1997) predict that most inner disc stars finish trapped inside the bulge, whereas a tiny fraction gains radial energy and expands through the thin outer disc. Secular increases of eccentricity and apocentric distance Ra, but without significant changes of scale height, are also predicted. After 4 Gyr evolution 0.1-0.2% of stars born at 2kpc from the GC, already populate the 7.5 - 8.5 kpc ring. This percentage increases with time as this thin disc component expands.

The bar action leads to a loss of memory about the bulge stars birthplaces. Inner disc stars trapped inside the bulge acquire its kinematics and share the peanut structure. They become kinematically indistinct from bulge stars formed earlier during the collapse.

The question whether SMRs are bulge-like stars or inner disc stars is then less pertinent. Nevertheless we have a clue about the of SMR origin, which is the observed velocity dispersion σW. Simulations show that σW is nearly conserved during the radial expansion if the effect of encounters is neglected. The present σW is then an upper limit for the initial $\sigma W_{\circ}$. The present flatness of the SMRs population is a strong indication of an origin in a thin disc close to the bulge, rather than in a bulge with a significant vertical velocity dispersion inherited from the collapse.

At the Sun distance, the bar modulates the radial motion of all stars with orbital eccentricity e > 0.2. The sign and amplitude of this U motion is function of the angle between the bar axis and the Sun - GC direction. Presently <U> = +25 km/s for V < -50 km/s stars. The asymmetry of the U distribution in the (U,V) plane was interpreted by Raboud et al.(1998) as an evidence of a local bar effect.

An additional effect, crucial for the SMR migration, is the de-stabilisation of galactic orbits for specific domains in the (U,V) plane. The area V -40 to -100 and U -20 to +120 km/s is a zone of chaotic orbits occupied by the majority of SMRs. Stars moving on chaotic orbits may wander erratically from inside the bar to outside corrotation with rapid changes of eccentricity and Rp. The available models do not predict the presence of SMRs in the quadrant V 0 to -50 and U -10 to -80. A combination of spiral arms and bar effects should be investigated.

The bar acts on all stars, irrespective to their ages and metallicities, provided they travel on orbits eccentric enough to bring them close to or inside the bar. A consequence of the orbital diffusion in the disc is a smoothing of the metallicity gradient for the oldest stars.

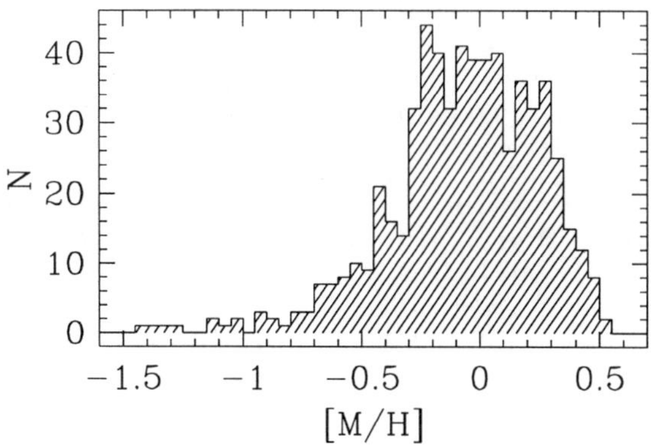

Figure 5. The metallicity distribution for 612 NLTT stars with Rp $\leq$ 5.5 kpc and absolute vertical velocity less than 20 km/s.

Figure 9 shows the metallicity distribution for the inner disc population defined by Rp < 5.5 kpc and |W+10| ≤ 20 km/s. The overall distribution is identical to that observed in the bulge when thick-disc stars are removed. In this selection SMRs with [M/H] ≥ +0.30 represent 11% of the sample. The mean metallicity is -0.02 dex, similar to that observed now in the solar vicinity for recently formed stars. The SMRs kinematical evolution is not at all distinct from that of other less metal-rich inner disc stars. Their peculiar metallicity, higher than that of the ISM now in the solar vicinity, makes them unambiguous tracers of the early inner galactic evolution.

10. Conclusions

1. Extreme SMRs with smallest Rp were formed early, about 9 Gyr ago, in the inner thin disc. They were scattered in and out mainly through the action of the galactic bar. 2. SRMs with intermediate Rp alike the metal-rich members of HR 1614 Eggen's group, were formed later in the inner disc, possibly 4 to 5 kpc away from the galactic center. The majority travel on chaotic orbits allowing a fast radial migration. 3. The SMR phenomenon for younger and evolved stars may result from line intensities enhancement in N-enriched atmospheres, for stars formed from metal-rich ISM as some members of the Hyades group and part of VSL giants.

References

1. Barbuy, B., Grenon, M., 1990, in *"Bulges of Galaxies"*, Jarvis et al. Eds., Garching, ESO/CTIO Workshop, pp. 83-87
2. Campbell, B., Lambert, D.L., Maillard, J.-P., 1990, *PASP*, **102**, pp. 79-87
3. Castro, S., Rich, R.M., Grenon, M., Barbuy, B. and Mc Carthy J.K., 1997, *Astron. J.*, **114**, pp. 376-387
4. Chavez, M. *et al.*,1998,*Revista Mexicana Conf. Ser.*, in press
5. Fux R., 1997, *Astron. Astrophys.*,**327**, pp. 983-1003
6. Grenon, M., 1972, in *"L'Age des Etoiles"*, G. Cayrel et al. Eds., pp. LV 1-6
7. Grenon, M., 1989, *Astrophysics and Space Sci.*, **156**, pp. 29-37
8. Luck, R.E.,Challener S.L., 1995, *Astron. J.*,**110**, pp. 2968-3008
9. Mishenina, T.V.,1996, *Astron.Astrophys.Suppl.*,**119**, pp. 321-328
10. Peterson, R., 1976, *Astrophys. J. Suppl.*, **30**, pp. 61-83
11. Raboud, D., Grenon, M., Martinet *et al.*, 1998, *Astron.Astrophys.*, **335**, pp. L61-64
12. Schaller, G., Schaerer, D. *et al.*, 1992, *Astron.Astrophys.Suppl.*, **96**, pp. 269-331
13. Spinrad, H., and Taylor, B. J., 1969, *Astrophys. J.*, **157**, 1279
14. Taylor, B.J., *Astrophys. J.Suppl.*, **102**, pp. 105-128
15. Tripicco, M.J., Bell, R.A. *et al.*, 1995, *Astron. J.*, **109**, pp. 1697-1705

Speaker: M. GRENON

M. TOSI: Do you think that the Sun could also have been affected by the bar action and have moved significantly from its birthplace, as suggested by some people (e.g. Wielen et al.)?

M. GRENON: The arguments to propose an inner disc birthplace for the Sun is based on a subsolar metallicity for the youngest stars in the solar vicinity as in Orion. At a given epoch there is a natural scatter in the gas at given R_{GC}, of ± 0.1 dex, thus there is no need for an exotic birthplace for the Sun. In addition, the present orbit of the Sun occupies an island of total stability with respect to the bar perturbation.

R. WYSE: As you said, the orbital diffusion will dilute abundance gradients. Have you estimated the amplitude of the effect? If so, how do the predictions agree with the observations of the gradient in e.g. K giants of Lewis-Freman?

M. GRENON: The gradient dilution is efficient only for old stars with eccentric orbits. The gradient for oldest disc stars is indeed zero inside the Sun orbit. If an unbiased sample of K stars is used (of low to intermediate ages) the gradient should be nearly the same as for the gas at the present time.

ABUNDANCES IN SUPERNOVAE

OBSERVATIONAL ASPECTS

I.J. DANZIGER
Osservatorio Astronomico di Trieste
Via G.B. Tiepolo, 11, I-34100, Trieste

1. Abstract

A review is presented of the various methods that have been used to determine the abundances of elements produced in supernovae. The methods range from spectroscopic analyses at early optically thick phases of the expanding envelope through the later optically thin nebular phases, for both of which optical and infrared spectra have been used. Since the luminosity of supernovae of all types are in general powered by radiocative decay of heavy elements, the study of light curves provides another effective method for determining the amounts of this radioactive material. Direct observations of the various gamma-ray lines produced in these decays provide other independent estimates of the mass of some radioactive elements. An attempt is made to fit some of these observations into a consistent framework of understanding progenitors, but the emergence of hypernovae in the last year demonstrates that there remain many surprises which could in the future affect our understanding of the processes of chemical enrichment of our and other galaxies.

2. Sources of Abundances

Here we briefly describe the various methods used for determining abundances in supernovae, together with comments elucidating the advantages and disadvantages of the methods. In general the methodologies can be applied to supernovae of all types be they thermonuclear Type Ia SNe or core collapse Type Ib/c and Type II, although specific attention will be drawn to examples of elements for which the method cannot be reliably used. I illustrate only with examples where quantitative results have been reported.

F. Matteucci and F. Giovannelli (eds.), The Evolution of the Milky Way, 59–74.

- 1. Photospheric analyses. This type of analysis is analogous to that used for the absorption line spectra of the Sun and stars, except that it is done in the framework of an expanding atmosphere which results in the redward shift of photons as they are absorbed and re-emitted by lines of different ions. As for stellar spectra one obtains information on abundances at or above the photosphere and information on total masss of any nuclear species would require an inward extrapolation of structure, both density and composition, used to model the outer part. At the present stage of supernova research such an extrapolation would not be considered a reliable procedure, at least to an observer. However as time progresses the effective photosphere of the envelope recedes inward relative to the outer boundary which means that analyses at successive times can give information on abundances at various depths. There should of course be consistency from one phase to another. This method is applicable to phases where an effective photosphere dominates the spectrum, that is from outburst up to about 20-30 days past maximum light, depending on the particular supernova. After that time one sees the gradual emergence of emission lines as a result of the expansion and its extension into an optically thin envelope. Spectra taken in the UV, optical and IR have been used for this purpose. Required ingredients for testing model results against observations are abundances and density gradients in the envelope, this latter parameter being the most intractable of the important ones.
- 2.Nebular analyses. This method is analogous to that used in analysing spectra of gaseous nebulae, and therefore requires in general, a knowledge of the density and temperatures in the line-emitting regions. These two quantities are never easy to determine with the limited availability of diagnostic lines, but in some cases fine structure transitions are detected and are useful. In addition it must always be remembered that in an extended expanding envelope neither temperature nor density are uniform. We also know that clumping of this expanding gas occurs, adding to the difficulty.
  Optical and near and mid-IR spectra have been used to good effect, but an enormous step forward would be achieved if the far IR (accessible with a sensitive space telescope) provided the important fine structure lines of many ions including those of C,N and O. Additional considerations include critical densities, optical depth effects, and states of ionization. Examples of optical photospheric and nebular spectra of SN1987A have been presented by Danziger et al.(1991) together with IR nebular spectra.
- 3.Bolometric Light Curves. These represent in principle the complete budget of energy released at any time. Since the radioactive β-decay

of ^{56}Ni into ^{56}Co and thence into ^{56}Fe is the main source of energy powering the light curves through γ-ray trapping and positron deposition with subsequent thermalisation of the electrons, one can determine the mass of this radioactive material by measuring the total energy released at any time. Observationally the determination of a bolometric light curve is made difficult by the requirement that the complete relevant wavelength range is monitored. In addition one needs to determine or have a good understanding of the distribution of radioactive material in the envelope and hence a reliable estimate of the fraction of radioactive energy deposited and the fraction that escapes. Results from light curves around maximum are more model dependent than those from later phases where one can see in many cases of Type II SNe an almost precise tracking of the exponential radioactive decay.
- 4.γ-Ray Emission Lines. The various radioactive species emit discrete lines which can and have been detected with appropriate space-borne detectors. To arrive at a total mass of emitting material one requires a model which provides an estimate of the fraction which escapes taking into account Comptonisation of γ-rays into X-rays.

3. Photospheric Spectra

Spectral synthesis codes have been used by a number of authors such as, for example, Harkness(1986), Branch et al.(1985), Hoeflich(1991). We discuss mostly results obtained with a Monte Carlo code originally devised by Lucy(1987) for analysing spectra of SN1987A and subsequently adapted by Mazzali and colleagues for work on other SNe, because this work has put most emphasis on abundance determinations.

The synthesis modelling of the UV spectrum of the Type II SN1987A by Lucy(1987) showed that LMC abundances in the outer envelope gave a satisfactory fit. With the exception of the BaII lines Fosbury et al.(1987) showed a reasonable fit in the optical region also. The BaII lines were seen to be very strong and narrower than other aborption features. Mazzali et al.(1992) modelled the early spectra to show that barium and another s-process element strontium were enhanced relative to LMC abundances by a factors of 5 and 2, and that barium was confined to velocities $<$7000ksec^{-1}, and therefore could not have extended to the surface of the progenitor star. Thus SN1987A provides evidence for s-processing in the interiors of massive stars.

Photospheric modelling of SNIa spectra has been able to contribute to our understanding that there is a range of properties among this class. The modelling of SN1990N by Mazzali et al. (1993) showed the power of the method. While it is not surprising that the W7 model of Nomoto et al.

(1984) should give a reasonable fit to the observations, since this model was adapted to meet the requirements of early observations of Type Ia SNe, it showed that mixing of the stratified layers was required and that elements of intermediate (Ca,Si) and heavy (Co,Fe) mass co-existed in the outer layers of the envelope.

Another SNIa modelled by Mazzali et al.(1995) the intrinsically bright SN1991T was of interest because of possible abundance differences seemingly apparent in the phases before maximum light when the SiII6355 lines were weak. This apparent anomaly was due mainly to a higher temperature causing much of the Si to be in the form of SiIII. Therefore whatever the reasons for the higher luminosity of this SN, it remains to demonstrate that significant abundance differences exist.

However an analysis of the intrinsically faint Type Ia SN1991bg by Mazzali et al.(1997), does point to a possible real difference from brighter SNIa. At maximum light a broad absorption feature is apparent near 7500A which has been well modelled with the OI 7772 triplet. Future work must try to demonstrate that this is a real abundance effect which differentiates it from brighter SNIa, and not a result of differences in the physical conditions obtaining in the smaller cooler envelope of this SN.

We can summarize this section on SNIa by pointing out that the analyses referred to above, as well as others for SN1994D (Patat et al.1996), suggest that the total mass of the envelope and the exploding mass correlate with the intrinsic luminosity. We return to this point using results from the nebular spectral analysis.

4. Nebular Spectra

With nebular lines one is in principle able to estimate the total mass of an ion provided the envelope is optically thin. The only Type II SN for which this has been done in an extensive way is SN1987A, because of the temporal and wavelength coverage and the quality of the data. But even so it has not been possible to derive an accurate mass of oxygen from the [OI]6300,63 lines because an excitation temperature could not be determined. This is unfortunate because oxygen is surely one of the main constituents of the ejecta.

The determination of the ^{56}Co and ^{57}Co masses was more successful with nebular spectra, because the ground state fine structure transitions for the dominant ion of CoII fall in an atmospheric window at 10.5μ, and could be observed for 500 days. The strength of these lines is insensitive to temperature. Because we know the half lives of these 2 unstable isotopes and they are quite different, being 77 and 270 days respectively, Danziger et al.(1991) were able to compare the observed line strength with that expected from the decreasing mass of these 2 isotopes. This resulted in a mass of $0.07 M_\odot$ of ^{56}Co and a $^{57}Co/^{56}Co$ ratio of 1.5 times the solar ratio of the corresponding Fe isotopes. Varani et al.(1990) used the ratio of FeII and CoII lines of higher but similar excitation to arrive at a similar conclusion to the one above.

A less direct method (Lundquist et al.1999) using IR nebular spectra of SN1987A taken with ISO provides a means of determining the original mass of ^{44}Ti assumed to be the energy source at day 3999 when the spectra were taken. This model allows for mixing of the radioactive ^{44}Ti (half life 61.7 years) out to a velocity of 2000ksec^{-1} in the envelope and by calculating the heating and cooling, estimates the strength of ground state fine structure lines of FeI, FeII, and FeIII. The strongest of these [FeII]25.99μ was not detected and an estimate of the upper limit to its strength provides an upper limit to the mass of ^{44}Ti=1.5x10$^{-4}M_\odot$.

The following table summarizes results obtained from nebular spectra of SN1987A at day 421 (Danziger et al. 1991).

TABLE 1. Nebular Abundances, SN1987A

Ion(line)	$M/M_\odot$	Factor
CI(9823,49)	0.036	(x2)
OI(6300,63)	0.2-3	
SiI(1.644μ)	0.034	(x3)
ArII(6.98μ)	0.0005	(x1.5)
CaII(7291,7324)	<0.007	
FeI(1.423,62μ)	0.043	
FeII(1.248,27μ)	0.098	
CoII(10.52μ)	0.0019	(x1.5)
NiI(3.12μ)	0.0007	(x3)
NiII(6.63μ)	0.0014	(1.5)

The third column is an approximate multiplication factor to apply for unseen stages of ionization to obtain total element abundance. Note that the value for Ca is an upper limit, and that a more sophisticated analysis by Li and McCray(1993) employing a 5-level atom and including clumping, suggests a mass of Ca an order of magnitude lower than this limit, and that therefore one is simply measuring the primordial mass of Ca in the progenitor star.

The nebular spectra of SNIa have many similarities to one another and notable differences. The similarities lie in the presence of lines of [FeII,III] and [CoIII] spread over the optical range. These identifications were convincingly demonstrated first by Meyerott(1980) and Axelrod (1980). Kuchner et al.(1994) have made a simpler independent test in which a broad feature near 4650A ascribed to [FeIII] has been measured relative to another broad feature at 5900A ascribed to [CoIII]. They have shown that this ratio increases with time in quantitative agreement with what is expected when Co decays into Fe. More detailed modelling by Mazzali and colleagues has quantified differences within the family of Type Ia. The immediate obvious difference is in the widths of the emission lines in different SNe observed at similar phases. Mazzali et al.(1998) have shown that these widths correlate with the absolute luminosity at maximum. Figure 1 taken from the work of Turatto et al.(1996) highlights the differences in the nebular spectra of Type Ia SNe taken at similar phases.

Infrared spectra of 4 SNIa have been modelled by Bowers et al.(1997) who obtain masses of ^{56}Ni in the range 0.37-0.86$M_{\odot}$. The full range in observed mass of Ni-Co-Fe produced in SNIa is provided by the work of Spyromilio et al.(1992) who determined from IR nebular spectra an estimate of $1M_{\odot}$ for the luminous SN1991T, and by Mazzali et al.(1997) who determined a mass of $0.1M_{\odot}$ or less for the sub-luminous SN1991bg. This agrees with the trends shown by the photospheric modelling. It also shows observationally that luminosity, velocity, shape of light curve, mass of radioactive material are related. And it demonstrates that the family of Type Ia SNe embraces a significant range of properties.

Strangely enough, although oxygen lines have been reported present in the photospheric spectra of Type Ia (SN1991bg discussed above), emission lines of this element have never been reported. If the progenitors are C-O white dwarfs, this demonstrates observationally that the thermonuclear reactions have been very efficient in converting intermediate mass elements into Fe. Present indications are that the mass of radioactive material is less than that of the envelopes (see next section) so observationally it is of interest to understand the composition of the remaining mass.

5. Light Curves

Light curves of SNe vary in luminosity and shape with type. SN1987A played a very valuable role in demonstrating that radioactive decay of Ni-Co-Fe was the main source of energy at least within the first 1000 days. This was because after 100 days or so the bolometric light curve started falling with a timescale similar to that given by the the half life of ^{56}Co. In fact the radioactive decay of ^{254}Cf was first proposed as the source of energy for the Type Ia SN1937C. The idea was correct but the radioactive isotope was wrong just because there is significant escape of γ-rays from

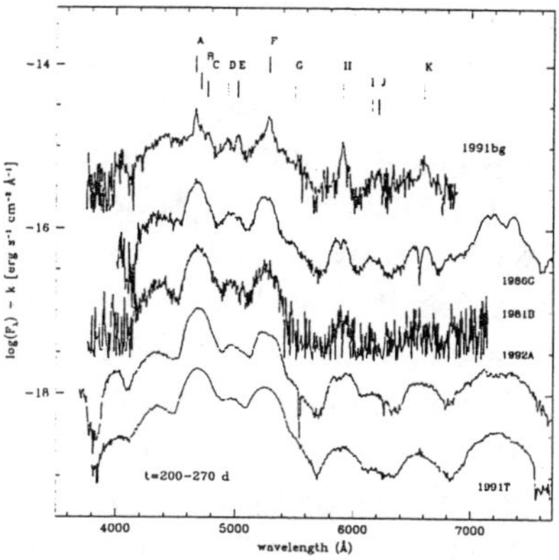

Figure 1. A comparison of the nebular spectra of 5 Type Ia SNe taken at similar phases. Most of the emission features are lines of [CoII,III] and [FeII,III] and systematic differences in line width can be correlated with luminosity.

Type Ia envelopes leading to a steeper decline than that suggested by ^{56}Co decay but consistent with ^{254}Cf.

Information concerning the masses of radioactive material is imprinted in the light curves at or near maximum light but more detailed modelling is required to extract an accurate measure of the mass. (See for example the discussion by Woosley(1988) for SN1987A). In what follows much more emphasis will be placed on results from obtained from measurements of the "tail" of the light curves which develop at approximately 100 days for Type II and earlier for Type I SNe.

Before discussing results it is worth recalling that there are hazards with this approach. In principle the tail of a bolometric light curve should lead directly with little modelling to a mass of the radioactive ^{56}Co first and then subsequently ^{57}Co and ^{44}Ti. However there are other factors which can affect an easy measurement of the radioactive luminosity. They include light echoes caused by reflection of SN light by interstellar material not on the line-of-sight. Echoes have been reported and studied in the case of SN1987A and SN1991T. Interaction of the SN envelope with circum-stellar material will result in some conversion of kinetic energy into radiative energy thus enhancing the luminosity. The moment at which this occurs depends on the distance of the CSM from the SN. This effect has been observed in more than 12 SNe and a large range in delay times after explosion has been recorded. Two outstanding examples of this type of interaction are SN1979C and SN1988Z. The modelling of this phenomenon though important is hardly relevant to our discussion of abundances.

Another effect internal to the supernova is "freeze out" whereby the energy stored in the envelope at an earlier phase through ionization is released later through recombination. This has been computed for SN1987A by Franssen and Kozma(1993), but did not affect the results for ^{56}Co, but could obscure the role of the succeeding isotopes.

If the SN exploded with a backgound of stars or within a group, more probable in the case of Type Ib,c and Type II than with Type Ia, at a sufficient distance these stars would not be seen or resolved from the SN. That being so they could contaminate the SN light increasingly as the SN faded. In the case of SN1987A one was able to see and subtract the effects of 2 neighbouring stars, but at the distance of the Virgo cluster they would have gone undetected except for the effect on the light curve. Lastly one knows that some SNe produce pulsars and these are capable of adding energy to the light curve as they spin down. In fact the Crab nebula is the outstanding example, but so far evidence for such an effect in recently observed SNe has been notably absent.

The results from SN1987A were particularly important because a real bolometric light curve was available through multi-colour observations. We refer to the work of Woosley (1988) for the early decline, and those by Bouchet and Danziger (1993) for the later epochs. During the interval 100-1000days the bolometric light curve of this SN yielded a mass of $^{56}Co = 0.07M_{\odot}$, consistent with the results from other methods. After that time the results for the succeeding isotopes using this method are less secure because of the competing effects of "freeze out", and the fact that the uncertainties in the bolometric magnitudes were increasing owing to the inability to monitor the thermal IR where most of the radiation was emerging as a result of thermalized dust emission.

It is now apparent that at maximum light there is a wide range of almost 5 magnitudes in luminosity of SN II. For a sample of these this does not mean a range of the same order in mass of radioactive material, but rather a large range in the size of the envelope of the exploding star. In a sample of Type II SNe where this range occurs (SNe1992am,1987A,1969L,1983K) one finds that all return to a similar level on the exponential tail, implying a mass of ^{56}Co similar to the well established amount in SN1987A. SNe for which interaction with the CSM is established must be excluded. Nevertheless there are still oddities such as SN1997D where Turatto et al.(1998) have shown that along with a low velocity of expansion and a low temperature of the envelope there occurred a dramatic drop in the luminosity after about 100days and the luminosity on the ensuing exponential tail implied a mass of ^{56}Co of $0.002M_{\odot}$. Although Zampieri et al.(1999) have proposed the formation of a black hole resulting from collapse, it is not clear observationally why this has occurred in this case. Clearly the question of the range of masses of radioactive material and hence Fe production is an important one and can only be answered with many more observations.

Modelling of the early light curve of the Type IIb SN1993J, embracing maximum light, has been done by Nomoto et al.(1993), Shigeyama et al.(1994), Woosley et al.(1994), and Young et al.(1995). The agreement among these various results is very good. A binary system is required to have transferred most of the hydrogen envelope from the progenitor star because only a little hydrogen was seen in the early phases. The star which exploded was a helium star of mass not exceeding $4M_{\odot}$ whose main sequence progenitor had a mass in the range 12-17$M_{\odot}$ and produced not less than $0.07M_{\odot}$ of ^{56}Ni or even twice that amount according to the last of the above authors.

The Type Ic core-collapse SN1994I has been modelled in a similar fashion by Nomoto et al.(1994), Iwamoto et al.(1994), and Young et al.(1995). Since normal Type Ic SNe do not show traces of helium, the favoured exploding star is a C-O star of about 3-4$M_{\odot}$, originating from a main sequence star of 13-20$M_{\odot}$ and producing 0.07-0.15$M_{\odot}$ of ^{56}Ni. Again the fitted data extends to about 50 days past maximum. However the recognition of helium via the HeI10830 in the spectrum of the hypernova SN1998bw, an apparently bright Type Ic event (see below) raises the question of just how important the presence of helium is in defining the progenitor star and whether it plays a significant role in the explosion.

Many papers have now been published elucidating the range in shapes and luminosities of Type Ia SNe. The Figure 2 taken from Turatto et al.(1996) shows clearly the difference in both shape and luminosity between the sub-luminous SN1991bg and the relatively normal SN1992A. In fact by overlaying light curves of SNIa covering the full range of luminosity at maximum one can see that the differences seen there persist and even increase out to the limits of observation on the exponential tails. Can these differences be quantified in terms of the mass of radioactive material produced. Cappellaro et al.(1997) have developed a simple Monte Carlo light curve model to simulate luminosity evolution of a sample of 5 SNIa spanning the full range of observed properties. Deposition of γ-rays and positrons is considered. The luminosity at maximum is sensitive to the mass of radioactive material and the mass of the envelope dictates the shape of the light curve at early and late stages depending on the deposition. The sample considered included the over-luminous SN1991T and the sub-luminous SN1991bg.Two interesting results emerge in a quantitative way even though bolometric light curves were not available. There is a range in mass of ^{56}Co required to match all examples, and a corresponding range of a factor of 2 in the masses of the ejected envelope. This therefore is consistent with results from previous methods and also raises 2 important questions. The first is of dir-

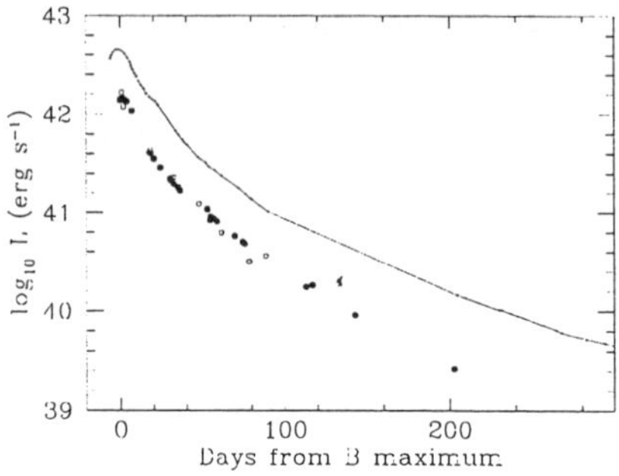

Figure 2. A comparison of pseudo bolometric light curves of SN1991bg (symbols) and SN1992A showing the difference in shape and luminosity.

ect concern to theories of chemical evolution of galaxies, namely what is the rate of occurrence of objects in the various ranges of luminosity. The second is whether we are seeing evidence for the explosion of sub-Chandrasekhar mass white dwarfs.

A final point concerns whether one is not making too much of the example of the sub-luminous SN1991bg. It is not unique and the example of the very similar SN1997cn discussed by Turatto et al.(1998) testifies to that. The full range is gradually being populated with more examples.

6. Gamma-rays

A discusion of γ-rays from supernovae are included in a review by Diehl and Timmes(1998) where more detailed references are available.The γ-ray lines at 847keV and 1238keV from radioactive decay of ^{56}Co and the one 122keV from ^{57}Co have been detected from SN1987A, while upper limits exist for the detection of the 1159,511keV lines from the decay of ^{44}Ti.In fact a light curve for the 847 and 1238keV lines extends over several hundred days, and the summation of all the data results in the detection of even weaker γ-ray lines at other energies. These measurements result in the following masses: ^{56}Co - $0.07M_\odot$; ^{57}Co - 2.5x10$^{-3}M_\odot$;^{44}Ti - <1.5x10$^{-4}M_\odot$. These values are in good agreement with those derived by other means discussed previously. Parenthetically it should be noted that the radioactive decay of ^{56}Ni results in γ-ray line emission at 3 energies in the keV range. But the short half-life of this isotope ensured that observations could not be undertaken before it had decayed to negligible amounts.

In fact the bright Type Ia SN1991T is the only other SN for which detection of the 847,1238 keV lines has been reported by Morris et al.(1998) using COMPTEL. The mass of ^{56}Co computed from this observation lies in the range 1.3-2.3$M_\odot$ depending on the assumed distance 13-17Mpc. Even the lower value exceeds determinations by other methods and is uncomfortably close to the Chandrasekhar limit for a white dwarf commonly believed to be the progenitor of such SNe. There is still some discussion concerning the calibration of this data.

Iyudin et al.(1997) using COMPTEL have reported the detection of the 1159keV of ^{44}Ti from the young remnant Cas A, with a resulting mass of this isotope derived to be 2.4x$10^{-4}M_{\odot}$. Although this is higher than in SN1987A there is some discussion that this measurement may be too high by a factor 2.5. Although a mass of ^{44}Ti has not been determined there has been a detection of the same line in the direction of an apparently new Xray SNR superimposed on a section of the Vela SNR by Iyudin et al.(1998).

7. Hypernovae

A recent development in SN research has been the discovery of a very luminous SN1998bw associated with a γ-ray burst source GRB980425. This SN lies in the spiral arm of a sub-luminous galaxy, and the coincidences in time and position make the association convincing. When discovered it had characteristics that apart from the GRB association made it seem unique. It was the brightest radio SN ever observed. The optical spectrum was peculiar and made classification difficult. However it has since been classified as Ic but with a much larger velocity than normally occurs in such SNe. It is mainly this huge velocity that affected the appearance of the spectrum. The earliest spectra taken 8 days after the explosion showed that relativistic velocities exceeding 60000kmsec^{-1} were apparent. Identification of absorption features due to SiII and CaII made this determination of large velocities possible.

In this case the early light curve and the spectra have been used together to converge on a model which also specifies a mass of ^{56}Ni. There are a whole range of models with different combinations of explosion energy and ejecta mass which can reproduce the light curve. This is because, as shown by Iwamoto et al.(1998), the photon diffusion time is dependent on different powers of both these quantities. Thus the expression for kinetic energy provides a second equation involving the energy and mass with the velocity determined observationally. Energy and mass are then uniquely determined and of course these parameters must lead to a synthetic spectrum that is in accord with the observed spectrum resulting from a judicious choice of abundances of intermediate mass elements. The result was presented by the above authors who obtained a mass of $^{56}Ni = 0.7M_{\odot}$, a value dictated by the luminosity at maximum light.

This modelling work differs from that previously described in employing a combination of light curve and spectra to arrive at a unique solution.

The final conclusion was that the mass of the exploding star was $13.8M_{\odot}$ and that of the ejecta was $10.9M_{\odot}$. This object looks like the C-O core of a much more massive star. It may have left behind a compact object. The identification of the HeI10830 line as an outer layer of the expanding envelope may complicate this picture a bit. A strong [OI]6300 line develops in the nebular phase but a mass determination is hindered by the lack of a well determined temperature.

Of additional interest is the fact that at 2 epochs in the first weeks linear polarization was measured in the spectra. See Danziger et al.(2000). Though small (<1 percent), it could still represent an indication of sizeable departures from spherical symmetry depending on the viewing angle. Polarization of other Type Ib/c SNe has been reported, and so in the future there will be increasing emphasis on modelling asymmetric explosions induced by the effects of angular momentum in the exploding star, a matter stressed by the work of Woosley et al.(1993), MacFayden and Woosley(1999), and Khokhlov et al.(1999). So far the emission line profiles do not unambiguously indicate departures from spherical symmetry. Nevertheless it is worth noting that many oxygen-rich young supernova remnants reveal structures that are ring-like rather than spherically symmetric.

Following this discovery other SNe with somewhat similar characteristics have been identified. The association with GRBs is not as secure but there is plausability to an association in the case of SN1997ef. Modelling by Iwamoto et al.(2000, in press) suggests that all the above parameters derived for SN1998bw are less extreme for this object.

Another class of SN which is very energetic is exemplified by SN1997cy whose association with a GRB is much less secure. In addition this is a Type II SN where the evidence for interaction with the CSM is strong, even though the observations would also allow the presence of $0.7M_{\odot}$ of ^{56}Ni in addition to the interaction energy. While its luminosity at maximum places it in the class of hypernovae and the current modelling with interaction (Turatto et al.2000, in press) is suggestive of a massive progenitor of $25M_{\odot}$, the lack of lines of oxygen and magnesium at any phase poses problems for a model requiring a high mass progenitor.

Why are these types of objects important in the study of chemical evolution of galaxies? The rate of occurrence of such objects in the nearby universe seems quite low. However the progenitor masses seem very high, which suggests that at much earlier epochs, when star formation was more pronounced, there could have been a much higher rate of occurrence. Since they appear to produce prodigious amounts of iron as well as oxygen this might have an impact on how the O/Fe ratio evolved with time.

8. Conclusion

The above discussion concerning the mass of ^{56}Ni determined from the light curves of core-collapse SNe allows one to begin to assemble a relationship between this mass and the mass of the main sequence progenitor. This has been done by Iwamoto et al.(2000) where the results for SN1993J, SN1994I, SN1987A, SN1997ef, and SN1998bw reveal a trend going from $0.07M_{\odot}$ of ^{56}Ni at a progenitor mass of $15M_{\odot}$ to $0.7M_{\odot}$ at $40M_{\odot}$. The Type II SN1997D with its low mass of ^{56}Ni is an obvious deviant point regardless of its progenitor mass. At the moment with only 5 points and a trend dominated by Type Ic SNe, one is not in a position to generalize nor to confidently extend this trend to lower progenitor masses. But it exemplifies an aim of future work. And although the work so far emphasizes quantitative results for Type Ib/c SNe, it raises the question whether most or all core collapse SNe originating from main sequence stars with masses $>20M_{\odot}$ explode after a He-C-O core has been laid bare by some form of mass loss. Also to be fair, it should be noted that in the case of the first hypernova SN1998bw (GRB980425), later observations of the light curve by McKenzie and Schaefer(1999) and modelling of asymmetric effects in explosions by Wheeler et al.(1999,in press) indicate that the mass of Fe produced could be significantly lower than that referred to earlier. Clearly this is an unfinished story.

From an observational point of view the quantitative determination of relative abundances and masses of elements is still sparse though gradually increasing.

Our lack of quantitative information on intermediate mass elements such as oxygen and magnesium is particularly frustrating because of the key role they play in the diagnostics of chemical evolution of our and other galaxies. Ultimately we may have to rely on modelling methods outlined by Franssen et al.(1996)who, by modelling the non-thermal excitation of [OI]6300 emitted by SN1987A after day 750, when thermal excitation has become unimportant, obtain a mass of oxygen $\geq 1.5\ M_{\odot}$.

The most reliable estimates are for the radioactive elements that lead to Fe production. This is true for SNe of all types. But even here we lack information on the composition of the residual mass particularly for Type Ia SNe.

Also what is lacking there are reliable statistics on the relative rates of different types even in our local universe. A much bigger gap occurs in our knowledge of how these rates change with redshift or look-back time. The occurrence of hypernovae with different characteristics as well as the occurrence of such objects as SN1997D, the Type II low Fe producer warns us that there are probably still new varieties of SNe out there remaining to be identified. Consequently the case for continuing to find and study in detail virtually all SNe over as wide a redshift range as possible is stronger than it ever was!

I thank E.Cappellaro for bringing my attention to some important references.

References

1. Axelrod, T.S. 1980, PhD thesis, UCRL-52994, Univ. California, Santa Cruz
2. Bouchet, P. et al. 1991, *AJ*,102,1135
3. Bouchet, P., Danziger, I.J. 1993, *A&A*,273,451
4. Bowers, E.J.C. et al. 1997, *MNRAS*,290,663
5. Branch, D. et al. 1985, *ApJ*,294,619
6. Cappellaro, E. et al. 1997, *A&A*,328,203
7. Danziger, I.J. et al. 1988, in Supernova 1987A in the Large Magellanic Cloud, Proc.of the fourth George Mason Astrophysic Workshop, ed. M.Kafatos and A.G.Michalitsianos, Cambridge Uni. Press,37
8. Danziger, I.J. et al. 1991, in Supernovae, The Tenth Santa Cruz Workshop in Astronomy and Astrophysics, ed. S.E.Woosley, Springer-Verlag,69
9. Danziger, I.J. et al. 2000, in Supernovae and Gamma-Ray Sources ed. M.Livio, in press
10. Diehl, R., Timmes, F.X. 1998, *PASP*,110,637
11. Fosbury, R.A.E.F. et al. 1987, in The ESO Workshop on the SN1987A, ed. I.J.Danziger, (ESO),139
12. Franssen, C., Kozma, C. 1993, *ApJ*,408,L25
13. Franssen, C. et al. 1996, in Supernovae and Supernova Remnants, IAU Coll. 145, Cambridge Uni. Press,211
14. Harkness, R. 1986, in Radiation Hydrodynamics in Stars and Compact Objects, ed. D.Mihalas and K.-H.A.Winkler, Springer-Verlag
15. Hoeflich, P. 1991, in Supernovae, The Tenth Santa Cruz Workshop in Astronomy and Astrophysics, ed. S.E.Woosley, Springer-Verlag,415
16. Iwamoto, K. et al. 1994, *ApJ*,437,L115
17. Iwamoto, K. et al. 1998, *Nature*,395,672
18. Iwamoto, K. et al. 2000, *ApJ*,in press
19. Iyudin, A.F. et al. 1997, in Proc.Second INTEGRAL Workshop, The Transparent Universe, ed. C.Winkler, T. J.-L.Courvoisier, & P.Durouchoux (ESA SP-382),37
20. Iyudin, A.F. et al. 1998, *Nature*,396,142
21. Khokhlov, A.M. et al. 1999,*ApJ*,524,L107
22. Kuchner, M.J. et al. 1994, *ApJ*,426,L89
23. Kurfess, J.D. et al. 1992, *ApJ*,399,L37
24. Li, H.-W., McCray, R. 1993, *ApJ*,405,730
25. Lucy, L.B. 1987, *A&A*,182,L31
26. Lundquist, P. et al. 1999, *A&A*,347,500
27. MacFayden, A.I., Woosley, S.E. et al. 1999, *ApJ*,524,262
28. Mazzali, P.A. et al. 1992, *A&A*,258,399
29. Mazzali, P.A. et al. 1993, *A&A*,269,423
30. Mazzali, P.A. et al. 1995, *A&A*,297,509
31. Mazzali, P.A. 1997, *MNRAS*,284,151
32. Mazzali, P.A., et al. 1998, *ApJ*,499,L49
33. McKenzie, E.H., Schaefer, B.E. 1999, *PASP*,111,964
34. Meyerott, R.E. 1980, *ApJ*,239,257
35. Morris, D.J. et al. 1998, in AIP Conf. Proc. 410,,Fourth Compton Symp. on Gamma-Ray Astronomy and Astrophysics, ed. C.Dermer,J.Kurfess, & M.Strickman (New York: AIP), 1084.

36. Nomoto, K. et al. 1984, *ApJ*,286,644
37. Nomoto, K. et al. 1993, *Nature*,364,507
38. Nomoto, K. et al. 1994, *Nature*,371,227
39. Patat, F. et al. 1996, *MNRAS*,278,111
40. Shigeyama, T. et al. 1994, *ApJ*,420,341
41. Spyromilio, J. et al. 1992, *MNRAS*,258,53P
42. Turatto, M. et al. 1996, *MNRAS*,283,1
43. Turatto, M. et al. 1998, *ApJ*,498,L129
44. Turatto, M. et al. 1998, *AJ*,116,2431
45. Turatto, M. et al. 2000, *ApJ*,in press
46. Varani, G.-F. et al. 1990, *MNRAS*,245,570
47. Wheeler, J.C. et al. 1999, *ApJ*,in press
48. Woosley, S.E. 1988, *ApJ*,330,218
49. Woosley, S.E. et al. 1993, *ApJ*,405,273
50. Woosley, S.E. et al. 1994, *ApJ*,429,300
51. Young, T.R. et al. 1995, *ApJ*,449,L51

Speaker: I.J. DANZIGER

G. ISRAELIAN: Observations show that many SN remnants are not chemically homogeneous and that the explosions are not spherically symmetric. I wonder how accurate are the abundances derived from spherically symmetric and homogeneous models of expanding photospheres.

I.J. DANZIGER: There is certainly growing observational evidence from polarization measures of light from early epochs that some SN-Ib/c and SN-II are asymmetric. Apparent ring structures particularly in young oxygen rich remnants also indicate asymmetries originating at the explosion. If we confine our determinations to gamma-rays or nebular spectra (provided we know physical conditions, a caveat that obtains whether asymmetries exist or not) the determination of masses should be independent of the existence of asymmetries.

OBSERVATIONS OF CANDIDATE SMR STARS

M.L. MALAGNINI[1]
Dipartimento di Astronomia, Università degli Studi di Trieste
Via G.B. Tiepolo 11, 34131 Trieste, Italy
C. MOROSSI[2]
Osservatorio Astronomico
Via G.B. Tiepolo 11, 34131 Trieste, Italy
A. BUZZONI[3]
Osservatorio Astronomico di Brera
Via Bianchi 46, 23807 Merate (Lc), Italy
AND
M. CHAVEZ[4]
Instituto Nacional de Astrofísica, Optica y Electrónica
A.P. 51 y 216, 72000 Puebla, Mexico

1. Introduction

The main objectives of our project on super metal rich stars (SMR) are the following:

1. to perform population synthesis of stellar aggregates;
2. to provide a complete stellar library as input for 1.;
3. to calibrate SMR stars in terms of atmospheric parameters;
4. to compute synthetic spectral indices.

In line with previous work on theoretical synthesis of spectral features in stars (Gulati et al. 1993; Malagnini et al. 1994; Chavez 1995; Chavez et al. 1995, 1996a,b,c), here we present preliminary results related to the analysis of candidate SMR stars in the solar neighborhood.

Our analysis is based on new spectroscopic observations of a large sample of candidate SMR stars, selected according to the [Fe/H] determinations reported in the literature. We especially discuss here the problems to be faced aiming to accurate and homogeneous determinations of the atmospheric parameters (namely surface gravity, effective temperature and overall chemical composition) for each star in the sample. The method we apply is based on model flux fitting: the observed spectra are compared

F. Matteucci and F. Giovannelli (eds.), The Evolution of the Milky Way, 75–79.

Figure 1. Sample of observed spectra

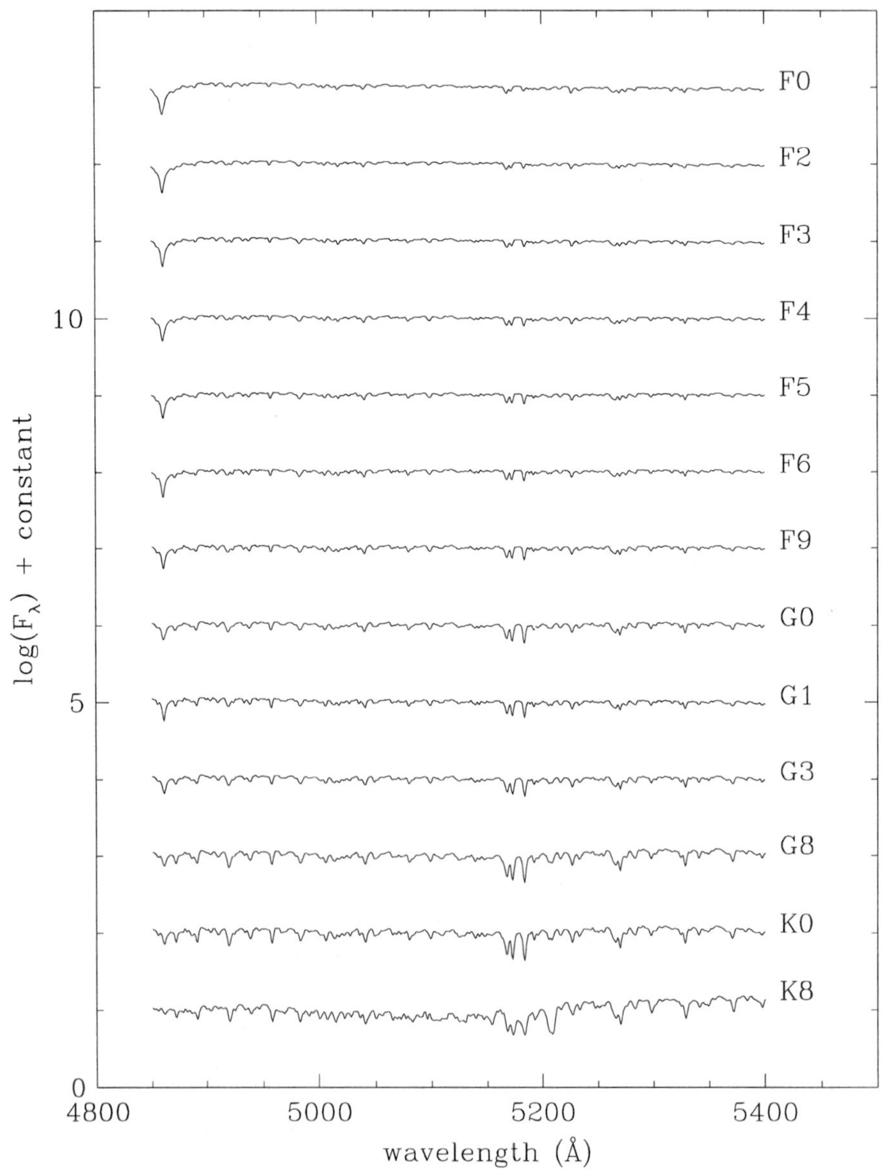

with the synthetic ones (Chavez et al. 1997) and the best parameter set is obtained by applying a least square fit criterion. A comprehensive analysis of the related problems is in fact mandatory to any further investigation given the coarse and heterogeneous estimates of the stellar distinctive parameters available in the literature.

2. The Observations

The observations cover 139 Pop I stars for a total of about 300 spectra in the range 4600–5500 Å and resolution $R = \lambda/\Delta\lambda = 2000$. The 2.12 m telescope of the INAOE "G. Haro" Observatory in Cananea (Mexico) equipped with a Böller & Chivens spectrograph was used. Figure 1 shows, as an example, a sequence of spectra of dwarf stars ordered according to spectral type.

3. The Synthetic Spectra

The synthetic spectra used for the present work belong to the collection of 711 high resolution synthetic spectra based on Kurucz (1993) atmosphere models and codes. The collection was constructed for stellar population studies and chemical abundance analysis and a complete description can be found in Chavez et al. (1997). Figure 2 shows, as an example, a sequence of synthetic spectra of solar chemical composition computed at different effective temperatures for fixed surface gravity ($\log g = 4.0$ dex).

4. Preliminary Results

The straightforward application of the flux fitting method described in Malagnini et al. (1999) to derive simultaneously the three parameters ($T_{\rm eff}$, $\log g$, $[M/H]$) from our observed spectra was not always satisfactory. Indeed, a major problem is a sort of "degeneracy" of the fitting parameters. In particular, our spectra cover a too narrow wavelength range to determine the stellar effective temperature with an accuracy as good as needed to avoid large uncertainties in the determination of metallicity and gravity. Much better results are obtained by fixing the $T_{\rm eff}$ values from broad band photometry.

As an example, Figure 3 shows the superposition of the observed spectrum (solid line) and of the synthetic one (dashed line) for the G8IV star HD 182572. The model fluxes were computed with the fixed $T_{\rm eff}$, obtained from Alonso et al. (1996) calibration of the Johnson $V - I$ color index, and the best fit $\log g$ and [M/H] values.

This work is in progress and the final results for the whole stellar sample will be presented in a forthcoming paper (Malagnini et al., 1999).

Figure 2. Sample of synthetic spectra

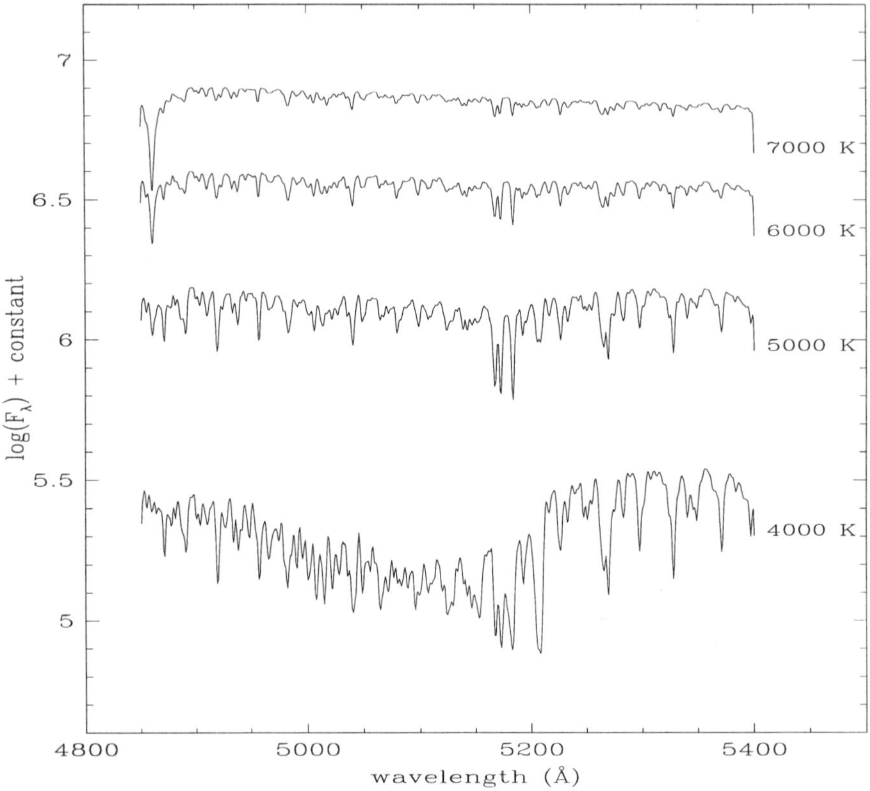

5. Acknowledgements

This work was partially supported by the Italian MURST (60% and 40% grants, Università degli Studi di Trieste, Osservatorio Astronomico di Trieste and Osservatorio Astronomico di Brera), and by the Consiglio Nazionale delle Ricerche (CNR–GNA). Partial support from Mexican CONACyT is also acknowledged via grant E9506 - 0099P-E

References

1. Alonso, A., Arribas, S., Martínez-Roger, C. 1996, A&A 313, 873
2. Chavez, M. 1995, PhD Thesis, SISSA
3. Chavez, M., Malagnini, M.L., Morossi, C. 1995, Ap.J. 440, 210
4. Chavez, M., Buzzoni, A., Malagnini, M.L., Morossi, C. 1996a, in "Fresh Views of Elliptical Galaxies", ASP Conf. Ser. 86 (eds. A.Buzzoni, A. Renzini, A. Serrano) p. 249

Figure 3. Comparison between the observed spectrum (solid line) and the synthetic one (dashed line) for HD 182572

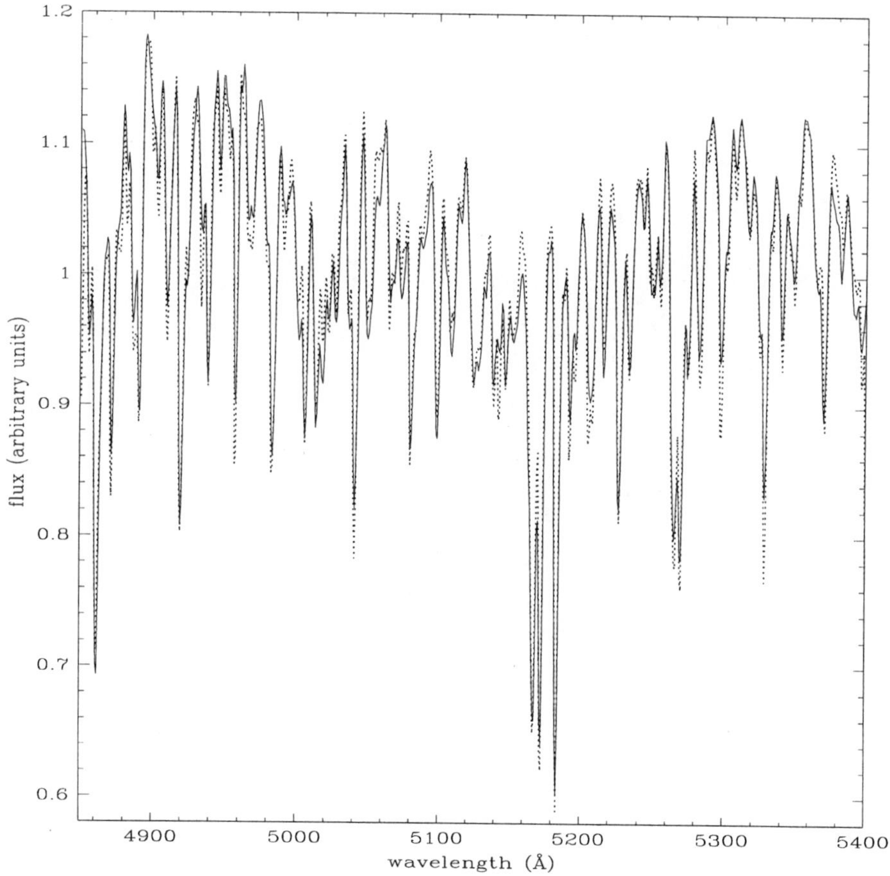

5. Chavez, M., Malagnini, M.L., Morossi, C. 1996b, Ap.J. 471, 726
6. Chavez, M., Malagnini, M.L., Morossi, C., Buzzoni, A. 1996c, in "From Stars to Galaxies", ASP Conf. Ser. 98 (eds. C. Leitherer, U. Fritze and J. Huchra) p.100
7. Chavez, M., Malagnini, M.L., Morossi, C. 1997, A&AS 126, 267
8. Gulati, R.K., Malagnini, M.L., Morossi, C. 1993, Ap.J. 413, 166
9. Kurucz, R.L. 1993, CD–ROM 13, ATLAS9 Stellar Atmosphere Programs and 2 km/s Grid (Cambridge: Smithsonian Astrophys. Obs.) (K93)
10. Malagnini, M.L., Morossi, C., Buzzoni, A., Mantegazza, L. 1994, Mem. SAIt, 65, 803
11. Malagnini, M.L., Morossi, C., Buzzoni, A., Chavez, M. 1999, A&A (submitted)

PLANETARY NEBULAE: ABUNDANCES AND ABUNDANCE GRADIENTS

W.J. MACIEL
IAG/USP
São Paulo, Brazil

Abstract. In this work, a review is given of some recent results and problems involved in the determination of chemical abundances of galactic planetary nebulae, particularly regarding disk and bulge objects.

1. Introduction

Chemical evolution models for the Galaxy can be included in one of four main classes: (i) analytical models, (ii) classical models, (iii) multiphase models and (iv) chemodynamical models (see for example Pagel 1997, Matteucci 1996). These models share a common characteristic, that is, they have to satisfy a series of observational constraints, most of which are related to the determination of chemical abundances. Some examples are the average gas metallicity in the disk, the age-metallicity relation and the existence of abundance gradients.

Planetary nebulae (PN) have an important role in the establishment of these constraints. As it is well known, PN are formed from the late evolutionary stages of intermediate mass stars ($1 \leq M/M_{\odot} \leq 8$), and their chemical composition reflects the nucleosynthetical processes that occur in these stars. Elements such as He and N can be significantly enhanced by the stellar chemical evolution, while S and Ar are probably unaltered during the evolution of the progenitor star, so that the nebular abundances of these elements can be taken as the interstellar abundances at the time when the progenitor star was formed.

In the framework of the Peimbert (1978) classification scheme PN can be classified as: Type I (disk objects with massive progenitors), Type II (disk objects with average mass progenitors), Type III (thick disk objects,

F. Matteucci and F. Giovannelli (eds.), The Evolution of the Milky Way, 81–92.

kinematically detached), Type IV (halo objects), and Type V (bulge objects) (see also Maciel 1989, Torres-Peimbert and Peimbert 1997).

In this work, some recent results concerning the abundance determinations of PN are considered, both regarding the abundances themselves and their spatial and temporal variations. The following sections discuss PN of types I–III (section 2), type IV (section 3) and type V (section 4).

2. The galactic disk

2.1. ABUNDANCES

Several independent groups have obtained abundances of disk PN, mainly from spectroscopic observations of visible and ultraviolet emission lines coupled with ionization correction factors or photoionization models. Some recent results are given by Kwitter and Henry (1998), Hajian et al. (1997) and Costa et al. (1996a).

Typical errors of the line intensities are of the order of 10–20%, leading to similar uncertainties in the derived electron temperatures. The abundances themselves have generally errors of 0.1–0.2 dex for the best measured elements, such as O/H and S/H, and somewhat higher for Ar/H and Cl/H. Geometrical effects as well as temperature fluctuations may increase these figures, especially for PN with massive central stars (Gruenwald and Viegas 1998). A discussion on the accuracy of the derived chemical abundances based on hydrodynamical models has been recently given by Perinotto et al. (1998).

Table 1 shows average values of the abundances of disk PN (Costa et al. 1996a). Type II has been subdivided into types IIa and IIb, according to Faúndez-Abans and Maciel (1987), and the table also lists types IV and V PN, which will be commented upon later. Only the best observed elements are given, though some recent work includes less abundant elements such as Si, Mg, Na and Fe (Perinotto et al. 1999, Pottasch and Beintema 1999).

The first group of abundances gives $\epsilon(X/H) = \log(X/H) + 12$, where X/H is the element abundance relative to hydrogen by number of atoms. For helium, the table gives the He/H ratio directly. The last two rows of the table give the average height z above the galactic plane (pc) and the peculiar velocity Δv (km/s). It can be seen that the He/H ratio increases in the disk along the sequence III–II–I, similarly to N/H, S/H, Ar/H and Cl/H. For O/H and Ne/H, there is an increase from type III to type II, but the average abundances of type I PN are not clearly higher than for type II, which may be partially due to ON cycling in the progenitor stars.

The second group of abundances in Table 1 shows abundances relative to oxygen of those elements (Ne, S, Ar, Cl) that are not produced during the evolution of the progenitor stars, so that they can be considered as

TABLE 1.

	I	IIa	IIb	III	IV	V
He/H	0.138	0.106	0.104	0.099	0.104	0.104
ϵ(O/H)	8.68	8.78	8.58	8.42	8.08	8.71
ϵ(N/H)	8.57	8.29	7.78	7.74	7.41	8.16
ϵ(S/H)	7.04	7.02	6.83	6.74	5.64	6.87
ϵ(C/H)	8.67	8.78	8.73	8.48	8.54	
ϵ(Ne/H)	8.03	8.06	7.87	7.71	7.27	
ϵ(Ar/H)	6.61	6.47	6.26	6.07	5.22	6.22
ϵ(Cl/H)	5.43	5.32	5.00	4.99		5.05
log(N/O)	−0.11	−0.49	−0.80	−0.68	−0.67	−0.55
log(Ne/O)	−0.65	−0.72	−0.71	−0.71	−0.81	
log(S/O)	−1.64	−1.76	−1.75	−1.68	−2.44	−1.84
log(Ar/O)	−2.07	−2.31	−2.32	−2.35	−2.86	−2.49
log(Cl/O)	−3.25	−3.46	−3.58	−3.43		−3.66
z (pc)	150	280	420	660	7200	
Δv (km/s)	20.5	21.3	22.1	64.0	172.8	

representative of the interstellar medium at the time of formation of these stars. Except for halo (type IV) PN, all objects have essentially constant ratios, a result that has been confirmed for S, Ne and Ar for HII regions both galactic and extragalactic (Henry and Worthey 1999). The ratio N/O is also included, as it is used to distinguish between type I and non-type I nebulae.

Abundance correlations are particularly important in order to understand the nucleosynthetical processes ocurring in the progenitor stars, since they do not depend on the often uncertain distances. Examples are the S/H $\times$ O/H and Ne/H $\times$ O/H as well as N/O $\times$ He/H correlations, which can in principle be used to separate the different PN types according to the mass of the progenitor star (Costa et al. 1996a, Henry 1998).

A potentially interesting source of information on PN abundances is their morphology, particularly since the publication of detailed high resolution images by Schwarz et al. (1992), Manchado et al. (1996) and Górny et al. (1999). Recent results based on HST observations show a clear evidence of N/O enhancements in bipolar nebulae (class B), and a corresponding underabundance for round PN (class R) (Stanghellini et al. 1999, Stanghellini 1999). On the other hand, elliptical PN (class E), which form the majority of galactic PN, show a wider range of N/O abundances. The immediate conclusion is that B nebulae have relatively massive progenitors, in agreement with their being generally closer to the galactic plane, while R nebulae are ejected from stars near the lower bracket of the intermediate mass stars. E nebulae are probably formed by stars in the whole mass interval. As an example, the abundances of 13 bipolar nebulae studied by Perinotto and Corradi (1998) agree very well with those of type I objects given in Table 1. Previous evolutionary models for AGB stars did not predict the He and N enhancements observed in these objects, a situation that has recently changed with the extension of the theoretical calculations to stars with $M \simeq 6M_{\odot}$ with overshooting (Marigo, this conference).

Abundance variations inside the nebula may be important, and are presently poorly known (Pottasch 1997). A recent study of bipolar nebulae (Perinotto and Corradi 1998) suggests that their observed sample is chemically homogeneous.

2.2. ABUNDANCE GRADIENTS

The study of radial abundance gradients includes basically (i) the average magnitudes of the gradients, (ii) the possible change of slope along the galactic disk, and (iii) the time variation of the gradients. The first item is relatively well established (cf. Maciel 1996, 1997, Henry and Worthey 1999), and it seems clear that an average gradient of −0.05 to −0.07 dex/kpc can be observed in the Galaxy for O/H, S/H, Ne/H and Ar/H. The PN derived gradient (Maciel and Quireza 1999, Maciel and Köppen 1994, Amnuel 1993, Pasquali and Perinotto 1993, Köppen et al. 1991) is close to – and slightly lower than – the well known gradient observed from HII regions in the Galaxy (Shaver et al. 1983, Afflerbach et al. 1997) and in other spiral galaxies (Kennicutt and Garnett 1996, Ferguson et al. 1998). Recent work on open cluster stars confirms these results (Friel 1999, Phelps, this conference), and also data on B stars, as recently discussed by Smartt and Rolleston (1997), Gummersbach et al. (1998), and Smartt (this conference), in contradiction with earlier work on these objects, which reported essentially flat gradients. The radial variations of O/H, S/H, Ne/H and Ar/H are consistent with essentially constant ratios of S/O, Ne/O and Ar/O from PN and HII regions, as can be seen from Table 1 and from a recent discussion on the gradients in the Galaxy and in other galaxies (Henry and Worthey 1999).

As an illustration, figure 1 shows the O/H gradient derived from HII regions in several spiral galaxies (Henry et al. 1994), along with the average gradient for disk PN in the Galaxy (straight line) from Maciel and Quireza (1999). In this figure, the abscissa gives the distance to the centre of the galaxy in terms of the effective radius, defined as the radius where half of

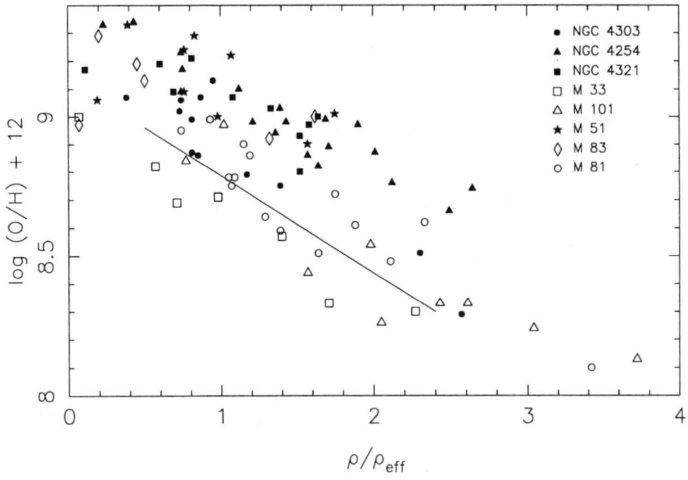

Figure 1. O/H radial gradient in spiral galaxies and PN (straight line).

the optical emission is contained (cf. Maciel 1997). The similarity of the PN and HII region gradients suggests that orbital diffusion is probably mild in the radial direction, since the PN population is at least a few Gyr older than the HII regions. Azimuthal diffusion could be much higher, however, as the PN are sampled all over the galactic disk. Therefore, figure 1 suggests that the gas is homogeneously distributed in the azimuthal direction. This can be confirmed by plotting the O/H abundances of PN in a concentric ring of width ΔR around the galactic centre as a function of the longitude. No correlations are observed, even if relatively wide rings ($\Delta R \simeq 1 - 2$ kpc) are considered. On the other hand, according to Wielen et al. (1996) the gradients are not much affected by orbital diffusion provided the surface density of stars falls off exponentially with the galactocentric distance R, and both the gradient and the diffusion coefficient do not appreciably change with R.

Radial variations of the gradients of O/H, S/H, Ne/H and Ar/H have been studied by Maciel and Quireza (1999). As an example, figure 2 shows the O/H gradient, where some recent data for planetary nebulae in the outer galactic disk from Costa et al. (1997) have been included. The dots are disk PN; the straight line gives a linear fit corresponding to -0.058 dex/kpc; the curve is a second-order polynomial fit corresponding to a gradient of -0.058 dex/kpc at the sun, where $R_0 = 7.6$ kpc, and the step function has been fitted with 1 kpc bins. A flattening at large radial distances can be observed, in agreement with some recent work on HII regions (Vílchez and Esteban 1996).

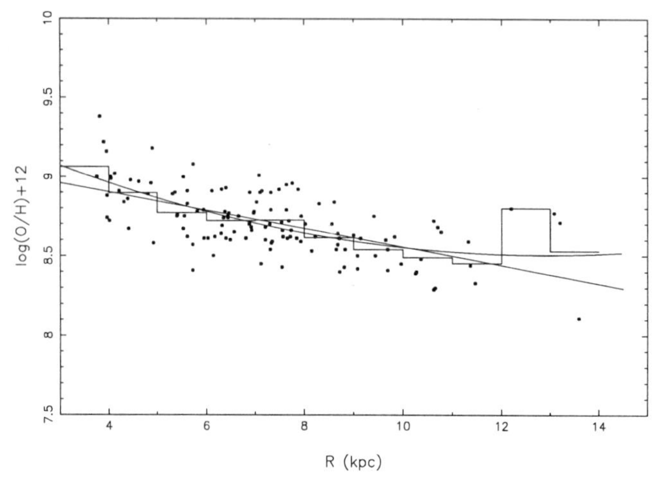

Figure 2. O/H gradient from disk PN (Maciel and Quireza 1999).

The time variation of the abundance gradients is not as well defined, basically due to the fact that the ages attributed to PN are not sufficiently well known and the age difference between HII regions and B stars is relatively small. Type III PN are generally older and show flatter gradients, but part of this may be due to orbital diffusion. Including HII regions, B stars and the different types of PN, the present results are consistent with some steepening of the gradients, in agreement with earlier suggestions by Maciel and Köppen (1994), Golovaty and Malkov (1997) and Maciel and Quireza (1999). The latter obtained a rough estimate of the average steepening rate as -0.004 dex kpc^{-1} Gyr^{-1}. Despite the large uncertainty of this result, it clearly shows that any time variation in the gradients is probably small, so that the possibility of a constant gradient cannot be ruled out.

Radial gradients and their variations are extremely important as constraints to chemical evolution models. Recent classical models predict similar gradients as observed in the framework of an inside-out scenario, showing some flattening near the outer Galaxy and a time steepening (Chiappini et al. 1997, Matteucci and Chiappini 1999). Chemodynamical models (Samland et al. 1997, Hensler 1999) also predict some flattening at large R, and time steepening gradients are predicted by theoretical models by Götz and Köppen (1992) and Tosi (1988). On the other hand, multiphase models (Ferrini et al. 1994, Mollá et al. 1997) suggest just the opposite behaviour, as also models developed by Prantzos and Silk (1998), Boissier and Prantzos (1999) and Allen et al. (1998), so that new independent observations and estimates of the gradients are needed to settle this question.

3. The halo

Only an handful of halo PN are known, and in some cases it is not clear whether the nebula is a true halo object or belongs to the thick disk. These PN have very low metal abundances as can be seen in Table 1, which shows very clearly halo–disk variations ranging from 0.10 dex to an order of magnitude, therefore much higher than the expected abundance uncertainties. A recent analysis of the halo object DDDM1 (Kwitter and Henry 1998) shows that the He, N, C, O, and Ne abundances are similar to those of type IV PN in Table 1. These abundances suggest that type IV PN reflect the metal poor conditions in the halo at the time of the formation of their progenitor stars, displaying a halo–disk vertical gradient, as found years ago by Faúndez-Abans and Maciel (1988) and confirmed by Cuisinier (1993), in contrast with an essentially flat vertical gradient for disk PN (Faúndez-Abans and Maciel 1988, Pasquali and Perinotto 1993, Cuisinier 1993).

From the second set of abundances of Table 1, it can be seen that halo PN have much lower S/O, Ne/O and Ar/O ratios than the remaining types, apart from being more metal-poor. According to Henry and Worthey (1999), these discrepancies could be caused by local abundance fluctuations due to unmixed ejecta from recent supernova events.

An additional problem is that several of the PN associated with the halo and thick disk (types III and IV) have probably low mass progenitors with strong stellar winds (Maciel et al. 1990, Maciel 1993). In fact, some of these nebulae apparently have central stars with masses under 0.55 $M_{\odot}$, in contradiction with most evolutionary tracks for intermediate mass stars.

4. The bulge

Many bulge, or type V PN (Maciel 1989) are known (see for example Beaulieu et al. 1999), but only recently accurate abundances have been obtained (Ratag et al. 1997, Costa et al. 1996b). Recent work by Cuisinier et al. (1999) and Costa and Maciel (1999) has led to He, O, N, Ar and S abundances to about 40 bulge PN, with an uncertainty comparable to disk objects. However, van Hoof and van de Steene (1999) have found a larger scatter among the recent determinations for a sample of 5 bulge PN, which was attributed to uncertainties in the electron temperature.

A comparison of He/H and N/O abundances in the bulge and in the disk shows that objects with higher ratios are less present in the bulge, suggesting that it contains an excess of older progenitor stars, since younger, massive objects are generally overabundant in these elements. On the other hand, the bulge metal abundances are generally comparable with those of the disk, and the O/H, Ar/H and S/H ratios can be higher than the disk

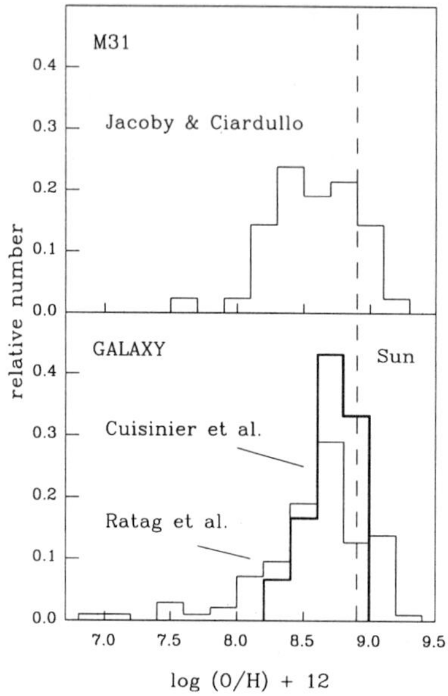

Figure 3. O/H distributions of bulge PN in the Galaxy and in M31.

counterparts even though very metal rich PN are missing in the bulge. Since underabundant nebulae are also present, these results suggest that the bulge contains a mixed population, so that star formation in the bulge probably spans a wide time interval.

Chemical abundances of PN in the bulge of M31 have been recently studied by Jacoby and Ciardullo (1999), who have included in their analysis some results by Stasińska et al. (1998) and Richer et al. (1999). Figure 3 shows a comparison of this sample (top panel) with the galactic bulge objects from Ratag et al. (1997) (lower panel, thin line) and Cuisinier et al. (1999) (lower panel, thick line). It can ben seen that the O/H abundance distributions are very similar, peaking around 8.7 dex, and showing very few if any super metal rich objects with supersolar abundances.

This fact has been interpreted by Jacoby and Ciardullo (1999) as an inability of metal rich stars to produce sufficiently bright PN and to a possible gradient in the bulge of M31. Coupled with the He/H and N/O abundances, it would suggest again that the youngest and most metal rich PN are less frequent in the bulge.

It is interesting to compare the PN metallicity distribution with the [Fe/H] distribution of stars in the bulge. A direct comparison depends on the adopted [O/Fe] × [Fe/H] relationship, as Fe abundances are known for a limited sample of PN only. Adopting the [O/Fe] × [Fe/H] relation

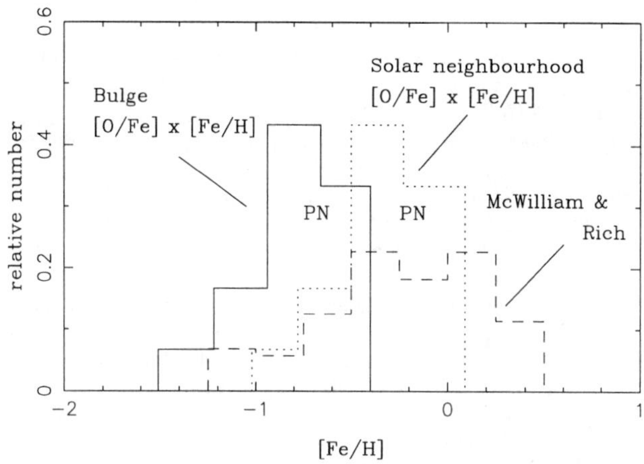

Figure 4. Metallicity distribution of bulge PN and K giants.

for the *solar neighbourhood* (Matteucci et al. 1999), it can be seen from figure 4 that the observed peak in the PN metallicity distribution for the bulge (dotted line) looks similar to the distribution of K giants in Baade's Window (McWilliam and Rich 1994, dashed line), but depleted of the very metal rich objects. In this case, the peak of the distribution occurs for [Fe/H] close to zero, which is also similar to the Mira variables metallicity distribution shown by Feast (this conference). Taking into account the predicted relationship of Matteucci et al. (1999) for the *bulge*, which assumes a faster evolution of the bulge compared to the halo, the observed peak of the PN metallicity distribution is displaced by about 0.5 dex towards lower metallicities (solid line), as in this case a given [Fe/H] metallicity implies a larger [O/Fe] ratio. As a result, the derived distribution shows an even stronger depletion of the very metal rich objects.

Several reasons can be considered to explain the differences in the metallicity distributions of figure 4, namely (i) systematic errors in the PN abundances, (ii) ON cycling in the PN progenitor stars, (iii) statistical uncertainties, and (iv) uncertainties in the adopted [O/Fe] $\times$ [Fe/H] relationships. As discussed by Maciel (1999), the first three are probably not important enough to account for a 0.5 dex discrepancy as shown in figure 4, so that we suggest that the main reason is an excess of oxygen enhancement relative to iron in the bulge [O/Fe] $\times$ [Fe/H] relation from Matteucci et al. (1999). This is supported by recent results by Barbuy (1999) for the bulge globular cluster NGC 6528 and for the six giants in the McWilliam and Rich (1994) sample for which [OI] lines have been measured.

Acknowledgements. This work was partially supported by CNPq and FAPESP.

References

1. Afflerbach, A., Churchwell, E., Werner, M.W.: 1997, *ApJ* **478**, 190
2. Allen, C., Carigi, L., Peimbert, M.: 1998, *ApJ* **494**, 247
3. Amnuel, P.R.: 1993, *MNRAS* **261**, 263
4. Barbuy, B.: 1999, *Galaxy evolution*, ed. M. Spite, F. Crifo, Kluwer (in press)
5. Beaulieu, S.F., Dopita, M.A., Freeman, K.C.: 1999, *ApJ* **515**, 610
6. Boissier, S., Prantzos, N.: 1999, *MNRAS* **307**, 857
7. Chiappini, C., Matteucci, F., Gratton, R.: 1997, *ApJ* **477**, 765
8. Costa, R.D.D., Chiappini, C., Maciel, W. J., Freitas-Pacheco, J.A.: 1996a, *A&AS* **116**, 249
9. Costa, R.D.D., Freitas-Pacheco, J.A., França, J.A.: 1996b, *A&A* **313**, 924
10. Costa, R.D.D., Chiappini, C., Maciel, W. J., Freitas-Pacheco, J.A.: 1997, *Advances in stellar evolution*, ed. R.T. Rood, A. Renzini, Cambridge, 159
11. Costa, R.D.D., Maciel, W.J.: 1999, *Galaxy evolution*, ed. M. Spite, F. Crifo, Kluwer (in press)
12. Cousinier, F.: 1993, *Acta Astron.* **43**, 455
13. Cousinier, F., Maciel, W.J., Köppen, J., Acker, A., Stenholm, B.: 1999 *A&A* (in press)
14. Faúndez-Abans, M., Maciel, W.J.: 1987, *A&A* **183**, 324
15. Faúndez-Abans, M., Maciel, W.J.: 1988, *Rev. Mex. A & A* **16**, 105
16. Ferrini, F., Mollá, M., Pardi, C., Díaz, A.I.: 1994, *ApJ* **427**, 745
17. Ferguson, A.M.N., Gallagher, J.S., Wyse, R.F.G.: 1998, *AJ* **116**, 673
18. Friel, E.D.: 1999, *Galaxy evolution*, ed. M. Spite, F. Crifo, Kluwer (in press)
19. Götz, M., Köppen, J.: 1992, *A&A* **262**, 455
20. Golovaty, V.V., Malkov, Yu.F.: 1997, *IAU Symp. 180*, ed. H.J. Habing, H.J.G.L.M. Lamers, Kluwer, 408
21. Górny, S.K., Schwarz, H.E., Corradi, R.L.M., van Winckel, H.: 1999, *A&AS* **136**, 145
22. Gruenwald, R.B., Viegas, S.M.M.: 1998, *ApJ* **501**, 221
23. Gummersbach, C.A., Kaufer, A., Schaefer, D.R., Szeifert, T., Wolf, B.: 1998, *A&A* **338**, 881
24. Hajian, A.R., Balick, B., Terzian, Y., Perinotto, M.: 1997, *ApJ* **487**, 313
25. Henry, R.B.C.: 1998, *ASP Conference Series vol. 147*, ed. D. Friedli, M. Edmunds, C. Robert, L. Drissen, 59
26. Henry, R.B.C., Pagel, B.E.J., Chincarini, G.L.: 1994, *MNRAS* **266**, 421
27. Henry, R.B.C., Worthey, G.: 1999, *PASP* **111**, 919
28. Hensler, G.: 1999, *Galaxy evolution*, ed. M. Spite, F. Crifo, Kluwer (in press)
29. Jacoby, G.H., Ciardullo, R.: 1999, *ApJ* **515**, 169
30. Kennicutt, R.C., Garnett, D.R.: 1996, *ApJ* **456**, 504
31. Köppen, J., Acker, A., Stenholm, B.: 1991, *A&A* **248**, 197
32. Kwitter, K.B., Henry, R.B.C.: 1998, *ApJ* **493**, 247
33. Maciel, W.J.: 1989, *IAU Symp. 131*, ed. S. Torres-Peimbert, Kluwer, 73
34. Maciel, W.J.: 1993, *A&SS* **209**, 65
35. Maciel, W.J.: 1996, *Stellar abundances*, ed. B. Barbuy, W.J. Maciel, J.C.Gregório-Hetem, IAG/USP, 79
36. Maciel, W.J.: 1997, *IAU Symp. 180*, ed. H.J. Habing, H.J.G.L.M. Lamers, Kluwer, 397
37. Maciel, W.J.: 1999, *A&A* (in press)
38. Maciel, W.J., Freitas-Pacheco, J.A., Codina, S.J.: 1990, *A&A* **239**, 301
39. Maciel, W.J., Köppen, J.: 1994, *A&A* **282**, 436
40. Maciel, W.J., Quireza, C.: 1999, *A&A* **345**, 629
41. Manchado, A., Guerrero, M., Stanghellini, L., Serra-Ricart, M.: 1996, *The IAC Morphological Catalog of Northern Galactic Planetary Nebulae*, IAC
42. Matteucci, F., 1996: *Fund. Cosm. Phys.* **17**, 283

43. Matteucci, F., Chiappini, C.: 1999, *Chemical evolution from zero to high redshift*, ed. J.R. Walsh, M.R. Rosa, ESO (in press)
44. Matteucci, F., Romano, D., Molaro, P.: 1999, *A&A* **341**, 458
45. McWilliam, A., Rich, R.M.: 1994, *ApJS* **91**, 749
46. Mollá, M., Ferrini, F., Díaz, A.I.: 1997, *ApJ* **75**, 519
47. Pagel, B.E.J.: 1997, *Nucleosynthesis and chemical evolution of galaxies*, Cambridge
48. Pasquali, A., Perinotto, M.: 1993, *A&A* **280**, 581
49. Peimbert, M.: 1978, *IAU Symp. 76*, ed. Y. Terzian, Reidel, 215
50. Perinotto, M., Bencini, C.G., Pasquali, A., Manchado, A., Rodriguez Espinoza, J.M., Stanga, R.: 1999, *A&A* **347**, 967
51. Perinotto, M., Corradi, R.L.M.: 1998, *A&A* **332**, 721
52. Perinotto, M., Kifonidis, K., Schönberner, D., Marten, H.: 1998, *A&A* **332**, 1044
53. Pottasch, S.R.: 1997, *IAU Symp. 180*, ed. H.J. Habing, H.J.G.L.M. Lamers, Kluwer, 483
54. Pottasch, S.R., Beintema, D.A.: 1999, *A&A* **347**, 975
55. Prantzos, N., Silk, J.: 1998, *ApJ* **507**, 229
56. Ratag, M.A., Pottasch, S.R., Dennefeld, M., Menzies, J.W.: 1997, *A&AS* **126**, 297
57. Richer, M.G., Stasinska, G., McCall, M.L.: 1999, *A&AS* **135**, 203
58. Samland, M., Hensler, G., Theis, C.: 1997, *ApJ* **476**, 544
59. Schwarz, H.E., Corradi, R.L.M., Melnick, J.: 1992, *A&AS* **96**, 23
60. Shaver, P.A., McGee, R.X., Newton, L.M., Danks, A.C., Pottasch, S.R.: 1983, *MNRAS* **204**, 53
61. Smartt, S.J., Rolleston, W.R.J.: 1997, *ApJ* **481**, L47
62. Stanghellini, L.: 1999, *Ap&SS* (in press, astro-ph/9906188)
63. Stanghellini, L., Blades, C.J., Osmer, S.J., Barlow, M.J., Liu, X.W.: 1999, *ApJ* **510**, 687
64. Stasinska, G., Richer, M.G., McCall, M.L.: 1998, *A&A* **336**, 667
65. Torres-Peimbert, S., Peimbert, M.: 1997, *IAU Symp. 180*, ed. H.J. Habing, H.J.G.L.M. Lamers, Kluwer, 175
66. Tosi, M.: 1988, *A&A* **197**, 47
67. van Hoof, P.A.M., van de Steene, G.C.: 1999, *MNRAS* **308**, 623
68. Vílchez J.M., Esteban, C.: 1996, *MNRAS* **280**, 720
69. Wielen, R., Fuchs, B., Dettbarn, C.: 1996, *A&A* **314**, 438

Speaker: W. MACIEL

A. MAEDER: It is clear that the abundances of some elements are not modified by internal nucleosynthesis; this is certainly true for Ar, S, Cl... However, for oxygen, this is much less certain. In this context, can you account for CNO processing?

W. MACIEL: In fact, only type I planetary nebulae show some evidences of ON cycling. The gradients that have been shown were derived from type II planetary nebulae, so that type I objects are not included.

PLANETARY NEBULAE AS PROBES OF STELLAR EVOLUTION AND POPULATIONS

LETIZIA STANGHELLINI
Space Telescope Science Institute
Affiliated to the Astrophysics Division,
Space Science Department of ESA
e-mail: lstanghe@stsci.edu

Abstract. Planetary Nebulae (PNe) have been used satisfactory to test the effects of stellar evolution on the Galactic chemical environment. Moreover, a link exists between nebular morphology and stellar populations and evolution. We present the latest results on Galactic PN morphology, and an extension to a distance unbiased and homogeneous sample of Large Magellanic Cloud PNe. We show that PNe and their morphology may be successfully used as probes of stellar evolution and populations.

1. Relevance of Planetary Nebulae

Planetary Nebulae represent the fate of most stars of masses in the range 1–10 $M_{\odot}$. The effects and consequences of their evolution are of great importance in stellar population studies, since PNe represent the progeny of the most common stars in galaxies.

PNe have been observed for centuries in the Galaxy, and more recently in other galaxies such as the Magellanic Clouds, and between galaxies. They have been identified as far as in the Virgo group. Planetary Nebulae studied in different galaxies are of exceptional interest to probe stellar populations in a variety of environments. These studies would also allow the direct comparison of similar stellar populations in the different host galaxies (see also Maciel, this volume).

The ejection of the stellar envelopes of Asymptotic Giant Branch (AGB) stars yields to the PN formation, and also marks the beginning of the *cosmical recycling* process. In fact, PNe are among the most important cosmic sources of carbon and other elements whose abundance is modified in the

F. Matteucci and F. Giovannelli (eds.), The Evolution of the Milky Way, 93–100.

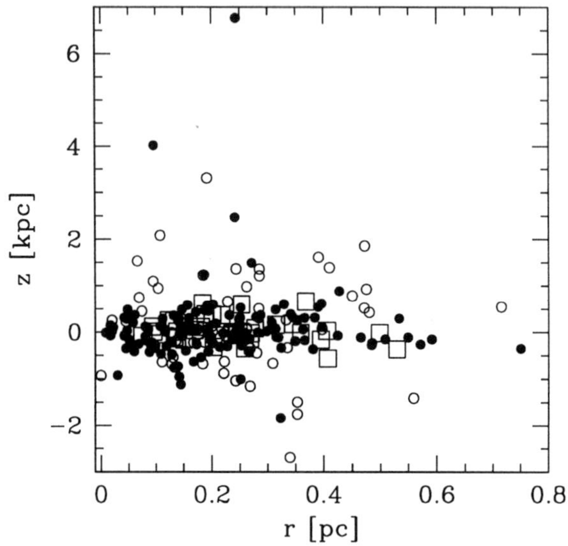

Figure 1. Distribution of sizes and location within the Galaxy of round (open circles), elliptical (filled circles), and bipolar (squares) PNe.

interiors of low- and intermediate-mass stars. The carbon, oxygen, nitrogen, and helium abundances of PNe are excellent tracers of the evolutionary paths of their progenitors, and their initial mass (Renzini & Voli 1981, Iben & Renzini 1983). It is worth mentioning that the chemical yields for AGB stars are at variation with respect to the initial chemical environment (van den Hoek & Groenewegen 1997), and that rotation and other stellar effects may severely affect the *classical* view of elemental production in AGB stars (e.g., Charbonnel 1995).

Planetary Nebulae also eject into the interstellar medium material that has not been processed during the life of the low-mass star, such as argon, neon, and sulphur. These elements are uncorrupted by the evolution of low- and intermediate-mass stars, thus their abundance mix allows to infer the chemical conditions of the interstellar medium (ISM) at the time of the formation of the PN progenitor.

Planetary Nebulae appear in a variety of morphologies, which have been related to their chemical content (Peimbert 1978, Maciel 1989, Torres-Peimbert & Peimbert 1997), and to stellar populations in the Galaxy (e.g., Stanghellini et al. 1993). It is thus important to study the morphology of PNe for an independent test on stellar evolution and populations.

Most of relations between PN morphology versus the evolution of the progenitor stars, and stellar populations, have been obtained for Galactic PNe. Only the use of the HST cameras allows the spatial resolution of extragalactic PNe. In §2 we discuss the work on Galactic PN morphology, based on contemporary databases. These results may be biased by the large uncertainties of the Galactic PN distance scale (Terzian 1993), and with the selection against PNe of low surface brightnesses, due to Galactic dust absorption. To circumvent these difficulties, and to study PNe in a different chemical environment, we have embarked a Large Magellanic Cloud PN survey, of which we give a preview in §3. Section 4 includes a summary and drafts future developments.

2. Morphology of Galactic Planetary Nebulae

Galactic PN morphology has been related to nebular and stellar physical properties in the past few decades (a review in Stanghellini 1995). The importance of these investigations is two-fold. First, the correlations between PN morphology and central star evolution allow us to use morphology to infer the evolutionary paths of the stellar progenitors (Stanghellini et al. 1993; Stanghellini & Pasquali 1995). Second, by relating the PN morphology to kinematic, spatial, and chemical quantities it is possible to use morphology as tracer of stellar populations (Peimbert 1978, Stanghellini et al. 1993).

In Figure 1 we show the distribution of Galactic PN as a function of their physical radii and the altitude above (or below) the Galactic plane. The various morphologies are coded with different symbols. It is evident that most bipolar PNe (sample from Manchado et al. 1996) lie closer to the Galactic plane than the round and elliptical PNe, leading to the conclusion that bipolar PNe are the progeny of a young stellar population. This result is not new; since the seventies Greig (1972), found that *binebulous* PNe constitute a younger PN population (Peimbert's *type I* PNe). The importance of this plot is that it is based on a homogeneous and complete PN sample, placing this conclusion on firmer ground.

In Figure 2 we examine the mass distribution of central stars of PNe across morphology. This plot is based on the same sample as Figure 1, and on new Zanstra analysis (Stanghellini et al. in preparation), and it confirms what already found by Stanghellini et al. (1993): bipolar PNe host stars in a wide range of masses, while very few high-mass central stars are found within the round PNe. One again, this is indicative of two different stellar

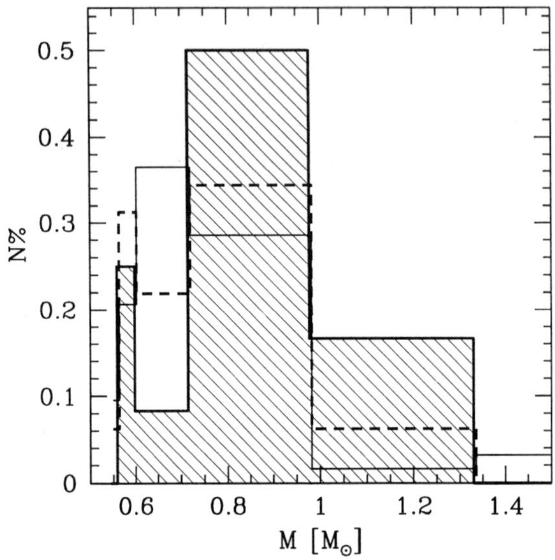

Figure 2. Mass histogram for the central stars of round (thin line), elliptical (dashed line), and bipolar (thick line and shaded histogram) planetary nebulae.

populations, and it is not in disagreement with the findings by Peimbert and his collaborators.

3. Morphology of Magellanic Cloud Planetary Nebulae

The importance of the results discussed in the previous session is somewhat dimmed by the difficulty to establish a Galactic distance scale to PNe. Even in the most optimistic view, the distance to Galactic PNe is known to within 50%, thus only the statistical distances to the selected samples have a scientific significance, while the individual distances are very strongly biased.

To alleviate this problem we are observing a large sample of Large Magellanic Cloud (LMC) PNe (Cycle 8; Program ID: 8271; PI: Stanghellini; Co-Is: Shaw, Blades, Balick). The main goal of our program is to detect LMC PN morphology and find the connection to the physical conditions of the nebulae and their central stars. We obtain the STIS/HST data in slitless mode. The advantage of slitless spectroscopy can be summarized in the following points: (1) each exposure gives spatial and spectral information at once, obtaining *narrow band* PN images plus excitation/ionization information and velocity information all at the same time; (2) we also contemporary obtain the spectrum of the central star, if bright enough to be visible. The caveat of slitless spectroscopy is that spatial and spectral information may overlap for PNe larger than the spectral segregation between nearby lines. For example, this technique can not be used for PNe larger than 1.5 arcsec in diameter. Interestingly enough, these observations are almost ideally suited for LMC PNe, which typically span 0.8 arcsec in size.

From the space-based observations we can thus derive (a) the PN shape and size in the prominent emission lines (Hα, Hβ, [O III] λ5007, [N II], etc); (b) the stellar and nebular spectrum, stellar continuum, central star identification, luminosity and temperature. Our HST program is about half completed, the other half to be finished after the third HST servicing mission (SM3a). Among the LMC PNe observed so far, we were able to recognize all the morphological classes that we were familiar with in Galactic PN studies.

From a related ground-based program (with NTT/EMMI at ESO) we will derive detailed abundance and plasma diagnostics. At the time of writing, the ground-based observations (acquired in October 1999) are completed but not analyzed yet, thus we use the literature abundances (main source: Leisy & Dennefeld 1996) for the preliminary analysis shown in this paper.

By correlating the PN morphology and chemical abundance we find that the yields of the products of stellar evolution change across morphology. We know that the carbon, nitrogen, and oxygen abundance of PNe vary with progenitor stellar mass. In particular, carbon is produced during core helium burning, but it is partially converted into nitrogen only in the massive progenitors (e.g. van den Hoek & Groenewegen 1997). The oxygen yield also varies with mass and other evolutionary parameters, but it seem to be consistently stable at the LMC metallicities (see Fig. 1 in van den Hoek & Groenewegen 1997). We thus can use oxygen as a reference element for the discussion on the evolutionary enrichment.

Figure 3 shows the evolution of the C/O versus N/O in LMC PNe across morphological types. We see that almost all targets are enriched in N/O with respect to the local environment. It is also evident that most symmetric PNe have high C/O and low N/O ratios, and asymmetric PNe behaving the opposite way. This figure give us a lot of confidence in what already found in Galactic PNe: asymmetric PNe may have more massive progenitors that symmetric PNe.

Figure 4 shows the marked difference of population-tracking elements

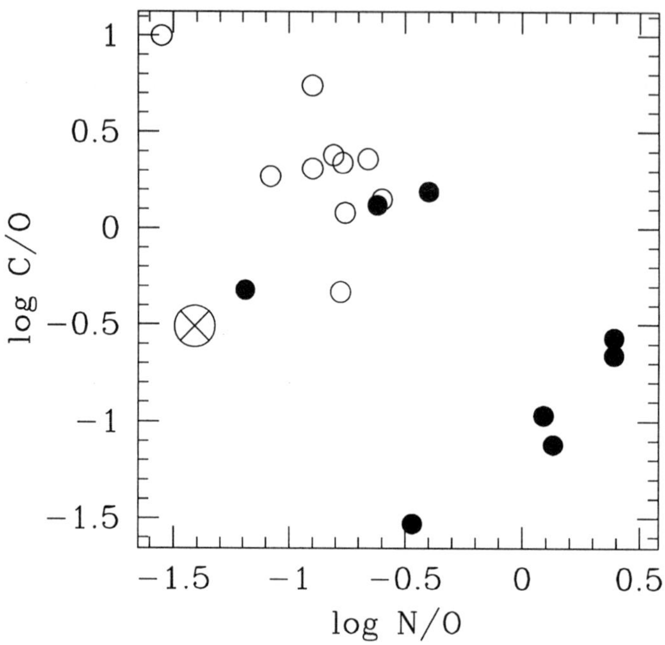

Figure 3. Log C/O versus log N/O in LMC PNe. Filled symbols: asymmetric PNe (bipolar and quadrupolar); open symbols: symmetric PNe (round and elliptical). The crossed larger circle indicates the LMC average of the H II regions.

between symmetric and asymmetric PNe. Since both oxygen and neon should not change during evolution through the PN phase, both elements describe the chemical environment prior to the formation of the PN progenitors. Even with relatively small number statistics, we can infer that asymmetric PN progenitors have formed in an environment that is chemically enriched with respect to where the progenitor of symmetric PN originate. Most asymmetric PNe have a higher neon content than the local H II regions, while only one symmetric PN is markedly enriched in neon with respect to the H II regions. Most of the symmetric PNe have lower oxygen abundance than their surroundings, while more than half asymmetric PNe are oxygen rich, again with respect to the local average. Since there is such a sharp contrast in the abundance of *primordial* elements across morphological classes, we also test whether there are spatial segregations of either morphological class or chemistry in the LMC. We did not find any convincing segregation. Our result indicates that the (low- and intermediate-mass)

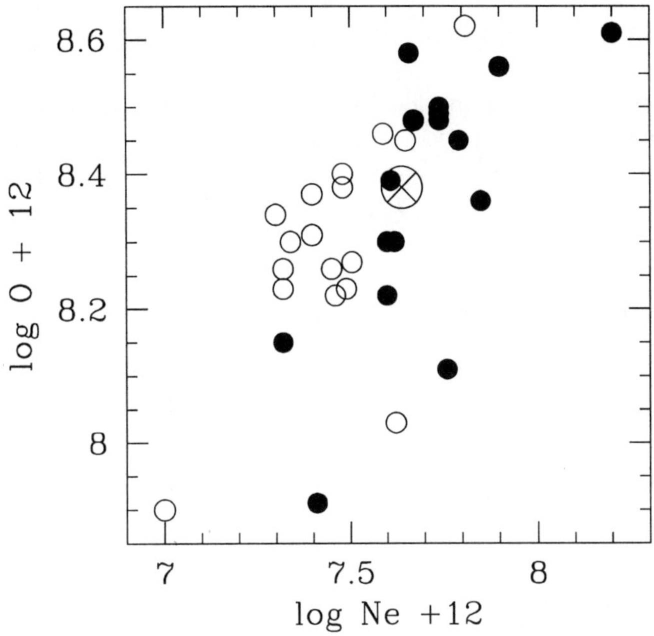

Figure 4. Oxygen versus Neon abundances in LMC PNe. Symbols as in Figure 3.

stellar populations are mixed within the LMC.

4. Summary

By analyzing Galactic and LMC PNe, we searched for the relations between morphological class and chemical and spatial characteristics to test whether PNe can be used as probes of stellar evolution and populations. The previously known correlations of Galactic PN show a marked segregation of high-mass stars to form asymmetric PNe, which is in agreement with the distribution of these PNe within the Galactic disk. Novel studies in the Large Magellanic Clouds hint to connection between morphology and stellar evolution, and formation/population as well. The extragalactic studies are based on a database in the making. The preliminary results lead to the existence of two PN populations, distinct in their masses and chemical contents, that can be resolved though morphological analysis.

Ours future plans include the completion of our Cycle 8 HST program, and of the ground-based spectroscopy, to constitute a new, homogeneous, chemical/morphological database. We will derive the morphological sequences and the dynamical ages, the stellar masses and evolutionary ages, the abundances, and finally we will model the surface brightness decline for each morphological class, as described in Stanghellini et al. (1999).

We also hope to embark a similar study in the Small Magellanic Cloud (SMC), in order to extend the *chemical baseline* for PN formation to a low-metallicity environment.

5. Acknowledgements

I thank Dick Shaw, Arturo Manchado, and Eva Villaver for contributing substantially to the science included in this paper, and Max Mutchler for the HST data analysis. I whish to express my gratitude to Francesca Matteucci for inviting me to an outstanding meeting.

References

1. Charbonnel, C., 1995, ApJ 453, L41
2. Greig, W. E., 1972, A&A 18, 70
3. Iben, I. Jr., & Renzini, A., 1983, ARA&A 21, 271
4. Leisy, P., & Dennefeld, M., 1996, A&AS 116, 95
5. Maciel, W. J., 1989, IAU Symp. 131, ed. S. Torres-Peimbert, Kluwer , 73
6. Manchado, A., Guerrero, M., Stanghellini, L., & Serra-Ricart, M. 1996, "The IAC Morphological Catalog of Northern Galactic Planetary Nebulae" (Tenerife: IAC)
7. Peimbert, M., 1978, IAU Symp. 76, ed. Y. Terzian, Reidel, 215
8. Renzini, A., & Voli, M., 1981, A&A
9. Stanghellini, L., Corradi, R. L. M., & Schwarz, H. E., 1993, A&A 279, 521
10. Stanghellini, L., & Pasquali, A., 1995, ApJ 452, 286
11. Stanghellini, L. 1995, Asymmetric Planetary Nebulae, Annales of the Israel Physical Society, Vol. 11, eds. A. Harpaz & N. Soker, p. 17
12. Stanghellini, L., Blades, J. C., Osmer, S. J., Barlow, M. J., & Liu, X.-W., 1999, ApJ 510, 687
13. Terzian, Y., 1993, IAY Symp. 155, ed. R. Weinberger & A. Acker, Kluwer, 109
14. Torres-Peimbert, S., & Peimbert, M., 1997, IAU Symp. 180, ed. H. J. Habing & H. J. G. L. M. Lamers, Kluwer, 175
15. van den Hoek, L. B., & Groenewegen, M. A. T., 1997, A&AS 123, 305

CLUMP STARS IN THE SOLAR NEIGHBOURHOOD

LÉO GIRARDI
Dipartimento di Astronomia, Università di Padova
Vicolo dell'Osservatorio 5, I-35122 Padova, Italy

Abstract: *Hipparcos* data has allowed the identification of a large number of clump stars in the Solar Neighbourhood. We discuss our present knowledge about their distributions of masses, ages, colours, magnitudes, and metallicities. We point out that the age distribution of clump stars is "biased" towards intermediate-ages. Therefore, the metallicity information they contain is different from that provided by the local G dwarfs. Since accurate abundance determinations are about to become available, these may provide useful constraints to chemical evolution models of the local disc.

1. The clump after *Hipparcos*

The clump of core-helium burning stars has been known for decades in the colour-magnitude diagrams (CMD) of open clusters and nearby galaxies. However, only recently we have been able to identify a significant number of clump stars in the Solar Neighbourhood, thanks to the *Hipparcos* mission (Perryman et al. 1997). In fact, the ESA (1997) catalog contains ~ 600 clump stars with parallax error lower than 10 %, and hence an error in absolute magnitude lower than 0.12 mag. This accuracy limit corresponds to a distance of ~ 125 pc within which the sample of clump stars is complete. Moreover, accurate BV photometry (and I for $\sim 1/3$ of the sample), is available, and interstellar absorption is small enough to be neglected.

The *Hipparcos* M_V versus $B - V$ CMD is illustrated on the left panel of Fig. 1. In this plot, we exclude binaries and limit the sample to stars with $\pi < 0.007$ arcsec and $V < 8.5$. The clump is well evident at $M_V \simeq 1$, $B - V \simeq 1$.

So far, most of the emphasis on nearby clump stars has been on their possible use as standard candles, since their mean absolute magnitude can be determined with an accuracy of hundredths of magnitude from *Hipparcos* database (cf. Paczyński & Stanek 1998; Stanek et al. 1998). This application requires a good understanding of the basic characteristics of clump stars, in

F. Matteucci and F. Giovannelli (eds.), The Evolution of the Milky Way, 101–107.

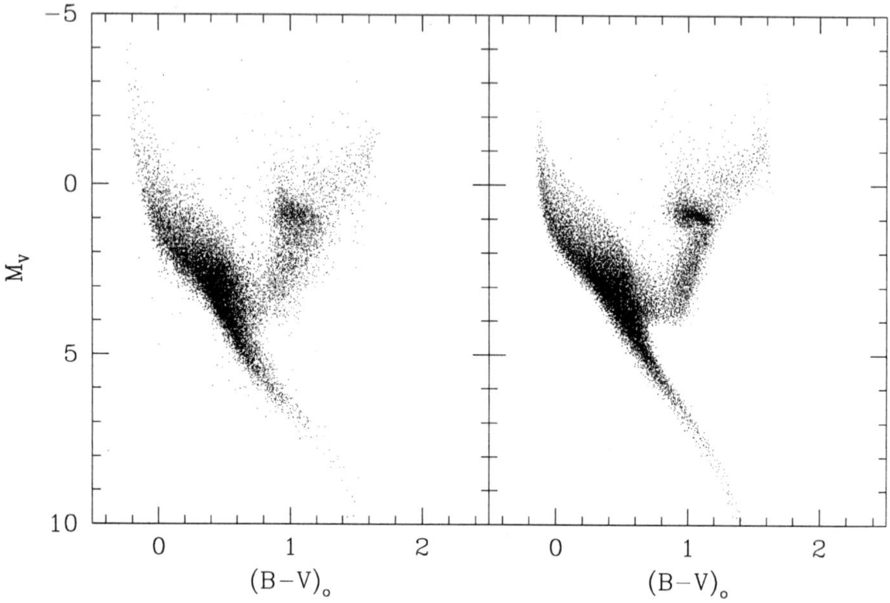

Figure 1. Left panel: *Hipparcos* data. Right panel: theoretical simulation (see text).

order to access the systematic changes of the clump luminosity in different populations, as those pointed out by Cole (1998) and Girardi et al. (1998).

2. The mass distribution

A common idea is that the mass distribution of clump stars roughly follows the IMF, presenting then a single and sharp peak at the lowest possible masses, i.e. with $0.8 - 1.2\ M_{\odot}$. This reasoning, however, is not valid for evolved stars such as core-helium burners. From simple arguments we can derive their mass distribution, in a galaxy of age T, as being proportional to the IMF ϕ_M, to the core-helium burning lifetime $t_{\rm He}$, and to the star formation rate (SFR) at the epoch of birth $\psi[T - t(M)]$. Since $t_{\rm He}$ presents a peak at about $2\ M_{\odot}$ (the transition from low- to intermediate masses), and the IMF shows a peak at the lowest masses, a *double-peaked mass distribution* appears (Girardi 1999). This is shown at the left panel of Fig. 2, for the case of a constant SFR from 0.1 to 10 Gyr ago, and a Salpeter IMF.

Moreover, it turns out that some intermediate-mass stars, from say 2 to $2.5\ M_{\odot}$, are not severely under-represented with respect to the low-mass ones. They are close enough to the clump region in the CMD to be considered as genuine clump stars. For the case of a constant SFR, they make about 20 % of the clump.

However, is the hypothesis of a constant SFR a realistic one for the

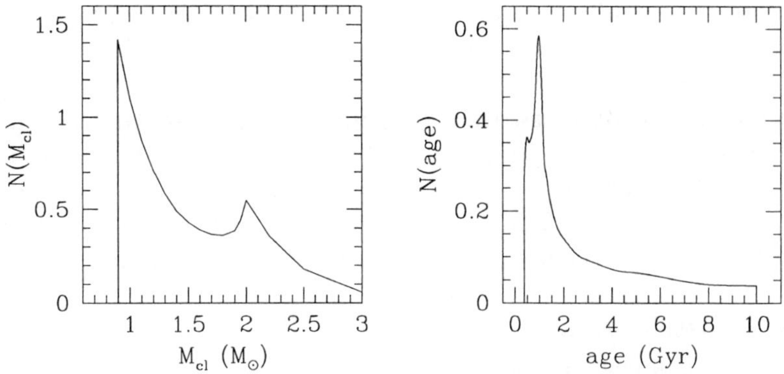

Figure 2. Mass (left panel) and age (right panel) distribution of clump stars for the case of a constant star formation rate (see text).

local disc? From careful analyses of the *Hipparcos* CMD, Bertelli & Nasi (1999) favour an almost-constant SFR with *increasing* rates in the last few Gyr, whereas Gilmore et al. (this meeting) find several episodes of star formation, without any marked long-term trend of decreasing or increasing SFR. An independent indication for a roughly-constant SFR (over time-scales of several Gyr) comes from the simulation we present in the right panel of Fig. 1. Using a population synthesis code, we generate an uniform distribution of stars, and limit it to $\pi < 0.007$ arcsec and $V < 8.5$. The stars are picked up from the Girardi et al. (2000) set of evolutionary tracks and isochrones, and distributed according to a constant SFR and with a metallicity distribution $[\mathrm{Fe/H}] = -0.12 \pm 0.18$ (see § 6 below). Although this was intended to be a *crude* initial model for the local stars, the derived synthetic CMD reproduces quite well the observed number of stars in several key regions of the *Hipparcos* CMD (Girardi et al., in preparation).

3. The age distribution

A constant SFR does not mean a constant age distribution, at least not for evolved stars. In fact, from the mass distribution of clump stars obtained with a constant SFR, we can easily derive the age distribution shown in the right panel of Fig. 2. The latter is far from being constant; it peaks at an age of 1 Gyr, and decreases monotonically afterwards. Half of the clump stars have ages lower than 2 Gyr! This result comes, essentially, from the continuous decrease, with age, of the birth rate of post-main sequence stars.

Therefore, the age distribution of clump stars (K giants) is very different from that of low-main sequence stars (G dwarfs). The long-lived G dwarfs have an age distribution simply proportional to the SFR, whereas K giants have it "biased" towards intermediate-ages ($1-3$ Gyr). This difference reflects also into the metallicities: since younger stars are more metal rich, giants tend to have a relatively narrow metallicity distribution, if compared to dwarfs (see § 6 below).

These results are in contrast with a common prejudice, i.e. that red giants sample equally well stellar populations of all ages, from the intermediate-age (metal rich) to the oldest (metal-poor) ones in a galaxy. This is not the case. Only in galaxies which formed few stars at intermediate-ages, can the age distribution of red giants be expected to be moderately flat.

4. The colour distribution

Red giant stars are known to become redder at higher metallicities. Jimenez et al. (1998) applied this concept to derive, solely from the colour range of *Hipparcos* clump stars, an estimate of metallicity range covered by them, obtaining $-0.7 < [\mathrm{Fe/H}] < 0.0$. Although this result has little effect on their estimate of the age of the Galactic disc (Carraro, this meeting), the approach is essentially misleading. In fact, red giants become redder also at lower masses, or equivalently, at higher ages. Girardi et al. (1998) demonstrated that a galaxy model with mean solar metallicity, a *very small* metallicity dispersion ($\sigma_{[\mathrm{Fe/H}]} = 0.1$ dex), and constant SFR up to 10 Gyr ago, shows a clump as wide in colour as the observed *Hipparcos* one – i.e. with $\Delta(V - I) \simeq 0.2$ mag (see also Fig. 1). It follows that a significant fraction of the colour spread of the local clump might be due to an age spread, rather than to a metallicity spread.

5. The magnitude distribution

The correct interpretation of the colour spread of clump stars is also crucial for distance determinations. Were the clump colour determined by metallicity only, the observed constancy of the I-band magnitude with colour inside the clump, in different galaxies, would be indicating that M_I is virtually independent of metallicity, and hence an excelent standard candle (cf. Paczyński & Stanek 1998; Stanek et al. 1998; Udalski 1998a).

This conclusion is not supported by theoretical models (Cole 1998; Girardi et al. 1998), which predict a brighter clump at lower metallicities, and a more complex dependence on age. The observational works by Udalski (1998a,b), instead, conclude for a very modest dependence of M_I on both metallicity and age. However, the data analyses contain several uncertain assumptions, like the large "geometric corrections" applied to the distances of LMC and SMC clusters in Udalski's (1998b) study. More recently, Twarog et al. (1999) and Sarajedini (1999) concluded for a larger age and metallicity dependence of the clump magnitude, using data from disc open clusters.

Models predict that the clump magnitude depends on age in a non-monotonic way, being fainter than the mean (by up to 0.4 mag) at both the ~ 1 Gyr and $\gtrsim 10$ Gyr age intervals. These two groups with the lowest luminosities correspond also to the bluest and reddest clump stars one finds for a given metallicity. The effect is such that, summing up stars of all ages, one can easily obtain a nearly-horizontal clump in the M_I versus $V-I$ CMD, similar to the *Hipparcos* one (Girardi et al. 1998). Almost-horizontal structures can also be obtained in galaxy models with decreasing SFRs and a reasonable age-metallicity relation (cf. Girardi 1999). Thus, the observed constancy of M_I with colour in the clump of different galaxies can be reasonably reproduced by models. Nonetheles, there is no reason for M_I being constant from galaxy to galaxy. In fact, the LMC clump should be intrinsically brighter than the local one by $0.2-0.3$ mag in the I-band (Cole 1998; Girardi et al. 1998).

A fine structure in the CMD – namely a fainter secondary clump plus a low-density bright plume – comes out in galaxy models containing $\sim$ 1-Gyr old populations with metallicities $Z \gtrsim 0.004$ (Girardi et al. 1998; Girardi 1999). These features are due to the intermediate-mass clump stars, i.e. those just massive enough for starting to burn helium in non-degenerate conditions. A nice confirmation of the predictions is provided by Bica et al. (1998) and Piatti et al. (1999), who detected a group of faint clump stars in several fields over the LMC. Similar structures are also present in the *Hipparcos* CMD (Girardi et al. 1998; Beaulieau & Sackett 1998).

6. The metallicity distribution

The metallicities of 581 nearby K giants (mainly clump stars) have been derived by Høg & Flynn (1998), based on DDO photometry. From their data it turns out that the $V-I$ colour does not correlate with [Fe/H] (cf. Paczyński 1998), contrarily to what expected, for intance, from Jimenez et al. (1998) analysis of local clump stars. On the basis of the arguments discussed in § 4 above, however, this result is not surprising: it might simply reflect that the colour spread of the local clump is mostly due to an age spread (from $\sim$ 1 to 10 Gyr), rather than due to a metallicity spread.

Girardi & Salaris (in preparation) selected a sample of clump stars with spectroscopic abundance determinations. The resulting metallicity distribution is Gaussian-like with a very small dispersion, i.e. $[\mathrm{Fe/H}] = -0.12 \pm 0.18$ dex. This disagrees with the metallicity range ($-0.7 \le [\mathrm{Fe/H}] \le 0.0$) derived from the colours by Jimenez et al. (1998). Again, the spectroscopic [Fe/H] determinations indicate that the colour spread in the local clump cannot be simply due to a spread of metallicity.

Another evident aspect in the metallicity distribution of clump stars is the anallogous of the G-dwarf problem: only 15 out of 334 clump stars have $[\mathrm{Fe/H}] < -0.5$, whereas *at least* 43 are expected from a simple closed-box model of chemical evolution. This new "K-giant problem" is quantitatively different from the G-dwarf one, since it is offered by stars which are, on the mean, younger than G dwarfs. Therefore, the chemical information provided by clump stars should be considered as additional to that provided by the dwarfs, and more suitable to probe the chemical conditions at intermediate-ages (from say 1 to 4 Gyr ago).

7. Concluding remarks

The *Hipparcos* catalog provides us an extremely interesting sample of clump stars, complete up to 125 pc, and representing a well-understood evolutionary stage. Thus far, it has provided interesting checks to the theory of stellar evolution and population synthesis. A good example is the finding of a fine structure in the clump, which represents a sort of "lower main sequence" for core-helium burning stars. Theoretical modelling of the clump in galaxies also brings some unexpected results, such as the wide and double-peaked mass distribution, and the intrinsic bias towards intermediate-ages. These aspects should be taken into account when interpreting the observational data related to clump stars, and to red giants in general.

The perspectives of using the clump data for probing the chemical evolution of the Galactic disc are also very promising. Grenon and Morossi (both in this meeting) announced more reliable and homogeneous chemical abundance determinations for nearby red giants. With these new data, the sample of nearby clump stars may render additional (and quantitative) constraints to chemical evolution models of the Solar Neighbourhood. In this respect, we should also consider the tight constraints to the disc SFR provided by the detailed analyses of the *Hipparcos* CMD.

References

1. Beaulieu J.-P., Sackett P.D., 1998, AJ 116, 209
2. Bertelli G., Nasi E., 1999, A&A in press
3. Bica E., Geisler D., Dottori H., et al., 1998, AJ 116, 723
4. Cole A.A., 1998, ApJ 500, L137
5. ESA, 1997, The Hipparcos and Tycho Catalogues, ESA SP-1200
6. Girardi L., Groenewegen M.A.T., Weiss A., Salaris M., 1998, MNRAS 301, 149
7. Girardi L., 1999, MNRAS 308, 818
8. Girardi L., Bressan A., Bertelli G., Chiosi C., 2000, A&AS in press
9. Høg E., Flynn C., 1998, MNRAS 294, 28
10. Jimenez R., Flynn C., Kotoneva E., 1998, MNRAS 299, 515
11. Paczyński B., 1998, Acta Ast. 48, 405
12. Paczyński B., Stanek K.Z., 1998, ApJ 494, L219
13. Perryman M.A.C., Lindegren L., Kovalevsky J., et al., 1997, A&A 323, L49
14. Piatti A., Geisler D., Bica E., et al., 1999, AJ in press (astro-ph/9909475)
15. Sarajedini A., 1999, AJ in press
16. Stanek K.Z., Zaritsky D., Harris J., 1998, ApJ 500, L141
17. Udalski A., 1998a, Acta Astr. 48, 113
18. Udalski A., 1998b, Acta Astr. 48, 383

Speaker: L. GIRARDI

M. TOSI: Have you included binaries in your simulations? Could the fact that your MS width is smaller than the observed one be due to the lack of binaries in the model?

L. GIRARDI: Binaries have not been included in the models, simply because they were eliminated from the Hipparcos sample we are using. In fact the Hypparcos and Tycho catalog includes enough information to allow the identification of the bulk of nearby binaries. I remark, however, that the MS width in the models is very similar to the observed one, when a metallicity dispersion of ~ 0.2 dex is considered.

M. GRENON's Comments:

1) The Hipparcos sample of clump stars is complete by construction of the survey, provided the selection is made according to the survey limits function of the galactic latitude.

2) There is a real spread in metallicity both in the clump and in the subgiant branch. The width of the GB is essentially in to [M/H] scatter and exceeds by a factor of ~ 2 that component in your simulation adopting a unique [Fe/H].

3) The (V-I) as published in the Hipparcos catalogue are mostly non genuine, but derived from (B-V), (or other indices more sensitive to metallicity than the true (V-I)).

L. GIRARDI

In fact, both the completeness and metallicity distributions adopted in the simulations (your points 1 and 2) should be improved, in order to make a more detailed analysis of the *Hipparcos* CMD. This is work in progress. With respect to (3), we selected from the *Hipparcos* catalog the I-band data corresponding to real photometric measures. When we refer to the behaviour of $V-I$ colours, we consider only this limited sample.

MIXING ALONG THE RED GIANT BRANCH IN METAL-POOR FIELD STARS

E. CARRETTA[1] AND R.G. GRATTON[1]
Osservatorio Astronomico di Padova
Vicolo dell'Osservatorio 5, I-35122 Padova, ITALY

C. SNEDEN[2]
Department of Astronomy and McDonald Observatory
The University of Texas at Austin, USA

AND

A. BRAGAGLIA[3]
Osservatorio Astronomico di Bologna
via Ranzani 1, I-40127 Bologna, ITALY

1. Introduction

Stellar models predict that as a small mass star evolves up the RGB, the outer convective envelope expands inward and penetrates into the CN-cycle processed interior regions (*first dredge-up*). Approximately, the outer 50% of the star by mass is involved in this mixing that brings to the surface mainly ^{13}C and ^{14}N, while the primordial ^{12}C and fragile, light elements like Li, Be, and B are transported from the surface to the interior. Results for old disk field giants (Shetrone et al. 1993) and metal-poor stars (Sneden et al. 1986) show that the first dredge-up occurs at the predicted luminosities; however, mixing in bright giants is much more extreme than predicted by evolutionary models.

On the other side, some further mixing is possible in the latest phases of the RGB (see e.g., Charbonnel 1994, 1995): in fact, after the end of the dredge-up phase is reached, the convective envelope begins to recede, leaving behind a chemical discontinuity. This is subsequently contacted by the H-burning shell, giving rise to the so-called RGB bump. Before this contact is made, the H-burning shell is advancing through a region where there are appreciable composition gradients, which should inhibit any (rotational induced) mixing. Thereafter, since there is no mean molecular

F. Matteucci and F. Giovannelli (eds.), The Evolution of the Milky Way, 109–114.

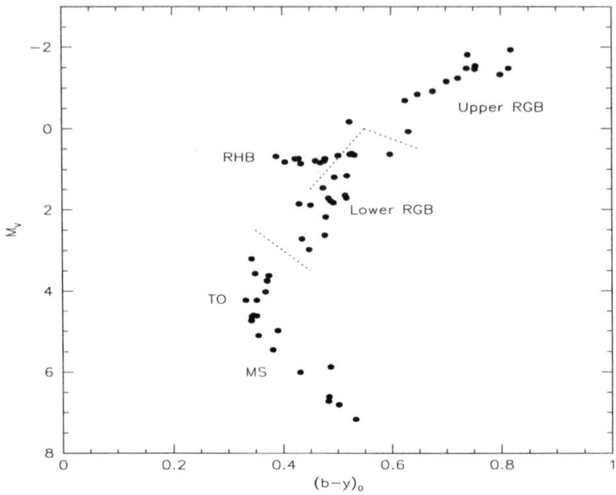

Figure 1. Selected sample: we define as lower red giant branch (lower-RGB) the evolutionary phase between first dredge-up completion and the RGB bump: small mass stars on the lower-RGB should have a well defined set of CNO and Li abundances, distinct from that observed in main sequence stars (MS stars, which have not yet experienced the first dredge up) and upper-RGB stars (hereinafter, upper-RGB stars are those stars first ascending the red giant branch, which are brighter than the RGB bump).

weight gradient between the convective envelope and the near vicinity of the shell, it is possible that circulation currents, perhaps driven by meridional circulations activated by core rotation (Sweigart & Mengel 1979) give rise to further mixing.

In principle, globular clusters offer a unique opportunity to verify this scheme, since they provide a large number of stars located at the same distance (thus allowing an accurate definition of the evolutionary status of individual stars), having the same age (thus similar masses evolving along the RGB), and, hopefully, the same initial chemical composition. However, surface abundances of globular cluster RGB stars have revealed a complex phenomenology (for a summary, see Kraft 1994), that has defied insofar any attempt of a detailed explanation. This is likely due to the fact that the surface abundances of these stars are significantly affected by several major factors: deep mixing within individual stars, primordial inhomogeneities within a cluster, and perhaps accretion of nuclearly processed material during the early phases of the cluster evolution. Furthermore, it is becoming increasingly clear that the dense environment plays an important role on determining other basic cluster features (like the colour of the horizontal branch), either by causing systematic variations in the basic stellar properties (like e.g., the initial angular momentum), or in favouring pollution of the surface layers of stars by ejecta from other stars, or both.

To solve these issues, it is necessary to first understand the evolution of single undisturbed small mass stars in the field, in a restricted range of mass and metal abundance (both of them affecting the luminosity of the RGB bump).

2. Sample selection and observations

To this purpose, we selected a sample from Anthony-Twarog & Twarog (1994, ATT) for the evolved stars) and Schuster & Nissen (1989) for the main sequence and turn-off stars, plus a few local subdwarfs from Clementini et al. (1998). We restricted to metallicities in the range $-2 <$[Fe/H]< -1, in order to: (i) avoid possible massive interlopers (thin disk stars), (ii) have a large number of moderately bright stars with accurate absolute magnitudes, hence clear evolutionary phases, and (iii) avoid complications from large star-to-star abundance variations present among the most metal-poor stars.

The resulting sample (62 stars) is shown in Figure 1; evolutionary phases are derived from the position on the $(b-y)_0 - c_{10}$ and the calassical colour magnitude diagram; M_V values are from ATT (giants and subgiants) or from Hipparcos (dwarfs).

High S/N (> 100), high resolution ($R > 50,000$) spectra were acquired at the McDonald Observatory (2.7m telescope + "2d-coudé" echelle spectrometer, spectral range $3800 \leq \lambda \leq 9000$ Å) and at ESO (CAT+CES spectrograph, spectral regions centered at 4230, 4380, 5680, 6300, 6700, and 7780 Å), to measure abundance indicators for Li, C, N, O, Na and Fe.

3. Atmospheric parameters and abundance analysis

Effective temperatures ($T_{\rm eff}$'s) were derived from the dereddened $(B-V)_0$ and $(b-y)_0$ colours using colour-$T_{\rm eff}$ transformations from Kurucz (1995) models with no overshooting, and reddenings from ATT or Carretta, Gratton & Sneden (1999). Note that no empirical correction was required for these calibrations (see e.g., Castelli, Gratton & Kurucz 1997). Gravities were derived from absolute magnitudes and $T_{\rm eff}$'s, assuming a mass M = 0.85 $M_\odot$ for all stars and B.C.'s from Kurucz (1995). Eliminating trends of abundances derived from Fe I lines versus EWs for stars with McDonald spectra, we obtained microturbulent velocities v_t's for this subsample and a tight relation $v_t = f(T_{\rm eff}, \log g)$. This was used to derive accurate v_t values for stars with CAT spectra (having very few Fe lines measured).

The abundance analysis was performed using Kurucz (1995) model atmospheres with no overshooting. Further details are in Gratton et al. (2000).

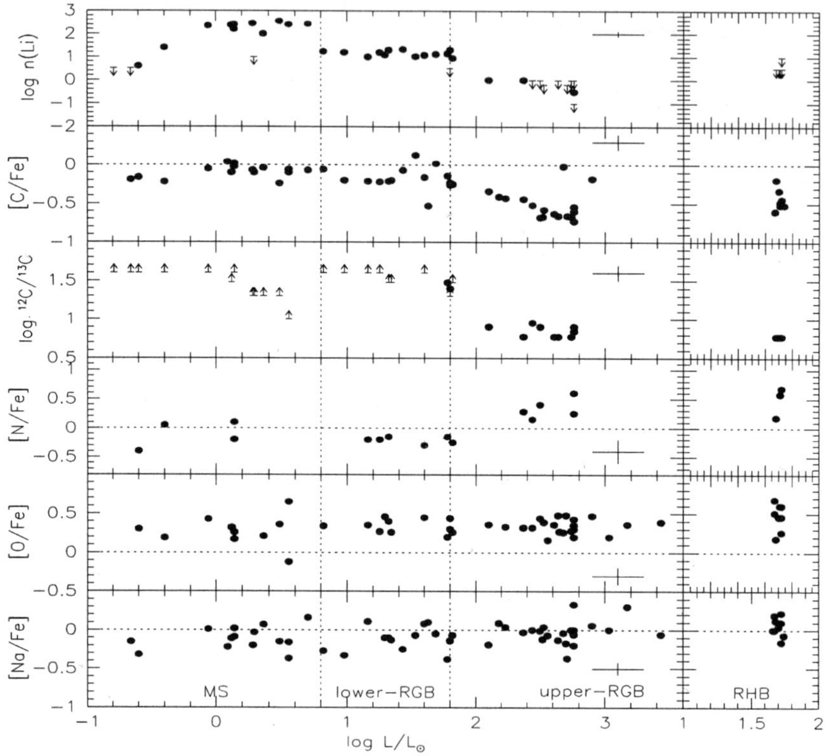

Figure 2. Run of the abundance of Li, of the abundance ratios [C/Fe], [N/Fe], [O/Fe] and [Na/Fe], and of the isotopic ratio $^{12}C/^{13}C$ with luminosity for stars with $-2 <$[Fe/H]< -1. Filled symbols are actual measures; arrows are upper (for Li) or lower (for $^{12}C/^{13}C$) limits. Typical error bars for the various quantities are shown. Dashed lines separate various evolutionary phases. Results for red HB (RHB) stars are plotted separately for clarity.

Abundances for Fe I, Fe II, O I, Na I, $\alpha-$elements and heavy elements were derived from measured equivalent widths (EWs). Whenever possible, we complemented the rather few atomic lines measured on CAT spectra with lists of accurate (errors ≤ 3 mÅ) and homogeneous EWs from literature (see Gratton, Carretta & Castelli 1997 for references).

O abundances were obtained from both forbidden and permitted lines. Abundances from these last (as well as the Na abundances from the 5682-88 Å and 6154-60 Å doublets) include corrections for departures from LTE (Gratton et al. 1999). Average [O/Fe] ratios were computed after a small offset of 0.08 dex between the abundances from [O I] and O I lines was accounted for.

C abundances as well $^{12}C/^{13}C$ isotopic ratios were derived from spectral synthesis of the G-band around 4300 Å. Synthetic spectra computations of

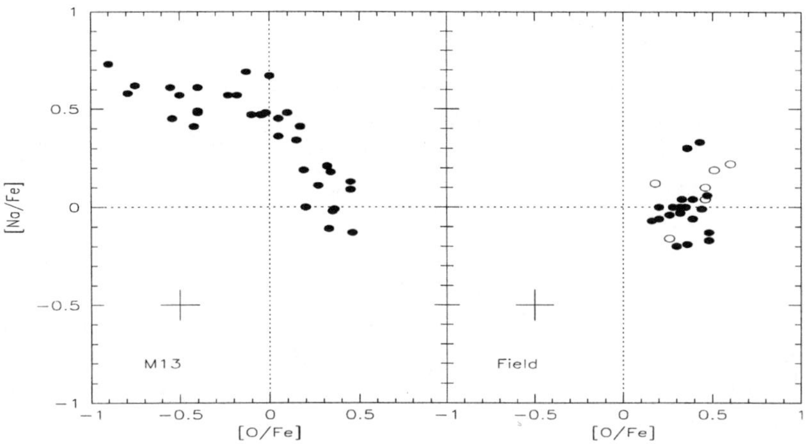

Figure 3. The Na-O anticorrelation for field stars and for stars in M13. Filled symbols are stars first ascending the RGB (only upper-RGB stars are shown); open symbols are RHB stars. Error bars are shown at the bottom left.

the (0.0) and (1,1) bandheads of the violet band of CN were used to derive N abundances for the 24 program stars with McDonald spectra.

Li abundances have been derived from comparison to synthetic spectra, and complemented with data from Pilachowski, Sneden & Booth (1993, PSB), after Fe and Li abundances from their analysis were increased to compensate the offsets (0.09 and 0.21 dex, respectively) with respect to our results, due to our higher T_{eff}'s (112 ± 32 K, from 14 stars in common with PSB).

4. Results

Results for Li, C, N, O, Na and $^{12}C/^{13}C$ isotopic ratios are summarized in Figure 2 for the 62 program stars with accurate evolutionary phase, plus 43 stars more having accurate abundances for some of these elements and similarly well defined luminosities from the literature. ¿From Figure 2 we can see that small mass lower-RGB stars (i.e., stars brighter than the first dredge-up luminosity and fainter than the RGB bump) have abundances of light elements in agreement with predictions from classical evolutionary models: only marginal changes occur for CNO elements, while dilution within the convective envelope causes the surface Li abundance to decrease by a factor of ~ 20.

A second, distinct mixing episode occurs in most (perhaps all) small mass metal-poor stars just after the RGB bump, when the molecular weight barrier left by the maximum inward penetration of the convective shell is canceled by the outward expansion of the H-burning shell, in agreement with recent theoretical predictions.

In field stars, this second mixing episode only reaches regions of incomplete CNO burning: it causes a depletion of the surface ^{12}C abundance by about a factor of 2.5, and a corresponding increase in the N abundance by about a factor of 4. The ^{12}C/^{13}C is lowered to about 6 to 10 (close to, but distinctly larger than the equilibrium value of 3.5), while practically all remaining Li is burnt.

However, an O-Na anti-correlation such as typically observed among globular cluster stars (see Figure 3) is not present in field stars. None of the 29 field stars more evolved than the RGB bump (including 8 RHB stars) shows any sign of O depletion or Na enhancement. This means that in field stars the second mixing episode is not deep enough to reach regions were ON-burning occurs.

References

1. Anthony-Twarog, B.J., Twarog, B.A. (1994) Reddening estimation for halo red giants using uvby photometry, *Astron. J.* **107**, 1577
2. Carretta, E., Gratton, R.G., Sneden, C. (1999), submitted to *Astron. Astrophys*
3. Castelli, F., Gratton, R.G., Kurucz, R.L. (1997), Notes on the convection in the ATLAS9 model atmospheres, *Astron. Astrophys* **318**, 841
4. Charbonnel, C. (1994), Clues for non-standard mixing on the red giant branch from C-12/C-13 and C-12/N-14 ratios in evolved star, *Astron. Astrophys* **282**, 811
5. Charbonnel, C. (1995) A Consistent Explanation for 12C/ 13C, 7Li and 3He Anomalies in Red Giant Stars *Ap. J.* **453**, L41
6. Clementini, G., Gratton, R.G., Carretta, E., Sneden, C. (1998) Homogeneous photometry and metal abundances for a large sample of Hipparcos metal-poor stars, *Mon. Not. R. Astr. Soc* **302**, 22
7. Gratton, R.G., Carretta, E., Castelli, F. (1997) Abundances of light elements in metal-poor stars. I. Atmospheric parameters and a new Teff scale, *Astron. Astrophys.* **314**, 191
8. Gratton, R.G., Gustafsson, B., Carretta, E., Eriksson, K. (1999) Abundances of light elements in metal-poor stars. II. Non-LTE abundance corrections *Astron. Astrophys,* **350**, 955
9. Gratton, R.G., Sneden, C., Carretta, E., Bragaglia, A. 2000, *Astron. Astrophys.*, in press
10. Kraft, R.P. (1994) Abundance differences among globular-cluster giants: Primordial versus evolutionary scenarios, *Pub. Astr. Soc. Pacific* **106**, 553
11. Kurucz, R.L. (1995), CD-ROM 13
12. Pilachowski, C.A., Sneden, C., Booth, J. (1993) The abundance of lithium in metal-poor subgiant stars, *Ap. J.* **407**, 699
13. Schuster, W.J., Nissen, P.E. (1989) ubvy-beta photometry of high-velocity and metal-poor stars. III - Metallicities and ages of the halo star, *Astron. Astrophys.* **222**, 69
14. Shetrone, M.D., Sneden, C., Pilachowski, C.A. (1993) Carbon isotope ratios and lithium abundances in old disk giants, *Pub. Astr. Soc. Pacific* **105**, 337
15. Sneden, C., Pilachowski, C.A., VandenBerg, D.A. (1986) Carbon isotope ratios in field Population II giant star, *Ap. J.* **311**, 826
16. Sweigart, A.V., Mengel, J.G. (1979) Meridional circulation and CNO anomalies in red giant stars, *Ap. J.* **229**, 624

THE METAL ABUNDANCE OF GALACTIC CEPHEIDS AS DERIVED FROM THEIR PULSATIONAL PROPERTIES

M. MARCONI, I. MUSELLA AND V. RIPEPI
Osservatorio Astronomico di Capodimonte, Via Moiariello 16, I-80131

AND

F. CAPUTO
Osservatorio Astronomico di Roma, Via di Frascati 33, Monteporzio Catone, I-00040

1. Introduction

Classical Cepheids play a fundamental role in the determination of distances to Local Group galaxies, in the calibration of various secondary distance indicators and finally in the evaluation of the Hubble constant H_0. A considerable amount of observational studies has been devoted to the calibration of their characteristic Period-Luminosity (PL) and Period-Luminosity-Color (PLC) relations and to the application of these relations to extragalactic samples.

In this context an important question to be answered is whether the PL and PLC relations depend on the chemical abundance of the variables. This is a crucial point since if there is a metallicity effect the distances to galaxies with metallicities different from the calibrating Cepheids, should be systematically underestimated or overestimated. In recent times, the need of assessing the magnitude and the sign of a possible metallicity effect on Cepheid PL and PLC relations has become more and more evident, with the observational tests of this effect suggesting that metal-rich Cepheids are brighter than metal-poor Cepheids at fixed period (see Kochanek 1997; Beaulieu et al. 1997; Sasselov et al. 1997; Kennicutt et al. 1998).

As for theoretical estimates, models based on *linear* pulsation calculations suggest a very low effect (Chiosi et al. 1993; Saio & Gautschy 1998; Alibert et al. 1999).

F. Matteucci and F. Giovannelli (eds.), The Evolution of the Milky Way, 115–121.

In the last few years we have been producing an extensive grid of *nonlinear*, nonlocal and time-dependent convective pulsating models for different values of the stellar mass and three different chemical compositions, taken as representative of Cepheids in the Galaxy and in the Magellanic Clouds, and the results of this investigation have already been presented in a series of papers (Bono et al. 1998; Bono et al. 1999a,b; Bono et al. 2000; Caputo et al. 1999, Caputo et al. 2000). We only remind here that the main advantages of nonlinear convective models are: 1) the possibility of evaluating the long term full-amplitude behavior of models and in turn of predicting the light curve morphology, the amplitudes for the various pulsation observables and the mean magnitudes and colors over a pulsation cycle; 2) the capability of deriving the location of both the boundaries (blue and red) of the instability strip through the nonlinear coupling between pulsation and convection.

The transformation of theoretical bolometric light curves into the various photometric bands, through the adoption of updated model atmospheres, allowed us to evaluate magnitude-weighted and intensity-weighted mean magnitudes and colors, to be coupled with the period information. The resulting least square solutions through models led to both statistical (connected with the instability strip topology, such as the PL and the Period-Color (PC)), and punctual (i.e. individually valid for each pulsator, such as the PLC and the Color-Color (CC)) relations in the BVRIJK bands. These relations turned out to depend, even if with different extents, on the adopted metal content (see Bono et al. 1999b, Caputo et al. 1999, Caputo et al. 2000) and this dependence was found to be the direct consequence of the dependence of the theoretical instability strip on metallicity.

As discussed in Bono et al. (1999b, 2000), the fundamental mode instability strip based on nonlinear convective models gets redder as the metallicity increases from 0.004 to 0.02, so that the average period for each fixed luminosity level increases with metallicity and, at the same time, the luminosity of metal reach Cepheids, at fixed period, is fainter than for metal poor Cepheids. This effect, which is already evident in bolometric PL relations, increases at the visual passbands due to the additional sensitivity of bolometric corrections. On the other hand, near infrared PL relations are slightly affected by both the metallicity variations and the spurious effect related to the finite width of the instability strip (see Bono et al. 1999b, Caputo et al. 1999, Caputo et al. 2000).

As for PLC relations, that are individually valid for each pulsator, thus not requiring any assumption on the way Cepheids are distributed within the instability strip, we also find a sensitivity to the adopted metal content.

Indeed, on the basis of the predicted BV and VR PLC relations, metal rich Cepheids should be brighter than the metal poor ones at fixed period and color, whereas the opposite behavior occurs for the VJ and VK PLC. As a natural consequence the VI PLC relation turns out to be the less affected by the adopted metal abundance thus representing a powerful tool for extragalactic distance determinations (provided that colors are well known).

2. Metallicity effect on the Cepheids distance scale

If, as suggested by nonlinear convective models, PL and PLC relations do depend on metallicity, it is important to discuss the effect of this sensitivity on the galactic and extragalactic distance evaluations. In fact, in this case, we have to deal with three unknown quantities: distance, reddening and metallicity.
This means that we need data in at least three bands in order to disentangle the effects of metallicity and reddening on our distance estimate with the help of theoretical relations.
A first application of this three-filter method to the Magellanic Cloud Cepheids (BVK data) has already been presented by Caputo et al. (1999) and we refer the interested reader to that paper for details on the adopted procedure and results.
In the next section we will apply a similar method to two samples of galactic Cepheids for deriving accurate individual estimates of reddening, distance and metallicity.

3. Application to galactic Cepheids

We considered the BVI data for the nearby galactic Cepheids analysed by Gieren et al. (1998, GFG98) and the outer-disk pulsators presented by Pont et al. (1997, P97) and, by adopting the theoretical CC, PL and PLC relations in the BVI bands, we self-consistently derived individual values of reddening, distance and metallicity.
First of all we derived the reddening E(B-V), for each pulsator in both samples and for each assumed metallicity, from the application of theoretical (B-V) vs (V-I) relations at Z=0.004, 0.008, 0.02 to the BVI observations. The comparison in the (B-V) vs (V-I) plane between theoretical predictions and data from GFG98 is shown in Fig. 1.
Using the metallicity-dependent reddening estimates derived from Fig. 1 we provided, for each assumed metal abundance, the distance moduli based both on the I band PL relation ($\mu_0 PL_I$) and on the VI PLC relation ($\mu_0 PLC_I$).
Then we required consistency between $\mu_0 PL_I$ and $\mu_0 PLC_I$. To this aim, in Fig. 2 the difference $\delta\mu_0(PLC_I - PL_I)$ is plotted as a function of the pulsation period, at varying Z, for the GFG98 (left panels) and the P97 (right panels) samples respectively. The solid lines in both figures are the

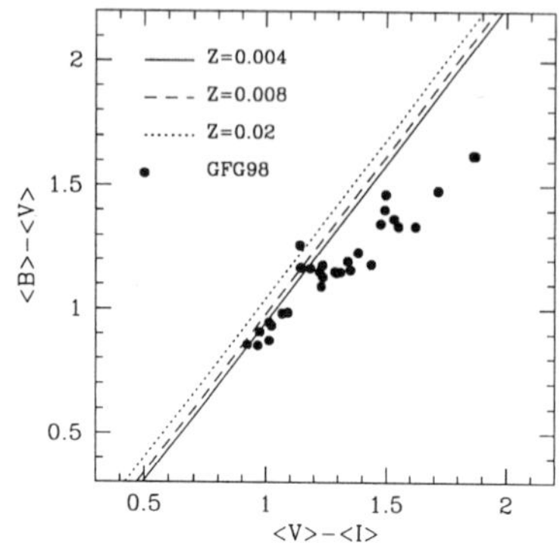

Figure 1. Comparison between theoretical Color-Color relations (lines) for the labeled metal abundances and the BVI data (filled circles) from GFG98

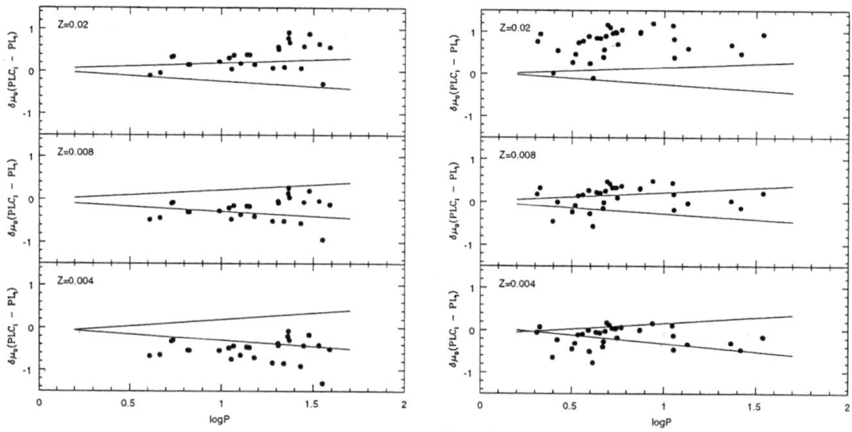

Figure 2. The difference between the moduli obtained using the PL and the PLC relations for each selected metal abundance as a function of the pulsation period for GFG98 (left) and the P97 (right) samples respectively. The the two solid lines in each panel represent the boundaries of the fundamental instability strip at the corresponding labeled metal abundances.

instability strip boundaries in this plane for the corresponding labeled metal abundances.

We notice that $\delta\mu_0(PLC_I - PL_I)$ is expected to be zero only in the middle of the instability strip due to the statistical nature of the PL relation. But each individual pulsator has to be located between the blue and the red boundary so that the comparison in Fig. 2 enables us to evaluate both the mean value of metallicity and the associated error.

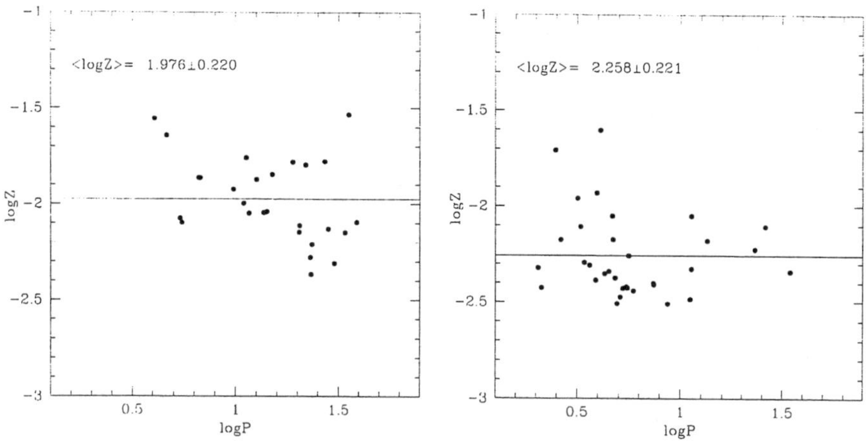

Figure 3. The resulting metallicity distribution for the GFG98 (left) and P97 (right) Cepheids as a function of the pulsation period.

The resulting metallicity distribution for the GFG98 and P97 Cepheids is shown in Fig. 3 and the mean values for both samples are labeled with the associated errors.

From inspection of this figure we notice that the mean metallicity value of the outer disk Cepheids by P97 is significantly lower than the mean abundance of nearby Cepheids by GFG98, suggesting a possible metallicty gradient among Galactic Cepheids (see Sect. 4).

4. The metallicity gradient

Since the three-filter method followed in the previous section allows the avaluation of individual distance, reddening and metallicity values for all the pulsators in our samples, we can assume a distance to the Galactic center and try to find a correlation between the galactocentric distances and the metallicities.

This is shown in Fig. 4 in which we have assumed a distance to the Galactic center equal to 8.5 Kpc (Feast & Withelock 1997) and we have plotted the metal abundances versus the Galactocentric distances for both the GFG98 Cepheids (squares) and the P97 outer disk pulsators (circles). The dashed line in this figure is the metallicity gradient (-0.07 dex Kpc^{-1}) suggested by Harris (1981). From this Figure one may reasonably conclude that theoretical pulsation models predict a metallicity gradient among galactic Cepheids with a slope which is in good agreement also with other estimates in the literature (e.g. Maciel & Quireza 1999).

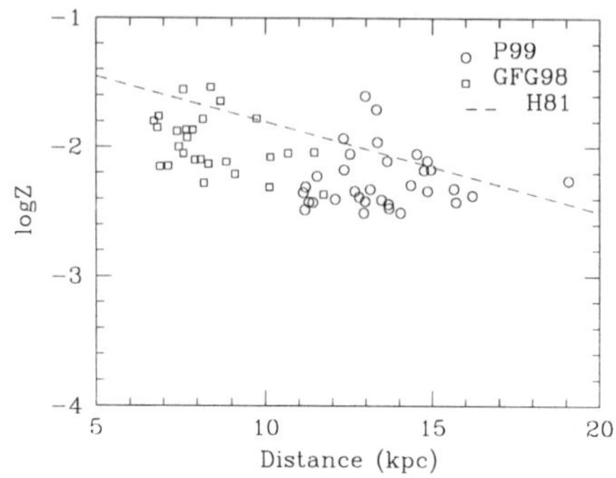

Figure 4. The metal abundances versus the Galactocentric distances for both the GFG98 Cepheids (squares) and the P97 outer disk pulsators (circles). The dashed line is the metallicity gradient (-0.07 dex Kpc^{-1}) suggested by Harris (1981).

5. Summary

Nonlinear nonlocal time-dependent convective models predict that Cepheid PL, PC, PLC and CC relations depend on metallicity. As a consequence self consistent estimates of E(B-V), μ_0 and Z may be derived if three filter data are taken into account. The application of this three-filter method to Galactic Cepheids (BVI data) provides individual values of reddening, distance and metallicity. When these metal abundances are plotted as a function of the Galactocentric distances a metallicity gradient appears, that is in good agreement with current empirical evaluations.

References

1. Alibert Y., Baraffe I., Hauschildt P., Allard, F. 1999, A&A 344, 551
2. Beaulieu J. P., Krockenberger M., Sasselov D. D. et al. 1997, A&A 318, L47
3. Bono G., Caputo F., Marconi M. 1998, ApJ 497, L43
4. Bono G., Marconi M., Stellingwerf R. F. 1999a, ApJS 122, 167
5. Bono G., Caputo F., Castellani V., Marconi M. 1999b, ApJ 512, 711
6. Bono G., Castellani V., Marconi M. 2000, ApJ, in press
7. Caputo F., Marconi M., Ripepi V. 1999, ApJ 525, 784
8. Caputo F., Marconi M., Musella I. 2000, A&A in press
9. Chiosi C., Wood P. R., Capitanio N. 1993, ApJS 86, 541
10. Feast, M. Whitelock, P., 1997, MNRAS, 291, 683
11. Gieren W. P., Fouqué P., Gómez M. 1998, ApJ 496, 17 [GFG98]
12. Harris, H.C. 1981, AJ, 86,707 [H81]
13. Kennicutt R. C., Stetson P. B., Saha A. et al. 1998, ApJ 498, 181
14. Kochanek C. S. 1997, ApJ 491, 13
15. Maciel, W.J., Quireza, C. 1999, A&A, 345, 629
16. Pont, F., Queloz, D., Bratschi, P., Mayor, M., 1997, A&A, 318, 416 [P97]
17. Saio H., Gautschy A. 1998, ApJ 498, 360
18. Sasselov D. D., Beaulieu J. P., Renault, C. at al. 1997, A&A 324, 471

Speaker: M. MARCONI

S. SMARTT: Regarding the galactic abundance gradient, is this representative of a particular element (e.g. O or Fe)? Can you disentangle the effects from different elements, or is this just a "global" value?

M. MARCONI: The derived abundance gradient is not representative of a particular element but of the global metallicity. With the outlined method we can only derive the global metal abundance gradient.

R. WYSE: How does the amplitude of the metallicity gradient you find for Cepheids compare with that for other tracers like HII regions, old open clusters etc.?

M. MARCONI: The amplitude of metallicity gradient we find on the basis of Cepheid theoretical models is in agreement within the errors with that for HII regions (e. g. Afflerbach et al., 1997), open clusters (e.g. Twarog et al., 1997), planetary nebulae (e.g. Maciel and Quireza, 1999).

II-BINARIES

The Volcano Stromboli

WOLF-RAYET STARS, BINARIES, BLACK HOLES

KAREL A. VAN DER HUCHT
Space Research Organization Netherlands,
Sorbonnelaan 2, NL-3584 CA Utrecht, the Netherlands

Abstract. We review the literature of the past two decades on the possible evolution from the massive star Wolf-Rayet phase to the black hole phase, and on the masses of WR stars observed in binary systems. A list of 55 WR binaries is presented. The 19 WR binaries with RV solutions have WN star masses in the range 2–55 $M_\odot$ and WC star masses in the range 9–16 $M_\odot$. We discuss four candidate WR+BH systems: WR 145a (Cyg X-3), WR 148 (HD 197406), BAT99-102 (R 140a2), and BAT99-116 (Mk 34).

1. Introduction

Although the notion that massive stars may evolve into black holes is over three decades old, it was not until the late 70-ies that it was realized that Wolf-Rayet (WR) stars represent a normal phase in the evolution of massive stars, *i.e.*, the phase just before the supernova explosion. We review here the literature on the relation between WR stars and black holes, and present a list of 55 currently known galactic WR binaries, 19 of which have RV-solutions and mass determinations, against which evolutionary models of massive stars can be tested.

2. Single Wolf-Rayet stars

Maeder (1981) presented evolutionary models with mass loss for initially massive ($M_\mathrm{i} > 30\,M_\odot$) stars, showing that WR stars are likely to be SN precursors. With WR stars having lost their hydrogen envelope, such explosions will have a spectrum showing no hydrogen lines and thus correspond to Type I SNe. Maeder & Lequeux (1982) determined the fraction of supernovae originating from WR progenitors in three different ways: (*i*) based on an estimated number of 1200 WR stars in the Galaxy and mean WR

F. Matteucci and F. Giovannelli (eds.), The Evolution of the Milky Way, 125–138.

lifetimes of $\lesssim 5\times10^5$ yr; (*ii*) based on SN remnants; and (*iii*) based on the mass spectrum and mass limits, with $M_i = 8\,M_\odot$ as the lower limit initial mass of a star evolving toward a SN explosion and $M_i = 23\,M_\odot$ as the lower mass limit to form a SN from a WR progenitor. All three estimates indicate that one out of 3–7 SN events may result from a WR progenitor. This agrees with the fact that of the five galactic SNe of the past millennium, at least one, Cas A (Langer & El Eid 1986; Langer 1987; Fesen *et al.* 1987), may have originated from a WR star. El Eid & Langer (1986) proposed for Cas A a SN explosion by $e^{\pm}$-pair creation instability in a $45\,M_\odot$ WN star (with $M_i = 100\,M_\odot$), its final fate being complete disruption by explosive oxygen burning. Maeder (1983) and Maeder & Meynet (1987) presented three scenarios to form SNe:

$40\,M_\odot < M_i$	O - Of - BSG - LBV - WNL - WNE - WC - (WO) - SN
$35\,M_\odot < M_i < 40\,M_\odot$	O - BSG - YSG - RSG - (BSG) - WNE - WC - SN
$15\,M_\odot < M_i < 35\,M_\odot$	O - (BSG) - RSG - YSG/Cepheid - RSG - SN

Maeder & Meynet (1989) concluded from further evolutionary models for massive stars with mass loss by stellar wind, that $M_i(\mathrm{WR}) = 20$–$40\,M_\odot$ is the minimum initial mass allowing the evolution to the WR phase. The minimum initial mass allowing evolution to the black hole phase is: (*i*) $M_i(\mathrm{BH}) = 25\,M_\odot$ from theoretical models (*e.g.*, Woosley 1986); (*ii*) $M_i(\mathrm{BH}) = 40$–$80\,M_\odot$ from two observed X-ray binaries (van den Heuvel & Habets 1984); and (*iii*) $M_i(\mathrm{BH}) = 50\,M_\odot$ from pulsars in young associations (Schild & Maeder 1985). Thus they concluded that the direct progenitors of black holes should be WR stars, notably WNL and WCE stars. Maeder (1991) specified massive star evolution as a function of both initial mass and metallicity.

Langer (1989) calculated models with mass-dependent mass loss rates and concluded that WNE and WC stars have masses in the range 5–$20\,M_\odot$ and that the mean final $M_{\mathrm{WR}} = 5$–$10\,M_\odot$. This makes them candidates for SN Ib progenitors, since Ensman & Woosley (1988) derived that light-curves of SN Ib explosions require 4–$7\,M_\odot$ progenitors with $M_i = 15$–$20\,M_\odot$. Langer (1991) concluded that: (*i*) stars with $M_i > 100\,M_\odot$ are supposed to explode during their WNL phase as peculiar SNe II due to the pair-creation mechanism, accounting for $\sim 1\%$ of the observed SNe; and (*ii*) the bulk of the stars with $35\,M_\odot < M_i < 100\,M_\odot$ is supposed to end their evolution as WC stars with $M_{\mathrm{WC}} < 10\,M_\odot$. Some of them explode due to the core-collapse mechanism, to become visible as possible SNe Ib. Langer & Woosley (1991) and Woosley *et al.* (1993) calculated a synthetic light-curve of an exploding $4.25\,M_\odot$ WC star, evolved from a $M_i = 60\,M_\odot$ star with mass-dependent mass loss rate, revealing similarities to observed SN Ib light-curves. Woosley *et al.* (1995) narrowed down the final mass range of helium

(WR) stars, with $M_\mathrm{i} = 4$–$20\,\mathrm{M}_\odot$ and mass-dependent mass loss rates, to 2.3–3.6 $\mathrm{M}_\odot$, and identified these as progenitors of SNe Ib (no hydrogen lines) and, perhaps, SNe Ic (not even helium lines). Up to 25% of all SNe are of Type Ib or Ic (typical light-curves are shown by Vanbeveren *et al.* 1998, p. 45), *i.e.*, powered by iron-core collapse in a hydrogen-free star.

An important parameter in the evolutionary models mentioned above is the mass-loss rate during the WR phase. WR mass loss rate determinations have come down in recent years when accounting for clumping effects, by factors of 3–4 with respect to homogeneous atmosphere models (to typical rates $\dot{M} \simeq 0.5$–5×10^{-5}), as examplified by, *e.g.*, Hillier (1996), Hamann & Koesterke (1998), Nugis *et al.* (1998), and reviewed by Crowther (1999). Therefore, perhaps, SNe Ic originate only from mass-transfer binaries, rather than from single massive stars.

3. Wolf-Rayet binaries

A significant fraction of the WR stars is contained in binaries, mostly SB2 WR+OB objects, in a few cases possibly as SB1 WR+c objects (c: neutron star or black hole). Van der Hucht *et al.* (1988) determined a WR binary frequency of 37 % within 2.5 kpc from the Sun, or 34 % within 3 kpc. Moffat (1995) counted 42 %. Incompleteness of both WR star detection and binary classification in these volumes renders these numbers only lower limits.

The occurrence of WR+c systems was proposed from evolutionary considerations for massive early-type binaries (van den Heuvel & Heise 1972). However, van den Heuvel (1994) noted that the evolution beyond the HMXB phase is uncertain due to lack of observed WR+c objects, apart from Cyg X-3 (see below). The most dependable test to check whether a given close binary does comprises a black hole is a large mass combined with strong X-ray emission. Indirect evidence for black holes may be obtained for WR stars with a massive unseen companion (SB1), even if the object is not a strong X-ray source, but has a ring nebula or lies high above the galactic plane (Tutokov & Cherepashchuk 1985). Stevens & Willis (1988) calculated the expected X-ray luminosity of the proposed WR+c system WR 6 (HD 50896, WN4) and found that, while the dense WR wind will absorb a large fraction of the X-rays originating from accretion on to the neutron star, the observed X-ray luminosity should be considerably higher than is actually observed, suggesting that an accreting neutron star is not present in WR 6. However, Vanbeveren *et al.* (1998, p. 238–245) argued that lack of X-rays can be explained if the compact companion is a rapidly spinning neutron star.

Mosqueda & Koenigsberger (1990) derived constraints that restrict the number of close massive binary systems in which black hole formation is allowed. They showed that in co-rotating massive close binaries, provided that no more than 40 % of the specific angular momentum of the exploding star is lost, black holes can form only from binary systems with pre-explosion orbital periods larger than 15 d. De Donder *et al.* (1997) predicted that (*i*) about 75 % of the observed WR+OB binaries will disrupt due the SN explosion of the WR component. The majority of the OB components of these disrupted binaries will evolve into WR stars, *i.e.*, single WRs with a binary history of possible accretion; and (*ii*) less than 3 % of the observed WR+OB binaries will form WR+c binaries, with periods in the range 0.05–3.6 d.

Woosley & MacFadyen (1999) and MacFadyen & Woosley (1999), in discussing the association of GRB 980425 with the SN 1998bw, argued that a black hole formed by incomplete explosion of a rapidly rotating massive star — a collapsar — produces a Type Ib/c supernova like SN 1998bw, but that 99% of SNe Ib/c are made in 3–4 $M_{\odot}$ helium cores (WR stars) that leave neutron stars and make no GRBs. Langer & Heger (1999) concluded that the majority of SNe Ib/c occur in post-mass-transfer close binaries. Yet the observed galactic binary frequency is $< 50\%$. Thus, although SN statistics show that there must be many more WR+OB binaries than single WR stars, we have not yet detected all those binaries.

The VIIth Catalogue of Galactic Wolf-Rayet stars (van der Hucht 2000, replacing the VIth Catalogue of van der Hucht *et al.* 1981), lists 227 WR stars. Among those are at least 55 WR binaries, 28 more than in the WR binary census of Smith & Maeder (1989). Among the WR binaries all WR subtypes are represented, apart from WN9-11 types. Tables 1 and 2 list the 55 WR binaries, including:
(*i*) double-line spectroscopic binaries (SB2);
(*ii*) single-line spectroscopic binaries (SB1);
(*iii*) long-period binaries deduced from high-spatial-resolution IR imaging of spiral dust formation around systems like WR 98a and WR 104 (Tuthill *et al.* 1999, 2000; Monnier *et al.* 1999), some of which may never get a RV solution because of low orbital inclination, although their composite spectra could be noted (*e.g.*, Williams & van der Hucht 1996; 2000);
(*iv*) very-long-period binaries inferred from episodic/periodic dust formation, like WR 19, WR 48a, WR 125, WR 137, WR 140 (Williams 1999), requiring monitoring campaigns of a few decades;
(*v*) very-long- and extreme-long-period binaries inferred from episodic/periodic non-thermal radio excesses like WR 140 (Williams *et al.* 1990, 1994), WR 146 and WR 147 (Setia Gunawan *et al.* 2000a,b) requiring monitoring campaigns of a few decades;
(*vi*) extreme-long-period binaries deduced from high-spatial-resolution non-thermal radio and IR/optical imaging, like WR 146 and WR 147 (Dougherty *et al.* 1996; Williams *et al.* 1997; Niemela *et al.* 1998); and

(*vii*) binaries inferred from composite spectra like, *e.g.*, WR 19, WR 27, WR 50, WR 69, WR 77, WR 104, WR 125, WR 137, and WR 146.
In the period characterization we adopt as terminology:

$P < 1$ d	:	very-short-period binary
1 d $< P <$ 10 d	:	short-period binary
10 d $< P <$ 100 d	:	medium-period binary
100 d $< P <$ 1000 d	:	long-period binary
1000 d $< P <$ 10000 d	:	very-long-period binary
10000 d $< P$	:	extreme-long-period binary

The 19 WR double-line spectroscopic binaries (SB2) have RV solutions and reliable masses. Of those 19, the 13 WN stars have masses in the range 2.3–55 $M_\odot$ with $\overline{M_{\rm WN}} = 22 \pm 17\, M_\odot$. Three WN5–7 stars (WR 22, WR 47 and WR 141) have masses $> 40\, M_\odot$. The 6 WC stars have masses in the range 9–16 $M_\odot$ with $\overline{M_{\rm WC}} = 12 \pm 3\, M_\odot$. In agreement with the evolutionary scenarios referred to above, $\overline{M_{\rm WC}} < \overline{M_{\rm WN}}$. For these WR binaries the mass ratio $q = M_{\rm WR}/M_{\rm O}$ ranges as $q_{\rm WN} = 0.2$–3 (four WN stars have $q > 1$), $q_{\rm WC} = 0.3$–0.6, and $q_{\rm WO} = 0.15$.

The orbital periods among the 55 WR binaries range from 0.2 d (WR 109, WR 145a) to 4700 d (WR 137) and larger (lower limits of 6600 d and 7800 d for, respectively, WR 125 and WR 48a). Especially the census of very-short-period WR binaries ($P < 1$ d) and long-period WR binaries ($P > 100$ d) is suffering from observational bias.

Two WR objects of which a possible binary status has been suggested in the literature, but which have not been included in Tables 1 and 2, are:
- WR 6 (HD 50896, WN4, $P = 3.7650$ d): earmarked as a possible SB1 by Firmani *et al.* (1980), its binary nature has been questioned over the years, *e.g.*, lately by Morel *et al.* (1998), and promoted, *e.g.*, by Duijsens *et al.* (1996) and Georgiev *et al.* (1999);
- WR 25 (HD 93162, WN6ha, excess L_X): shows optical polarization variability on a time-scale of years (Drissen *et al.* 1992). Its excess X-ray luminosity has been associated with a possible very-long-period colliding wind binary (CWB) status by, *e.g.*, van der Hucht *et al.* (1992) and Corcoran *et al.* (2000).

4. Four Wolf-Rayet + black hole candidate binaries

WR 145a (Cygnus X-3, WN4-7+c, $P = 0.1997$ d)
At a distance of 10 kpc, Cyg X-3 is one of the brightest X-ray sources in the Galaxy: $\log(L_X/L_\odot) \simeq 3$–5 (van der Klis 1993). The companion of Cyg X-3 was predicted (van den Heuvel & de Loore 1973) and shown (van Kerkwijk *et al.* 1992, 1996; Fender *et al.* 1999) to be a WR star. Schmutz *et al.* (1996) found $f(M) = 2.3\, M_\odot$. Assuming $M_{\rm WR} \simeq 13\, M_\odot$ they derived $M_{\rm c} \simeq 17\, M_\odot$,

suggesting the companion be a black hole. However, Mitra (1996, 1998) objected against the WR classification for the visible component of Cyg X-3, stating that a WR wind should be opaque to low-energy X-rays (; see also Vanbeveren *et al.* 1998, p. 245–251; Vanbeveren 1999). Fender *et al.* (1999) proposed that the WR wind of Cyg X-3 is significantly flattened, reconciling a massive WR secondary with the relatively small optical depth to X-rays. Verbunt & van den Heuvel (1995) argued that Cyg X-3 is an ideal progenitor for a close double neutron star, and that after spiral-in the binary will consist of a helium star and a black hole. Upon exploding, the WR star may leave a neutron star or a black hole. In the first case, an eccentric close binary will result, consisting of a young pulsar and a black hole. Ergma & Yungelson (1998) argued that for a model with two massive helium stars as immediate progenitors of the Cyg X-3 system, the requirement of having two WR stars in the post-Common-Envelope orbit, each with strong mass loss, would prevent formation of WR+BH systems with orbital periods less than several days.

WR 148 (HD 197406, WN8h+B3IV/c, P = 4.32 d)
Drissen *et al.* (1986) derived from $f(m) = 0.28\,M_\odot$, $i = 66\overset{\circ}{.}7$ and an assumed $M_{WN} = 60\,M_\odot$, that the mass of the unseen companion is $12.4\,M_\odot$. Marchenko *et al.* (1996) estimate the mass of the unseen companion in the range $5–10\,M_\odot$. The system is a runaway star with a net peculiar velocity of $52\,\mathrm{km\,s^{-1}}$ (Moffat & Seggewiss 1980) and located 870 pc above the galactic plane (van der Hucht *et al.* 1988). If the unseen companion of WR 148 is a black hole, then the recoil from a preceding supernova explosion may be the cause of its present runaway status. We would expect WR+BH systems to be strong X-ray sources. However, $L_X/L_\odot = 0.16$ (Pollock *et al.* 1995), *i.e.*, only three times the average WR X-ray luminosity. Yet, Marchenko *et al.* (1996) argued in favor of a black hole scenario, based on timing arguments of the orbiting ionized cavity (; see also Vanbeveren 1998, pp. 245, 286; Vanbeveren 1999).

R 140a2 (WN6+O, P = 2.76 d) and Mk 34 (WN5h) in 30 Doradus (LMC)
Wang (1995) found with *ROSAT*-HRI two X-ray sources in 30 Doradus in the LMC, with $L_X \simeq 260\,L_\odot$. One is associated with the R 140a2 (BAT99-102, WN6+O, $f(M) = 0.10\,M_\odot$), with a mass of the unseen companion in the range $2.4–15\,M_\odot$. The other is associated with Mk 34 (BAT99-116, WN5h, Breysacher *et al.* 1999) and of similar nature. Thus these two X-ray sources may represent WR+BH binaries.

Acknowledgments. Comments and suggestions by N. Langer and P.M. Williams are gratefully acknowledged.

Table 1. 29 galactic WN-type binaries arranged by subtype (from van der Hucht 2000)

WR	HD/name	WR-type OB/c-type	binary status	P (d)	e	q $(\frac{M_{WR}}{M_{OB/c}})$	$f(M)$ $(M_\odot)$	$a \sin i$ $(R_\odot)$	$M \sin^3 i$ $(M_\odot)$	i (source)	M_{WR} $(M_\odot)$	$M_{OB/c}$ $(M_\odot)$	references
3	HD 9974	WN3+O4	SB1	46.85 ± 0.02			~0.2				(20)	(5–12)	MLS86,SH89,SM89,SS96
46	HD 104994	WN3p+?	SB1	0.2824 – 0.2725				4 – 7					VG95,NB95,MM96,VGC00,VGH00
127	HD 186943	WN3	SB2	9.5550	0.07 ± 0.04	0.47		39 ± 6	9.3	55.3 ± 8.2	17		Ma81,MS86,SM89,LM96
		O9.5V						18.6 ± 0.7	20	(phot)		36	
97	HD 320102	WN3-4	SB2	12.595		0.56		28.0 ± 1.2	2.3 ± 0.3	85.4 ± 2.0	2.3		SM89,NC95,NR96,LM96
		O5-7						14.8 ± 0.5	4.1 ± 0.6	(pol'n)		4.1	
31	HD 94546	WN4	SB2	4.831 ± 0.003	0.0	0.43		17.2 ±0.7	2.7	61.6 ± 1.7	4		NM85,LM87,SM89,LM96
		O8V						7.6 ±0.4	6.3	(phot)		9	
128	HD 187282	WN4(h)+?	SB1	3.56 / 3.871									La83,AC85,MS86
151	CX Cep	WN4	SB2	2.12687		0.71		14.9 ± 0.5	17.8 ± 1.4	74.5 ± 5	20		SM89,SH89,LM93
		O5V						10.5 ± 0.4	25.2 ± 1.9	(pol'n)		28	
145a	Cyg X-3	WN4-7	SB1	0.1997	0.0	0.76	2.3	3 – 6		~ 50	~ 13		KB89,KC92,Ke93,CM94,
		c										~17	KG96,MM96,SG96
21	HD 90657	WN5	SB2	8.2545	0.04 ± 0.03	0.52		33	8.4	49.6 ± 3.7	19		NM82,SM89,LM96,Ni99
		O4-6						17	16.3	(phot)		37	
109	V617 Sgr	WN5h+?	SB2	0.207		0.92							SC88,SD98,CD99,SC99
133	HD 190918	WN5	SB2	112.4 ± 0.2	0.39 ± 0.07	0.49		70.3	0.40 ± 0.01	16.8	(17)		SM89,Be94,UH94,Be95,
		O9I						34.5	0.82 ± 0.02			(34)	MMH98,Ni99
138	HD 193077	WN5o+B:	SB2	1538	0.28 ± 0.07		4.7±0.8	930 ± 60					LM82,An91,Ni99
139	HD 193576	WN5	SB2	4.212435	0.04 ± 0.01	0.34		25.6	8.8 ± 0.4	78.7 ± 0.5	9		SM89,RM90,MMK94,Ni99
		O6III-V						10.2	26.3 ± 1.0	(pol'n)		28	
141	HD 193928	WN5°	SB2	21.6895 ± 0.0003	0.02 ± 0.03	1.38		50 ± 2	28.9 ± 4.2	68 ± 12	45		SM89,GM91,LM96,MMEe98,
		O5V-III						69 ± 5	21.0 ± 3.1			30	IV99
21a	Th35-42	WN6+OB	sp	...									MB94,Re99
43b	HD 97950-B	WN6h+a	SB1	3.7720			0.15±0.03						MN84,MS85,DM95
47	HD 311884	WN6	SB2	6.2395		0.85			40	67.1 ± 1.9	51		LM87,SM89,MD90,LM96
		O5V							47			60	
153ab	GP Cep	WN6/WCE	SB4	6.6884 + 3.4698	0.0 + 0.000	0.54			13	74.0 ± 0.7	15		Ma81,SM89,PS90,An94,
		O6I (+WN+O)							24	(phot)		27	LM96,Ni99
155	CQ Cep	WN6	SB2	1.6412436	0.01 ± 0.01	1.24			20.9 ± 1.1	72.5 ± 0.4	24		SB84,SM89,MM95,HH97,
		O9V							25.9 ± 2.7	(phot)		30	Ni99
8	HD 62910	WN7/WCE+?	SB1	38.4 / 115 ± 13				17.3-54.3	0.8-2.6				Ni91,MMH98
22	HD 92740	WN7ha	SB2	80.336 ± 0.001	0.60 ± 0.01	2.69		62.5 ± 0.9	55.3 ± 7.3	90	55		SM89,RV96,Ni99,SSS99
		O9III-V						168.2 ± 9.0	20.6 ± 1.7			21	
29	MS 3	WN7h+OB	SB2	3.164			3±1	13 ± 2					Ni99,NG00
63	LSS 3289	WN7+OB	sp	...									SS96
145	V1923 Cyg	WN7/WCE	SB1	20.0 ± 0.4									RC89,SM89
158	AS 513	WN7h+Be?	sp	...									AV92
12	Ve 5-5	WN8h(+O)	SB1	23.9			5.5	61		78.8 ± 2.2			Ni82,SM89,LM96
98	HD 318016	WN8/WC7	SB1	48.7			1.9 ± 0.4						Ni91,MMEv98
147	AS 431	WN8(h)+B0.5V	CWB	ELPB + 2550 ± 50									WD97,NS98,SB00b
148	HD 197406	WN8h	SB1	4.317364	0.08 ± 0.03	~ 3		7.5 ± 0.2		66.7	(30)		DL86,SM89,Mo92,MML96,
		B3IV/BH					0.28 ± 0.06	3.1 ± 0.8		(pol'n)		(10)	MMEv98

Table 2. 26 galactic WC- and WO-type binaries arranged by subtype (from van der Hucht 2000)

WR	HD/name	WR-type OB-type	binary status	P (d)	e	q $(\frac{M_{WR}}{M_{OB}})$	$f(M)$ $(M_\odot)$	$a \sin i$ $(R_\odot)$	$M \sin^3 i$ $(M_\odot)$	i (source)	M_{WR} $(M_\odot)$	M_{OB} $(M_\odot)$	references
30a	MS 4	WO4+O4	SB2	4.62		0.15	0.015						Ni95,GR99
19	LS 3	WC4+O9.5-9.7	CWB	3690									VH98
4	HD 16523	WC5+?	SB1	2.4096	0.0		0.007-0.016						RC89
9	HD 63099	WC5 O7	SB2	14.305	0.0	0.28		35 22	5.3 18.8	56.8 ± 2.0 (phot)	9	 32	NM84,SM89,Ni95,LM96
27	LS 4	WC6+a	sp	...									SS90
30	HD 94305	WC6 O6-8	SB2	18.82		0.48		73 35	15.4 31.9	78.3 ± 5.8 (phot)	16	 34	NM83,SM89,Ni95,LM96
48	θ Mus	WC6+O9.5/B0Iab	SB2	18.341 ± 0.008			triple?	63					MS77,SD87,SM89,HM99,Ni99
132	HD 190002	WC6+?	SB1	8.16									BF83
146	HM19-3	WC6+O8	CWB	ELPB + 1235 ± 10									DW96,NS98,SB00a
14	HD 76536	WC7+?	SB1	2.42									Sh90,Ni99
42	HD 97152	WC7 O7V	SB2	7.886 ± 0.003	0.0	0.60			3.7 6.2	40.3 ± 2.9 (phot)	14	 23	DM81,SD87,SM89,LM96,Ni99
50	V864 Cen	WC7+a	SB1	1.06									GH90,SS90,GV91
79	HD 152270	WC7 O5-8	SB2	8.8908 ± 0.0005		0.37			1.8 4.9	33.6 ± 2.3 (phot)	11	 29	SD87,SM89,LM96,Ni99
93	HD 157504	WC7+O7-9	sp	...									LS84,LT84,SS90
125	V378 Vul	WC7+O9III	SB2	> 6600									Wi97,Wi00
137	HD 192641	WC7+OB	SB2	4700 ± 100	0.07		16.6 ± 1.7						An95,Wi97,Ni99,WH99,Wi00
140	HD 193793	WC7 O4-5	SB2	2900 ± 10	0.85 ± 0.01	0.37		2750 ± 144 1022 ± 46	23.2 ± 3.4 62.7 ± 9.6				WH87,ML87,SM89,WH90, An95,Ni99
11	γ^2 Vel	WC8 O7.5III-V	SB2	78.53 ± 0.01	0.33 ± 0.01	0.31		180 ± 3 56 ± 3	6.8 ± 0.6 21.6 ± 1.1	63 ± 8	9	 30	Mo77,NS80,MV86,SD87,SM89, SSS97,Ni99,SSK99,TC99,DS99
48a	Danks 1	WC8+OB	CWB	> 7800									Wi97,Wi00
77	He3-1239	WC8(+a)	sp	...									CV90
113	HD 168206	WC8 O8-9IV	SB2	29.704 ± 0.002	0.19 ± 0.03	0.48		49 ± 1	10.6 22.3	70.4 ± 2.3 (phot)	13	 27	SM89,LM96,NM96,Ni99
98a	IRAS 17380-3031	WC8-9+OB	CWB	565 ± 50						35 ± 6			MT99
69	HD 136488	WC9+OB	SB2	2.293 ± 0.005									MMH98,Ni99,WH00
70	HD 137603	WC9 B0I	SB2	> 4000		0.45		34 7	0.2 1.1	< 20	(> 5)	 (> 27)	Go87,SM89,Ni95,Wi97, Ni99,Wi00
103	HD 164270	WC9+?	SB1	1.7556									IM81,GH86,MLC86,SD87, BE89,SH89,Ni99
104	Ve2-45	WC9+B0.5V	SB2	243.5 ± 3.0						20 ± 5			TM99,MT99,WH00

Notes to Tables 1 and 2:
() : estimate; ELPB: extreme long-period binary, i.e., > 10000 d.
Binary status:
SB2 : double-lined spectroscopic binary; SB1 : single-lined spectroscopic binary; CWB : colliding wind binary effects discovered, not yet a RV solution; sp : absorption lines indicative of OB companion.
References:
AC85 : Antokhin & Cherepashchuk 1985; An91 : Annuk 1991; An94 : Annuk 1994; An95 : Annuk 1995; AV92 : Andrillat & Vreux 1992; BE89 : Balona et al. 1989; Be94 : Bertrand 1994; Be95 : Bertrand 1995; BF83 : Bisiacchi et al. 1983; CD99 : Cieslinski et al. 1999; CM94 : Cherepashchuk & Moffat 1994; CV90 : Conti & Vacca 1990; DL86 : Drissen et al. 1986; DM81 : Davis et al. 1981; DM95 : Drissen et al. 1995; DS99 : De Marco & Schmutz; DW96 : Dougherty et al. 1996; GH86 : van Genderen & van der Hucht 1986; GH90 : van Genderen et al. 1990; GM91 : Grandchamps & Moffat 1991; Go87 : Golombek 1987; GR99 : Gosset et al. 1999; GV91 : van Genderen et al. 1991; HH88 : van der Hucht et al. 1988; HH97 : Harries & Hilditch 1997; HM99 : Hartkopf et al. 1999; IM81 : Isserstedt & Moffat 1981; IV99 : Ivanov et al. 1999; KB89 : van der Klis & Bonnet-Bidaud 1989; KC92 : van Kerkwijk et al. 1992; Ke93 : van Kerkwijk 1993; KG96 : van Kerkwijk et al. 1996; La83 : Lamomtagne 1983; LM82 : Lamontagne et al. 1982; LM87 : Lamontagne & Moffat 1987; LM93 : Lewis et al. 1993; LM96 : Lamontagne et al. 1996; LS84 : Lundström & Stenholm 1984; LT84 : Lortet et al. 1984; Ma81 : Massey 1981; MB94 : Mereghetti et al. 1994; MD90 : Moffat et al. 1990; ML87 : Moffat et al. 1987; MLC86 : Moffat et al. 1986a; MLS86 : Moffat et al. 1986b; MM95 : Marchenko et al. 1995; MM96 : Moffat & Marchenko 1996; MMEe98 : Marchenko et al. 1998b; MMEv98 : Marchenko et al. 1998a; MMH98 : Marchenko et al. 1998c; MMK94 : Marchenko et al. 1994; MML96 : Marchenko et al. 1996; MN84 : Moffat & Niemela 1984; Mo77 : Moffat 1977; Mo92 : Moffat 1992; MS77 : Moffat & Seggewiss 1977; MS85 : Moffat et al. 1985; MS86 : Moffat & Shara 1986; MT99 : Monnier et al. 1999; MV86 : Moffat et al. 1986c; NB95 : Niemela et al. 1995a; NC95 : Niemela et al. 1995b; NG00 : Niemela & Gamen 2000; Ni82 : Niemela 1982; Ni91 : Niemela 1991; Ni95 : Niemela 1995; Ni99 : Niemela 1999; NM82 : Niemela & Moffat 1982; NM83 : Niemela et al. 1983; NM84 : Niemela et al. 1984; NM85 : Niemela et al. 1985; NM96 : Niemela et al. 1996a; NR96 : Niemela et al. 1996b; NS80 : Niemela & Sahade 1980; NS98 : Niemela et al. 1998; PS90 : Panov & Seggewiss 1990; RC89 : Rustamov & Cherepashchuk 1989; Re99 : Reig 1999; RM90 : Robert et al. 1990; RV96 : Rauw et al. 1996; SB84 : Stickland et al. 1984; SB00a : Setia Gunawan et al. 2000a; SB00b : Setia Gunawan et al. 2000b; SC88 : Steiner et al. 1988; SC99 : Steiner et al. 1999; SD87 : St-Louis et al. 1987; SD98 : Steiner & Diaz 1998; SG96 : Schmutz et al. 1996; SH89 : Schulte-Ladbeck & van der Hucht 1989; Sh90 : Shylaja 1990; SM89 : compilation of Smith & Maeder 1989; SS90 : Smith et al. 1990; SS96 : Smith et al. 1996; SSK99 : Schweickhardt et al. 1999b; SSS97 : Schmutz et al. 1997; SSS99 : Schweickhardt et al. 1999a; TC99 : Tokovinin et al. 1999; TM99 : Tuthill et al. 1999, 2000; UH94 : Underhill & Hill 1994; VG95 : Veen et al. 1995; VGC00 : Veen et al. 2000b; VGH00 : Veen et al. 2000a; VH98 : Veen et al. 98; WD97 : Williams et al. 1997; WH87 : Williams et al. 1987; WH90 : Williams et al. 1990; WH99 : Williams & van der Hucht 1999; WH00 : Williams & van der Hucht 2000; Wi97 : Williams 1997; Wi00 : Williams, private communication.

References

1. Andrillat, Y., Vreux, J.-M. 1992, A&A (Letters) 253, L37.
2. Annuk, K. 1991, in: K.A. van der Hucht & B. Hidayat (eds.), Wolf-Rayet Stars and Interrelations with Other Massive Stars in Galaxies, Proc. IAU Symp. No. 143 (Dordrecht: Kluwer), p. 245
3. Annuk, K. 1994, A&A 282, 137
4. Annuk, K. 1995, in: K.A. van der Hucht & P.M. Williams (eds.), Wolf-Rayet Stars: Binaries, Colliding Winds, Evolution, Proc. IAU Symp. No. 163 (Dordrecht: Kluwer), p. 231
5. Antokhin, I.I., Cherepashchuk, A.M. 1985, Pis'ma Astron. Zh. 11, 838 (= Sov. Astron. (Letters) 11, 355)
6. Balona, L.A., Egan, J., Marang, F. 1989, MNRAS 240, 103
7. Bertrand, J.-F. 1994, M.Sc. thesis, Univ. de Montréal
8. Bertrand, J.-F. 1995, in: K.A. van der Hucht & P.M. Williams (eds.), Wolf-Rayet Stars: Binaries, Colliding Winds, Evolution, Proc. IAU Symp. No. 163 (Dordrecht: Kluwer), p. 248
9. Bisiacchi, G.F., Firmani, C., de Lara, E. 1983, RMAA 5, 285
10. Breysacher, J., Azzopardi, M., Testor, G. 1999, A&AS 137, 117
11. Cherepashchuk, A.M., Moffat, A.F.J. 1994, ApJ (Letters) 424, L53
12. Cieslinski, D., Diaz, M.P., Steiner, J.E. 1999, AJ 117, 534
13. Conti, P.S., Vacca, W.D. 1990, AJ 100, 431
14. Corcoran, M.F. 2000, in: N.I. Morrell & V.S. Niemela (eds.), Hot Stars in Open Clusters of the Galaxy and the Magellanic Clouds, Proc. La Plata Workshop, December 1997, RMAA-SC 8, 131
15. Crowther, P.A. 1999, in: K.A. van der Hucht, G. Koenigsberger & P.R.J. Eenens (eds.), Wolf-Rayet Phenomena in Massive Stars and Starburst Galaxies, Proc. IAU

Symp. No. 193 (San Francisco: ASP), p. 116
16. Davis, A.B., Moffat, A.F.J., Niemela, V.S. 1981, ApJ 244, 528
17. De Donder, E., Vanbeveren, D., Van Bever, J. 1997, A&A 318, 812
18. De Marco, O., Schmutz, W. 1999, A&A 345, 163
19. Dougherty, S.M., Williams, P.M., van der Hucht, K.A., Bode, M.F., Davis, R.J. 1996, MNRAS 280, 963
20. Drissen, L., Lamontagne, R., Moffat, A.F.J., Bastien, P., Seguin, M. 1986, ApJ 304, 188
21. Drissen, L., Robert, C., Moffat, A.F.J. 1992, ApJ 386, 288
22. Drissen, L., Moffat, A.F.J., Walborn, N.R., Shara, M.M. 1995, AJ 110, 2235
23. Duijsens, M.F.J., van der Hucht, K.A., van Genderen, A.M., Schwarz, H.E., Linders, H.P.J., Kolkman, O.M., 1996, A&AS 119, 37
24. El Eid, M.F., Langer, N. 1986, A&A 167, 275
25. Ensman, L.M., Woosley, S.E. 1988, ApJ 333, 754
26. Ergma, E., Yungelson, L.R. 1998, A&A 333, 151
27. Fender, R.P., Hanson, M.M., Pooley, G.G. 1999, MNRAS 308, 473
28. Fesen, R.A., Becker, R.H., Blair, W.P. 1987, ApJ 313, 388
29. Firmani, C., Koenigsberger, G., Bisiacchi, G.F., Moffat, A.F.J., Isserstedt, J. 1980, ApJ 239, 607
30. van Genderen, A.M., van der Hucht, K.A. 1986, A&A 162, 109
31. van Genderen, A.M., van der Hucht, K.A., Larsen, I. 1990, A&A 229, 123
32. van Genderen, A.M., Verheijen, M.A.W., van der Hucht, K.A., de Loore, C.W.H., Schwarz, H.E., van Esch, B.P.M., Greidanus, H., van der Heiden, R., van Kampen, E., Kuulkers, E., le Poole, R.S., Reijns, R.A., Robijn, F.H.A., Spijkstra. L. 1991, in: K.A. van der Hucht & B. Hidayat (eds.), Wolf-Rayet Stars and Interrelations with Other Massive Stars in Galaxies, Proc. IAU Symp. No. 143 (Dordrecht: Kluwer), p. 129
33. Georgiev, L., Koenigsberger, G., Ivanov, M.M., St-Louis, N., Cardona, O. 1999, A&A 347, 583
34. Golombek, D.A. 1987, Thesis University of Buenos Aires
35. Gosset, E., Royer, P., Rauw, G., Manfroid, J., Vreux, J.M. 1999, in: K.A. van der Hucht, G. Koenigsberger & P.R.J. Eenens (eds.), Wolf-Rayet Phenomena in Massive Stars and Starburst Galaxies, Proc. IAU Symp. No. 193 (San Francisco: ASP), p. 354
36. Grandchamps, A., Moffat, A.F.J. 1991, in: K.A. van der Hucht & B. Hidayat (eds.), Wolf-Rayet Stars and Interrelations with Other Massive Stars in Galaxies, Proc. IAU Symp. No. 143 (Dordrecht: Kluwer), p. 258
37. Hamann, W.-R., Koesterke, L. 1998, A&A 335, 1003
38. Harries, T.J., Hilditch, R.W. 1997, MNRAS 291, 544
39. Hartkopf, W.I., Mason, B.D., Gies, D.R., ten Brummelaar, T., McAlister, H.A., Moffat, A.F.J., Shara, M.M., Wallace, D.J. 1999, AJ 118, 509
40. van den Heuvel, E.P.J., Heise, J. 1972, Nature 239, 67
41. van den Heuvel, E.P.J., de Loore, B. 1973, A&A 25, 387
42. van den Heuvel, E.P.J., Habets, G.M.H.J. 1984, Nature 309, 598
43. van den Heuvel, E.P.J. 1994, in: D. Vanbeveren, W. van Rensbergen & C. de Loore (eds.), Evolution of Massive Stars: A Confrontation between Theory and Observation, Proc. Brussels Workshop, SSR 66, 309
44. Hillier, D.J. 1996, in: J.-M. Vreux, A. Detal, D. Fraipont-Caro, E. Gosset & G. Rauw (eds.), Wolf-Rayet Stars in the Framework of Stellar Evolution, Proc. 33rd Liège Int. Astroph. Coll. (Liège: Univ. of Liège), p. 509
45. van der Hucht, K.A., Conti, P.S., Lundström, I., Stenholm, B. 1981, SSR 28, 227
46. van der Hucht, K.A., Hidayat, B., Admiranto, A.G., Supelli, K.R., Doom, C. 1988, A&A 199, 217
47. van der Hucht, K.A. 1992, The A&A Rev. 4, 123
48. van der Hucht, K.A., Williams, P.M., Spoelstra, T.A.Th., de Bruyn, A.G. 1992, in: L. Drissen, C. Leitherer & A. Nota (eds.), Non-isotropic and Variable Outflows from Stars, Proc. STScI Workshop, ASP-CS 22, 253

49. van der Hucht, K.A. 2000, New Astronomy Reviews to be submitted
50. Isserstedt, J., Moffat, A.F.J. 1981, A&A 96, 133
51. Ivanov, M.M., Valchev, T.S., Georgiev, L.N., Barbá, R., Iliev, I.Kh. 1999, RMAA 35, 25
52. van Kerkwijk, M.H., Charles, P.A., Geballe, T.R., King, D.L., Miley, G.K., Molnar, L.A., van den Heuvel, E.P.J., van der Klis, M., van Paradijs, J. 1992, Nature 355, 703
53. van Kerkwijk, M.H. 1993, A&A (Letters) 276, L9
54. van Kerkwijk, M.H., Geballe, T.R., King, D.L., van der Klis, M., van Paradijs, J. 1996, A&A 314, 521
55. van der Klis, M., Bonnet-Bidaud, J.M. 1989, A&A 214, 203
56. van der Klis, M. 1993, SSR 62, 173
57. Lamontagne, R., Moffat, A.F.J., Koenigsberger, G., Seggewiss, W. 1982, AJ 253, 230
58. Lamontagne, R. 1983, Thesis, Université de Montréal
59. Lamontagne, R., Moffat, A.F.J. 1987, AJ 94, 1008
60. Lamontagne, R., Moffat, A.F.J., Drissen, L., Robert, C., Matthews, J.M. 1996, AJ 112, 2227
61. Langer, N., El Eid, M.F. 1986, A&A 167, 265
62. Langer, N. 1987, A&A (Letters) 171, L1
63. Langer, N. 1989, A&A 220, 135
64. Langer, N. 1991, in: S. Woosley (ed.), Supernovae, Proc. 10th Santa Cruz Workshop (New York: Springer), p. 549.
65. Langer, N., Woosley, S.E. 1991, in: K.A. van der Hucht & B. Hidayat (eds.), Wolf-Rayet Stars and Interrelations with Other Massive Stars in Galaxies, Proc. IAU Symp. No. 143 (Dordrecht: Kluwer), p. 566
66. Langer, N., Heger, A. 1999, in: K.A. van der Hucht, G. Koenigsberger & P.R.J. Eenens (eds.), Wolf-Rayet Phenomena in Massive Stars and Starburst Galaxies, Proc. IAU Symp. No. 193 (San Francisco: ASP), p. 187
67. Lewis, D., Moffat, A.F.J., Matthews, J.M., Robert, C., Marchenko, S.V. 1993, ApJ 405, 312
68. Lortet, M.C., Testor, G., Niemela, V.S. 1984, A&A 140, 24
69. Lundström, I., Stenholm, B. 1984, A&AS 58, 163
70. MacFadyen, A.I., Woosley, S.E. 1999, ApJ 524, 262
71. Maeder, A. 1981, A&A 99, 97
72. Maeder, A., Lequeux, J. 1982, A&A 114, 409
73. Maeder, A. 1983, A&A 120, 113
74. Maeder, A., Meynet, G. 1987, A&A 182, 243
75. Maeder, A., Meynet, G. 1989, A&A 210, 155
76. Maeder, A. 1991, A&A 242, 93
77. Maeder, A. 1992, A&A 264, 105 (Erratum: 1993, A&A 268, 833)
78. Marchenko, S.V., Moffat, A.F.J., Koenigsberger, G. 1994, ApJ 422, 810
79. Marchenko, S.V., Moffat, A.F.J., Eenens, P.R.J., Hill, G.M., Grandchamps, A. 1995, ApJ 450, 811
80. Marchenko, S.V., Moffat, A.F.J., Lamontagne, R., Tovmassian, G.H. 1996, ApJ 461, 386
81. Marchenko, S.V., Moffat, A.F.J., Eversberg, T., Hill, G.M., Tovmassian, G.H., Morel, T., Seggewiss, W. 1998a, MNRAS 294, 642
82. Marchenko, S.V., Moffat, A.F.J., Eenens, P.R.J. 1998b, PASP 110, 1416
83. Marchenko, S.V., Moffat, A.F.J., van der Hucht, K.A., Seggewiss, W., Schrijver, H., Stenholm, B., Lundström, I., Setia Gunawan, D.Y.A., Sutantyo, W., van den Heuvel, E.P.J., De Cuyper, J.-P., Gómez, A.E. 1998c, A&A 331, 1022
84. Massey, P. 1981, ApJ 244, 157
85. Mereghetti, S., Belloni, T., Shara, M., Drissen, L. 1994, ApJ 424, 943
86. Mitra, A. 1996, MNRAS 280, 953
87. Mitra, A. 1998, ApJ 499, 385

88. Moffat, A.F.J. 1977, A&A 57, 151
89. Moffat, A.F.J., Seggewiss, W. 1977, A&A 54, 607
90. Moffat, A.F.J., Seggewiss, W. 1980, A&A 86, 87
91. Moffat, A.F.J., Niemela, V.S. 1984, ApJ 284, 631
92. Moffat, A.F.J., Seggewiss, W., Shara, M.M. 1985, ApJ 295, 109
93. Moffat, A.F.J., Shara, M.M. 1986, AJ 92, 952
94. Moffat, A.F.J., Lamontagne, R., Cerruti, M. 1986a, PASP 98, 1170
95. Moffat, A.F.J., Lamontagne, R., Shara, M.M., McAlister, H.A. 1986b, AJ 91, 1392
96. Moffat, A.F.J., Vogt, H., Paquin, G., Lamontagne, R., Barrera, L.H. 1986c, AJ 91, 1386
97. Moffat, A.F.J., Lamontagne, R., Williams, P.M., Horn, J., Seggewiss, W. 1987, ApJ 312, 807
98. Moffat, A.F.J., Drissen, L., Robert, C., Lamontagne, R., Coziol, R., Mousseau, N., Niemela, V.S., Cerruti, M.A, Seggewiss, W., van Weeren, N. 1990, ApJ 350, 767
99. Moffat, A.F.J. 1992, A&A 253, 425
100. Moffat, A.F.J. 1995, in: K.A. van der Hucht & P.M. Williams (eds.), Wolf-Rayet Stars: Binaries, Colliding Winds, Evolution, Proc. IAU Symp. No. 163 (Dordrecht: Kluwer), p. 213
101. Moffat, A.F.J., Marchenko, S.V. 1996, A&A (Letters) 305, L29
102. Monnier, J.D., Tuthill, P.G., Danchi, W.C. 1999, ApJ (Letters) 525, L97
103. Morel, T., St-Louis, N., Moffat, A.F.J., Cardona, O., Koenigsberger, G., Hill, G.M. 1998, ApJ 498, 413
104. Mosqueda, A., Koenigsberger, G. 1990, RMAA 20, 79
105. Niemela, V.S., Sahade, J. 1980, ApJ 238, 244
106. Niemela, V.S. 1982, in: C. de Loore & A.J. Willis (eds.), Wolf-Rayet Stars: Observations, Physics, Evolution, Proc. IAU Symp. No. 99 (Dordrecht: Reidel), p. 299
107. Niemela, V.S., Moffat, A.F.J. 1982, ApJ 259, 213
108. Niemela, V.S., Méndez, R., Moffat, A.J. 1983, ApJ 272, 190
109. Niemela, V.S., Massey, P., Conti, P.S. 1984, PASP 96, 549
110. Niemela, V.S., Mandrini, C.H., Méndez, R.H. 1985, RMAA 11, 143
111. Niemela, V.S. 1991, in: K.A. van der Hucht & B. Hidayat (eds.), Wolf-Rayet Stars and Interrelations with Other Massive Stars in Galaxies, Proc. IAU Symp. No. 143 (Dordrecht: Kluwer), p. 201
112. Niemela, V.S. 1995, in: K.A. van der Hucht & P.M. Williams (eds.), Wolf-Rayet Stars: Binaries, Colliding Winds, Evolution, Proc. IAU Symp. No. 163 (Dordrecht: Kluwer), p. 223
113. Niemela, V.S., Barbá, R.H., Shara, M.M. 1995a, in: K.A. van der Hucht & P.M. Williams (eds.), Wolf-Rayet Stars: Binaries, Colliding Winds, Evolution, Proc. IAU Symp. No. 163 (Dordrecht: Kluwer), p. 245
114. Niemela, V.S., Cabanne, M.L., Bassino, L.P. 1995b, RMAA 31, 45
115. Niemela, V.S., Morrell, N.I., Barbá, R.H., Bosch, G.L. 1996a, in: V.S. Niemela & N.I. Morrell (eds.), Colliding Winds in Binary Stars, Proc. Workshop celebrating the 80th birthday of Jorge Sahade, RMAA-SC 5, 100
116. Niemela, V.S., Rovero, A., Cerruti, M.A. 1996b, in: V.S. Niemela & N.I. Morrell (eds.), Colliding Winds in Binary Stars, Proc. Workshop celebrating 80th birthday Jorge Sahade, RMAA-SC 5, 126
117. Niemela, V.S., Shara, M.M., Wallace, D.J., Zurek, D.R., Moffat, A.F.J. 1998, AJ 115, 2047
118. Niemela, V.S. 1999, in: K.A. van der Hucht, G. Koenigsberger & P.R.J. Eenens (eds.), Wolf-Rayet Phenomena in Massive Stars and Starburst Galaxies, Proc. IAU Symp. No. 193 (San Francisco: ASP), p. 26
119. Niemela, V.S., Gamen, R. 2000, in: N.I. Morrell & V.S. Niemela (eds.), Hot Stars in Open Clusters of the Galaxy and the Magellanic Clouds, Proc. La Plata Workshop, December 1997, RMAA-SC 8, 59
120. Nugis, T., Crowther, P.A., Willis, A.J. 1998, A&A 333, 956

121. Panov, K.P., Seggewiss, W. 1990, A&A 222, 117
122. Pollock, A.M.T., Haberl, F., Corcoran, M.F. 1995, in: K.A. van der Hucht & P.M. Williams (eds.), Wolf-Rayet Stars: Binaries, Colliding Winds, Evolution, Proc. IAU Symp. No. 163 (Dordrecht: Kluwer), p. 512
123. Rauw, G., Vreux, J.-M., Gosset, E., Hutsemékers, D., Magain, P., Rochowicz, K. 1996, A&A 306, 771
124. Reig, P. 1999, A&A 345, 576
125. Robert, C., Moffat, A.F.J., Bastien, P., St-Louis, N., Drissen, L. 1990, ApJ 359, 211
126. Rustamov, D.N., Cherepashchuk, A.M. 1989, Astron. Zh. 66, 67 (= Sov. Astron. 33, 36)
127. Schild, H., Maeder, A. 1985, A&A (Letters) 143, L7
128. Schmutz, W., Geballe, T.R., Schild, H. 1996, A&A (Letters) 311, L25
129. Schmutz, W., Schweickhardt, J., Stahl, O., Wolf, B., Dumm, T., Gäng, Th., Jankovics, I., Kaufer, A., Lehmann, H., Mandel, H., Peitz, J., Rivinius, Th. 1997, A&A 328, 219
130. Schulte-Ladbeck, R.E., van der Hucht, K.A. 1989, ApJ , 872
131. Schweickhardt, J., Schmutz, W., Stahl, O., Szeifert, Th., Wolf, B. 1999a, A&A 346, 127
132. Schweickhardt, J., Schmutz, W., Kaufer, A., Stahl, O., Wolf, B. 1999b, in: K.A. van der Hucht, G. Koenigsberger & P.R.J. Eenens (eds.), Wolf-Rayet Phenomena in Massive Stars and Starburst Galaxies, Proc. IAU Symp. No. 193 (San Francisco: ASP), p. 98
133. Setia Gunawan, D.Y.A., de Bruyn, A.G., van der Hucht, K.A., Williams, P.M. 2000a, A&A in press
134. Setia Gunawan, D.Y.A., de Bruyn, A.G., van der Hucht, K.A., Williams, P.M. 2000b, in preparation
135. Shylaja, B.S. 1990, ApSS 164, 63
136. Smith, L.F., Maeder, A. 1989, A&A 211, 71
137. Smith, L.F., Shara, M.M., Moffat, A.F.J. 1990, ApJ 358, 229
138. Smith, L.F., Shara, M.M., Moffat, A.F.J. 1996, MNRAS 281, 163
139. Steiner, J.E., Cieslinski, D., Jablonski, F.J. 1988, in: V.M. Blanco & M.M. Phillips (eds.), Progress and Opportunities in Southern Hemisphere Optical Astronomy, Proc. *CTIO* 25th Anniversary Symposium, ASP-CS 1, 67
140. Steiner, J.E., Diaz, M.P. 1998, PASP 110, 276
141. Steiner, J.E., Cieslinski, D., Jablonski, F.J., Williams, R.E. 1999, A&A 351, 1021
142. Stevens, I.R., Willis, A.J. 1988, MNRAS 234, 783
143. Stickland, D.J., Bromage, G.E., Budding, E., Burton, W.M., Howarth, I.D., Jameson, R., Sherrington, M.R., Willis, A.J. 1984, A&A 135, 45
144. St-Louis, H., Drissen, L., Moffat, A.F.J., Bastien, P. 1987, ApJ 322, 870
145. Tokovinin, A.A., Chalabaev, A., Shatsky, N.I., Beuzit, J.L. 1999, A&A 346, 481
146. Tuthill, P.G., Monnier, J.D., Danchi, W.C. 1999, Nature 398, 487
147. Tuthill, P.G., Monnier, J.D., Danchi, W.C. 2000, in: S.C. Unwin & R.V. Stachnik (eds.), "Working on the Fringe", Optical and IR Interferometry from Ground and Space, Proc. Intern. Conf. JPL, Dana Point (CA), 24-27 May 1999, ASP-CS 194, ...
148. Tutokov, A.V., Cherepashchuk, A.M. 1985, Astron. Zh. 62, 1124 (= Sov. Astron. 29, 654)
149. Underhill, A.B., Hill, G.M. 1994, ApJ 432, 770
150. Vanbeveren, D., van Rensbergen, W., de Loore, C. 1998, The Brightest Binaries, Astrophysics & Space Science Library Vol. 232 (Dordrecht: Kluwer)
151. Vanbeveren, D. 1999, in: K.A. van der Hucht, G. Koenigsberger & P.R.J. Eenens (eds.), Wolf-Rayet Phenomena in Massive Stars and Starburst Galaxies, Proc. IAU Symp. No. 193 (San Francisco: ASP), p. 196
152. Veen, P.M., van Genderen, A.M., Verheijen, M.A.W., van der Hucht, K.A. 1995, in: K.A. van der Hucht & P.M. Williams (eds.), Wolf-Rayet Stars: Binaries, Colliding

Winds, Evolution, Proc. IAU Symp. No. 163 (Dordrecht: Kluwer), p. 243
153. Veen, P.M., van der Hucht, K.A., Williams, P.M., Catchpole, R.M., Duijsens, M.F.J., Glass, I.S., Setia Gunawan, D.Y.A. 1998, A&A (Letters) 339, L45
154. Veen, P.M., van Genderen, A.M., Crowther, P.A., van der Hucht, K.A. 2000a, A&A (to be submitted)
155. Veen, P.M., van Genderen, A.M., van der Hucht, K.A., Jones, A.F., Arentoft, T. 2000b, A&A (to be submitted)
156. Verbunt, F., van den Heuvel, E.P.J. 1995, in: W.H.G. Lewin, J. van Paradijs & E.P.J. van den Heuvel (eds.), X-Ray Binaries (Cambridge: CUP), p. 457
157. Wang, Q.D. 1995, ApJ 453, 783
158. Williams, P.M., van der Hucht, K.A., van der Woerd, H., Wamsteker, W.M., Geballe, T.R., Garmany, C.D., Pollock, A.M.T. 1987, in: H. Lamers & C.W.H. de Loore (eds.), Instabilities in Luminous Early Type Stars, Proc. of a Workshop in honour of Cornelis de Jager (Dordrecht: Reidel), p. 221
159. Williams P.M., van der Hucht, K.A., Pollock, A.M.T., Florkowski, D.R., van der Woerd, H., Wamsteker, W.M. 1990, MNRAS 243, 662
160. Williams, P.M., van der Hucht, K.A., Spoelstra, T.A.Th. 1994, A&A 291, 805
161. Williams, P.M., van der Hucht, K.A. 1996, in: J.-M. Vreux, A. Detal, D. Fraipont-Caro, E. Gosset & G. Rauw (eds.), Wolf-Rayet Stars in the Framework of Stellar Evolution, *Proc. 33rd Liège Int. Astroph. Coll.* (Liège: Univ. de Liège), p. 353
162. Williams, P.M. 1997, in: I. Cherchneff & T.J. Millar (eds.), Dust and Molecules in Evolved Stars, Proc. UMIST/CCP7 Workshop, Manchester, March 1997, ApSS 251, 321
163. Williams, P.M., Dougherty, S.M., Davis, R.J., van der Hucht, K.A., Bode, M.F., Setia Gunawan, D.Y.A. 1997, MNRAS 289, 10
164. Williams, P.M. 1999, in: K.A. van der Hucht, G. Koenigsberger & P.R.J. Eenens (eds.), Wolf-Rayet Phenomena in Massive Stars and Starburst Galaxies, Proc. IAU Symp. No. 193 (San Francisco: ASP), p. 267
165. Williams, P.M., van der Hucht, K.A. 1999, *UKIRT Newsletter* No. 5 (September 1999), p. 3
166. Williams, P.M., van der Hucht, K.A. 2000, MNRAS in press
167. Woosley, S.E. 1986, in: B. Hauck *et al.* (eds.), Nucleosynthesis and Chemical Evolution, 16th Saas Fee Conf., p. 1
168. Woosley, S.E., Langer, N., Weaver, T.A. 1993, ApJ 411, 823
169. Woosley, S.E., Langer, N., Weaver, T.A. 1995, ApJ 448, 315
170. Woosley, S.E., Eastman, R.G., Schmidt, B.P. 1999, ApJ 516, 788
171. Woosley, S.E., MacFadyen, A.I. 1999, A&AS 138, 499

THE EVOLUTION OF BINARIES

D. VANBEVEREN
Astrophysical Institute, Vrije Universiteit Brussel
Pleinlaan 2, 1050 Brussels, Belgium

1. Abstract

We discuss the evolution of intermediate mass and massive binaries and highlight their importance for population synthesis. The latter is a first step to study the effect of binaries of all types on the chemical evolution of galaxies in general, the Milky Way in particular.

2. Introduction

In the present paper we summarise binary star research and we focus on items that are important for population studies and galactic evolution. The next session deals with the observations of the binary population. The evolution of binaries can not be studied separately from the evolution of single stars. Section 4 therefore considers single stars whereas binaries are discussed in section 5. In sections 6 and 7 we present a few examples of population number synthesis and galactic chemical evolutionary calculations including a realistic fraction of binaries.

3. The interacting binary frequency

Let us first remark that interaction (i.e. Roche lobe overflow = **RLOF** and mass transfer, common envelope = **CE** evolution, spiral-in = **SpI** effects) occurs in binaries with period up to 10 years and more.

Interacting binary frequencies and binary statistics has been studied for the Solar neighbourhood but incompleteness is still significant. The Galactic situation at larger distances is much worse and the extra-galactic binary population is totally unknown.

Facts: among the bright O-type and early B-type stars 33% are primary of a spectroscopic binary (Garmany et al., 1980; Vanbeveren et al., 1998a); more than 50% of the stars in the young cluster Tr 16 are primary of an

F. Matteucci and F. Giovannelli (eds.), The Evolution of the Milky Way, 139–150.

interacting binary (Levato et. al., 1991); 74% of the B2-B5 dwarfs are binary components (Abt et al., 1990); 52% of the B-type stars in the cluster NGC 6231 and 74% of the early type stars in the Sco OB2 association are binary components (Raboud, 1996; Verschueren et al., 1996).

These percentages are lower limits. Due to observational selection, most of the known interacting binaries have orbital period smaller than 100 days (Mason et al., 1998). Moreover, there are very few spectroscopic binaries with small mass ratio. Obviously, also this effect may not be real since small mass ratio binaries are very hard to detect. A quantitative study of all possible selection effects (Hogeveen, 1991, 1992; Halbwachs, 1987) reveals that the real interacting binary frequency may be much larger than presently observed; primarily small mass ratio systems and systems with orbital period $P > 100$ days are lacking.

As a consequence we are forced to conclude that

Population studies (the chemistry, stellar content etc) where the effect of binaries has not been investigated in detail, have an academic value but may be far from reality.

4. The evolution of single stars

The evolution of intermediate mass single stars has been discussed by different authors in the present proceedings. Rotation and mass loss by stellar wind (**SW**) affect significantly massive star evolution. The first process has been reviewed here by A. Maeder. The following subsection summarises the present state of SW mass loss.

4.1. STELLAR WIND (SW) MASS LOSS

Three mass loss phases deserve special attention: the red supergiant phase (**RSG SW**), the luminous blue variable phase (**LBV SW**) and the Wolf-Rayet phase (**WR SW**).

There are no RSGs observed brighter than $M_{bol} = -9.5$ (corresponding to stars with initial mass larger than 40 $M_{\odot}$). In the same luminosity range, LBVs are hydrogen shell burning stars or stars close to the end of the core hydrogen burning phase losing mass by stellar wind at very large rates. As a working hypothesis for stellar evolution, it is logical to adopt a model where the LBV SW (possibly followed by a RSG SW) is large enough to assure a RSG phase which is short enough to explain the lack of observed RSGs with $M_{bol} < -9.5$.

The SW rates of RSGs are poorly known. Equation 1 gives a rate relation as a function of luminosity for 16 RSGs in the Large Magellanic Cloud (Reid et. al., 1990; Jura, 1987).

$$\log(-\dot{M}) = 0.8 \log L - 8.7 \tag{1}$$

(the rate and L in Solar units). This relation was used in Vanbeveren et al. (1998a, b) to calculate the evolution of massive single stars. The conclusions resulting from these calculations were confirmed by Salasnich et al. (1999) who used a very similar SW formalism. Remark that the rates predicted by equation (1) differ significantly from those used in the standard massive single star models of the Geneva group.

Using a NLTE atmosphere code where the SW is assumed to be homogeneous, Hamann et al. (1995) determined rates for a large number of WR stars. However, the assumption of homogeneity overestimates these values, typically by a factor of two (Hamann & Koesterke, 1998). A mass loss rate relation that closely fits most present day observational constraints is given by equation 2 (the rate and L in Solar units).

$$\log(-\dot{M}) = \log L - 10 \tag{2}$$

Remark that equation (2) predicts values which are up to a factor 4 smaller than the equation proposed by Langer (1989).

4.2. THE MASS OF A STAR PRIOR TO FE-NI CORE COLLAPSE

The effect of smaller WR SW rates on massive star evolution has been discussed in Vanbeveren et al. (1998 a,b,c). One of the major effects is of course the much larger mass of a massive star prior to Fe-Ni core collapse which favours the formation of massive black holes (**BH**). BH formation is very important for models that predict the galactic evolution of chemical elements that are formed primarily in the most massive stars, like oxygen (see section 7).

4.3. SINGLE STAR EVOLUTIONARY SCENARIO

The overall intermediate mass and massive single star evolutionary scenario is sketched in figure 1. Apart from the abbreviations that were introduced earlier, we use **Min(WR)** as the minimum initial mass for a star to evolve into a WR star. In the case of single stars, its value depends on the RSG SW and the LBV SW. Using the SW formalism discussed in subsection 4.1 and adopting a metallicity dependence predicted by the radiatively driven wind theory (Castor et al., 1975; Pauldrach et al., 1994; Puls et al., 1996), Min(WR) $\approx$ 20-25 $M_\odot$ for the Galaxy and larger than 40 $M_\odot$ for small metallicity regions (like the Small Magellanic Cloud).

Whether BH formation is accompanied by a 'normal' SN explosion is uncertain. The presence of O, Mg, Si and S in the atmosphere of the optical components of the BH X-ray binary Nova Scorpii 1994 suggests a SN explosion of the BH progenitor (Israelian et al., 1999). The standard X-ray binary Cyg X-1, with a massive BH candidate, is a runaway. This could suggest that a SN explosion occurred. Cluster ejection to make a runaway can not be excluded although in the case of Cyg X-1, the progenitor runaway must have been a binary and as shown by Mason et al. (1999) the binary frequency among the runaways is extremely small. Wherever we use 'no SN' in the proposed evolutionary scenario's, we add a question mark.

5. The evolution of intermediate and massive interacting binaries

Figures 2-6 illustrate the overall evolutionary scenario of intermediate mass and massive interacting binaries. We use the classical definitions of $CaseA/Cas$ and $CaseB_c/CaseC$ binaries.

5.1. BINARY EVOLUTIONARY PROCESSES

Depending on the binary mass ratio and/or period, the following processes can happen (for more details, we like to refer to Van den Heuvel, 1993; Vanbeveren et al., 1998a):

RLOF: the evolution of case B_r binaries with initial mass ratio > 0.2 is governed by the Roche lobe overflow process of the primary (= mass loser) accompanied by mass transfer and mass accretion of the secondary (= mass gainer). For population studies, it is important to know quantitatively the evolution of the mass gainer.

CE: when during RLOF both binary components come into contact, further mass loss and mass transfer will make a common envelope; CE evolution will very likely happen in a Case B_c or case C binary when the primary has a convective envelope at the onset of RLOF. CE is accompanied by large mass loss from the system but the efficiency of the process is uncertain and has to be parameterised.

SpI: when the mass ratio of a binary is smaller than 0.2, RLOF does not happen. Rather, the small mass component is dragged into the envelope of the higher mass star and the small mass star starts spiralling-in. Similarly as the CE process, SpI may result into large mass loss but also here, the efficiency is rather poorly known and is parameterised as well.

Merger process: a non-conservative RLOF, a CE evolution or the SpI process is always accompanied by large orbital angular momentum loss from the system. This may imply a drastic shrinking of the distance between both components and they may merge. When one of the components is a neutron star (**NS**) or a BH, merging leads to the formation of a Thorne-Zytkow object (**TZO**). Evolutionary computations for TZOs were presented by Cannon et al. (1992). The merging of a BH may result in a gamma ray burst (**GRB**, Fryer and Woosley, 1999).

SN explosion in a binary: when one of the components of a binary explodes and leaves a compact remnant (**SN $I_{b/c}$ or SN II**), depending on the pre-SN orbital parameters and on the asymmetry of the explosion, the binary is disrupted or not. Knowing the degree of asymmetry is crucial. The latter can be linked to the observed space velocities of pulsars which seems to be very large (Lorimer et al., 1997) although the real distribution is still uncertain. Of particular importance are **SN Ia** supernovae and their progenitors.

5.2. THE PROGENITORS OF SN IA SUPERNOVAE

It is widely accepted that a Type Ia SN is the thermonuclear explosion of a mass-accreting white dwarf (**WD**) (Nomoto et al., 1997 for a recent review). A most plausible scenario is the accretion of matter lost by a main sequence star or red giant filling its Roche lobe. Accounting for the reaction of a WD when it accretes hydrogen rich matter (Nomoto et al., 1979) and the possible effect of a stellar wind when due to accretion, the luminosity of the WD becomes comparable with the critical Eddington luminosity (Hachisu et al., 1996), Li and van den Heuvel (1997) calculated the possible WD + normal companion binary configurations where the WD mass can grow and reach the Chandrasekhar mass. Improvements were presented by Hachisu et al. (1999). The maximum possible mass of the WD companion is 3.7 $M_{\odot}$. A 3.7 $M_{\odot}$ star reaches its Roche lobe after some 300 million years. This means that in a galactic evolutionary model, the first SN Ia event happens 300 million years after the onset of star formation.

Using a population synthesis code, it is then straightforward to compute the initial binary population that produces the observed SN Ia rate. This will be discussed in section 6.3.

6. Population synthesis

Population number synthesis (PNS) of stars can be defined as a comparison between the observed number of stars with common observational characteristics and a theoretical number. To calculate the latter, one needs

•An evolutionary model for single stars; in the case of massive stars, it is essential to use the evolutionary calculations with the most recent SW mass loss rate formalisms

•An evolutionary model for interacting close binaries where the evolution of the mass gainer is considered simultaneously; the binary period evolution (the formalism is different for different types of interacting binaries) must be followed in detail; in the case of massive binaries, similarly as for massive single stars, evolutionary computations must be considered where the most recent SW formalisms are included; one has to account for the effect of an asymmetrical SN explosion on binary parameters; to study the X-ray binary population, a model is needed describing the formation of hard X-rays when matter is accreted onto a compact star

•An unambiguous definition of when a star calculated with an evolutionary code will be observed as a star belonging to the class of stars that is studied, and the lifetime of single stars and of binary components during such an evolutionary phase

•Input parameter distributions for stellar objects at birth, i.e.

a. The IMF of single stars and of primaries of binaries

b. The binary frequency

c. The period and mass ratio distribution of binaries

d. A parameter distribution describing the asymmetry of the SN explosion

e. The distribution of the spin period of the NSs at birth and of the magnetic field of NSs.

Using a PNS code, it is then possible to study the population of WDs, cataclysmic variables, Algol type binaries, Be binaries, X-ray binaries, WR stars, starbursts, pulsars, SN rates, GRBs, super-soft X-ray sources etc. The number of publications that appeared the last two decades is very large. Too large to include them here but we strongly advise an interested scientist to make a detailed literature search before starting to work in the field. Here, we consider a few topics that were studied in Brussels.

6.1. THE WR-O-RSG POPULATION

The WR-O-RSG population including binaries has been discussed by Vanbeveren et al. (1998b). A few conclusions:

•The observed WR/O and WR+OB/WR number ratios in the Solar neighbourhood can be reproduced only if the overall massive close binary formation rate is larger than 0.7, i.e. the majority of the massive stars are born as a component of an interacting close binary

•To explain the WR/O and WR+OB/WR number ratios in the Magellanic Clouds, we must adopt an interacting binary frequency which is smaller than in the Solar neighbourhood;

Could it be that the binary frequency is not the same everywhere in the cosmos, but may depend on the metallicity?

•The interpretation of the integrated spectra of starbursts depend critically on the effects of interacting binaries whereas the age determination becomes ambiguous (see also Van Bever and Vanbeveren, 1998; Van Bever et al., 1999).

6.2. THE SN II AND $I_{B/C}$ RATES

The effect of binaries on SN II and $I_{b/c}$ rates has been discussed by De Donder and Vanbeveren (1998). The following conclusions hold:

•Depending on the minimum mass of BH formation, it is conceivable that only few WR stars contribute to the SN $I_{b/c}$ population; the latter then originate from binaries with primary mass smaller than 20 $M_\odot$

•The number ratio ($II/I_{b/c}$) strongly depends on the adopted binary formation rate, the binary mass ratio and period distribution

The following remark is appropriate: type II SN are usually related to massive single red supergiants belonging to population I environments. There is however a possible class of type II that deserves extra attention. Consider a 5+4 $M_\odot$ case B_r close binary. After quasi-conservative RLOF, the binary consists of a 0.7-0.8 $M_\odot$ subdwarf or WD and a 8 $M_\odot$ (rejuvenated) main sequence star. The further evolution of the binary will be characterised by the SpI process and, in most of the cases, the WD will merge with the He-core of the mass gainer. When this new core is massive enough, an SN II may happen. Our PNS code predicts that a significant fraction (up to 30%) of all SN II may have such an intermediate mass binary merger origin (De Donder and Vanbeveren, 2000).

6.3. THE SN IA RATES

•SN Ia progenitors are binaries and accounting for the model discussed in section 4.2 it is obvious that the estimated number of SN Ia depends on the overall binary frequency (linearly), the binary mass ratio and period distribution.

•In the Solar neighbourhood, 15% of all SN are expected to be type Ia. With the model of section 5.2 and a binary population with a mass ratio and period distribution corresponding to observations, to obtain a 15% SN Ia frequency one has to adopt an overall interacting (intermediate mass) binary frequency (at birth) $\geq$ 50%.

6.4. OVERALL CONCLUSION CONCERNING THE SN RATES

To estimate (relative) SN rates in one galaxy, data are collected of a large sample of galaxies of the same type. The average value is then considered as representative for all of them (Turatto, present proceedings). However, accounting for the dependency of SN ratios on binary statistics, it follows that

Estimating SN number ratios in a particular galaxy by using the average ratio in a large set of galaxies of the same type, is meaningful only if the

binary formation rate, the binary mass ratio and period distributions are the same in all the galaxies where the average value is based on.

Whether the conditions mentioned above are valid is questionable.

7. Chemical evolution and binaries

In the previous sections, we have demonstrated that it can not be excluded that a significant fraction of all stars are formed as component of an interacting binary. Therefore, predicting the chemical evolution of galaxies with single stars only and comparing with observation has only academic value if the effect of binaries is not known. Present day chemical evolutionary models simulate the effect of type Ia SN. However, it is important to realise that assuming a 15% SN Ia rate implies an interacting binary frequency $\geq$ 50%, i.e. for more than 50% of the stars, the chemical yields may be different from single star yields. Furthermore, it is uncertain whether the binary frequency is everywhere the same, i.e. whether a 15% SN Ia rate can be applied independent from metallicity.

In a forthcoming paper (De Donder and Vanbeveren, 2000), the effect of binaries on chemical evolutionary models will be discussed in detail. Here we raise a corner of the veil and consider the effect of BH formation on the evolution of oxygen. We constructed a galactic evolutionary model similar to the one of Chiappini et al.(1997) but we included a detailed scenario for all types of binaries with the appropriate yields. Remark that including binaries makes the galactic models orders of magnitude more complicated. To investigate the effect of BH formation, we used two schemes, i.e. scheme 1: all single stars (resp. case B_r primaries) with initial mass larger than 25 $M_\odot$ (resp. 40 $M_\odot$) form a BH and when a BH is formed, the BH mass = the final stellar mass (i.e. either no SN explosion occurs or, when a SN explosion happens, the mass of the SN shell is very small) and scheme 2: above 40 $M_\odot$, the mass of the BH = the final stellar mass whereas single stars with initial mass between 25 $M_\odot$ and 40 $M_\odot$ end their life in a SN explosion and, although a BH may be formed, the mass of the BH is not very different from the Chandrasekhar mass. We conclude:

A galactic chemical model, including a detailed model for the effect of binaries, reproduces a Solar-type oxygen abundance for the Solar neighbourhood only when scheme 2 is applied. With scheme 1, the predicted oxygen is a factor 2 smaller.

This means that, if the oxygen abundance of the Solar neighbourhood equals the one of the Sun, chemical evolutionary models promote 40 $M_\odot$ as minimum mass for the progenitors of BHs which are formed without SN explosion. Remark that this mass limit is similar to the one proposed by Fryer (1999) who studied BH formation in massive stars from first principles.

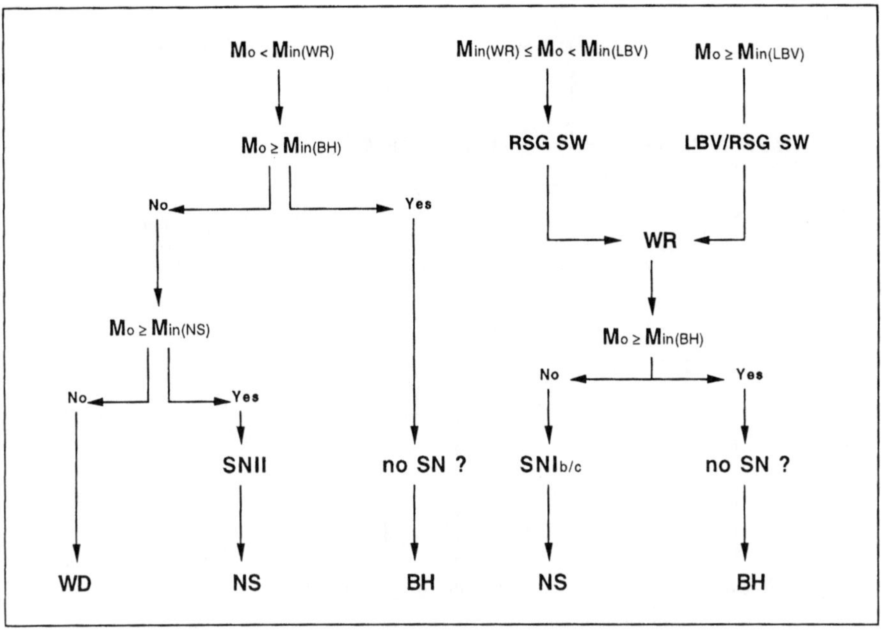

Figure 1. The overall evolutionary scenario of intermediate mass and massive single stars

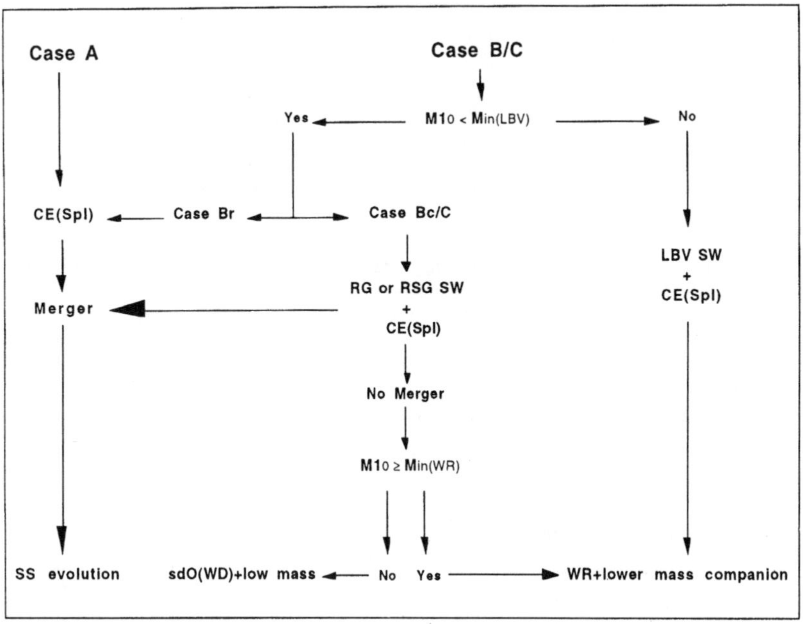

Figure 2. The overall evolutionary scenario of intermediate mass and massive binaries with $q < 0.2$

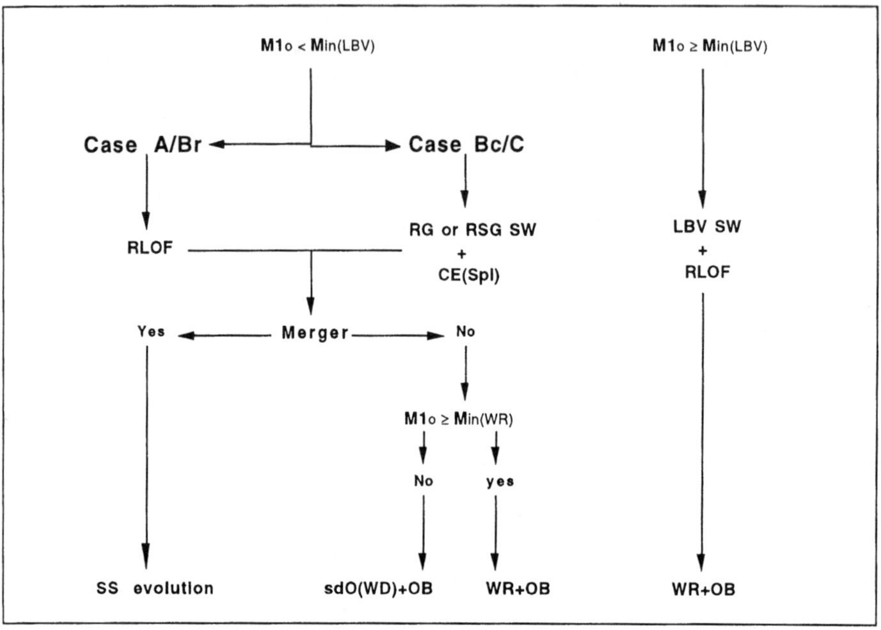

Figure 3. The evolutionary scenario of intermediate mass and massive close binaries with $q > 0.2$, up to the beginning of core helium burning

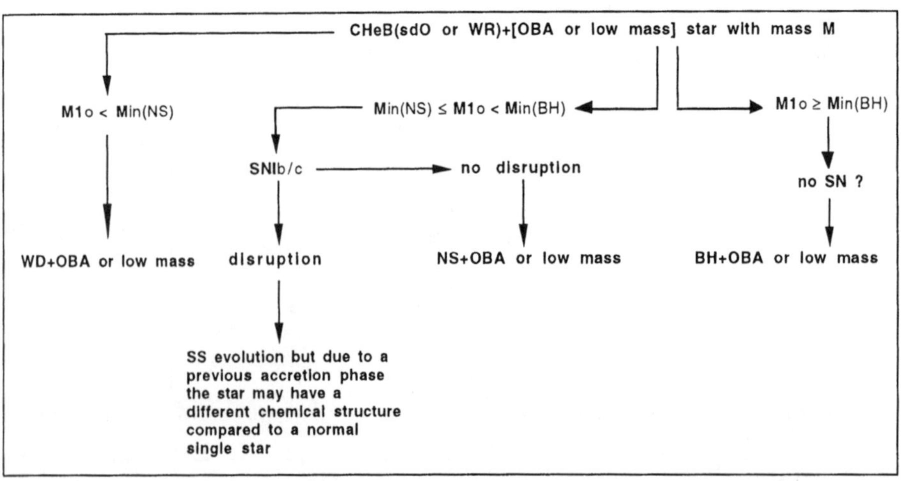

Figure 4. The evolution of intermediate mass and massive close binaries from the beginning of core helium burning up to the formation of a degenerate companion

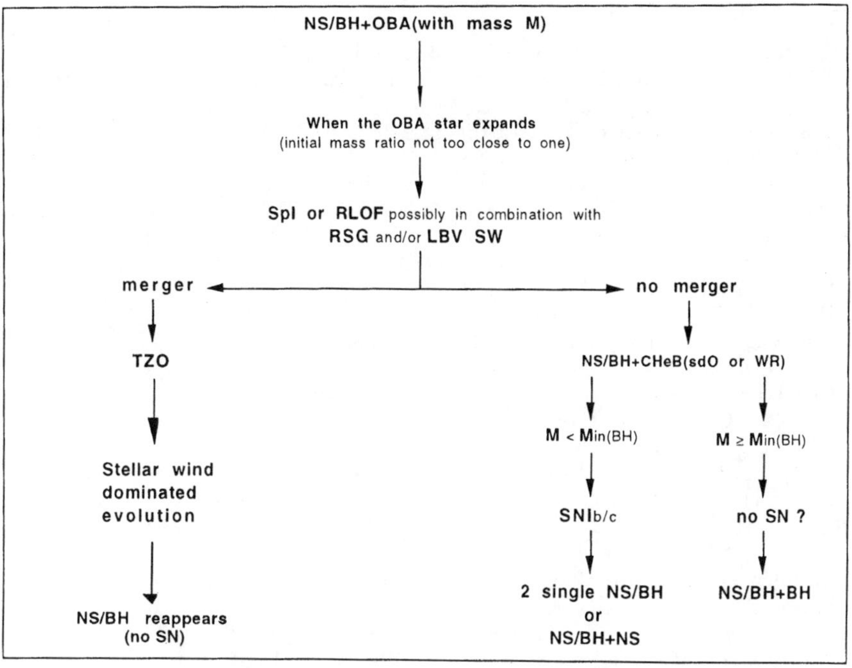

Figure 5. The evolution of OBA type stars with a NS or BH companion

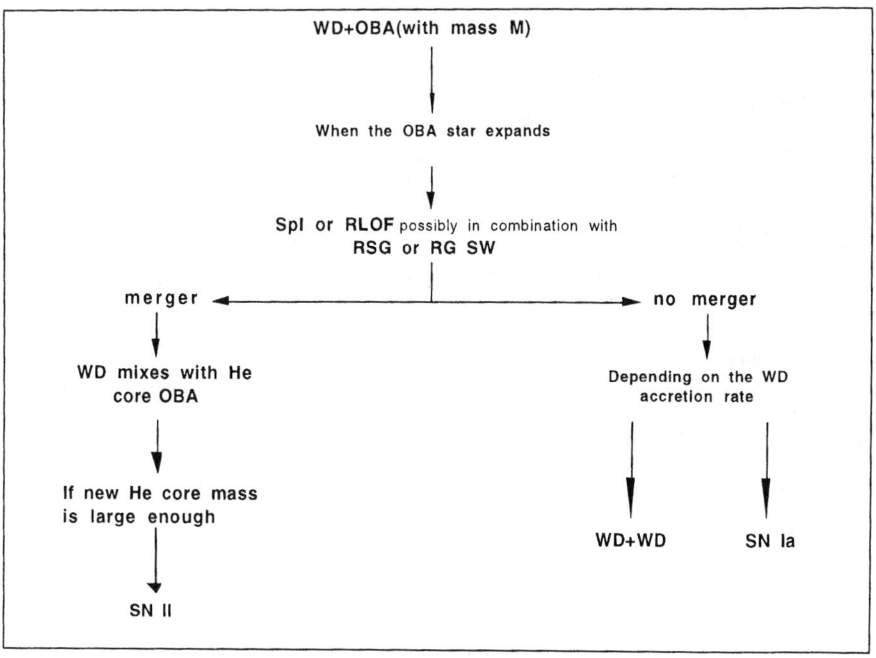

Figure 6. The evolution of OBA type stars with a WD companion

References

1. Abt, H.A., Gomez, A.E., Levy, S.G.: 1990, ApJS; 74, 551
2. Cannon, R., Eggleton, P.P., Zytkow, A.N., Podsiadlowski, Ph.: 1992, ApJ, 386, 206
3. Castorn J., Abbott, D.C., Klein, R.: 1975, ApJ, 195, 157
4. Chiappini, C., Matteucci, F., Gratton, R.: 1997, ApJ, 477, 765
5. De Donder, E., Vanbeveren, D.: 1998, A.A., 333, 557
6. De Donder, E., Vanbeveren, D.: 2000, A.A. (submitted)
7. Fryer, C.L., Woosley, S.E.: 1998, ApJ, 501, 780
8. Garmany, C.D., Conti, P.S., Massey, P.: 1980, ApJ, 242, 1063
9. Hachisu, I., Kato, M., Nomoto, K.: 1999, ApJ, 521 (in press).
10. Halbwachs, J.L.: 1987, A.A., 183, 234
11. Hamann, W.-R., Koesterke, L., Wessolowski, U.: 1995, A.A., 299, 151
12. Hamann, W.-R., Koesterke, L.: 1998, A.A., 335, 1003
13. Hogeveen, S.: 1991, Ph.D. Thesis, Univ. of Amsterdam
14. Hogeveen, S.: 1992, Astrophys. Space Sci., 196, 299
15. Israelian, G., Rebolo, R., et al.: 1999, Nature, 401, 142
16. Jura, M.: 1987, ApJ, 313, 743
17. Langer, N.: 1989, A.A., 210, 93
18. Levato, H., Malaroda, S., Morell, N., Garcia, B., Hernandez, C.: 1991, ApJS, 72, 869
19. Li, X.-D., van den Heuvel, E.P.J.: 1997, A.A., 322, L9
20. Lorimer, D.R., Bailes, M., Harrison, P.A.: 1997, MNRAS, 289, 592
21. Mason, B.D., Gies, D.R. et al.: 1998, ApJ, 115, 821
22. Nomoto, K., Iwamoto, K., Kishimoto, N.: 1997, Science, 276, 1378
23. Nomoto, K., Nariai, K., Sugimoto, D.: 1979, PASJ, 31, 287
24. Pauldrach, A.W., Kudritzki, R.P., Puls, J., Butler, K., Hunsinger, J.: 1994, A.A., 283, 525
25. Puls, J., Kudritzki, R.P., Herrer, A. et al.: 1996, A.A., 305, 171
26. Raboud, D.: 1996, A.A., 315, 384
27. Reid, N., Tinney, C., Mould, J.: 1990, ApJ, 348, 98
28. Salasnich, B., Bressan, A., Chiosi, C.: 1999, A.A., 342, 131
29. Van Bever, J., Belkus, H., Vanbeveren, D., van Rensbergen, W.: 1999, New Astronomy, 4, 173
30. Van Bever, J., Vanbeveren, D.: 1998, A.A., 334, 21
31. Van den Heuvel, E.P.J.: 1993, in 'Interacting Binaries', Saas-Fee Advanced Course 22, eds. H. Nussbaumer and A. Orr, Springer-Verlag: p. 263
32. Vanbeveren, D., De Donder, E., Van Bever, J., van Rensbergen, W., De Loore, C.: 1998b, New Astronomy, 3, 443
33. Vanbeveren, D., De Loore, C., van Rensbergen, W.: 1998c, Astron. Astrophys. Rev., 9, 63
34. Vanbeveren, D., van Rensbergen, W., De Loore, C.: 1998a, 'The Brightest Binaries', Kluwer Academic Publishers: Dordrecht
35. Verschueren, W., David, M., Brown, A.G.A.: 1996, in ASP Conf. Ser. 90, The Origins, Evolution and Destinies of Binary Stars in Clusters, eds. E.F. Milone and J.-C. Mermilliod, p.131

THE ORIGIN OF X-RAY PULSARS IN BINARY SYSTEMS

F. GIOVANNELLI
Istituto di Astrofisica Spaziale, CNR - Area di Ricerca di Tor Vergata
Via del Fosso del Cavaliere 100 - I 00133 Roma, Italy

AND

L. SABAU-GRAZIATI
División de Ciencias del Espacio, INTA
Cra. de Ajalvir Km 4 - E 28850 Madrid, Spain

1. Abstract

The evolution of binary systems predicts that the component faster evolving can explode as supernova (SN). Depending on the initial conditions and on the kind of SN explosion: i) the progenitor binary system cannot be disrupted and a remnant binary system, in which an X-ray pulsar orbiting around an OB/Be star, and the correspondent SNR itself are the secondary products of such an explosion; ii) the progenitor binary system can be disrupted: a pulsar and a SNR are the compact and diffuse remnants of the explosion.

A comparative analysis of the samples of X-ray pulsars and galactic SNRs shows that a significant number of objects of the two samples are correlated. We show the evidence that roughly 47% of the actual sample of X-ray pulsars can be the compact remnants of supernova explosions occurred in progenitor massive close binary systems, as suggested by us since 1993 with a smaller sample.

2. Introduction

X-ray binary systems are the best laboratory for the study of accreting processes thanks to their relative high luminosity in a large part of the electromagnetic spectrum. For this reason, multifrequency observations and relative analysis and interpretation are fundamental in understanding their morphology and the physics governing their behaviour.

F. Matteucci and F. Giovannelli (eds.), The Evolution of the Milky Way, 151–161.

One of the most exciting problems on knowledge of X-ray binaries is connected with their origin and evolution.

The evolutionary models (e.g. Rappaport and van den Heuvel, 1982, and de Loore et al., 1982) show that an X-ray/Be binary can be the result of the evolution of a moderate massive ($M_{1,2} \approx 10 - 20 M_{\odot}$) B-type close binary. However, this kind of evolution leads to systems with low orbital eccentricities ($e \approx 0.1 - 0.2$) (van den Heuvel and Rappaport, 1987), while the observed values in several X-ray/Be systems are $0.3 \leq e \leq 0.8$ (Hutchings et al., 1985). Such higher eccentricities can be reached either assuming an asymmetric supernova explosion (van den Heuvel and Rappaport, 1987) or quite massive close binary precursor 1988) ($M_{1,2} \geq 50 M_{\odot}$), which underwent highly non-conservative evolution (Habets, 1987). Indeed, most of the radio pulsars tend to be runaway objects, as well as massive X-ray binaries (van Oijen, 1989); this fact can be an indication that the neutron stars receive a kick velocity of $\approx 100 - 200$ Km s^{-1}, due to supernova mass ejection, when they are born (Dewey and Cordes, 1987).

Brandt and Podsiadlowski (1995) simulated the sky distribution of 54 HMXBs (High Mass X-ray Binaries) and 36 LMXBs (Low Mass X-ray Binaries), which reproduces very well the distribution of the actual systems. They also computed the distribution of final eccentricities versus the final orbital period for typical HMXBs, assuming an isotropic kick-velocity distribution. The results are in agreement with the experimental values of orbital periods and eccentricities of known systems.

Iben and Tutukov (1998) comment on the intrinsic kicks at birth to explain the observed properties of single and binary neutron stars. They conclude that, surely, kicks do occur. However, it is necessary to construct a probability distribution function for kicks (for magnitude and direction) and take into account a probable dependence on presupernova binary configuration.

Then, the problem on the possible origin of the observed motion of pulsars is still open. However, two different models of neutrino emission from a cooling neutron star are in good quantitative agreement and predict the same order of magnitude for the pulsar kick velocity, consistent with the data (Kusenko and Segre, 1999).

On the basis of positional coincidence, distance and age, the association of the 104-s X-ray pulsar A 0535+26 with the SNR S147 has been claimed by Giovannelli et al. (1993, 1994). The analysis of the samples of SNRs (Green, 1988; 1991) and 38 X-ray pulsars, then known, revealed the possible association of nine more binary systems with SNRs.

The purpose of this paper is to demonstrate the correlation between X-ray pulsars and SNRs, by using the whole updated samples of both classes. In order to show the general criteria used, we will start by using the example of A 0535+26/HDE 245770, associated with the SNR S147, which triggered this work.

3. Evidence of correlations between A 0535+26/HDE 245770 and S 147

A 0535+26/HDE 245770 is a typical X-ray/Be runaway system (van Oijen, 1989), whose components are a 104-s X-ray pulsar (Rosenberg et al., 1975) and an O9.7 IIIe star (Giangrande et al., 1980). The X-ray pulsar is spinning-up with a pulse period derivative $\dot{P} = -4.9 \times 10^{-4}\ yr^{-1}$ (Coe et al., 1990) with alternating phases of spin-down during X-ray outbursts. The rotational equilibrium period formula, derived by Bisnovatyi-Kogan (1991) for long-periodic X-ray pulsars, sets A 0535+26 very close to its rotational equilibrium, as already suggested by Giovannelli and Ziolkowski (1990). The O9.7 IIIe companion does not fill its Roche lobe (de Loore et al., 1984), although a temporary accretion disk can be formed around the neutron star when it is close to the periastron (Giovannelli and Ziolkowski, 1990). The orbital period of the system is $\sim$ 111 days (Priedhorsky and Terrell, 1983) and the eccentricity is 0.47 (Finger et al., 1994). Complete reviews of this system can be found in the papers by Giovannelli et al. (1985), Giovannelli and Sabau Graziati (1992), and Burger et al. (1996).

De Loore et al. (1984) computed the simultaneous evolution of the two components of a system of $15M_\odot + 10M_\odot$, with an initial period of 8 days. The final system, consisting of a primary of 3.42 $M_\odot$ and a secondary of 20.5 $M_\odot$, has an orbital period of 68 days. 1.08 $M_\odot$ was lost by the system during the contact phase. The accreting component (initially 10 $M_\odot$) becomes as large as 20.5 $M_\odot$. The helium remnant of the primary meanwhile continued its evolution and exploded. The system in this phase could represent an X-ray/Be system, very similar to A 0535+26/HDE 245770, with a $\sim$ 1.5 $M_\odot$ neutron star orbiting around a $\sim$ 20 $M_\odot$ Be star with a period of 81 days. The system A 0535+26/HDE 245770 lies within the extended radio source S 147, which is undoubtedly an intermediate age supernova remnant (G 180.0-1.7), showing the presence of a well developed radio shell. S 147 is one of the oldest optically visible SNRs in the Galaxy (Minkowski, 1958; Kirshner and Arnold, 1979; Lozinskaya, 1980). The high resolution 6 cm observations of selected portions of S 147 show good positional agreement between radio and optical emission structures within the remnant (Sofue and Furst, 1980), as well as with the first X-ray map obtained with the EXOSAT satellite. The SNR might be the source detected at medium energy (2-6 KeV band), even if the spectrum is somewhat too hard, which indicates that a part of this emission is expected to come from the pulsar A 0535+26, which was contained in the observed field (Sauvageot, et al., 1990). Giovannelli

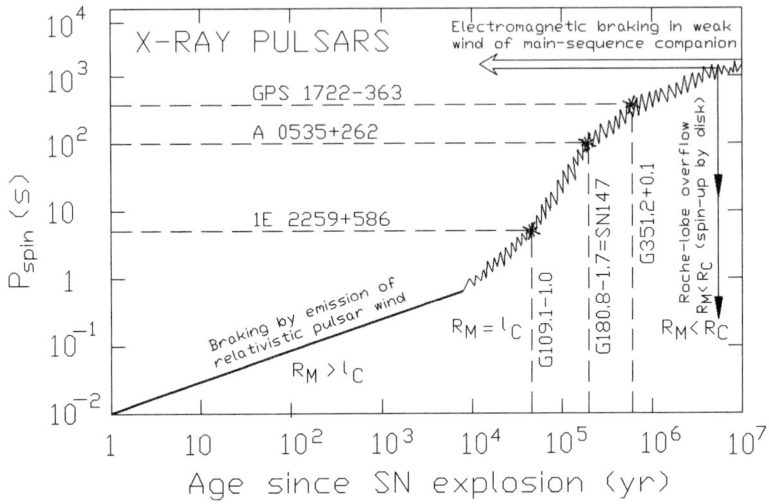

Figure 1. Spin period of X-ray pulsars vs. the age since SN explosion. Suggested phases in the rotational evolution of a pulsar that is born in a massive close binary system. R_M, R_C, and l_C are the radius of the magnetosphere, of co-rotation, and of the light cylinder, respectively (after van den Heuvel, 1981). The position of the X-ray/Be systems A 0535+26/HDE 245770, GPS 1722-36 and the accreting neutron star 1E 2259+586, probably in a LMXB are reported too.

et al. (1993) suggested that the SNR S 147 could have been originated by an asymmetric supernova explosion in a moderate massive B-type binary system which produced the A 0535+26 pulsar, eccentrically orbiting around the Be star HDE 245770. Indeed:

i) Following the rotational history of the pulsars in massive X-ray binaries (Savonije and van den Heuvel, 1977) and the relationship between the pulse period and the time since the SN explosion, which originated the binary system (van den Heuvel, 1981), a 104-s X-ray pulsar can be the collapsed remnant of a supernova explosion occurred roughly 1.5×10^5 years ago. Figure 1, adapted from van den Heuvel (1981), shows the suggested phases in the rotational evolution of a pulsar that is born in a massive binary system and superimposed the representative points of the X-ray/Be systems A 0535+26/HDE 245770, GPS 1722-36 and the accreting neutron star 1E 2259+586, probably in a LMXB (see later). The ages of the three X-ray pulsars are roughly in agreement with the ages of the correspondent SNRs, derived from radio measurements. For example, the derived age of S 147 is $\sim 110,000$ yr (Kundu et al., 1980), compatible with that derived by using the curve reported in Fig. 1.

ii) The SNR S 147 is placed at 1.4 ± 0.3 Kpc (Clark and Caswell, 1976; Sofue and Furst, 1980) and HDE 245770 is placed at 1.8 ± 0.6 Kpc (Giangrande et al., 1980). The SNR S 147 southern rim reaches the position of

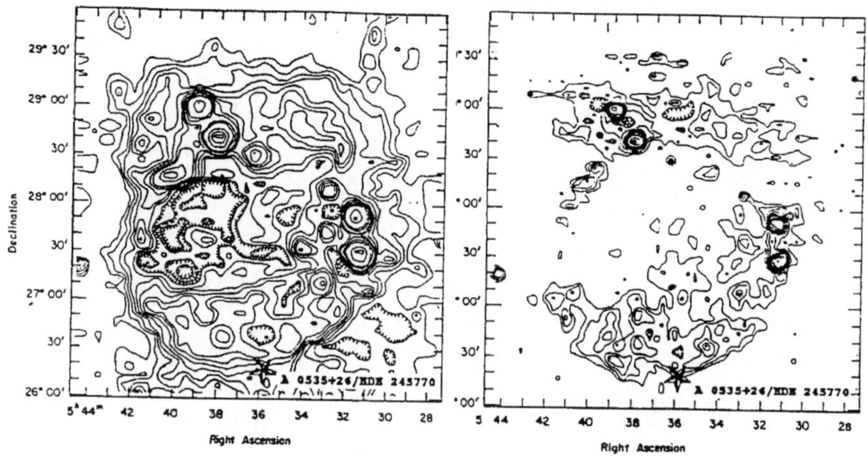

Figure 2. Radio maps at 18.2 cm (left panel) and 11.1 cm (right panel) of the SNR S 147 (Kundu et al., 1980). The stars indicate the position of the binary system A 0535+26/HDE 245770 in both maps.

A 0535+26. This fact is supported by the radio maps of the SNR made at 18.2 cm and 11.1 cm (Kundu et al., 1980), as shown in Figure 2. Therefore, the space positions of the X-ray/Be system and SNR are fully compatible.

iii) The proper motion of HDE 245770 is poorly determined (SAO catalogue) and only upper limits of 2×10^{-2} arcsec yr^{-1} in right ascension and declination are given. Assuming these upper limits as the components of the proper motion of the Be star, at a distance of 1.8 Kpc, the HDE 245770/A 0535+26 system has travelled $\sim$ 25 pc in $\sim 10^5$ yr. This figure is compatible with the estimated radius of the SNR S 147, from radio measurements (Kundu et al., 1980). The X-ray/Be system in order to cross 25 pc in $\sim 1.5 \times 10^5$ yr (the age of S 147 derived from the Figure 1) would have had a velocity of $\sim$ 160 Km s^{-1}, which is a value compatible with the possible kick velocity in an asymmetric supernova explosion (Dewey and Cordes, 1987; Brandt and Podsiadlowski (1995); Vanbeveren, this workshop), and lower than the tangential orbital velocity of the system, which is $\sim$ 210 Km s^{-1} (Motch et al. 1991).

4. The samples of galactic SNRs and X-ray pulsars

The catalog of Galactic SNRs (Green, 1998), containing 220 SNRs, is an updated version of those presented in detail in Green (1984; 1988). SNRs are concentrated on the galactic plane (Green, 1991). The sample of HMXB transient sources (HMXBTSs) - reported in Table 1 of Giovannelli et al. (1994) paper - contained 51 objects at that time known, among which 38 systems were known as X-ray pulsars. The updated list of X-ray pulsars (compiled by us) contains 88 galactic objects. Most of these X-ray pulsars have a Be star companion. The 5-9 second pulsars probably belong to LMXB systems (Mereghetti and Stella, 1995).

Few X-ray pulsars, e.g. Vela X1, Cen X3, 4U 1538-522, OAO 1657-415, RX J0648.1-4419, 1E 1145-6141, GX 301-2, and 4U 1909+097 have a B supergiant companion rather than a Be star companion.

For other few of them the optical companion has not yet been found. Two sources, Cir X1 (1M 1516-569) and SS 433 (4U 1909+05), are black hole candidates, whose companions are OB stars.

Evidence for a supernova origin for the black hole in the system GRO J1655-40 has been shown by Israelian et al. (1999) on the basis of large overabundances of the α-elements oxygen, magnesium, silicon and sulfur in the atmosphere of the star orbiting a black hole in the binary system, also known as Nova Scorpii 1994. However, the relative SNR is not visible, probably because the SN explosion occurred roughly 10^6 yr ago.

All these binary systems are concentrated on the galactic plane in a narrow band of galactic latitude (3.9 Deg) and are uniformly distributed in galactic longitude (van Paradijs, 1989). Therefore, their spatial density is:

$D = 88/(360 \times 3.9) = 0.0627\ Deg^{-2}$

Supposing, in the first approximation, that the distribution is gaussian, the standard deviation is $\sigma = 7.9 \times 10^{-3}\ Deg^{-2}$.

In the same band of 3.9 Deg around the galactic plane, the SNRs of the entire sample lie, with the exception of few objects. Taking into account the dimensions of each SNR (Green, 1998), the SNRs of such a band keep occupied $\sim 100\ Deg^2$. Supposing the two distributions independent, the casual coincidence between X-ray pulsars and SNRs is:

$$N = (62.7 \pm 7.9) \times 10^{-3} \times 100 \sim 6.3 \pm 0.8$$

Table 1 shows 41 X-ray pulsars which lie within SNRs.

This number of coincidences exceeds roughly seven times what is expected from chance coincidence.

5. Discussion and conclusions

The number of coincidences found demonstrates that X-ray pulsars in binary systems, as well as isolated, can originate from SN explosions. A better knowledge of the distances and ages of the objects of both classes could facilitate the search for direct experimental evidences of such associations, that indeed recently they are coming out for several objects, like commented later. This number of coincidences, however, represents a lower limit to the actual number of associations between X-ray pulsars and SNRs, since some of the oldest SNRs, associated with the slowest pulsars are not visible any more; moreover, some sources in very crowded regions, such as the Galactic Center, has been removed in order to avoid uncertain associations: e.g. the four transients 1,2,3 and 4 discovered by GINGA (Koyama et al., 1989a) Furthermore, some X-ray pulsars can have left the apparent radius of the remnant, which is, however, energy dependent and difficult to be determined. It is also possible that some SNRs have not yet been recognized. Among these objects placed within SNRs, we can remark that:

i) The space coincidence of the SNR S 147 and HDE 245770/A 0535+26 system has been proved. Indeed, their distances to the Earth are 1.4 ± 0.3 Kpc and 1.8 ± 0.6 Kpc, respectively. The 104-s X-ray pulse period of A 0535+26 is just that one expects if the system originated from a supernova explosion in a medium mass binary system occurred $\sim 10^5$ years ago. This age is compatible with that of the SNR S 147. So that, the transient X-ray/Be system can be now considered physically, and not only prospectically, within the SNR S 147, at the southern rim. Thus, the runaway X-ray/Be system A 0535+26/HDE 245770 and the SNR S 147 are still interacting. This possibility could be supported by some direct experimental evidences. Indeed, the stellar wind from the binary system has a terminal velocity of ~ 630 Km s^{-1} (Giovannelli et al., 1982; de Loore et al., 1984), which is a supersonic velocity if we admit that it is outflowing through the material of the SNR, having a temperature of 4.7×10^7 K (Sauvageot et al., 1990). Therefore, a strong interaction due to the shock could be detectable.

ii) The space coincidence of 4U 0115+634 with G 126.2+1.6 is supported also by the evaluation of the distances. Indeed the 3.6-s X-ray pulsar is placed at ~ 3.5 Kpc (White et al., 1983) and the SNR is placed at ~ 3 Kpc with an age of $\sim 5 \times 10^4$ yr (Furst et al., 1984). We note that this age is roughly in agreement with that derived for the X-ray/Be system by using the diagram of Figure 1.

iii) The coincidence of the 6.9 X-ray pulsar 1E 2259+586 with the SNR G 109.1-1.0 is very probable. Indeed, Koyama et al. (1989b) suggested a distance to the pulsar of 3.6 Kpc, and the distance to G 109.1-1.0 is between 3.6 and 5.2 Kpc (Gregory and Fahlam 1980; Hughes et al., 1981). The derived age of the SNR is $(1.2-1.7) \times 10^4$ yr, which is not incompatible with the age of the pulsar derived from the X-ray pulse period. This probable association was already noted by Hughes et al. (1984), who gave a distance of the X-ray pulsar to the center of the SNR equal to 3.6 arcmin.

iv) The two non-pulsing objects, 1M 1516-569 (Cir X1) and 4U 1909+05 (SS 433), black hole candidates, have ages and distances difficult to evaluate. However, Cir X1 appears to be at similar distance to that of G 321.9-0.3, whose age is ranging from 2×10^4 yr and 10^5 yr (Clark et al., 1975); indeed, the distances of Cir X1 is ~ 8 Kpc (Goss and Mebold, 1977) and

$\sim$ 5.5 Kpc for G 321.9-0.3 (Clark et al., 1975). Considering the uncertainties of both estimates, the two are in rather good agreement. Stewart et al. (1993) reported the same distance (6.5 Kpc) of both systems and a common age of 28,600 yr. This fact, strongly supports our conclusions on the association between X-ray/OB systems and SNRs. The physical interaction of SS 433 with the SNR G 39.7-2.0 has been already suggested in literature (e.g. Velusamy and Pramesh Rao, 1983).

v) Morini et al. (1988) reported the EXOSAT observations of the SNR G109.1-1.0. with a deduced age of the order of 10,000 yr. They noted the central 6.9-s X-ray pulsar 1E 2259+586 but their data did not allow to argue any association between the pulsar and SNR. Recently, Parmar et al. (1998) from BeppoSAX observations of the same objects 1E 2259+586 and the SNR G109.1-1.0 deduced that probably the pulsar lies some distance beyond the SNR and then the two objects are unrelated, or there is a presence of absorbing material local to the pulsar which would reconcile the higher value of its N_H column with respect to that of the SNR. The evaluated age of the SNR is about 3,000 yr.

vi) Evidence for a low-luminosity 5.5-s X-ray pulsar (RXJ 1838.4-0301) associated with the SNR G28.8+1.5 has been presented by Schwentker (1994) on the basis of the positional coincidence and the compatible estimated distances.

vii) Vasisht and Gotthelf (1997) report the discovery of 12-s pulsed X-ray emission from the compact source 1E 1841-045. The X-ray source is located in the center of the small-diameter SNR Kes 73. Then, 1E 1841-045 is very likely the compact stellar remnant of the supernova explosion that formed Kes 73. This association was already suggested by Giovannelli et al. (1993, 1994).

viii) Recently, a new SNR (G29.6+01) has been found to be coincident with the slow X-ray pulsar AX J1845-0258 (Gaensler et al., 1999). The young age ($<$ 8,000 yr) of G29.6+01 is consistent with the interpretation that the anomalous X-ray pulsars (AXPs) are isolated, highly magnetized neutron stars (*magnetars*) (Thompson and Duncan, 1995).

Therefore, we can conclude that X-ray pulsars in binary systems, or isolated, can be formed in SN explosions. Among the known X-ray pulsars, 47 % of them are associated with known SNRs. Further investigations in X-ray, radio and optical energy ranges of X-ray pulsars and SNRs will provide direct experimental proves of all the association presented in this paper, and maybe more.

Aknowledgements

We would like to thank Mr Juan Antonio Balsalobre (DCE-INTA) and Mr Massimo Frutti (IAS-CNR) for helping in graphics.

References

1. Bisnovatyi-Kogan, G.S. 1991 *A&A*, 245, 528
2. Bradt, H.V.D. & McClintock, J.E. 1983, *Ann. Rev. A&A*, 21, 13
3. Brandt, N. & Podsiadlowski, P. 1995, *MNRAS*, 274, 461
4. Burger, M., van Dessel, E.L., Giovannelli, F., Sabau-Graziati, L., Bartolini, C., Guarnieri, A. & Piccioni, A. 1996, in *Multifrequency Behaviour of High Energy Cosmic Sources*, F. Giovannelli & L. Sabau-Graziati (eds.), *Mem. S.A.It*, 67, 365
5. Clark, D.H., Caswell, J.L. & Green, A.J. 1975, *Aust. J. Phys. Astrophys. Suppl.*, No. 37, 1
6. Clark, D.H. & Caswell, J.L. 1976, *MNRAS*, 174, 274
7. Coe, M.J., Carpenter, G.F., Engel, A.R. & Quenby, J.J. 1975, *Nature*, 256, 630
8. Coe, M.J. et al. 1990, *MNRAS*, 243, 475
9. Dewey, R.J. & Cordes, J.M. 1987, *ApJ* 321, 780
10. Finger, M.H., Cominsky, L.R., Wilson, R.B., Harmon, B.A. & Fishman, G.J. 1994, in *The Evolution of X-ray Binaries*, S.S. Holt & C.S. Day (eds.), *AIP Conf. Proc.*, 308, 459
11. Furst, E., Reich, W. & Steube, R. 1984, *A&A*, 133, 11
12. Gaensler, B.M., Gotthelf, E.V. & Vasisht, G. 1999, *ApJL*, 526, L37
13. Giangrande, A., Giovannelli, F., Bartolini, C., Guarnieri, A. & Piccioni, A. 1980, *A&A Suppl. Ser.*, 40, 289
14. Giovannelli, F. et al. 1982, *Proc. of the Third European IUE Conf.*, ESA SP-176, p. 233
15. Giovannelli, F. et al. 1985, in *Multifrequency Behaviour of Galactic Accreting Sources*, F. Giovannelli (ed.), Siderea, Roma, Italy, p. 284
16. Giovannelli, F. & Ziolkowski, J. 1990, *Acta Astron.*, 40, 95
17. Giovannelli, F. & Sabau-Graziati, L. 1992, *Space Sci. Rev.*, 59, 1
18. Giovannelli, F., Polcaro, V.F., Sabau-Graziati, L, & Norci, L. 1993, in *Frontier Objects in Astrophysics and Particle Physics*, F. Giovannelli & G. Mannocchi (eds.), Italian Physical Society, Editrice Compositori Bologna, Italy 40, 323
19. Giovannelli, F., Norci, L., Polcaro, V.F. & Sabau-Graziati, L 1994, in *Evolutionary Links in the Zoo of Interacting Binaries*, F. D'Antona, V. Caloi,C. Maceroni & F. Giovannelli (eds.), *Mem. S.A.It.*, 65, 373
20. Goss, W.M. & Mebold, U. 1977, *MNRAS*, 181, 255
21. Green, D.A. 1984, MNRAS, 209, 449
22. Green, D.A. 1988, *Ap&SS*, 148, 3
23. Green, D.A. 1991, *PASP*, 103, 209
24. Green, D.A. 1998, *A Catalog of Galactic Supernova Remnants*, Mullard Radio Astronomy Observatory, Cambridge, UK
25. Gregory, P.C. & Fahlman, G.G. 1980, *Nature*, 287, 805
26. Habets, G.M.H.J. 1987, in *Physics of Be Stars*, A. Slettebak & T.P. Snow (eds.), Cambridge University Press, NY, USA, p. 509
27. van den Heuvel, E.P.J. 1981, *Space Sci. Rev.*, 30, 623
28. van den Heuvel, E.P.J. & Rappaport, S. 1987, in *Physics of Be Stars*, A. Slettebak & T.P. Snow (eds.), Cambridge University Press, NY, USA, p. 291
29. Hughes, V.A., Harten, R.H. & van den Bergh, S. 1981, *ApJL*, 246, L127
30. Hughes, V.A., Harten, R.H., Costain, C.H., Nelson, L.A. & Viner, M.R. 1984, ApJ, 283, 147
31. Hutchings, J.B., Crampton, D., Cowley, A.P., Olszweski, E., Thompson, I.B. & Suntzeff, N. 1985, *PASP*, 97, 418
32. Iben, I.Jr. & Tutukov, A.V. 1998, *ApJ*, 501, 263
33. Israelian, G., Rebolo, R., Basri, G., Casares, J. & Martin, E.L. 1999, *Nature*, 400, 142
34. Janot Pacheco, E., Motch, C. & Mouchet, M. 1987, *A&A* 177, 91
35. Kirshner, R.P. & Arnold, C.N. 1979, *ApJ*, 229, 147

36. Koyama, K. et al. 1989a, *PASJ*, 41, 483
37. Koyama, K. et al. 1989b, *PASJ*, 41, 461
38. Koyama, K., Kunieda, H., Takeuchi, Y. & Tawara, Y. 1990, *PASJL*, 42, L59
39. Koyama, K. et al. 1991a, *ApJL*, 366, L19
40. Koyama, K, Kunieda, H., Takeuchi, Y. & Tawara, Y. 1991b, ApJL 370, L77
41. Kundu, M.R., Angerhofer, P.E., Furst, E. & Hirth, W. 1980, *A&A*, 92, 225
42. Kusenko, A. & Segre, G. 1999, *PhRv*, 590, 76
43. de Loore, C., Burger, M., van Dessel, E.L. & Mouchet, M. 1982, in *Be Stars*, M. Jaschek & H.G. Groth (eds.), Reidel Publ. Co., Dordrecht, Holland, p. 347
44. de Loore, C., Giovannelli, F., van Dessel, E.L. et al. 1984, *A&A*, 141, 279
45. Lozinskaya, T.A. 1980, *A&A*, 84, 26
46. Mereghetti, S. & Stella, L. 1995, *ApJL*, 442, L17
47. Minkowski, R. 1958, *Rev. Mod. Phys.*, 30, 1048
48. Motch, C., Stella, L., Janot Pacheco, E. & Mouchet, M. 1991, *ApJ*, 369, 490
49. Morini, M., Robba, N.R., Smith, A. & van der Klis, M. 1988, *ApJ*, 333, 777
50. Nagase, F. 1989, *PASJ*, 41, 1
51. Nagase, F. et al. 1982, *ApJ*, 263, 814
52. van Oijen, J.G.J. 1989, *A&A*, 217, 115
53. van Paradijs, J. 1989, in *Timing Neutron Stars*, H. Ogelman & E.P.J. van den Heuvel (eds.), Kluwer Academic Publishers, Dordrecht, Holland, p. 191
54. Parmar, A.N., Oosterbroek, T., Favata, F., Pightling, S., Coe, M.J., Mereghetti, S. & Israel, G.L. 1998, *A&A*, 330, 175
55. Priedhorsky, W.C. & Terrell, J. 1983, *Nature*, 303, 681
56. Rappaport, S. & van den Heuvel, E.P.J. 1982, in *Be Stars*, M. Jaschek & H.G. Groth (eds.), Reidel Publ. Co., Dordrecht, Holland, p. 327
57. Rappaport, S. & Joss, P.C. 1981, in *X-Ray Astronomy*, R. Giacconi (ed.), D. Reidel Publ. Co., Dordrecht, Holland, p. 123
58. Rosenberg, F.D., Eyles, C.J., Skinner, G.K. & Willmore, A.P. 1975, *Nature*, 256, 628
59. Sauvageot, J.L., Ballet, J. & Rothenflug, R. 1990, *A&A*, 227, 183
60. Savonije, G.J. & van den Heuvel, E.P.J. 1977, *ApJL*, 214, L19
61. Schwentker, O. 1994, *A&AL*, 286, L47
62. Shull, J.M. 1983, in *Supernova Remnants and their X-Ray Emission*, J. Danzinger & P. Gorenstein (eds.), D. Reidel Publ. Co., Dordrecht, Holland, p. 99
63. Sofue, Y. & Furst, E. 1980, *PASJ*, 32, 1
64. Stewart, R.T., Caswell, J.L., Haynes, R.F. & Nelson, G.J. 1993, *MNRAS*, 261, 593
65. Tawara, Y., Yamauchi, S., Awaki, H., Kii, T., Koyama, K. & Nagase, F. 1989, *PASJ*, 41, 473
66. Thompson, C. & Duncan, R.C. 1995, *MNRAS*, 275, 255
67. Vanbeveren, D. 2000, it This workshop
68. Vasisht, G. & Gotthelf, E.V. 1997, *ApJL*, 486, L129
69. Velusamy, T. & Pramesh Rao, A. 1983, in *Supernova Remnants and their X-Ray Emission*, J. Danziger & P. Gorenstein (eds.), D. Reidel Publ. Co., Dordrecht, Holland, p. 465
70. White, N.E., Swank, J.H. & Holt, S.S: 1983, *ApJ*, 270, 711
71. Yamauchi, S. & Koyama, K. 1990, *PASJL*, 42, L83

TABLE 1. X-ray pulsars associated with SNRs. The two pulsars marked with * are associated to the same SNR. The absence of spin period is marked with +: the two relative objects are black hole candidates. Symbol ? in third and fourth columns means doubt in the association and uncertainty in the size of the SNR, respectively.

X-Ray Pulsar	Spin Period (s)	SNR	SNR Diameter (arcmin)@1 GHz
4U 0115+634	3.6	G126.2+1.6	70
4U 0142+614	8.69	G127.1+0.5	45
RX J 0146.9+6121	1413	G132.7+1.3	80
RX J 0440.9+4431	220.5	G160.9+2.6	140x120
A 0535+262	104	G180.0-1.7	180
GS 0834-430	12.3	G260.4-3.4	60x50
1E 1048.1-5937	6.44	G289.7-0.3	18x14
A 1118-616	405	G294.1-0.0	40
4U 1145.1-6141*	297	G296.1-0.5	37x25
4U 1145-619 *	292	G296.1-0.5	37x25
4U 1258-61	272	G302.3+0.7	17
1SAXJ 1324.4-6200	170.84	G308.1-0.1	20x30?
2S 1417-624	17.6	G312.4-0.4	38
1SAXJ 1452.8-5949	437.4	G317.3-0.2	11
1M 1516-569	+	G321.9-0.3	31x23
2S 1553-54	9.26	G327.4+0.4	21
AXS J 161730-505505	0.069	G332.4-0.4	10
OAO 1657-415	38	G344.7-0.1	10
1RXSJ 170849.0-400910	10.997	G346.6-0.2	8
GPS 1722-363	414	G351.2+0.1	7
4U 1728-247	120	G5.4-1.2	35
XRE J 1739-302	>300	G357.7+0.3	24
GRO J 1744-28	0.467	G0.9+0.1	8
GRO 1750-27	4.45	G4.2-3.5	28
AX J 1749.1-2725	220	G1.4-0.1	10
AX J 1820.5-1434	152.26	G16.8-1.1	30x24
X 1833-076	111	G24.7+0.6	30x15
RX 1838.4-0301	5.45	G28.8+1.5	100?
1E 1841-045	11.76	G27.4+0.0	4
GS 1843-024	94.9	G29.7-0.3	3
AX J 1844.8-0258	6.97	G29.6+0.1	4.5x5.0
AX J 1845.0-0433	7	G27.8+0.6	50x30
XTE J 1855-026	361	G30.7-2.0	16
XTE J 1858+034	221	G36.6-07	25?
XTE J 1906+09	437	G42.8+0.6	24
4U 1909+05	+	G39.7-2.0	120x60
J 1946+274	15.8	G63.7+1.1	8
J 1948+32	18.7	G69.0+2.7	80?
1WGAJ 1958.2+3232	721	G69.7+1.0	16
1SAXJ 2103.5+4545	358.61	G82.2+5.3?	95x65
1E 2259+586	6.98	G109.1-1.0	5.4

III-GALACTIC CLUSTERS

Conference dinner

METAL ABUNDANCES IN GLOBULAR CLUSTERS

R.G. GRATTON
Osservatorio Astronomico di Padova, ITALY
Vicolo dell'Osservatorio, 5, 35122 Padova, ITALY

1. THE METALLICITY SCALE OF GLOBULAR CLUSTERS

Traditionally, the most used scale of abundances for globular clusters is that prepared by Zinn & West (1984); however, this is now superseded by the most recent one by Carretta & Gratton (1997). This is based on homogenous high dispersion ($R \sim 30,000$) abundance derivations for stars in 24 globular clusters; the abundances derived from these spectra are compared with the results of consistent solar analysis with Kurucz (1992) model atmospheres. A comparison between the two scales shows that while they agree at the extremes of the scale (both for metal-poor clusters like M92, and for metal-rich clusters as 47 Tuc), quite large differences exists for clusters of intermediate metallicity (like M3 or M13), where the new Carretta & Gratton scale gives larger metal abundances (by about 0.2 to 0.3 dex). Abundances for a larger set of clusters (71) derived from the Ca II triplet, calibrated against the Zinn & West as well as the Carretta & Gratton scales is given by Rutledge et al. (1997); they may be used to have quite accurate abundances for those clusters lacking high dispersion data.

When comparing the two scales, we should remind that the effects on e.g. age derivations are not negligible. They depend on the adopted technique used to derive distances to globular clusters; for instance, when using the subdwarf fitting method they are ad large as $\Delta Age/\Delta$[Fe/H]$\sim$ 1 Gyr/0.1 dex.

Recently, the Carretta & Gratton abundance scale has been extended to very high metal abundances by the analysis of a number of red horizontal branch stars in NGC 6553 (Cohen et al. 1999), and even more recently in NGC 6528 (Carretta et al. 2000), based on spectra gathered with HIRES at Keck I (see Figure 1). Both these clusters have [Fe/H]~ -0.2; these metal abundances agree well with those given in the compilation by Harris (1996), but are slightly larger than estimated by Barbuy et al. (1999) from

F. Matteucci and F. Giovannelli (eds.), The Evolution of the Milky Way, 165–183.

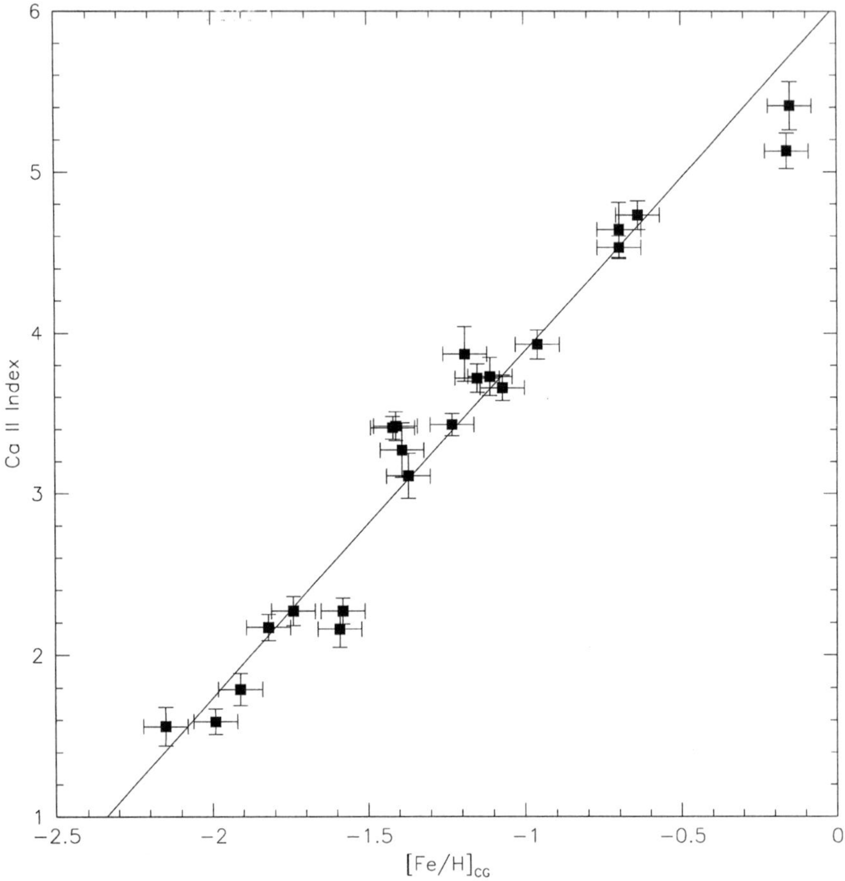

Figure 1. Calibration of the Ca II index (Rutledge et al. 1997) using the Carretta & Gratton (1997) high dispersion metal abundances; also plotted are the abundances for NGC 6553 (Cohen et al. 1999) and NGC 6528 (Carretta et al. 2000), which extend the calibration at high metal abundances ([Fe/H]> -0.5)

spectra of cool red giants. This difference seems due to adoption of lower temperatures in the Barbuy et al. analysis, and may reflect problems in the analysis of very cool, metal-rich stars with presently available model atmospheres. Analysis of warmer subgiants in these two clusters would be highly welcomed.

While the Carretta & Gratton scale is as homogeneous as presently possible, and uses updated model atmospheres, a concern still exists, because it is not obvious that dwarfs and giants give the same abundances. In fact systematic differences in the analysis of dwarfs and giants might be due either to problems in the model atmospheres and in the $T_{\rm eff}$ scale. This point is discussed at some length in Gratton (1998).

While the issue of the metallicity is still not completely settled, I think that we may now consider an even more interesting question, that is: how globular clusters got their metals?

2. CLUSTER CHEMICAL EVOLUTION

Scenarios of cluster chemical evolution have been considered by Cayrel (1986), Burkert et al. (1995), and Jehin et al. (1998). In order to understand the relevant mechanisms, we have to consider the formation of stars in clusters, to be used for the construction of a general scheme of chemical evolution. The paradigma is that in clusters low mass stars form first; when first massive stars form and explode as SNe, star formation is stopped. The remaining (metal enriched) gas is lost from the system. Within this paradigma (where a substantial fraction of the initial mass is lost), it is difficult to have tied systems; violent relaxation should also be considered. Furthermore, the role of winds from massive stars is not clear.

Observations of R136 (=30 Doradus: a $10^4\ M_\odot$ cluster a few 10^6 yrs old in the LMC, with no detected SN remnant; likely the best example of a very young, massive cluster we may observe in detail; Rubio et al. 1998) indicate that the strong winds from the most massive stars are able to sweep out all the gas from the cluster core. The shock due to the interaction between the fast wind and the original nebula create a superbubble where new stars form. Winds from these massive stars, as well as from core collapse SNe, do not have the original composition; stars forming in the superbubble might then have at birth a chemical composition different from those formed in the central regions of the cluster. However, it is not obvious that stars forming in the superbubble will remain tied to the cluster.

It is not clear if core collapse SNe indeed participate into this game. A part from the obvious case of ω Cen (the most massive cluster, that might even be the nucleus of a now dispersed dwarf galaxy), evidences for inhomogeneities in Fe-peak elements (that would be a clear indication of a second generation of stars, polluted by the ejecta from core collapse SNe) are not very clean in globular clusters. Langer et al. (1998) found small star-to-star differences at $\sim$ 0.1 dex level in M92. Star-to-star differences have been found also for neutron capture elements with an $r-$ signature, also likely connected to core-collapse SNe, in M15 (Sneden et al. 1997). Upper limits to chemical inhomogeneities (at least for what concern heavy elements) are given by the narrow cluster subgiant branches observed in most clusters when using broad band photometry, since the dependence of colours on metal abundances for these stars is quite strong: $\delta(B-V)/\delta[\mathrm{Fe/H}] \sim 0.2$. In some clusters, this set an upper limit of ~ 0.15 dex for star-to-star variations in [Fe/H]. Finally, recent observations of the Orion association by Cunha et al. (1998), which show large variations in the O abundances with no variation of Fe, may also be relevant, although this a very loose system.

It is not even obvious that star-to-star abundance variations are indeed to be expected in self enrichment scenarios. Following Burkert et al. (1995) and Jehin et al. (1998), the stars we presently observe might all be second generation stars, stars of the first generation being all lost due to severe mass loss occurring during the evolution of the most massive stars. These authors argue that stars in the second generation might then collapse to form a new cluster. The second generation will show signatures of enrichment by core collapse SNe alone; turbulent mixing keep the supershell chemical homogeneous.

3. RELATION BETWEEN GLOBULAR CLUSTERS AND FIELD STARS: HALO INHOMOGENEITIES

There are however stronger evidences for individual chemical evolution for globular clusters. Significant (0.2 dex peak to valley) star-to-star differences in the ratios between $\alpha-$elements and Fe (at a given metallicity) have been found for metal-rich halo field stars ([Fe/H]> -1: Nissen & Schuster 1997), while thick disk stars in the same metal abundance range shows only a small dispersion, with a general large overabundance of $\alpha-$elements. Similarly, a number of $\alpha-$poor (or rather Fe-rich) field stars are known to exist in the metal abundance range $-2 <$[Fe/H]< -1 (King 1997; Carney et al. 1997). A recent survey by Fulbright & Kraft (1998) revealed that these stars are rather uncommon (a similar earlier result was obtained by Gratton 1998). Much larger star-to-star differences have been found amongst the most metal-poor stars ([Fe/H]< -2: McWilliam et al. 1995), differences being largest for n-capture elements.

Amongst globular clusters, the most famous cases are those of Pal 12 and Rup 106, for which nearly solar [α/Fe] abundance ratios have been found by Brown et al. (1997). These clusters are known to be significantly younger than the bulk of the globular clusters (Gratton & Ortolani 1988; Stetson et al. 1989; Buonanno et al. 1993; according to Rosenberg et al. 1999, Pal 12 has about 60% the age of typical globular clusters), and are likely to be accreted objects. Of particular interest is the case of the couple M5-M4 recently considered by Ivans et al. (1999): while these two clusters have very similar Fe abundances, they have clearly different Si abundances (by 0.2 to 0.3 dex). This case is very interesting because M5 and M4 are rather metal-rich clusters ([Fe/H]~ -1.1, according to Carretta & Gratton

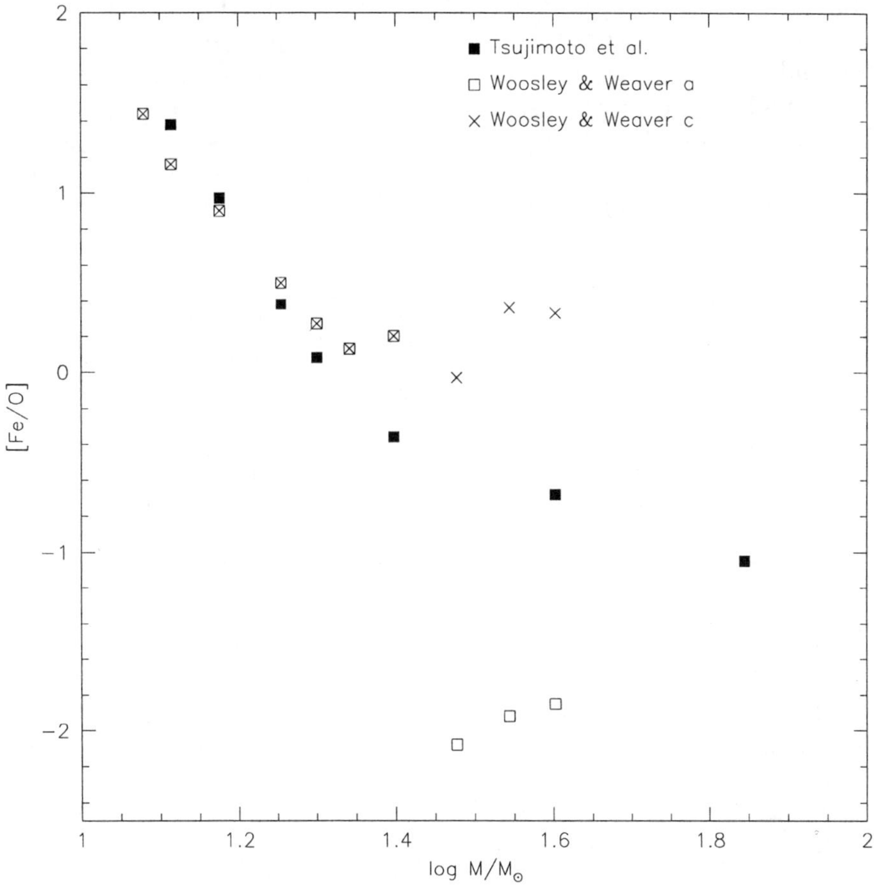

Figure 2. Ratios of the yields for Fe and O from core-collapse SNe as a function of mass according to various authors. The ratios are given with respect to the abundance ratio observed in the Sun

scale). Furthermore, relative ages from Rosenberg et al. (1999) indicate that these two clusters have the same age as the bulk of the globular clusters (within 5%, that is < 1 Gyr).

There are two interpretative schemes for the scatter in [α/Fe]: (i) younger ages (likely the case for Pal 12 and Rup 106); (ii) ISM not well mixed (stochastic variation: McWilliam 1997 scenario, or the more recent model by Tsujimoto et al. 1999). The last model is more appealing for several field stars, as well as for the M5-M4 case. It should be noticed that both scenarios imply a halo forming out of fragments, as proposed by Searle &

Zinn (1978), and are not mutually exclusive. However, while in the first case the low [α/Fe] ratios are attributed to the contribution by type Ia SNe, in the second case they may be simply explained by the stochastic effects of a small number of core collapse SNe, coupled with a different production of $\alpha-$elements and Fe in individual SNe (perhaps, but not necessarily, due to a difference in the progenitor mass; see Nakamura et al. 1999), and then do not require significant age differences.

TABLE 1. Expected scatter in [Fe/O] as a function of mass of fragments having independent chemical evolution

M_{star}	M_{total} [Fe/H]=-2	M_{total} [Fe/H]=-1	N(SNII)	[Fe/O] rms	O_{mass}
10^3	3×10^4	3×10^5	3	0.59	23
10^4	3×10^5	3×10^6	34	0.15	230
10^5	3×10^6	3×10^7	340	0.04	2300

It is interesting to explore the mass scale involved. To this purpose, I considered the SN yields as a function of mass (see Figure 2); I used the theoretical values by Thielemann et al. (1996), and by Woosley & Weaver (1995), although it should be reminded that these values are largely uncertain (due to poor knowledge of the explosion mechanism). Observations also indicate the presence of a wide scatter (Iwamoto et al. 1998; Turatto et al. 1998), but these also are very uncertain, save for SN1987A (see Danziger talk at this meeting). These yields may be coupled with some IMF (e.g. a Salpeter one) in a MonteCarlo model, in order to produce the expected scatter in the abundance ratios. This is done in Table 1 for fragments of different mass. From this table it is clear that signicant (observable) variations are expected when the chemical evolution of fragments with masses $\leq 10^7$ $M_\odot$ can be considered individually. While the value for the typical mass of the fragments depends on the assumed spread in the production from individual SNe and on the assumed IMF (and it is then premature to conclude a value for this quantity simply from the observed spread in [α/Fe] values), it is clear that independent chemical evolution of small fragments may result in observable variations of the abundance ratios, and then to a chemical individuality of each globular cluster. Small fragments (and hence larger inhomogeneities) are most likely to be present in the halo (where density is lower), while inhomogeneities should be much smaller in the thick disk (and in the bulge), where the higher density suggests that clouds having an individual chemical evolution should have a much larger size.

4. BACK TO OBSERVED INHOMOGENEITIES: THE CNO ANTICORRELATION

A quite different issue concerns the inhomogeneities in the abundances of several light elements in globular cluster stars. While first noted in the early '70s, the nature of these inhomogeneities is now becoming slowly cleaner thanks to the patient efforts of a number of investigators. Probably the most notable contribution has been done by the Texas-Lick group (Kraft and coworkers), that in a serie of paper has evidenced the existence of anticorrelations existing between the abundances of several couples of elements (C-N, O-Na, Mg-Al).

These anticorrelations clearly shows that products of the ON-cycle are present in the atmospheres of globular cluster stars (Langer et al. 1993). However, the origin of this material is still unclear (Kraft 1994). Kraft and coworkers noticed that stars exhibiting the O-Na anticorrelation are generally bright giants; they then suggested that the ON-cycle occurs within the same stars where abundance anomalies are observed, and brought to the surface by some deep mixing mechanism (e.g. meridional circulation activated by core rotation). Such a mechanism may work only after the molecular weight barrier created by the deepest inward penetration of the outer convection is canceled by the outward expansion of the H-burning shell (Sweigart & Mengel 1979; Charbonnel 1994). A recent careful work on field stars (Gratton et al. 1999) showed that indeed this mechanism is universally active in metal-poor, low-mass stars; however, at least in field stars such mixing is deep enough to reach the CN-burning region, but not the ON one, so that the abundances of O and Na are left undisturbed. If the correct explanation is indeed deep mixing, it is then necessary that this is systematically larger in cluster stars than in field ones, e.g. because they might rotate faster; however it seems worth to explore alternative mechanisms.

Further strong indications that additional mechanisms are involved come from observations of anomalous abundances in stars less evolved than the bump and down to the main sequence. C and N variations in these stars have been observed since several years (Hesser & Bell 1980; Cannon et al. 1998); very recently, variations in the abundances of other elements have been found in the analaysis of subgiants in M92 using Keck (King et al. 1998). The effect is not limited to a few stars, as shown by the Strömgren photometry data by Grundahl et al. (1998) in several clusters. Since these stars are not evolved enough for having produced themselves these anomalies, these should be due to some not in situ mechanisms.

There are two classes of such mechanisms. The first one consider primordial variations: however this is not appealing because the anomalous abundances of the light elements are not accompanied by corresponding spreads in the heavier elements, that would be expected by e.g. merging two clouds having an original different chemical composition. Much more interest has been raised by self pollution mechanism (Cottrell & Da Costa 1981; D'Antona et al. 1983; Jehin et al. 1998). Within these models, the most appealing donors are the largest amongst intermediate mass stars ($3 < M < 5{-}6\ M_\odot$, lifetime $2 \times 10^7 < t < 5 \times 10^8$ yr), because the outer layers of these stars may be polluted by the products of hot bottom burning, which roughly have the required abundance pattern. These stars produce in fact huge amounts of N, and possibly an O-Na anticorrelation. A further advantage of these objects is that they have low main sequence $T_{\rm eff}$'s and lifetimes as post-AGB stars so short that they do not ionize large amounts of gas and they end their lifes as PN, with massive low velocity winds. They will then inject low energy into the ISM. Also expected evolution of mass loss by stars within a cluster reaches a maximum after 10^8 yr, so that in this phase we have the highest probability of accretion (virtually no gas is presently observed in globular clusters). The main problem with these scenarios is due to scarce knowledge of the accretion mechanisms. In the rest of this talk, I will explore a little bit some of the expected features of these models.

4.1. SPHERICAL ACCRETION

Spherical accretion (Bondi accretion) has been considered by D'Antona et al. (1983); it requires low velocities and high densities. The basic formula giving spherical accretion (Lang 1974) is:

$$dm/dt = 6.4 \times 10^{-8}\ M^2\ \rho_g/v^3 \tag{1}$$

Where m is the accreted mass (in $M_\odot$); t is time in Myr; M is the stellar mass (in $M_\odot$); ρ_g is the gas density (in $M_\odot/{\rm pc}^3$); and v is the quadratic sum of the relative velocity of the star and gas, and of the sound velocity (of the order of 1-10 km/s), in units of 10 km/s.

This formula may explain while accretion is not active for field stars (the existence of extremely metal-poor dwarfs having tiny subatmospheric convection zones sets upper limits to accretion in these stars at $< 10^{-8}\ M_\odot/{\rm Myr}$), since in that case it may be stopped by stellar winds: for instance, the Sun looses through its wind $\sim 2\,10^{-8}\ M_\odot/{\rm Myr}$, at a velocity of $\sim 5\,10^2$ km/s. Momentum due to this wind should prevent accretion unless the accretion rate is higher than $\sim\ 10^{-6}\ M_\odot/{\rm Myr}$ (for a typical relative velocity of 10 km/s). This requires a quite high density of 15 $M_\odot/{\rm pc}^3$, much higher than the typical density of the local interstellar medium (ISM: $< 10^{-1}\ M_\odot/{\rm pc}^3$). The critical density for accretion scales with the square of the relative velocities between the accreting star and the ISM: it is then $\sim 10^3$ larger for field halo stars. Even a wind much weaker than the solar one should then be enough to prevent accretion on these objects.

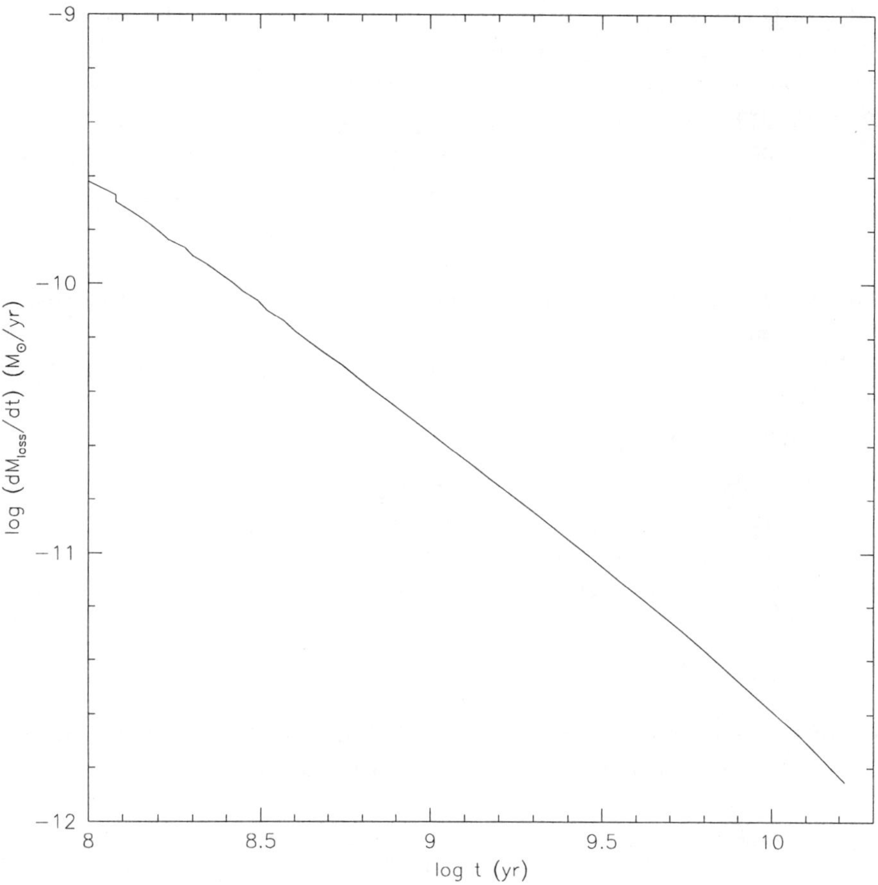

Figure 3. Mass loss rate from a stellar population having a Salpeter mass function, as a function of time. Note the strong decrease in the mass loss rate as the population ages. If we assume that all this mass ends up in the cluster core (having a volume of ~ 1 pc^3), and that is wiped out at every passage through the galactic disk (once every 10^8 yr), the maximum gas density is $\rho_g = 10^{13} dM_{loss}/dt\ M_\odot/\mathrm{pc}^3$ for a cluster of $10^5\ M_\odot$

However high densities coupled with low relative velocities between the stars and the ISM were possible in the core of globular clusters during the accretion epoch, perhaps allowing accretion to occur. In this case, gas density ρ_g may be estimated using the following equation:

$$\frac{d\rho_g}{dt} = \frac{dm_{lost}}{dt} - \frac{dm_{acc}}{dt} - \frac{dm_{wind}}{dt}. \tag{2}$$

On the right side of this equation, the first term represents the mass lost and the second the mass accreted by clusters stars, and the third term represents the gas lost by the cluster. When using this equation, I will assume that all gas lost by stars istantaneously ends up in the cluster core (that has a typical volume of $\sim 1\ \mathrm{pc}^3$).

Total mass lost m_{lost} may be written as:

$$\frac{dm_{lost}}{dt} = k(t)\, m_{tot} \tag{3}$$

where m_{tot} is the total mass in stars in the cluster. I estimated (using a Salpeter mass function) that after 10^8 yr of cluster life is $k \sim 2.5 \times 10^{-4}\ \mathrm{Myr}^{-1}$ (if t is in Myr); it scales approximately inverse with time (see Figure 3).

Total accreted mass is:

$$\frac{dm_{acc}}{dt} = 6.4 \times 10^{-8}\, m_{tot} \int \psi(M)\, M^2 \frac{\rho_g}{[v(M)]^3}\, dM, \tag{4}$$

where $\psi(M)$ is the mass function, normalized so that:

$$\int M\, \psi(M)\, dM = 1.$$

Assuming energy equipartition, and a central velocity dispersion σ_V (in units of 10 km/s) for a star of 1 $M_\odot$, it is possible to write:

$$\frac{dm_{acc}}{dt} = 6.4 \times 10^{-8}\, m_{tot} \frac{\rho_g}{\sigma_V^3} \int \psi(M)\, M^{7/2}\, dM, \tag{5}$$

Finally, the mass lost by the cluster dm_{wind}/dt may be obtained by e.g. assuming that all gas is swept out every time the cluster passes through the galactic plane (disk shocking).

Figure 4 describes the expected time evolution of ρ_g for a cluster of $10^5\ M_\odot$, in the hypothesis that all gas ends up in the cluster core, assumed to have a volume of 1 pc^3; in this simulation I assumed a central velocity dispersion of $\sigma_V = 10$ km/s for 1 $M_\odot$ stars. I assumed that all gas is swept out every time the cluster passes through the galactic plane and I plotted time averages over the interval between two consecutive passages. I considered two cases: in the first one (solid line) the interval between two consecutive passages is 10^8 yr; in the second one (dashed line) is half that. These are typical values for outer and inner halo clusters respectively. As it can be seen, for cluster of this size gas density decreases approximately linearly with time (this is not a general property of this model: for instance, $\rho_g \sim t^{-2/3}$ if the cluster mass is $10^6\ M_\odot$, having the same core diameter and velocity dispersion). During the first Gyr of cluster lifetime, the value

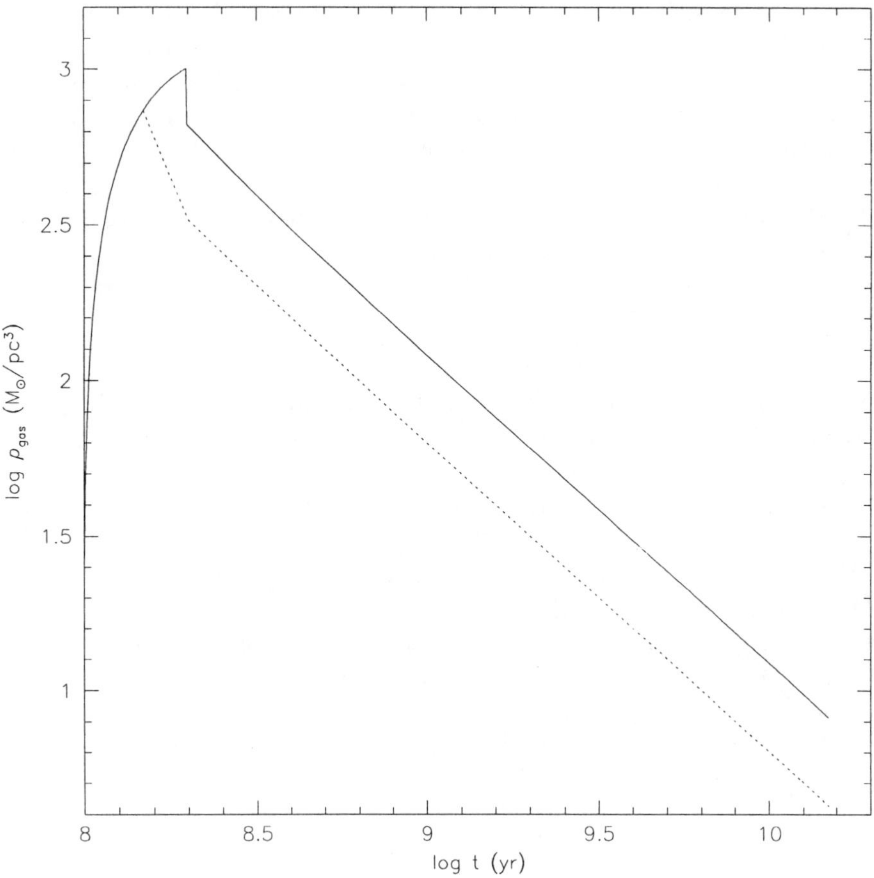

Figure 4. Expected run of ρ_g for a cluster of 10^5 $M_\odot$, in the hypothesis that all gas ends up in the cluster core, assumed to have a volume of 1 pc^3, as a function of time. I assumed a typical velocity of 10 km/s for 1 $M_\odot$ stars. I assumed that all gas is swept out every time the cluster passes through the galactic plane and (after this first passage) I plotted time averages over the interval between two consecutive passages. I considered two cases: in the first one (solid line) the interval between two consecutive passages is 10^8 yr; in the second one (dashed line) is half that

of gas density (in the cluster core) is much larger than the threshold value required to overcome the wind momentum: spherical accretion should then be possible even in the presence of a stellar wind stronger than the solar wind. However, it is not likely to be any more active at present time; in the following, I will assume that the accretion phase ends after about 1.5 Gyr (when 2 $M_\odot$ stars are in evolution; in our example this corresponds to threshold gas densities of $\rho_g \sim 50\ M_\odot \mathrm{pc}^3$, three times higher than expected for a wind identical to the solar one). Note also the low expected gas density after > 10 Gyr: this is due to the low mass loss rate predicted after such a long time; this may explain the lack of gas at present in globular clusters.

Note that Bondi accretion assumes only gravitational effects and is then valid only for neutral gas; however this seems a reasonable assumption insofar the gas density is high. In fact, we may write the fraction of gas f_{ion} that is ionized in the core of a cluster as:

$$f_{ion} = \frac{f(t)\, M_{tot}}{T_e\, \rho_g^2\, V}, \tag{6}$$

where T_e is the electron temperature (in K: $\sim 10^4$ for ionized material), V is the volume of the cluster core (~ 1 pc^3), and $f(t)$ is a function representing the evolution of the ionizing flux of the stellar population (per unit solar mass of material). f evolves approximately as t^{-3} and in the units we are considering is $\sim 10^{-4}$ after 10^8 yr. We then have $f_{ion} << 1$ when $\rho_g > 1\ M_\odot/\mathrm{pc}^3$.

Once the evolution of ρ_g with time is known, we may integrate equation (1) (assuming that accretion occurs during the range of ages from 100 Myr to 1.5 Gyr; note that these limits are not critical, since the total accreted mass depends only weakly on them), and get the total (average) accreted mass. For the typical cluster of $10^5\ M_\odot$, $\sigma_V = 10$ km/s for 1 $M_\odot$ stars, core volume 1 pc^3 we are considering, total (average) accretion is:

$$m_{acc} = 0.023\, (M/M_\odot)^{7/2} \tag{7}$$

if passages through the galactic plane occur every 10^8 yr, and half that if the passages occur every $5\,10^7$ yr (accretion rate is about 5 times larger for a cluster of $10^6\ M_\odot$). A typical star now evolving on the RGB, presently of $\sim 0.9\ M_\odot$, should have accreted on average $0.016\ M_\odot$ of gas. This value is comparable with the mass of the convective envelope of a TO star, while it is smaller then the mass of the convective envelope on the RGB: so we expect that if accretion is important, most stars at the TO of GCs should exhibit abundance anomalies, while only a few RGB stars should be anomalous. However, this is an average value: in the following we will see how different effects might change this simple prediction.

4.2. THE ROLE OF BINARIES: BIMODAL DISTRIBUTION

An important costraint in the accretion (as well as mixing) scenarios is that they should explain the observed spread in abundances. Accretion is a strong function of velocity (stellar mass), and hence most material will be accreted by the low velocity stars residing in the core. Then significant spread is expected if the relaxation time is longer than the accretion time ($\sim 10^9$ yr). But the two-body relaxation time is $\sim 10^8$ yr; then rather small spread in abundances are expected because all small mass stars have similar integrated accretion rates.

This problem may be circumvented considering binaries: in fact accretion rates on individual components of equal mass binaries is a factor $2^{5/2} = 5.7$ larger than on single stars, due to the larger mass of the system. Observations (Pryor et al. 1989; Kaluzny et al. 1998; Rubestein & Baylin 1997; Ferraro et al. 1997) indicate that $\sim 15\%$ of stars in clusters are binaries (may be $\sim 25\%$ in the cluster core where accretion occurs: Fischer et al. 1994; Cote et al. 1994): then binaries contribute more than half (and up to 90%) of the total accretion within a cluster. Binaries in cluster cores are likely not primordial, but rather result from dynamical evolution within the cluster. This favours the most massive (equal-mass) systems. The binary systems we are interested in are quite wide systems, easily disrupted within a cluster (we are interested in systems having typical lifetimes of a few $\sim 10^8$ yr; likely with separations of 1-10 AU); they probably constitutes the bulk of binaries in the cluster cores.

A present TO star that was member of a binary system during the accretion epoch in the typical cluster of $10^5\ M_\odot$, $\sigma_V = 10$ km/s, core volume 1 pc^3, should have accreted on average 0.09 $M_\odot$ if passages through the galactic plane occur every 10^8 yr, and half that if the passages occur every $5\,10^7$ yr (accretion may be as large as several tenths of $M_\odot$ for a cluster of $10^6\ M_\odot$ velocity dispersion of $\sigma_V = 10$ km/s, and passing through the galactic plane every 10^8 yr). This mass is of the same order of magnitude of the mass of convection envelopes in red giants, so that substantial modifications in the surface abundances of red giants should be expected.

Within each cluster we should then expect a roughly bimodal distribution:

- a group of heavily contaminated stars, those that were components of binary systems in the cluster core during the accretion phase. These stars have accreted a mass of the same order than the mass of the RGB convective envelope ($\sim 0.3\ M_\odot$). This group may include a substantial fraction of the cluster members, because several generations of binaries may be considered (however, the number of heavily contaminated stars may be substantially reduced by ejection of one binary component from the cluster during binary disruption)
- all the other stars will be about an order of magnitude less contaminated; dilution in the convective envelope at the base of RGB roughly bring their surface abundances back to the original values

Several factors may dilute this bimodal distribution: (i) binaries may be not stable over the whole accretion phase; (ii) mass ratio is likely represented by a distribution of values; (iii) stars dynamical parameters are distributed over a range of values. Furthermore, overall cluster parameters (mass, concentration, orbital period) are important factors modulating the relevance of accretion.

4.3. RECYCLING

A difficulty in the Bondi accretion scenario is that a significant fraction of the accretion on stars with present mass in the range 0.8-1.0 $M_\odot$ is due to material ejected from stars with 2-3 $M_\odot$, and occurs in a phase from 0.5 to 1.5 Gyr after cluster formation. Current models for these stars (and observations of MC clusters: see the plot by Marigo et al. 1996) indicate that these stars undergo triple-α reactions (producing large quantities of primary C) and significant amounts of primary s-elements. Accretion scenario may be saved if the threshold for accretion is rather large ($\rho_g \sim 100\ M_\odot \mathrm{pc}^3$) and/or mass lost by stars with 3-5 $M_\odot$, mostly accreted by stars with initial mass in the range 2-3 $M_\odot$, is recycled. Note also that within this scenario, a star with initial mass in the range 2-3 $M_\odot$ may easily accrete $\sim 1\ M_\odot$ of gas, so that its evolution (and likely composition of its wind) is greatly modified.

Main predictions of this scenario are:

- abundance anomalies will be seen only in clusters that (at this early epochs) were able to retain the ejecta of PNe and IMS winds, that is with escape velocities $> 10 - 20$ km/s, that is a few $10^4\ M_\odot$. No anomaly is expected in some very small outer halo clusters and in open clusters
- interaction with ISM (mainly in the disk) and disk shocking likely increase this threshold mass for clusters with abundance anomalies within the disk and in Magellanic Clouds
- since disruption of binaries may easily result in ejection from the cluster, a small fraction ($\sim 10^{-3}$) of field stars should exhibit the same abundance anomalies (CNO anticorrelation)
- the incidence of abundance anomalies should be correlated with cluster dynamical parameters, although more detailed modelling is required to make firmer predictions
- extreme cases of abundance anomalies should exhibit the abundance pattern typical of PNe, with most C and O converted into N, and the $^{12}C/^{13}C$ isotopic ratio at the equilibrium value
- predictions for Li and s-process elements are more difficult since they depend on the range of masses considered for the donors
- a fraction of the cluster stars should have deep He-rich, H-poor envelopes. This should influence their evolution on the RGB and on HB

(where we expect them to be bluer and brighter)
- since most cluster stars should accrete some H-poor material it is not obvious that the surface abundances for low accretion objects (interpreted as "single" stars during the accretion phase) reflects the original cluster abundance when the convective envelopes are tiny (as in TO stars). Lower RGB stars after the dilution episode due to inward expansion of the convective envelope should be closer to it
- safe data on original Li may be derived from TO stars only in the less massive clusters

4.4. ACCRETION BY PROTOSTELLAR DISKS

Suppose Bondi accretion does not works (e.g. due to strong stellar winds, or to reionization of the ISM). An alternative accretion scenario might consider proto-stellar disks. Size and lifetime of disks can be obtained by considering disruption by close passages of other stars (less than 10 times the disk radius). The number of encounters is:

$$n \sim 6 \times 10^{-5} a^2 t \rho_* v, \tag{8}$$

where a is closest encounter distance (in AU); t is time (in Myr); ρ_* is star density (in units of 10^5 star/pc^3); and v is velocity (in units of 10 km/s). In these conditions, we expect typically 1 encounters within 12 AU every 100 Myr. Typical proto-stellar disk sizes are $\sim 1-2$ AU. Also, disks of this size are likely to survive SN explosions, while larger disks are disrupted.

For protostellar disks, we will consider geometric accretion cross section:

$$dm/dt \sim 8 \times 10^{-10} a^2 v \rho_g \quad M_\odot/\mathrm{Myr} \tag{9}$$

with the same units as above, and ρ_g is the gas density (in units of $M_\odot$/pc^3)

In this case ρ_g cannot be computed from equilibrium conditions, that are never reached, because accretion is not efficient enough. We may rather assume that every 10^8 yr the residual gas is cleared by e.g. passage through the galactic disk (although it is not clear that a disk existed in this early phases). In this case, given the mass loss rate, average gas densities of $\rho_g \sim 4 \times 10^4$ $M_\odot$/pc^3 may be obtained in the core ($r_c = 0.5$ pc) of a cluster of 10^6 $M_\odot$ (values as large as 10^5 $M_\odot/pc^3$ seem feasible in the most massive clusters).

With this ρ_g, typical accretion rates for individual stars (assuming protodisks of ~ 2 AU) are:

$$dm_{acc}/dt \sim 1.3 \times 10^{-4} \quad M_\odot/\mathrm{Myr} \tag{10}$$

Typical accretion in 10^8 yr (the lifetime of protodisks) may then be 0.013 $M_\odot$ (this value is independent of the stellar mass), with values as large as

0.05 $M_\odot$ or even perhaps 0.1-0.2 $M_\odot$ possible in the most favourable and rare cases where a large protodisk survive long enough. Of course only a fraction of the stars should have retained even a small protodisk for such a long time

Since this mechanism only works in very early phases of cluster evolution, the donors are in this case stars with masses of about 5 $M_\odot$. Stars of this mass in the LMC never become carbon stars (likely due to hot bottom burning, transforming ^{12}C into ^{14}N): their abundance pattern matches quite well the abundance anomalies seen in GCs. However, average accretion is so low that no important changes are expected for typical RGB stars with massive convective envelopes.

Within this scenario, average accretion is a function of total cluster mass, although maybe limited by disturbances causing loss of gas.

5. SUMMARY

The chemical evolution history of clusters is a complex scenario. The interplay between star formation, stellar evolution (with mass loss), and cluster dynamical evolution must be considered. Critical tests would be to derive abundances in dwarfs and lower RGB stars, and compare them with those obtained for upper RGB stars; this may clear the role of mixing, and give precious information about dilution at the base of the RGB, that is important to understand if pollution and diffusion are indeed important. This is now possible thanks to UVES at VLT.

Accretion scenarios are interesting possibilities to explain abundance anomalies in GCs. Bondi accretion (if any) could be large enough to produce large abundance anomalies in GCs. The mechanism have a very strong mass dependence. Binaries may help to produce bimodal distributions. However, if accretion on more massive stars is not suppressed, most of material accreted by stars presently on the RGB should come from stars with masses in the range 2-3 $M_\odot$. Recycling of material may help to solve this concern An alternative, less appealing scenario considers accretion by proto-stellar disks.

References

1. Barbuy, B., Renzini, A., Ortolani, S., Bica, E., & Guarnieri, M.D. 1999, A&A, 341, 539
2. Brown, J.A., Wallerstein, G., & Zucker, D. 1997, AJ, 114, 180
3. Burkert, A., Brown, J., & Truran, J.W. 1995, ApJ, 440, 666
4. Buonanno, R., Corsi, C.E., Fusi Pecci, F., Richer, H.B., & Fahlman, G.G. 1993, AJ, 105, 184
5. Cannon, R.D., Croke, B.F.W., Bell, R.A., Hesser, J.E., & Stathakis, R.A. 1998, MNRAS, 298, 601

6. Carney, B.W., Wright, J.S., Sneden, C., Laird, J.B., Aguilar, L.A., & Latham, D.W. 1997, AJ, 114, 363
7. Carretta, E., & Gratton, R.G. 1997, A&AS, 121, 95
8. Carretta, E., Cohen, J.G., Gratton, R.G., Behr, 2000
9. Cayrel, R. 1986, A&A, 168, 81
10. Charbonnel, C., 1994, A&A, 282, 811
11. Cohen, J.G., Gratton, R.G., Behr, B.B., Carretta, E. 1999, ApJ, 523, 739
12. Cote, P. Welch, D.L., Fischer, P., Da Costa, G.S., Tamblyn, P. Seitzer, P., & Irwin, M.J. 1994, ApJS, 90, 83
13. Cottrell, P.L., & Da Costa, G.S. 1981, ApJ, 245, 79
14. Cunha, K., Smith, V.V., & Lambert, D.L. 1998, ApJ, 493, 195
15. D'Antona, F., Gratton, R., & Chieffi, A. 1983, MSAIt, 54, 173
16. Ferraro, F., Carretta, E., Fusi Pecci, F., Zamboni, A. 1997, A&A, 327, 528
17. Fischer, P., Welch, D.L., Mateo, M., & Cote, P. 1994, AJ, 106, 1508
18. Fulbright, J.P., & Kraft, R.P. 1998, AAS, 193, 6812
19. Gratton, R.G., 1998, MSAIt, 69, 145
20. Gratton, R.G., & Ortolani, S. 1988, A&AS, 73, 137
21. Gratton, R.G., Sneden, C., Carretta, E., Bragaglia, A. 1999, submitted to A&A
22. Grundahl, F., VandenBerg, D.A., Andersen, M.I. 1998, ApJL, 500, L179
23. Harris, W.E. 1996, AJ, 112, 1487
24. Hesser, J.E., & Bell, R.A. 1980, ApJ, 238, 149
25. Ivans, I.I., Sneden, C., Kraft, R.P., Suntzeff, N.B., Smith, V.V., Langer, G.E., & Fulbright, J.P. 1999, AJ, 118, 1273
26. Iwamoto et al. 1998, Nature, 395, 672
27. Jehin, E., Magain, P., Neuforge, C., Noels, A., & Thoul, A.A. 1998, A&A, 330, 33
28. Kaluzny, J., Hilditch, R.W., Clement, C., & Rucinski, M.S. 1998, MNRAS, 296, 347
29. King, J.R. 1997, AJ, 113, 2302
30. King, J.R., Stephens, A., Boesgaard, A.M. 1998, AJ, 115, 666
31. Kraft, R.P. 1994, PASP, 106, 553
32. Kurucz, R. 1992, CD-ROM 13
33. Lang, K.R., 1974, Astrophysical Formulae (Berlin: Springer)
34. Langer, G.E., Hoffma, R., & Sneden, C. 1993, PASP, 105, 301
35. Langer, G.E., Fischer, D., Sneden, C., & Bolte, M. 1998, AJ, 115, 685
36. Marigo, P., Bressan, A., & Chiosi, C. 1996, MSAIt, 67, 713
37. McWilliam, A. 1997, ARA&A, 35, 503
38. McWilliam, A., Preston, G.W., Sneden, C., Searle, L. 1995, AJ, 109, 2757
39. Nakamura, T., Umeda, H., Nomoto, K., Thielemann, F.-K., & Burrows, A., 1999, ApJ, 517, 193
40. Nissen, P.E., & Schuster, W.J. 1997, A&A, 326, 751
41. Pryor, C., McClure, R.D., Hesser, J.E., Fletcher, J.M. 1989, in Dynamics of Dense Stellar Systems (Cambridge: Cambridge University Press) p. 175
42. Rosenberg, A., Saviane, I., Piotto. G., Aparicio, A. 1999, AJ, in press
43. Rubestein, E.P., & Baylin, C.D. 1997, ApJ, 474, 701
44. Rubio, M., Barbá, R.H., Walborn, N.R., Probst, R.G., García, J, & Roth, M.R. 1998, AJ, 116, 1708
45. Rutledge, G.A., Hesser, J.E., & Stetson, P.B. 1997, PASP, 109, 907
46. Searle, L., & Zinn, R. 1978, ApJ, 225, 357
47. Sneden, C., Kraft, R.P., Shetrone, M.D., Smith, G.H., Langer, G.E., & Prosser, C.F. 1997, AJ 114, 1964
48. Stetson, P.B., Hesser, J.E., Smith, G.H., VandenBerg, D.A., Bolte, M. 1988, AJ, 97, 1360
49. Sweigart, A.V., & Mengel, J.G. 1979, ApJ, 229, 624
50. Thielemann, F.-K., Nomoto, K., & Hashimoto, M. 1996, ApJ, 460, 408
51. Tsujimoto, T., Shigeyama, T., & Yoshii, Y. 1999, ApJ, 519, 63

52. Turatto, M., et al. 1998, ApJL, 498, L129
53. Woosley, S.E., & Weaver, T.A. 1995, ApJS, 101, 181
54. Zinn, R., & West, M.J. 1984, ApJS, 55, 45

Speaker : R. GRATTON

S. CASSISI: The accretion mechanism could have been at work in high density clusters, so in which way can we explain the presence of high density clusters, like 47 Tuc, which do not show anomalies in the abundance patterns?

R. GRATTON: Abundance anomalies are clearly present amongst main sequence and turn-off stars of 47 Tuc.

ON THE GLOBULAR CLUSTER IMF BELOW 1 $M_\odot$

FRANCESCO PARESCE AND GUIDO DE MARCHI
European Southern Observatory
Karl-Schwarzschild Str. 2, D–85748 Garching, Germany

1. Abstract

Accurate luminosity functions (LF) for a dozen globular clusters have now been measured at or just beyond their half-light radius using HST. They span almost the entire cluster main sequence (MS) below $\sim 0.75\,M_\odot$. Transformation of the LF into mass functions (MF) by means of the available mass luminosity (ML) relations that are consistent with all presently available data on the physical properties of low mass, low metallicity stars shows that all the LF observed so far can be obtained from MF having the shape of a log-normal distribution with characteristic mass $m_c = 0.33 \pm 0.03\,M_\odot$ and standard deviation $\sigma = 0.34 \pm 0.04$. After correction for the effects of mass segregation, the variation of the ratio of the number of higher to lower mass stars with cluster mass or any simple orbital parameter or the expected time to disruption shows no statistically significant trend over a range of this last parameter of more than a factor of ~ 100. We conclude that the global MF of these clusters have not been measurably modified by evaporation and tidal interactions with the Galaxy and, thus, should reflect the initial distribution of stellar masses.

2. Introduction

The IMF is a critical ingredient in our understanding of a large number of basic astronomical phenomena. It plays a dominant role in any star formation theory as the end result of molecular cloud contraction and fragmentation.The actual measurement of a MF is a complex process whose ultimate precision and reliability rests heavily on a very careful quantitative analysis of all sources of possible random and systematic error. For many reasons, it has proven very difficult to pin down the shape of the IMF observationally with the required reliability and accuracy in a wide variety of stellar envir-

F. Matteucci and F. Giovannelli (eds.), The Evolution of the Milky Way, 185–196.

onments (Scalo 1998, 1999). The slope of the MF at the lowest mass end of the stellar MS and, in particular, whether or not there is a turn-over at the lowest masses before the H-burning limit and whether or not the IMF is universal or rather depends on the initial physical conditions in the natal environment are critical open issues at present.

Globular clusters represent, in principle, the ideal sample from which to deduce the stellar IMF and properly answer the above questions. They offer a large statistically significant sample of relatively bright, coeval, equidistant stars with, in most cases, relatively small variations of chemical composition and extinction within each cluster. They were all formed very early in the history of the Galaxy and there is no evidence of subsequent star formation episodes. The binary fraction outside the core is less than $10-15\,\%$ and has an insignificant effect on the measured LF. Mass segregation is a relatively straightforward and well understood phenomenon quantifiable by simple Michie–King models. The only potentially serious obstacle is related to the possible modification of the IMF by the effect of tidal interactions with the Galaxy potential. This interaction, integrated over the orbit and time, is expected to slowly decrease the slope of the global mass function of the cluster (Vesperini 1998) thereby effectively masking the original IMF from our present day observations, no matter how precise and detailed they are.

Since deep LF of a dozen globular clusters in our Galaxy have now been accurately measured, we are in a good position to address observationally the issue of if and, possibly, how the interaction history of these clusters, whose Galactic orbits are reasonably well known, affects their LF in the mass range where the signature is expected to be most significant. In this talk, we show that LF obtained at or just beyond the half-light radius of these clusters surveyed are completely insensitive to this history and that they can indeed be used to deduce an uncontaminated stellar IMF below $1\,M_\odot$ for these stars.

3. Observational Data

The main characteristics of the data used for this study are summarized in Table 1 and their relevant presently available physical parameters are listed in Table 2. All the listed clusters have well measured LF in the critical range $6.5 < M_I < 10$ and some even well beyond these limits. The LF of these clusters in number per 0.5 magnitude bins as determined by analysis of their color–magnitude diagrams are plotted with the relevant $1\,\sigma$ error-bar in Figure 1, as a function of the absolute I-band magnitude obtained using the distance moduli given in Table 2. The family of curves shown in Figure 1 represents a very homogeneous sample of objects all observed and analyzed

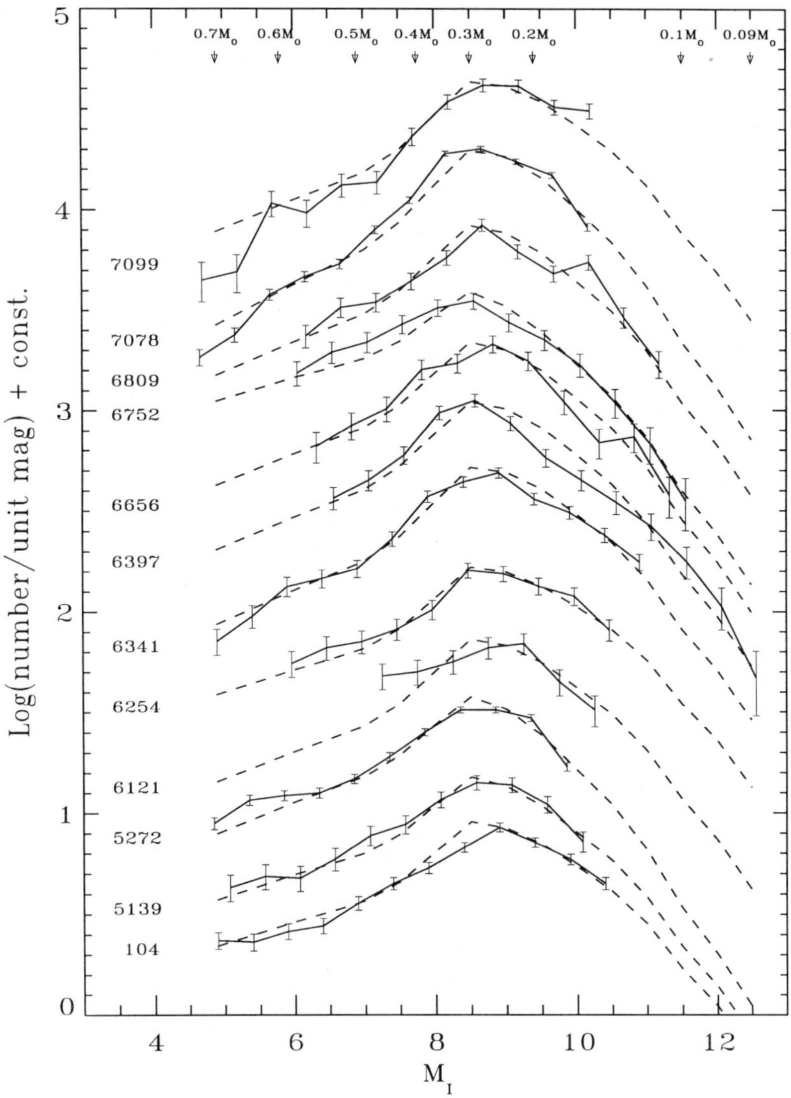

Figure 1. Luminosity functions of the clusters in our sample (see Table 1). The data have been shifted vertically by an arbitrary amount for enhanced visibility. The mass values shown at the top are taken from Baraffe et al. (1997) for $[Fe/H] = -1.5$. The dashed lines show the result of folding the log-normal MF of Figure 2 through the ML relation of Baraffe et al. (1997)

with the same basic techniques well outside the core in a region where the low-mass MS is well populated. The most obvious feature of the observed

TABLE 1. The clusters in our sample. Columns 2 and 3 show the radial distance at which the LF have been measured, respectively in arcmin and in units of the half-light radius (r_h; see Table 2). LF have been measured within $\pm 1\, r_h$ of the given average position r.

NGC	Name	r	r/r_h	Reference
104	47 Tuc	4.6	1.6	De Marchi & Paresce (1995b)
5139	ω Cen	4.6	0.9	De Marchi (1999)
5272	M 3	1.5	1.5	Marconi et al. (1997), Carretta et al. (1999)
6121	M 4	6.1	1.3	Pulone, De Marchi & Paresce (1999)
6254	M 10	2.4	1.3	De Marchi & Paresce (1996)
6341	M 92	4.6	4.5	Piotto, Cool & King (1997)
6397		4.6	1.8	Paresce, De Marchi & Romaniello (1995)
6656	M 22	2.6	0.8	De Marchi & Paresce (1997)
6752		3.1	1.5	Ferraro et al. (1997)
6809	M 55	2.5	0.9	De Marchi & Paresce (1996)
7078	M 15	4.6	4.6	De Marchi & Paresce (1995a)
7099	M 30	4.6	4.6	Piotto, Cool & King (1997)

LF is the peak located at $M_I \simeq 8.5 - 9$ with a rising and descending part on each side.

4. Conversion to a Mass Function

The observed local LF shown in Figure 1 can be converted into the corresponding MF by the application of a mass-luminosity relation (ML). The critical step here, therefore, is intimately connected to the proper realization of the appropriate ML relation for the age and metallicity of the cluster. A number of possibilities exists presently but the most reliable are the theoretical ML relations explicitly computed for the appropriate observational bandpasses by Baraffe et al. (1997, 1998) and Cassisi et al. (1999). The two models are,fortunately, practically indistinguishable from one another in the I band and our mass range. This consistency between independent determinations increases our confidence in the basic reliability of the M-L relation used here. We assess the validity of a model MF by converting it into the observational plane and comparing directly the resulting LF with the observed one.

In Figure 2, we show the log-normal distributions that accurately reproduce the LF plotted in Figure 1 over the whole magnitude range spanned by the observations. Solid lines mark the portion of the MF that have been fitted to the data, while the dashed lines represent the extrapolation of the

TABLE 2. Clusters' structural parameters. Columns are as follows: (1) NGC number; (2) distance modulus in the I band defined as $(m - M)_V + 0.48A_V$, with the latter two values taken from Djorgovski (1993); (3) half-light radius in arcmin (Djorgovski 1993); (4) and (5) core radius in arcmin and concentration ratio (Webbink 1985); (6) and (7) distance in kpc respectively from the Galactic plane and center (Djorgovski 1993); (8) perigalactic distance in kpc (Dinescu et al. 1999); (9) metallicity (Djorgovski 1993); (10) and (11) time to disruption in Gyr respectively from Gnedin & Ostriker (1997) and Dinescu et al. (1999); (12) and (13) average characteristic mass (solar units) and variancestandard deviation of the log-normal distribution that best fits the MF; (14) logarithmic ratio ΔN of lower to higher mass stars

NGC	$(m\text{-}M)_o$	r_h	r_c	c	Z_G	R_G	P	[Fe/H]	T_d	T_d	m_c	σ	ΔN
104	13.37	2.9	0.2	2.48	3.2	7.4	5.3	-0.71	88	131	0.34	0.40	0.13
5139	13.68	4.9	2.6	1.15	1.3	6.3	1.2	-1.59	40	16	0.37	0.37	0.12
5272	15.03	1.2	0.4	1.89	10.0	12.0	5.5	-1.66	213	275	0.36	0.30	0.25
6121	12.11	4.5	1.2	1.53	0.5	6.2	0.7	-1.33	13	2	0.32	0.35	0.21
6254	13.59	1.9	0.7	1.66	1.7	4.7	3.4	-1.60	22	23	0.33	0.41	0.14
6341	14.41	1.0	0.3	1.65	4.3	9.1	1.4	-2.24	30	33	0.30	0.33	0.10
6397	12.00	2.9	0.1	1.69	0.5	6.0	3.1	-1.91	4	4	0.32	0.33	0.23
6656	12.93	3.3	1.2	1.70	0.4	5.1	2.9	-1.75	31	29	0.33	0.33	0.21
6752	13.18	2.0	0.5	2.15	1.8	5.1	4.8	-1.54	35	96	0.42	0.35	0.09
6809	13.57	2.7	1.7	1.27	1.9	4.2	1.8	-1.82	14	11	0.32	0.32	0.25
7078	15.19	1.0	0.1	1.77	4.8	10.5	5.5	-2.17	48	155	0.30	0.28	0.14
7099	14.43	1.0	0.1	2.40	5.4	6.8	3.2	-2.13	23	40	0.30	0.35	0.16

same MF to fill in the range $0.09 - 0.75\,\mathrm{M}_\odot$. The dashed lines in Figure 1 represent the best fit LF obtained with the MF shown in Figure 2. The log-normal MF is characterized by only two parameters namely the characteristic mass m_c and the standard deviation σ and takes on the form:

$$\ln f\,(\log m) = A - \frac{[\log(m/m_c)]^2}{2\sigma^2} \tag{1}$$

where A is a normalization constant. The average values of the parameters for this sample of clusters are $< m_c >= 0.33 \pm 0.03$ and $< \sigma >= 0.34 \pm 0.04$. A simple, unbiased measure of the steepness of the rise to the maximum of the MF shown in Figure 2 that does not depend on any preconceived notion on the shape of the MF is ΔN, defined as the logarithmic ratio of the number of lower to higher mass stars taken from the MF in the mass range $m = m_c$ and $m = 0.7\,\mathrm{M}_\odot$. This is probably the most convenient parameter to describe the region of the mass distribution most likely to be affected by external and internal dynamics and is listed in Table 2 for each cluster. Another advantage of ΔN is that it is defined in a mass range where the stellar surface structure is best understood and

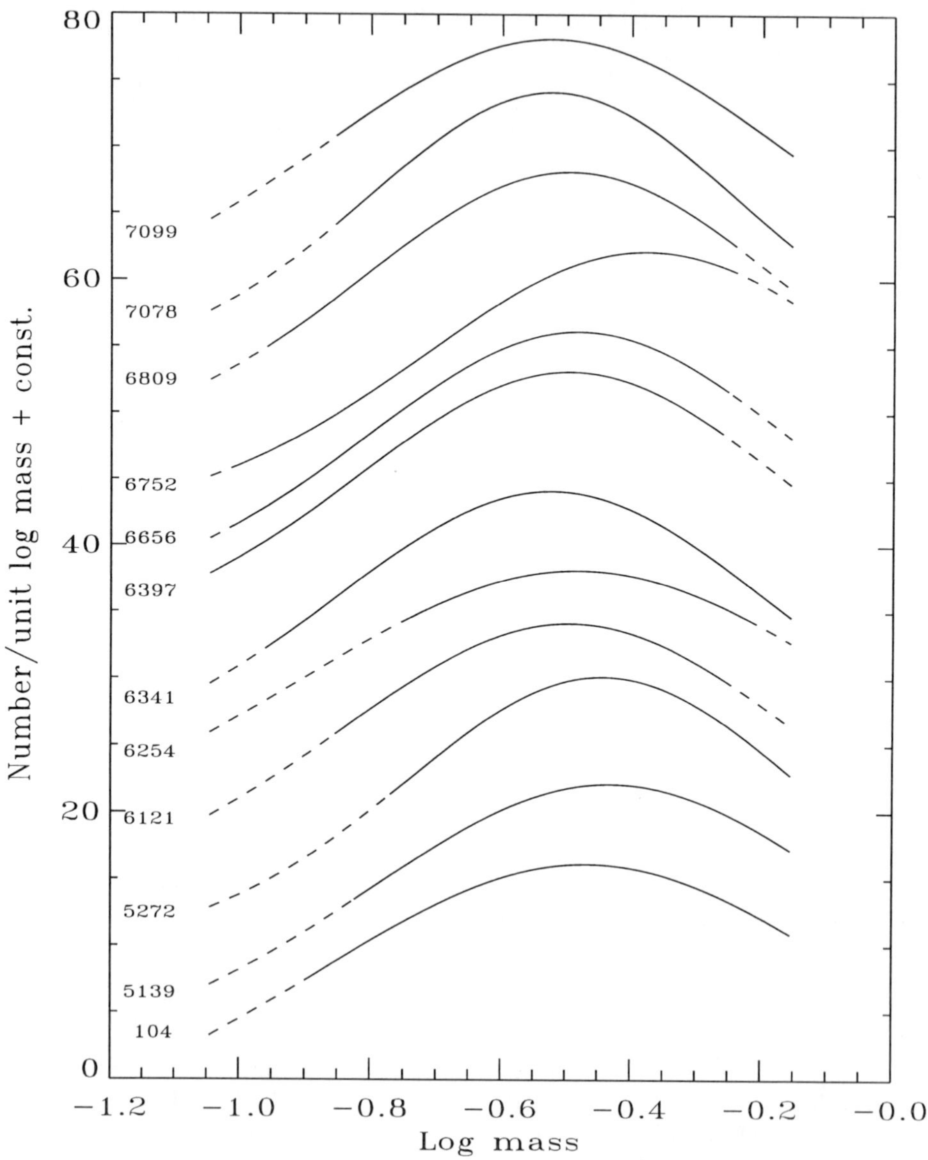

Figure 2. Log-normal mass functions that best fit the LF shown in Figure 1. Note that the scale is linear along the y axis.

all presently available models for the ML relation are in good agreement with each other (Silvestri et al. 1998) and is, in consequence, least likely to be subject to uncertainties due to the LF to MF conversion.

5. Correction for Mass Segregation

The MF shown in Figure 2 need to be corrected for the effects of mass segregation due to energy equipartition as has been extensively discussed in Pulone et al. (1999), for example, in the case of NGC 6121 (M 4). To assess the magnitude of the expected effect on the shape of the local MF, we have run standard multi-mass Michie–King models for the clusters in our sample. The corrected ΔN for these clusters obtained by using the appropriate Michie-King models are listed in Table 2. The effect of this correction is a decrease, as expected, of the value of ΔN for all clusters whose LF was determined far from the half mass radius.

6. Physical Data and Tidal Disruption

In Table 2, we list the main physical parameters of the clusters surveyed so far. Since our main objective is to search for a signature of the cluster's dynamical history on its low mass MF, we have included in this table whatever is known about its orbit in the Galactic tidal field. The space motion data were obtained from the work of Dauphole et al. (1996) and Odenkirchen et al. (1997). These data can be used in a theoretical model to determine the change with time of the cluster's main characteristics such as total mass, mass and luminosity functions, tidal radii, central concentrations, relaxation times, etc. Both N-body and Fokker-Planck models of increasing sophistication have been used recently to compute such evolution (Takahashi & Portegies Zwart 1999; Gnedin, Hernquist, & Ostriker 1999; Vesperini 1998, 1997; Vesperini & Heggie 1997; Gnedin & Ostriker 1997; Capriotti & Hawley 1996; Murali & Weinberg 1997). Although different authors use different initial conditions and approximations to the complex tidal interaction mechanisms, the generally physically plausible final result is a flattening of an assumed power-law low mass MF with time due to the preferential evaporation of lower mass stars forced by two-body relaxation out to the cluster periphery where the evaporation process is accelerated by tidal shocks.

A direct calculation of this phenomenon for a specific cluster orbit has not been carried out yet but an indirect indication at least of the magnitude of the effect can be gleaned from the recent calculations of the time to disruption T_d of specific clusters carried out by Gnedin & Ostriker (1997) and by Dinescu, Girard, & Van Altena (1999). These times are given in Gyr in Table 2 (assuming a value of 10 Gyr for a Hubble time) where the two values of the total destruction rate given by Gnedin & Ostriker (1997) for the two galactic models used in their calculations have been averaged in column (10). The observed clusters cover quite a large range of T_d from a minimum of 4 Gyr for NGC 6397 to 213 Gyr for NGC 5272 (using Gnedin

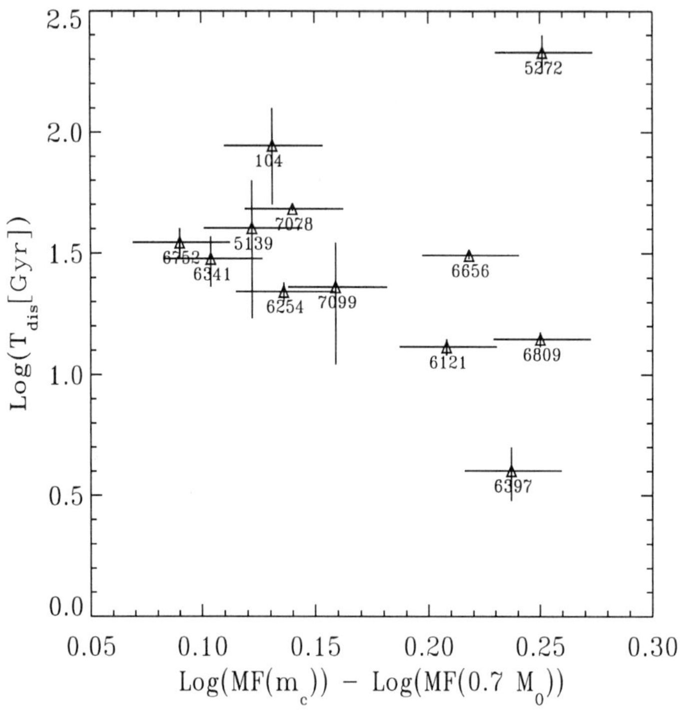

Figure 3. The time to disruption in Gyr as calculated by Gnedin & Ostriker (1997) is shown here as a function of the index ΔN describing the shape of the MF in Figure 3 (see text). Vertical error bars reflect the difference between the values of T_d obtained with two different models describing the distribution of stars in the Galaxy. Horizontal error bars mark the range in which ΔN can vary as a result of the uncertainty affecting the values of m_c and σ in the MF. As such, both error bars define in practice a $\pm 3\,\sigma$ uncertainty.

& Ostriker's values), or from 2 Gyr for NGC 6121 to 275 Gyr for NGC 5272 (following Dinescu et al. 1999). An inspection of Table 2 shows that no one particular orbital parameter or the cluster mass by itself is sufficient to foretell what the fate of the cluster will be. Even, for example, the cluster's perigalactic distance or its height above the plane are not well correlated with T_d. This means that the overall impact of the repeated bulge and disk shocks on the cluster over its lifetime is not easily predictable from a simple glance at the orbital parameters but only from the use of calculations over the entire orbit such as those referred to above.

All things being equal, then, we would expect that the clusters with the largest times to destruction T_d or those that have suffered the least tidal disruption to have the largest low to high mass number ratio ΔN.

The actual situation is shown in Figure 3 where we have plotted the time to disruption of Gnedin & Ostriker (1997) as a function of ΔN. Within the errors, there is no discernible trend in this direction and the conclusion at this point is, therefore, quite clear: *the global MF of the clusters in our sample show absolutely no evidence of evolution with time within the quoted errors.*

In light of the very small range spanned by m_c and σ deduced for all our clusters, the conclusion is, then, that a single form of the MF can easily reproduce all the 12 deep LF obtained so far and, since there is no obvious dependence on dynamical history over an extremely wide range of conditions, that this MF is most likely to represent the initial distribution of stellar masses in the cluster, namely the IMF.

7. Conclusions

We have analyzed in detail the implications for the IMF of our present reasonably good knowledge of the MS LF of a dozen Galactic globular clusters covering a wide range of physical and orbital characteristics. We have shown, first, that they can be converted to a MF by the application of a ML relation that incorporates all the relevant internal and atmospheric physics of low mass low metallicity stars appropriate to the cluster sample under investigation. We have, then, calculated the possible effect of mass segregation due to energy equipartition on the locally derived MF.

The MF obtained in this way could all descend from the same log-normal form of the global MF within a small range of mass scales and standard deviation. The MF of the four clusters of the sample whose LF extend significantly beyond the mass peak at $0.33\,M_\odot$ cannot be reproduced by a single power law throughout the MS mass range explored in this paper, but would require at least an unphysical double power law. We, then, explored the possible modification of these global MF with orbital history of the individual clusters by comparing the number gradient of their MF with theoretical estimates of their survivability in the Galactic potential. No statistically significant effect is found, no matter what particular model is used. We conclude that the effect, if present at all in this type of clusters, is completely obscured by the present observational uncertainties and that, therefore, the global MF we measure today must be, within those uncertainties, identical to the original MF namely the IMF.

It is clear that much remains to be done to clarify and establish the range of validity of this conclusion and to understand the origin of the possible deviations. This investigation should yield a bountiful harvest of information on the stellar IMF in the near future. Of particular importance in this endeavour, will be securing a sufficiently large, clean sample of stars of the same physical and kinematical type in a wide variety of environments and ages and to develop the most accurate models of their energy output as a function of mass.

References

1. Baraffe, I., Chabrier, G., Allard, F., and Hauschildt, P. 1997, A&A, 327, 1054
2. Baraffe, I., Chabrier, G., Allard, F., and Hauschildt, P. 1998, A&A, 337, 403
3. Capriotti, E.R., and Hawley, S.L. 1996, ApJ, 464, 765
4. Carretta, E., et al. 1999, in preparation
5. Cassisi, S. 1999, private communication
6. Dauphole, B., Geffert, M., Colin, J., Ducourant, C., Odenkirchen, M., and Tucholke, H.J. 1996, A&A, 313, 119
7. De Marchi, G. 1999, AJ, 117, 303
8. De Marchi, G., and Paresce, F. 1995a, A&A, 304, 202
9. De Marchi, G., and Paresce, F. 1995b, A&A, 304, 211
10. De Marchi, G., and Paresce, F. 1996, in Science with the Hubble Space Telescope - II, Eds. P. Benvenuti, F.D. Macchetto, & E. Schreier (Baltimore: STScI), 310
11. De Marchi, G., and Paresce, F. 1997, ApJ, 476, L19
12. Dinescu, D.I., Girard, T.M., and Van Altena, W.F. 1999, AJ, 117, 1792
13. Djorgovski, S. 1993, in Structure and Dynamics of Globular Clusters, ASP Conf. Ser. 50, ed. S. Djorgovski and G. Meylan (San Francisco: ASP), 373
14. Ferraro, F.R., Carretta, E., Bragaglia, A., Renzini, A., and Ortolani, S. 1997, MNRAS, 286, 1012
15. Gnedin, O.Y., Hernquist, L., and Ostriker, J. P. 1999, ApJ, 514, 109
16. Gnedin, O.Y., and Ostriker, J.P. 1997, ApJ, 474, 233
17. Marconi, G., Buonanno, R., Carretta, E., Ferraro, F.R., Fusi Pecci, F., Montergriffo, P., De Marchi, G., Paresce, F., and Laget, M. 1997, MNRAS, 293, 479
18. Murali, C., and Weinberg, M.D. 1997, MNRAS, 291, 717
19. Odenkirchen, M., Brosche, P., Geffert, Tucholke, H.J. 1997, NewA, 2, 477
20. Paresce, F., De Marchi, G., and Romaniello, M. 1995, ApJ, 440, 216
21. Piotto, G. , Cool, A. M., and King, I. R. 1997, AJ, 113, 1345
22. Pulone, L., De Marchi, G., and Paresce, F. 1999, A&A, 342, 440
23. Scalo, J.M. 1998, in The Stellar Initial Mass Function, ASP Conf. Ser. 142, eds. G. Gilmore and D. Howell (San Francisco: ASP), 201
24. Scalo, J.M. 1999, in The Birth of Galaxies, Blois, France, astro-ph/9811341
25. Silvestri, F., Ventura, P., D'Antona, F., and Mazzitelli, I. 1998, ApJ, 509,192
26. Takahashi, K., and Portegies Zwart, S. 1999, ApJ, submitted, astro-ph/9903366
27. Vesperini, E. 1998, MNRAS, 299, 1019
28. Vesperini, E. 1997, MNRAS, 287, 915
29. Vesperini, E., and Heggie, D.C. 1997, MNRAS, 289, 898
30. Webbink, R.F. 1985, IAU Symp, 113, 541

Speaker: F. PARESCE

S. CASSISI's Comments: I have few comments: I don't think that the Lyon models for very low mass stars are the only reliable ones. The "Teramo" models satisfy, within the current uncertainties on GC distances and metallicities, the observational constraints as well as the Lyon models. In addition, we have shown, in different papers that the use of model atmospheres for fixing the outer boundary conditions in stellar models introduces not negligible uncertainties in the position of the VLM models in the H-R diagram.

F. PARESCE: We were not aware of the existence of the newer "Teramo" models before this meeting. We have now used the Teramo models and find them also quite satisfactory. It is comforting to know that we can now rely on two different models of the low mass stars in globular clusters and obtain the same answer.

M. TOSI: I am puzzled by the fact that you use open cluster MFs to check whether or not the MFs derived in your globular clusters are plausibly similar to their IMF: aren't open clusters supposed to be heavily affected by mass segregation and/or evaporation? Can their MFs be taken as representative of their IMFs?

F. PARESCE: We selected these open clusters for comparison precisely because the measurements were made over very wide fields encompassing most or all of the known clusters. The very young ones like the ONC and IC348 are probably too you young for this effect to be significant. But, of course, there is no guarantee that they have been affected and so this should remain a caveat.

G. PIOTTO: Apart from the models, which predict a dynamical evolution induced by the Galactic potential, there are also empirical evidences that globular clusters loose a substantial amount of stars. E.G. Grillmair et al. found extended haloes far beyond (after times) the tidal radius in many clusters (including objects in your sample). It seems a strong assumption to say that the global present day mass functions you have are the IMF.

F. PARESCE: Yes, at first sight, this effect may seem to be inconsistent with our conclusions. But further reflection indicates that 1) the actual number of stars at or just beyond the tidal radius is very small (certainly less than 1% of the cluster) - see Combes, Meylan and Leon: 1999, A&A in press; 2) all these stars in any case are still bound to the clusters (see Testa et al., 1999 A&A in press); 3) these stars will stay bound to the clusters for a Hubble time (see Fukushige and Heggie: 1999, Astro-Ph/9910468).

THE EFFECT OF CNO AND ENHANCED α-ELEMENTS ON GLOBULAR CLUSTER STARS

ROBERT T. ROOD
University of Virginia

Abstract. While the idea of non-solar ratios for the heavy elements was new 20 years ago, enhancements in the abundances of the α-elements are now routinely included in stellar models. In general these lead to better agreement between the observations and theory. I give a personal history and simple review of the subject.

1. Introduction

My job in this presentation is simple—there is an excellent discussion of the subject in Salaris, Chieffi, & Straniero (1993) (SCS93). In addition shortly before writing this up I received a nice preprint on the subject by VandenBerg et al. (2000) (see also VandenBerg 2000). From the observational point of view there is the review of Kraft (1994). I have little to add to these papers other than perhaps a little history and some prejudices. I was a student of Icko Iben at the time he and his collaborators were first considering how the CNO cycle rate affected stellar structure. It was inevitable that an investigation of CNO effects would be part of my thesis. This early work was part of a larger effort to learn how all the components of the physical input affected stars.

In the 1970's the observational evidence, particularly that from Kraft and his coworkers, began to grow indicating that the CNO fraction might actually vary with [Fe/H]. There were even some hints that CNO might vary star to star within a single globular cluster. I was having some problems in modeling some horizontal branches (Rood 1973) and realized that having CNO could help with these. Thus, I began a more serious investigation including the opacity and energy generation effects of CNO separately. I think I was the first to do this. Most of this work was never published.

F. Matteucci and F. Giovannelli (eds.), The Evolution of the Milky Way, 197–203.

Much of my effort went into revising my code in response to the computer center billing algorithm. Like the CNO cycle I reached equilibrium; even though I worked on the projects they never got closer to completion.

Carbon, nitrogen, and oxygen play a special role among the heavy elements in stellar structure because they serve as a "catalyst" in determining the rate of hydrogen burning. The other heavies enter only through their contribution to the opacity, κ. In globular cluster stars hydrogen burning on the main sequence occurs via the p-p chains. However, even at the lowest metallicities, by the time of central hydrogen exhaustion most of energy output is occurring through the CNO cycles. In equilibrium the rate of hydrogen burning is determined by C + N + O (CNO). In the regions which dominate energy production in all post-turnoff stars the H-burning rate is set by the equilibrium CNO-cycles.

It is now more common to discuss the α-elements as a group (O, Ne, Mg, Si, S, Ca) rather than CNO. These elements are all products Type II supernovae, and it makes sense that their abundances should vary together. To a large extent observations support that view. McWilliam (1997) gives a detailed review of of the α abundances. However, even at this meeting we have seen some debate about how O varies as compared to the other α-elements. The prevailing view has been that [O/Fe] rises as [Fe/H] decreases but reaches a plateau of about 0.3–0.4 for [Fe/H] ≤ -1 (e.g., Kraft 1994). However Israelian et al. (1998) and Boesgaard et al. (1999) argue that at the lowest metallicities [O/Fe] continues to rise.

I will consider [O/Fe] and [α/Fe] separately here. There are a few cases where the [O/Fe] does not fit the normal pattern (references in McWilliam 1997). In addition, there is evidence that some very low metallicity stars are enriched my single supernova (McWilliam 1997; Burress et al. 2000). While the observations are for heavier elements one might find variations in O as well. In SN II models available when I was first thinking of this problem the yields of α elements increased with stellar mass. The more recent models of Woosley & Weaver (1995) seem consistent with this. In these stars the O lies to the exterior of the other elements. With some judicious fiddling with the mass cut between the ejecta and core one could imagine massive low metallicity SN that produce only O.

In addition to overall trends in [O/Fe] with [Fe/H] I want to remain open to the idea of star to star variations of [O/Fe] within a given globular cluster. In a massive cluster there could have been hundreds of SNe. The collapse time for a proto-globular-cluster is comparable to a pre-SN stellar life time. Particularly if the proto-cluster was much more massive than the current cluster, i.e., star formation efficiency was comparable to that now found in giant molecular clouds, it may take some time for the SNe to blow away the collapsing gas. Even one SN might make an interesting alteration to the O in the lowest metallicity clusters. This hypothesis was explored Truran, Brown & Burkert (1991) & Brown, Burkert & Truran (1991). While their models have not been widely accepted, I think they are still worth considering for possible anomalies at very low metallicities and special cases like the dwarf spheroidals.

I was attracted to [O/Fe] as a tool to deal with three problems associated with the horizontal branch: the second parameter problem (Rood 1973), bimodal HBs, and the Oosterhoff Dichotomy/Sandage Period Shift. O had the advantage of having a fairly strong effect on the HB while being poorly constrained observationly (at that time). O, as opposed to the α elements, does not affect the color of the red giant branch (RGB) which is set by the opacity of low ionization potential elements. Despite statements to the contrary, the color spread near the turnoff did not place significant constraints on a possible dispersion of [O/Fe] within a cluster. O could be observed spectroscopically only in the brighter cluster giants. These were known to have undergone CNO processing, and the mechanism which mixed this processed material to the surface was unknown. Thus, if the observed O disagreed with one's prejudices, the truly weaselly investigator could dismiss the observed abundance as due to mixing.

2. The Effect of Increased CNO

My ancient models gave the following results (Rood & Crocker 1997). I think these are all consistent with the more modern models of CSS93 and VandenBerg et al. (2000).

• *The Main Sequence:* The MS is shifted very slightly to the red. This is an opacity effect.

• *The MS Turnoff:* The MSTO is shifted to lower luminosity and to the red. This is mostly an opacity effect. The higher opacity leads to a higher T gradient and less fuel consumed. At the lowest metallicities opacity driven effects most go to zero.[1] Even at [Fe/H] $= -2.3$ half of the dependence is due to opacity (VandenBerg et al. 2000). A dispersion in [O/Fe] of 0.5 would give a dispersion in turnoff $\delta(B-V) < 0.01$ mag. The associated dispersion in the subgiant branch is $\delta V \sim 0.07$ mag.

• *The RGB luminosity function bump:* The luminosity of the bump is an opacity driven effect. The critical temperature range is 1–few 10^6 K. It should be well parameterized by the total metallicity mass fraction Z.

• *RGB Color:* Depends on low temperature opacity, hence low ionization potential abundances. Independent of [O/Fe]. Will vary with [α/Fe].

[1]This is why I often use $\mathrm{asinh}^{-1}(Z/\mathrm{constant})$ in analytic fits to opacity driven quantities.

• *RGB Tip:* The luminosity at helium flash depends on the rate that the He core is growing. This depends on rate of CNO cycle burning and thus CNO. The core mass at ignition ($M_{\rm core}$) will decrease as CNO increases. The flash L will increase as CNO increases.

• *The HB:* The behavior of the HB is complex. As CNO increases: (1) $M_{\rm core}$ decreases and L from helium burning decreases; (2) L from the hydrogen burning shell increases. The second factor leads to a decrease in $T_{\rm eff}$. The increase in opacity leads to a drop in L and $T_{\rm eff}$.

3. The Effect of $[\alpha/{\rm Fe}]$

As opposed to increases in just CNO, $[\alpha/{\rm Fe}] > 0$ increases low temperature opacity and affects RGB colors. SCS93 argued that enhanced $[\alpha/{\rm Fe}]$ could be well mimicked by using scaled solar abundances with $Z = Z_{\rm tot}$, the total mass fraction of the heavies with the enhanced α-elements included. In particular, α enhancements could be accurately reproduced using [M/H] with $[\alpha/{\rm Fe}] = 0$ instead of [Fe/H] with $[\alpha/{\rm Fe}] > 0$. $[M/{\rm Fe}]$ was computed assuming $[M/{\rm Fe}] = \log(Z/Z_\odot)$ where $Z = Z_0(0.638 f_\alpha + 0.362)$ with $[{\rm Fe/H}] = \log(Z_0/Z_\odot)$ and the α-enhancement factor defined as $[\alpha/{\rm Fe}] = \log f_\alpha$. While I was not surprised that this duplicated the shape of evolutionary tracks and isochrones well, I was surprised that it did such a good job with L and $M_{\rm core}$ at the helium flash and the HB properties which depend so strongly on $M_{\rm core}$. These factors should be driven almost exclusively by the CNO burning rate. However, looking more carefully I discovered that their prescription conserved CNO over the range of cluster abundances—very clever.

More recently VandenBerg et al. (2000) have found that as Z approaches solar values these simplifying approximations do not work so well. This is not surprising since the nuclear and opacity effects really are quite different.

4. Observational Results

Probably the cleanest observational results involving enhanced α abundances are the luminosities of the RGB luminosity function bump, an opacity driven parameter, and the luminosity at helium flash, $L_{\rm tip}$. Ferraro et al. (2000) have measured the bump in 8 clusters and the tip in 10 using IR photometry. Figure 1 shows their absolute K magnitude $M_K^{\rm bump}$ as a function of $[M/{\rm Fe}]$. The metallicity scale is that suggested by Carretta & Gratton (1997). Ferraro et al. assumed that $[\alpha/{\rm Fe}] = 0.28$ for $[{\rm Fe/H}] < -1$ decreasing linearly to zero over the interval $-1 < [{\rm Fe/H}] < 0$. $[M/{\rm Fe}]$ is from the SCS93 prescription. The theoretical relation of Straniero, Chieffi, & Limongi (1997) is shown as a dashed line. The fit between the observations and

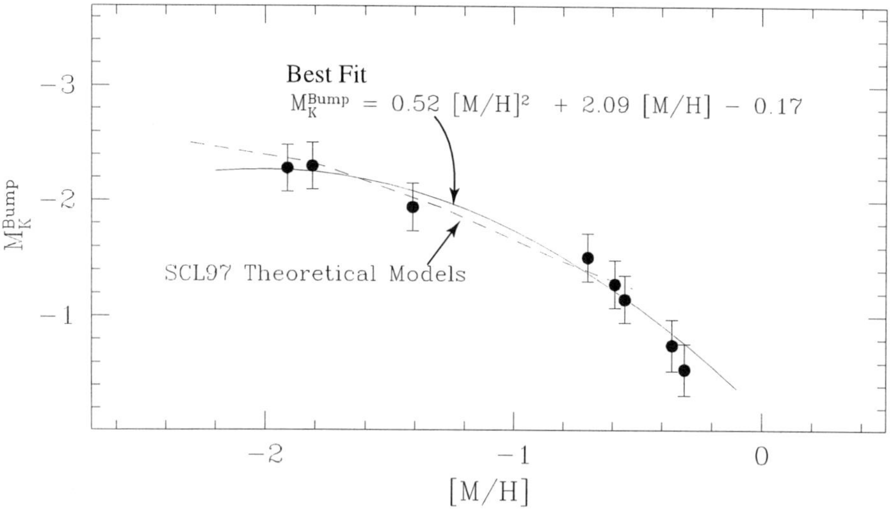

Figure 1. The absolute K magnitude of the LF-bump as a function of metallicity (from Ferraro et al. 2000).

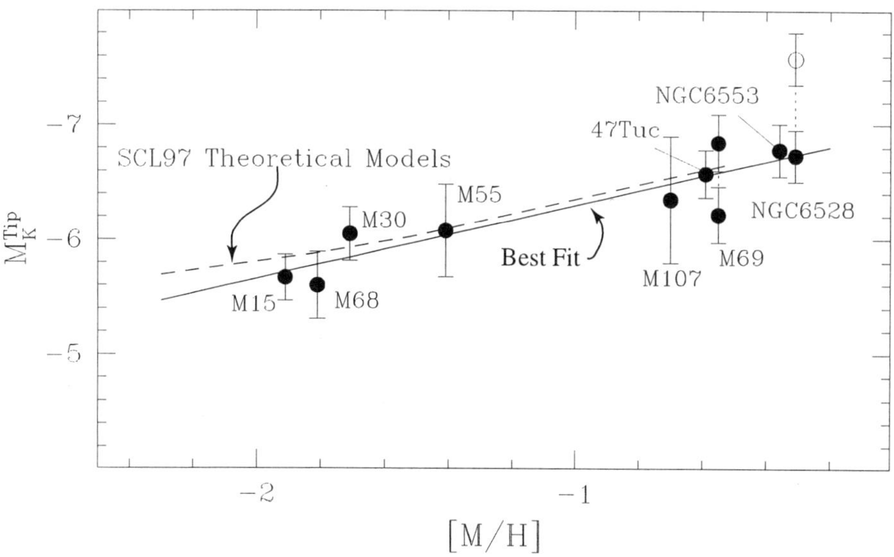

Figure 2. The absolute K magnitude of the RGB-tip as a function of metallicity (from Ferraro et al. 2000).

the theory is quite good. If the α elements were not enhanced $[M/\mathrm{Fe}]$ would overestimate the opacity and observations would lie above the theory at low metallicity.

Figure 2 shows the the Ferraro et al. (2000) $M_K^{\rm tip}$ as a function of $[M/\rm Fe]$. The fit between the observations and the theory is again quite good. If the CNO elements were not enhanced $[M/\rm Fe]$ would overestimate the energy generation rate and observations would lie below the theory at low metallicity.

5. Conclusions

In general enhanced $[\alpha/\rm Fe]$ or [O/Fe] has led to a better agreement between globular cluster color magnitude diagrams and the theory. In addition to the LF-bump and RGB-tip results presented above important consequences include:

- *Cluster ages:* Ages are reduced by 2–3 Gyr. This reduces the discrepancy between the stellar age scale and the expansion age scale.
- *HB Morphology:* It is very difficult to model systems with a significant number of red HB stars at low metalicity $[\rm Fe/H] \sim -2.3$ without enchanced CNO. Rood & Seitzer (1981) first noted this for the Draco dwarf spheroidal system. This might be a place where having a non-universal [O/Fe], [Fe/H]-relation might help. For example, it is non-trivial to understand the differences between the RR Lyrae variable and red HB populations of M92 & M15. If M15 has higher [O/Fe] than M92 things would be easier. If that were the case, then M92 would lie above the theoretical relation in Figure 2.

Of the other things that first drove me to consider variations in [O/Fe]: It was of no help in the Oosterhoff/Sandage Period Shift problem. The problem of HB gaps and bimodality is still with us. Intracluster variations in [O/Fe] could certainly produce multimodal HBs. However, there is no compelling reason to chose that the over other less than satisfying candidate mechanisms.

This research was supported by NASA Long Term Astrophysics Grant NAG 5-6403 and STScI grant GO-6607. I would like to especially thank Francesco Ferraro for providing the figures and many of the Bologna stellar astrophysics group for discussions over the years.

References

1. Boesgaard, A., King, J. R., Deliyannis, C. P., & Vogt, S. S. 1999, AJ, 117, 492
2. Brown, J., Burkert, A. & Truran, J. W., 1991, ApJ, 376, 115
3. Buress, D. L., Pilachowski, C. A., Armandroff, T. E., Sneden, C., Cowan, J. J., & Roe, H. (2000), preprint
4. Carretta, E. & Gratton, R. 1997, A&AS, 121, 95
5. Ferraro, F. R., Montegriffo, P., Origlia, L., & Fusi Pecci, F. 2000, AJ, in press (astro-ph/9912265)
6. Israelian, G., Garcia Lopez, R. J., & Rebolo, R. 1998, ApJ, 507, 805

7. Kraft, R. P. 1994, PASP, 106, 553
8. McWilliam, A. 1997, ARA&A, 35, 503
9. Rood, R. T. 1973, ApJ, 184, 815
10. Rood, R. T. & Crocker, D. A. 1997, Some Unpublished Results of Possible Interest, `http://www.astro.virginia.edu/ rtr/papers/`
11. Rood, R. T., & Seitzer, P. O. 1982, in Astrophysical Parameters of Globular Clusters, ed. A.G.D. Philip & D. S. Hayes (L. Davis: Schenectady), 369
12. Salaris, M., Chieffi, A. & Straniero, O. 1993, ApJ, 414, 580
13. Straniero, O., Chieffi, A., & Limongi, M. 1997, ApJ, 490, 425
14. Truran, J. W., Brown, J., & Burkert, A. 1991, in The Formation and Evolution of Star Clusters, ed. K. Janes (ASP: San Francisco), 78
15. VandenBerg, D. A. 2000, preprint
16. VandenBerg, D. A., Swenson, F., Rogers, F. J., Igelesias, C. A., & Alexander, D. R. 2000, preprint
17. Woosley, S. E. & Weaver, T. A. 1995, ApJS, 101, 181

HST–UV OBSERVATIONS OF STELLAR CLUSTERS: LOOKING FOR BSS

F. R. FERRARO
European Southern Observatory
Karl Schwarzschild Str. 2,
D–85748 Garching bei München, Germany
and
Osservatorio Astronomico di Bologna
Via Ranzani 1, I–40127 Bologna, ITALY

Abstract. HST–WFPC2 observations in the UV have been used to select Blue Straggler Stars in the central regions of four Galactic Globular Clusters (namely, M3, M13, M80, M92). Some recent results are presented and discussed.

1. Introduction

Globular Star Clusters are extremely important astrophysical objects since (1) they are prime laboratories for testing stellar evolution; (2) they are "fossils" from the epoch of galaxy formation, and thus important tools for studying the formation and the chemical evolution of the Galaxy; (3) their individual stars test models for stellar dynamics.

A few years ago we started a long–term project aimed to perform a global quantitative analysis of the stellar populations, in a sample of prototype clusters, by measuring **all** the evolved stars (i.e. Post Main Sequence Stars) at **any** radial distance from the cluster center in **the most suitable** photometric bands. In this contest, while optical and IR observations have been used to study the morphology of the Red Giant Branch (RGB) and its behaviour as a function of the cluster metallicity (see Ferraro et al. 1999a, Ferraro et al. 2000), UV observations, have been used to study blue sequences as the Horizontal Branch (HB) and the Blue Straggler Stars (BSS). These sequences are very sensitive to the environment conditions and the dynamical evolution of the cluster, and their morphology and Lu-

F. Matteucci and F. Giovannelli (eds.), The Evolution of the Milky Way, 205–211.

minosity Function (LF) can be used as a powerful diagnostics to quantify dynamical effects on stellar evolution.

2. The BSS population in GGCs from UV Observations

BSS were first discovered by Sandage (1953) in M3. In the color–magnitude diagram (CMD) they formed a sparsely populated sequence extending to higher luminosities than the turn–off point of normal hydrogen burning Main–Sequence (MS) stars. For this reason they are interpreted as MS stars of larger mass with respect to the bulk of the cluster stars. However, since there is no other indication of star formation after the burst which formed the bulk of the cluster stars, two main mechanisms for making BSS are favored: *(i)* the merger of two stars in a primordial binary system and *(ii)* collisions in regions of very high stellar density.

Therefore BSS in general and collisional BSS in particular, represent an important tool to study dynamical interactions inside star clusters (see Bailyn (1995) for a recent review on this subject).

UV CMDs turn to be the most suitable observational planes in which BSS can be detected and studied. In fact, since the BSS are one of the hottest sub–population in the cluster, they are easily separated from the cooler turn–off and SGB stars. And indeed they are the ideal targets for near– and far–UV imaging with HST, which is the only instrument that combines the necessary space resolution with the UV capabilities.

We successfully used HST–WFPC2 UV observations to survey the core of 4 GGCs, namely M3, M13, M80 and M92, looking for BSS candidates.

Figure 1 shows the (m_{255}, $m_{255} - U$) CMDs for the four clusters: more than 50,000 stars are plotted. Horizontal and vertical shifts have been applied to the M13, M80 and M92 sequences in order to match those in M13. The bright BSS ($m_{255} < 19$) are evidenced as big filled circles. The BSS population selected in such a way (bright BSS) has been compared to reference stellar populations (RGB and HB) observed in the same area of the cluster. The main results can be summarized as follows: (1) BSS are more centrally concentrated than the corresponding RGB stars in M80 and M92, whereas in M13 no significant difference has been found in the radial distribution between the two sequences. In M3 BSS have a very peculiar spatial distribution; (2) the BSS Specific Frequency (defined as $F_{BSS} = N_{BSS}/N_{HB}$) turns out to be largely variable in the four clusters, varying by a factor of 10 from M13 (the lowest) to M80 (the highest).

The case of M3 and M80 deserve a more accurate discussion.

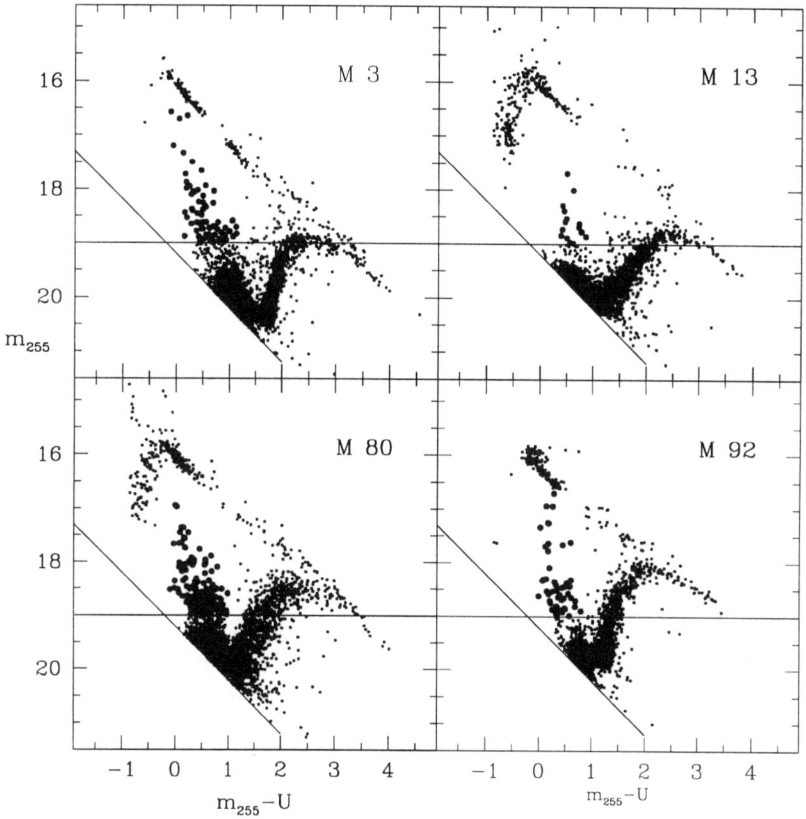

Figure 1. ($m_{255}, m_{255} - U$) CMDs for M3, M80, M92 and M13. Horizontal and vertical shifts have been applied to M13, M80, and M92 so that their sequences match those of M3. The horizontal solid line corresponds to $m_{255} = 19$. The bright BSS candidates are plotted as big asterisks.

2.1. THE ANOMALOUS RADIAL DISTRIBUTION OF BSS IN M3

As can be seen from Figure 1, the BSS population is a quite well defined sequence spanning ~ 3 mag in m_{255}. Moreover, they are clearly distinguishable from SGB–TO stars. In M3, we divide the global BSS sample in two sub–samples, the bright BSS with $m_{255} < 19$ and the faint BSS with $19 < m_{255} < 19.4$, in order to be consistent with the limiting magnitude adopted in the two more external previous surveys (Ferraro et al., 1993; Buonanno et al., 1994). These observations have been used to construct the *first complete catalog* of more than 114 bright BSS spanning *the whole cluster* from 0" to 360" from the cluster center. The radial distribution of the global bright sample of BSS compared with 1581 RGB stars turned to be quite unusual: the BSS in M3 displays a bimodal radial distribution with a dip within 100" $< r <$ 200".

Which is the origin of such a peculiar distribution? Sigurdsson et al. (1994) suggested that the observed distribution could be due to the dynamical evolution of collisional BSS originated in the central region of the cluster.

But another scenario is also possible, as first suggested by Ferraro et al. (1993): the bimodal BSS radial distribution naturally leads to hypothesize that the BSS located in different regions of the cluster could have a different origin: BSS in the center could have a collisional origin, while BSS in the outer regions could be originated by primordial binary merging. This hypothesis is also supported by comparison with theoretical BSS LFs by Bailyn & Pinsonneault (1995) (see Figure 11 in Ferraro et al., 1997).

2.2. THE SPECTACULAR POPULATION IN M80

The UV–CMD in the $(m_{255}, m_{255} - U)$ plane for more than 13,000 stars identified in the HST field of view of the central region of M80, is presented in Figure 2. In selecting the BSS sample we adopted the same criteria we used in M3: in order to insure the same BSS limiting absolute magnitude for M80 as we adopted in M3, we aligned the two $(m_{255},\ m_{255} - U)$ CMDs, using the bright portion of the HB as *normalization* region. The resulting fainter boundary of the BSS sequence in M80 is $m_{255} = 20.55$. By adopting this value M80 turns to have a spectacularly large population of BSS: 305 candidates have been found in the WFPC2 field of view. The number of BSS in M80 is huge. The previous record number was in M3 which has a population of $\sim$ 170 BSS (about half of the population in M80) in the WFPC2 field of view. Even the specific frequency of BSS in M80 is large ($\sim$ 1), moreover, considering only the field of view of the PC, the specific frequency of BSS in M80 rises to $\sim$ 1.7, i.e., the BSS are almost twice as abundant as the HB stars.

The BSS in this cluster turned to be far more concentrated towards the cluster center than either the HB or RGB stars. Note that while half of the BSS population is within $8''$ from the cluster center, only $\sim$ 20% of the HB or RGB can be found in the same region. The Kolmogorov–Smirnov test applied to the two distributions shows that the probability of drawing the two populations from the same distribution is very small, $\sim 10^{-4}$.

This suggest that the BSS population found in M80 is **the largest** and **most concentrated** population **ever** observed in a GGC.

Why M80 has such a large population of BSS? M80 has one of the highest central density ($Log\rho_0 \sim 5.4 M_{\odot}\ pc^{-3}$) of any GGC which has

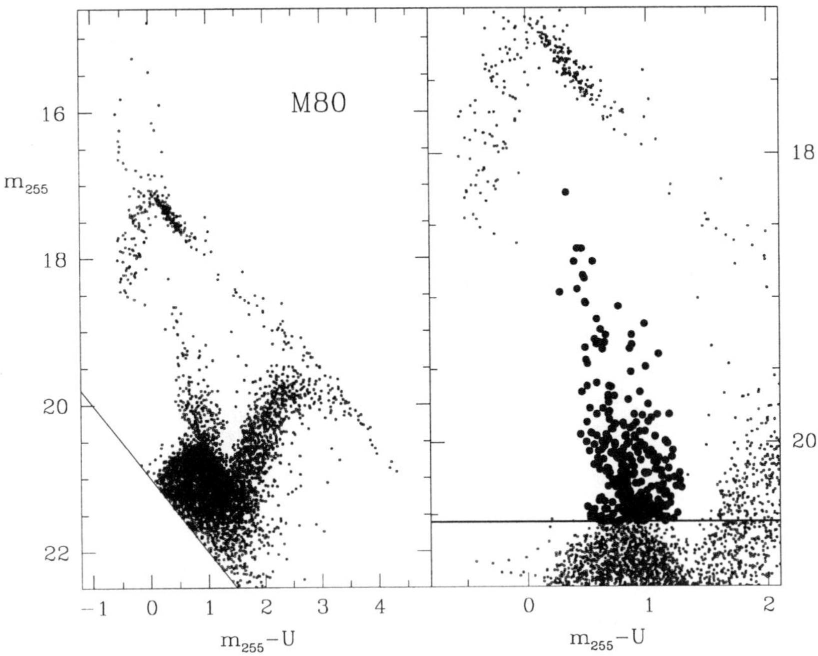

Figure 2. The $(m_{255}, m_{255}-U)$–CMD of the central region of M80. In the left panel the whole CMD is shown. The solid line corresponds to $U = 21$. The right panel shows the zoomed CMD in the BSS region. BSS are plotted as solid circles. The heavy horizontal line at $m_{255} = 20.55$ corresponds to the assumed limiting magnitude for the selected BSS.

shown no previous evidence for having undergone core collapse (Djorgovski 1993, Ferraro et al. 1999b).

However, the BSS population in M80 turns to be huge even with respect to clusters (observed with the WFPC2) with similar central density (as, for example, NGC6338 and NGC2808), and also with respect to Post Core Collapse (PCC) clusters (as 47Tuc and M30).

Thus neither the high density nor the PCC state can account for the large number of BSS in M80. We suggest that the most plausible hypothesis is that the BSS arise from the core collapse process. It is commonly thought that collisional binary formation plays an important role in the core collapse process, by delaying and eventually halting the collapse itself. With its high central density M80 is probably trying very hard to undergo core collapse but, ongoing binary formation, is preventing this from happening. If this scenario is correct the large population of collisional BSS arises because we have caught a cluster at a critical phase during its dynamical evolution, when the stellar interactions are delaying the collapse of the core and producing BSS (see Ferraro et al. 1999b).

3. Acknowledgments

This research was partially financed by the *Agenzia Spaziale Italiana* (ASI). The financial support of the *"Ministero della Università e della Ricerca Scientifica e Tecnologica"* (MURST) to the project *Stellar Evolution* is kindly acknowledged.

References

1. Bailyn, C.D.: 1995, 'Blue Stragglers and other stellar anomalies: implications for dynamics of globular clusters', *ARA&A*, **33**, 133
2. Bailyn, C.D., Pinsonneault, M.H.: 1995, 'On the LF, lifetimes and origin of BSS in GGCs', *ApJ*, **439**, 705
3. Buonanno, R., Corsi, C.E., Buzzoni, A., Cacciari,C., Ferraro F.R., Fusi Pecci, F.: 1994, 'The stellar population of the GGC M3. I. Photographic Photometry of 10,000 stars', *A&A*, **290**, 69
4. Djorgovski, S.: 1993, 'Physical parameters of Galactic Globular Clusters', in *Structure and Dynamics of Globular Clusters*, G.S. Djorgovski & G. Meylan eds., ASP, S. Francisco (CA), ASP Conf. Ser., **50**, 373
5. Ferraro F.R., Fusi Pecci, F., Cacciari, C., Corsi, C.E., Buonanno, R., Fahlman, G.G., Richer, H.B.: 1993, 'Blue Stragglers in the GGC M3: Evidence for two populations', *AJ*, **106**, 2324
6. Ferraro F.R. et al.: 1997, 'HST observations of blue straggler stars in the core of the globular cluster M3', itA&A, **324**, 915
7. Ferraro, F.R., Messineo, M., Fusi Pecci, F., DePalo, M.A., Straniero, O., Chieffi, A., Limongi, A.: 1999a, ' The giant, Horizontal and Asymptotic branches of GGCs. I. The catalog, photometric Observables and features', *AJ*, **118**, 1738
8. Ferraro, F.R., Paltrinieri, B., Rood, R.T., Dorman, B.: 1999b, 'Blue Straggler Stars: The Spectacular Population in M80', *ApJ*, **523**, 752
9. Ferraro, F.R., Montegriffo, P., Origlia, L., Fusi Pecci, F.: 2000, 'A new IR–Array photometric survey of Galactic Globular Clusters: A detailed study of the RGB sequence as a step towards the global testing of stellar models', *AJ*, in press
10. Sandage, A.R.: 1953, 'The colour–magnitude diagram for the cluster M3', *AJ*, **58**, 61.
11. Sigurdsson, S., Davies, M.B., Bolte, M.: 1994, 'Modeling the radial distribution of BSS in M3', *ApJ*,**431**, 115

Speaker: F. FERRARO

D. VANBEVEREN: Blue stragglers (BSs) are observed in many young clusters as well. They are frequently classified as rapid rotators (OBe). Do you see rotational properties in BSs of globular clusters?

F. FERRARO: We did not get any measure of velocity rotation for these BSs yet. However, the rotation feature cannot be considered a general characteristic for BSs. For instance, from a theoretical point of view Livio and Leonard (1996) showed that collisional BSs are not necessarily rapid rotators.

F. PARESCE: For a long time, it has been thought that 47Tuc was the archeotypical cluster "on the verge of collapse" just like M80. Why does it not have as many BSs? as M80?

F. FERRARO: I think that M80 is in a peculiar and possibly short dynamical phase. It is possible that in 47Tuc the production of collisional BSs has not reached the maximum efficiency. On the other hand M80 has the highest rate of stellar collisions much larger than 47Tuc and any other GGC.

HST - WFPC2 PHOTOMETRY OF THE GLOBULAR CLUSTER NGC 288: BINARY SYSTEMS, BLUE STRAGGLERS AND VERY BLUE STARS

M. BELLAZZINI AND M. MESSINEO
Osservatorio Astronomico di Bologna
Via Ranzani 1, 40127 Bologna, ITALY

Abstract. We report on new WFPC2 observations of the globular cluster NGC 288, focusing our attention on peculiar stars. A very pronounced binary sequence, paralleling the ordinary Main Sequence (MS) is clearly observed in the Color Magnitude Diagram (CMD) and a huge relative fraction of Blue Straggler Stars is measured. The dataset offers the opportunity of studying the evolution of a large population of binaries (and binary evolution by-products) in a low density environment, where the evolution of such systems is not dominated by collisions and encounters. Three (very) Extreme Horizontal Branch Stars have been found, all lying outside of the cluster core.

1. Introduction

As a part of a long term programme aimed to obtain complete samples of all the evolved stars in a few prototypical globular clusters[1], we have obtained multi-wavelenght observations of a couple of WFPC2 fields towards two very important globulars: NGC 2808 and NGC 288.

Here we report some preliminary result about NGC 288, a cluster that holds a key role in the literature about the *Second Parameter* phenomenon (see [?] for a review, and references therein).

[1]The programme is named *Large Population Studies of Globular Clusters (LPSGC)* and is a collaboration between scientists from the Bologna and Rome observatories, the Goddard Space Flight Center (Baltimore, MD) and the University of Virginia (Charlottsville), see Ferraro, this volume.

F. Matteucci and F. Giovannelli (eds.), The Evolution of the Milky Way, 213–219.

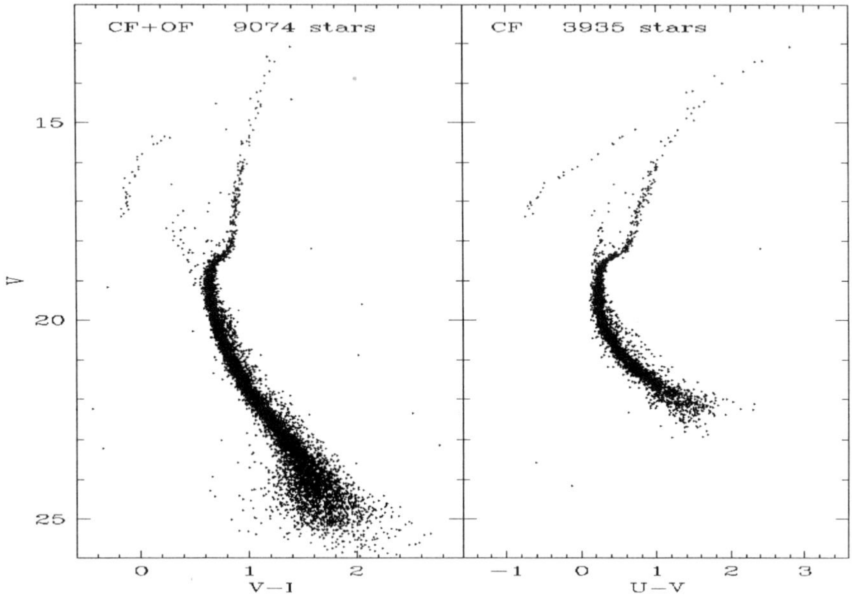

Figure 1. (V-I,V) and (U-V,V) Color magnitude diagrams of NGC 288.

2. Observations and data reductions

We observed a field centered on the core of the cluster (Central Field, hereafter CF) with the filters F255, F336, F555, F814 and a field a couple of arcmin apart (Outer Field, hereafter OF) with the filters F555 and F814.

The data reduction has been performed with DOPHOT [?], according to the general prescriptions adopted by [?]. The absolute calibration in the Johnson-Cousins system has been obtained adopting the Holtzman relations [?]. The final magnitudes are the average of various (2-4) measures on repeated exposures.

3. The CMD: a clear secondary sequence

In fig. 1 the V *vs.* V-I CMD of the whole sample and the V *vs.* U-V CMD of CF are shown. The most striking feature of these diagrams is the evident secondary sequence paralleling the Main Sequence that is naturally associated with a population of couples of MS stars measured as a single source. This occurrence can be due to chance superposition of otherwise unrelated stars or to the physical association of the couple (i.e. binary systems, see [?] for a review).

To evaluate the number of chance blending contaminating the observed secondary sequence and to obtain robust estimates of the true binary frac-

TABLE 1. Comparison between the F_{HB}^{BSS} and other relevant parameters for an homogeneous selected sample of Globular Clusters. The clusters are ordered with growing central density.

ID	[Fe/H]	$\log\rho_{0\,L_{\odot V}pc^{-3}}$	$t_{rc[10^9 yr]}$	F_{HB}^{BSS}
NGC 288	-1.40	1.80	1.230	1.35
NGC 6205 - M13	-1.65	3.32	0.871	0.17
NGC 5272 - M3	-1.66	3.54	0.602	0.67
NGC 6341 - M92	-2.24	4.38	0.069	0.55
NGC 6093 - M80	-1.64	4.79	0.040	1 - 1.7
NGC 7099 - M30	-2.13	5.07	0.002	0.49

tion, extensive artificial stars experiments are needed [?] and are presently in progress. However we are dealing with a very low density and relatively nearby cluster, so crowding has not to be a major concern in this case. The probability that two stars images overlap to within 2 FWHM is $P_{overlap} = \pi * FWHM^2 * N_s/A$, where N_s is the number of stars in the observed field and A is the area of the field. For our Central Field, the most "crowded" one, $P_{overlap} < 3\%$, thus the contamination by chance blending is expected to be negligible. So, it can be safely expected that most of the stars populating the secondary sequence are genuine binary systems and that the binary fraction in this cluster is remarkably high.

4. Blue Stragglers: a very high fraction

It is now widely accepted that Blue Straggler Stars (BBS) are products of the evolution of binary systems [?]. If the above discussed secondary sequence is indeed due to a large binary fraction in NGC 288, a significant population of BSS is also expected. This is actually the case. In fig. 2 the BSS are clearly identified in the F255 *vs.* F255-U CMD. While the absolute number of BSS is not particularly remarkable (as expected for such a low mass cluster), the specific frequency - defined by [?] as the ratio of the number of BSS and the number of HB stars - is one of the highest ever found in a globular, $F_{HB}^{BSS} = 1.35$. In table 1 [2] the F_{HB}^{BSS} ratio is reported together with other relevant parameters for a handful of clusters observed with HST-WFPC2 and with the adoption of the same selection criteria for the BSS sample (see [?]). The sample is very small but homogeneous and fully self-consistent for what concern the census of BSS.

[2]First three columns from [?], last column from [?] and the present work.

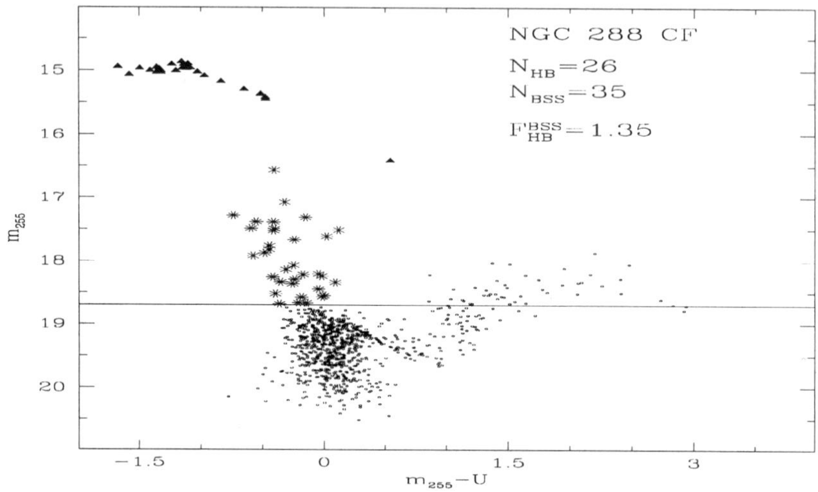

Figure 2. F_{HB}^{BSS} ratio from the selection operated in the F255W *vs.* F255W-U plane, according to the method adopted by Ferraro et al. (1999). BSS are indicated by asterisks and HB stars by triangles. The faint limit for BSS selection is indicated by the $m_{255} = 18.8$ line.

It is immediately evident that the only cluster with a comparable F_{HB}^{BSS} ratio is M80 which is a very dense cluster and that harbours most of its BSS in the very central regions. In particular the two entries for F_{HB}^{BSS} refer to the global and the central value respectively. [?] argues that the exceptional (and exceptionally concentrated) BSS population in M80 can be understood if it is assumed that the cluster is in a very special phase of its dynamical evolution, i.e. it is struggling against the unavoidable core collapse, a favorable condition for the formation of *collisional* BSS. However, while the collision time (see [?], eq. 8–125) for a MS star in the core of M80 is $t_{coll}(M80) \sim 5\ Gyr$, $t_{coll}(NGC288) \sim 1000\ Gyr$, so collisions between MS stars has to be exceedingly rare. Thus, it is very likely that the large majority of BSS in NGC 288 originated from a population of *primordial binaries* through a mechanism as efficient as the one operating in M80, but completely different.

5. HB morphology: Extreme Horizontal Branch stars?

In the right panel of fig. 3, a zoomed view of the HB stars in the (V, V-I) CMD is reported, while the left panel report the position of the same stars within the whole observed field. Three main facts can be remarked about the HB morphology:

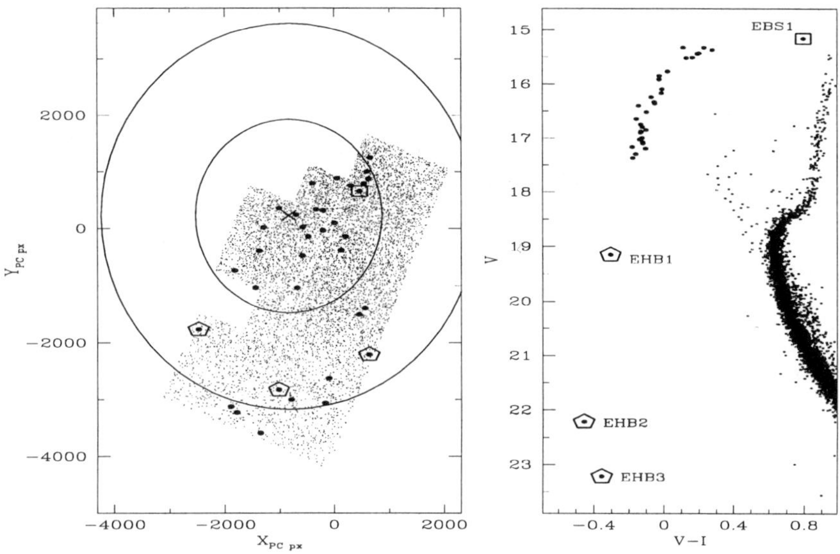

Figure 3. **Right panel:** HB morphology for the CF+OF sample. HB stars are indicated by larger dots, the candidate Evolved BSS by the open square and the candidates EHB stars by pentagons.**Left panel:** Map of the whole sample. North is up and East is Left, the symbols are the same as in the right panel. The cross marks the center of the cluster. The circles with radius of 1 r_c and 2 r_c are also reported.

- The overall HB morphology is the well known - almost purely blue - one of NGC 288. There is no possible candidate RR Lyrae star and a marginally significant gap around $V - I = 0.05$.
- There is a single red HB star. Since the red clump is the place where the Evolved BSS (EBSS, [?]) are expected to lie, during their core He burning phase, may be that this anomalous HB star is indeed an EB-SS. Based on crude evolutionary time consideration one would expect $N_{EBSS} \sim 0.6 * N_{BSS}$. In the present case $N_{EBSS} \sim 2$ is expected, in good agreement with the observed sole red HB star. Note that the EBSS candidate is found near the center of the cluster (fig. 3, left panel).
- There are three very blue stars on the ideal extension of the HB ridge line, much fainter than the end of the main HB distribution. All of these Extreme HB (EHB) candidates are genuine, isolated stars that have many consistent measurement in each filter from repeated exposures. All of them lie well beyond the cluster core, far from the densest region of the cluster. Unfortunately we have only V and I observations for these stars since they all fall in the OF. While EHB1 is fully compatible

with being a bona-fide HB star, the nature of EHB2 and EHB3 is much more uncertain. It may be that no reasonable HB model would be able to push stars at such high temperatures. Tests in this sense are in progress and other hypothesis are under consideration. Preliminary checks seems to exclude background quasars, field stars, white dwarfs and ordinary cataclysmic variables as viable possibilities.

6. Acknowledgments

The authors are grateful to all the members of the LPSGC team: F. Fusi Pecci, F.R. Ferraro, C. Cacciari, R. Buonanno, C.E. Corsi, G. Marconi, R.T. Rood, B. Dorman and W. Landsman. This research is partially funded by a MURST Grant assigned to the project "Stellar Evolution" (national coordinator prof. V. Castellani).

References

1. Fusi Pecci F. and Bellazzini M. (1997) HB morphology and Second Parameter Effect: faint stars in a big game, in *The Third Conference on Faint Blue Stars*, A.G.D. Davis Philip, J.W. Liebert & R. Saffer eds., Davis Press, Schenectady, p. 255
2. Schechter P.L., Mateo M. and Saha A. (1993) DOPHOT, a CCD photometry program: Description and tests,*PASP*, **105**, 1342
3. Olsen K.A., Hodge P.W., Mateo M., Olszewski E.W., Schommer R.A., Suntzeff N.B. and Walker A.R. (1998) HST colour-magnitude diagrams of six old globular clusters in the LMC,*MNRAS*, **300**, 665
4. Holtzman J.A. et al. (1995) The Photometric Performance and Calibration of WF-PC2, *PASP*, **107**, 1065
5. Hut P. et al. (1992) Binaries in globular clusters, *PASP*, **104**, 981
6. Rubenstein E.P and Bailyn C.D. (1997) Hubble Space Telescope Observations of the Post–Core-Collapse Globular Cluster NGC 6752. II. A Large Main-Sequence Binary Population, *AJ*, **112**, 2408
7. Bailyn C.D. (1995) Blue Stragglers And Other Stellar Anomalies: Implications for the Dynamics of Globular Clusters, *ARAA*, **33**, 133
8. Ferraro et al. (1997) HST observations of blue Straggler stars in the core of the globular cluster M 3, *A&A*, **324**, 915
9. Ferraro F.R., Paltrinieri B., Rood R.T. and Dorman B. (1999) Blue Straggler Stars: The Spectacular Population in M80, *ApJ*, **522**, 983
10. Binney J. and Tremaine S. (1987) *Galactic Dynamics*, Princeton Un. Press, Princeton (NJ), p. 539
11. Fusi Pecci F., Ferraro F.R., Corsi C.E., Cacciari C. and Buonanno R. (1992) On the blue stragglers and horizontal branch morphology in Galactic globular clusters - Some speculations and a new working scenario, *AJ*, **104**, 1831
12. Djorgovski S.G. (1993) Physical parameters of Galactic globular clusters, in *Structure and Dynamics of Globular Clusters*, G.S. Djorgovski & G. Meylan eds., ASP, S. Francisco (CA), ASP Conf. Ser., **50**, 373

Speaker: M. BELLAZZINI

M. LIMONGI: Did you check if the three faint stars could be white dwarfs?

M. BELLAZZINI: At the present time we only compared the position in the CMD of our blue faint stars with the white dwarf fiducial sequence of M4 by Richer et al. (1997) and we found that the M4 sequence starts some ~ 1 mag fainter.

INFRARED OBSERVATIONS OF STELLAR CLUSTERS

L. ORIGLIA
Osservatorio Astronomico di Bologna
Via Ranzani 1, 40127 Bologna, ITALY

Abstract.
Near infrared color–magnitude diagrams and absorption spectra are powerful instruments to study the cool stellar sequences also in very reddened and crowded environments. As an example, we discuss the case of Terzan 1, a globular cluster in the Bulge of our Galaxy. By coupling near infrared photometric and spectroscopic diagnostics we perform a preliminary abundance analysis.

1. Introduction

Cool stars with typical temperatures in the range 3000–5000 K, like massive red supergiants and low mass Asymptotic (AGB) and Red (RGB) Giant Branch stars, can dominate the infrared (IR) luminosity of integrated stellar systems in specific epochs. The near IR spectral range is particularly suitable to study these cool stellar systems, being very sensitive to low temperatures and also to high metallicities. In addition, compared to the visual range, the reddening corrections, the blanketing effects and the background contamination by MS stars are much less severe, allowing to properly characterize the red stellar sequences even though strongly affected by extinction and crowding, like for example the stellar clusters in the Bulge of our Galaxy.

Nevertheless, only recently we could start to fully benefit such a preferential spectral window, since only the last generation of IR detectors have performances close to the optical standards.

The current IR imager/spectrometers are equipped with 1k $\times$ 1k arrays and have longslit spectroscopic capabilities at low and medium resolution. Multiobject and integral field units should be available in the next few years together with larger array detectors.

F. Matteucci and F. Giovannelli (eds.), The Evolution of the Milky Way, 221–227.

2. Near IR observables

The most common observables to trace red stellar populations are IR colors and absorption spectra. Colors mainly work as thermometers, while absorption lines can provide detailed abundance information.

IR color–magnitude diagrams (CMDs) have been found to be the ideal instruments to properly define the overall physical and chemical properties of the AGB and RGB sequences in intermediate and old stellar systems (cf. e.g. Frogel et al. 1983;1990, Ferraro et al. 1995;2000), while more detailed abundance analysis can be performed by means of IR spectroscopic observations.

The near IR spectra of cool stars are dominated by absorption lines due to neutral metals and molecules, mainly CO and OH (cf. e.g. Lambert et al. 1984, Kleinmann & Hall 1986, Wallace & Hinkle, 1996). A reasonable number of these lines are well isolated, not affected by severe blending and can be measured even at relatively low resolution (cf. Origlia et al. 1993), which in principle make them powerful abundance tracers not only in stars but also in more distant stellar clusters and galaxies (cf. Origlia et al. 1997, Oliva & Origlia 1998). Also their modeling is relatively straightforward: the molecular lines are due to roto–vibrational transitions in the ground electronic state, so they can be safely treated under the LTE approximation. At these low temperatures the metals are poorly ionized, at most once, so also the computations of the ionization equilibria are not a crucial point, and the atomic oscillator strengths are the major source of uncertainty.

From the CO and OH lines around 1.62 μm one can infer quite reliable estimates of the C/O relative abundance, since for example a slight depletion of carbon or a slight enhancement of oxygen implies quite different spectral shapes even at low resolution. An estimate of the $^{12}C/^{13}C$ can be inferred using the (6–3) ^{12}CO and the (3–0) ^{13}CO bandheads at about 1.619 and 1.612 μm, respectively. The CO and OH molecular lines are also very sensitive to the stellar temperature and, in the case of globular cluster RGB stars, they also provide some constraints to the overall metal content, since the RGB location in temperature is strictly related to the cluster global metallicity.

3. An example: Terzan 1

As an example of the application of such an IR diagnostics to study the chemical properties of stellar clusters, by coupling photometric and spectroscopic information in the near IR, the case of Terzan 1 will be briefly discussed.

Terzan 1 is a globular cluster in the Bulge of our Galaxy with a high and quite uncertain reddening. Photometric estimates of its metal content

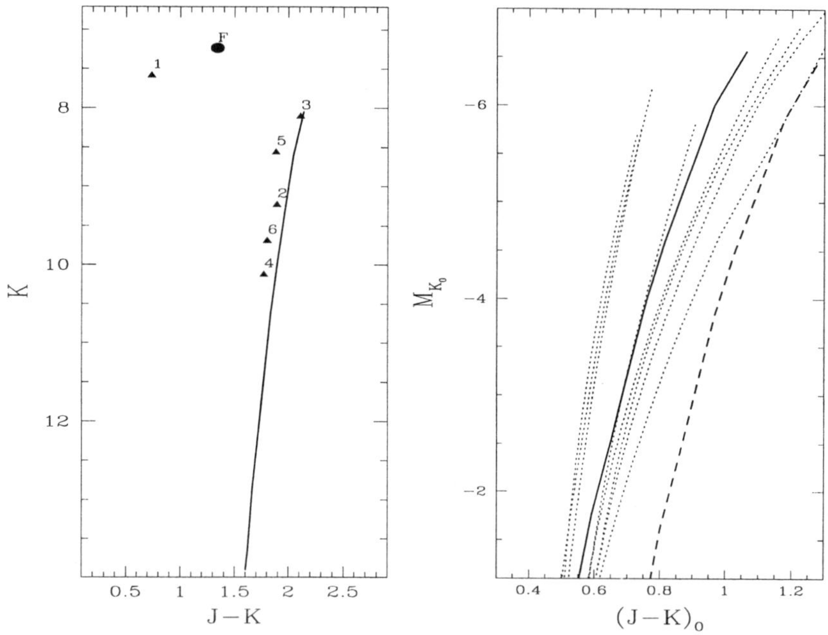

Figure 1. Left panel: RGB fiducial line of Terzan 1 in the K,J–K plane. The giants with observed near IR spectra are marked as triangles. The bright, intrinsically red field star in the cluster center is also indicated as a filled dot and labeled "F". Right panel: RGB fiducial line of Terzan 1 in the absolute M_{K_0},(J–K)$_0$ plane for two different reddening corrections: E(B–V)=1.64i (dashed line, Harris 1996) and E(B–V)=2.04 (full line, Peterson 1993). The mean ridge lines of other GGCs are plotted as dotted lines for comparison.

(Ortolani et al. 1993) and integrated spectroscopy of the very central region both in the visual (using the calcium triplet over a field of view of 10"×90", Bica et al. 1998) and in the IR (using the CO bandheads over a field of view of 5"×5", Origlia et al. 1997) suggest almost solar metallicity.

Nevertheless, very recently Ortolani et al. (1999) by means of high resolution HST images in the V and I bands suggested that Terzan 1 can be rather metal poor and the apparent metal richness is mainly due to field contamination and/or poor blanketing corrections.

In June 1993 we took J,K images of the central 4' using the ESO–MPI 2.2m telescope and the near IR camera IRAC2. In May 1999, as a backup programme at the ESO–NTT telescope, we took H and K band spectra of six bright stars using SOFI, the near IR imager/spectrometer.

The RGB slope in the K,J–K plane is one of the most robust (reddening–independent) indicator of the cluster metallicity and a linear correlation between these two quantities has been found (cf. e.g. Kuchinski 1995; Fer-

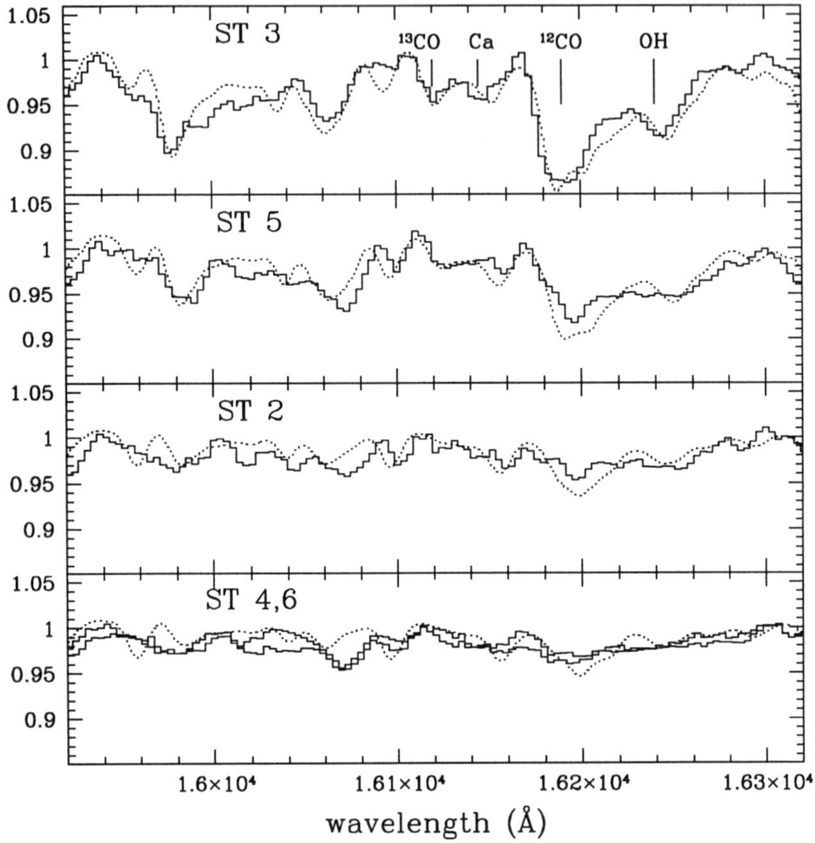

Figure 2. Observed H band spectra (full lines) of some bright giants in Terzan 1. Overimposed are our best fit synthetic spectra (dotted lines).

raro et al. 2000). The preliminary results of our photometric analysis of the upper RGB stars is reported in Fig.1. In the left panel we plot the observed RGB ridge line in the (K,J–K) CMD, and we also indicate the position of the six stars for which we collected IR spectra. Star #1 is most probably a field star, so we exclude it from the following analysis. All the other five stars lie along the upper RGB, where the temperature rapidly decreases, ascending the branch. In the right panel we locate the Terzan 1 RGB sequence in the absolute plane, assuming a true distance modulus $(m-M)_0=13.9$ (Harris 1996) and two different reddening corrections: a smaller correction (E(B–V)=1.64) as quoted by Harris (1996) and a larger one (E(B–V)=2.04) as suggested by Peterson (1993). The ridge lines of other clusters with different metallicities (Ferraro et al. 2000) are also preported for comparison.

Using the lower reddening correction, the mean ridge line of Terzan 1 would result extremely red, in the super–metal rich domain, but the overall RGB slope is somewhat too steep compared to the morphology of the other metal rich clusters. If a larger reddening correction is adopted, as suggested by Peterson (1993), a better location in the intermediate metallicity region is obtained. Larger reddening corrections would move the mean ridge line further to the blue, in the metal poor domain, but in this case the slope is too flat compared to the morphology of the other metal poor clusters.

Hence, our preliminary photometric analysis of the near IR CMD seems to indicate that Terzan 1 is a intermediate metallicity cluster, in good agreement with the recent findings by Ortolani et al. (1999).

In Fig.2 we plot the spectra around 1.6 μm of the five stars, candidate cluster members. A few major absorption features due to the CO and OH bandheads are present. The CO lines are very sharp thermometers since immediately below the RGB tip region, when the stellar temperature exceeds 4000K, as in the case of stars #2, #4, and #6, the CO lines become very shallow. Our preliminary abundance analysis indicates an overall metal content of about 1/10 solar, a C/O relative abundance a factor of 2 less than solar and a $^{12}CO/^{13}CO$ ratio of about 5.

In order to fit the observed CO and OH features with significative more metal poor atmospheres, one should need unlikely cooler stars or carbon over–abundance, while, on the contrary, significative more metal rich atmospheres would imply unlikely warmer stars or oxygen under – abundance. Moreover, it would be almost impossible to assume a unique carbon abundance for all the five stars, but what is truly unrealistic is the fact that more carbon depletion should be needed at lower RGB luminosities, contrary to what predicted by the nucleosynthesis calculations.

Why the most recent photometric and spectroscopic studies seem to indicate that Terzan 1 is an intermediate metallicity cluster, while the previous ones based on integrated spectroscopy of the central region suggest a very metal rich cluster?

About 5" from the nominal cluster center there is a very bright and intrinsically red field star, almost 1 magnitude brighter then the brightest cluster members (see Fig.1). This star strongly contaminates the integrated spectrum of the cluster central region. Bica et al. (1998) did not include such a bright star in their field of view but contamination by other red, field stars is possible.

In practice, integrated spectra of Bulge GGCs over large areas can lead to misleading metallicity estimates, also in the IR, because of the many red stars present in the Bulge. In this respect, Terzan 1 is a very unlucky example because it has a very red, field star in the central few arcsec.

4. Acknowledgments

The financial support of the *"Ministero della Università e della Ricerca Scientifica e Tecnologica"* (MURST) to the project *Stellar Evolution* is kindly acknowledged.

References

1. Ferraro, F.R., Fusi Pecci, F., Testa, V., Greggio, L., Corsi, C.E., Buonanno, R., Terndrup, D.M., Zinnecker, H.: 1995, 'Globular clusters in the Magellanic Clouds – II. IR–array photometry for 12 globular clusters and contributions to the integrated cluster light', *MNRAS*, **272**, 391
2. Ferraro, F.R., Montegriffo, P., Origlia, L., Fusi Pecci, F.: 2000, 'A new IR–Array photometric survey of Galactic Globular Clusters: A detailed study of the RGB sequence as a step towards the global testing of stellar models', *AJ*, in press
3. Frogel, J.A., Cohen, J.G., Persson, S.E.: 1983, 'Globular cluster giant branches and the metallicity scale', *ApJ*, **273**, 773
4. Frogel, J.A., Mould, J., Blanco, V.M.: 1990, 'The asymptotic giant branch of Magellanic Cloud clusters', *ApJ*, **357**, 453
5. Harris, W.E.: 1996, 'A Catalog of Parameters for Globular Clusters in the Milky Way', *AJ*, **112**, 1487.
6. Kleinmann, G., Hall, D.N.B.: 1986, 'Spectra of late–type standard stars in the region 2.0–2.5 μm', *ApJS*, **62**, 501
7. Kuchinski L.E., Frogel J.A., Terndrup D.M., Persson S.E.: 1995, 'Infrared array photometry of metal–rich globular clusters. I: Techniques and first results', *AJ*, **109**, 1131
8. Lambert, D.L., Brown, J.A., Hinkle, K.H., Johnson, H.R.: 1984, 'Carbon, nitrogen and oxygen abundances in Betelgeuse', *ApJ*, **284**, 223
9. Oliva, E., Origlia, L.: 1998, 'IR spectra of young Magellanic Cloud clusters and starburst galaxies: constraints on the temperature of red supergiants and new estimates of metallicity in young stellar populations', *A&A*, **332**, 46
10. Origlia, L., Moorwood, A.F.M., Oliva, E.: 1993, 'The 1.5–1.7 μm spectrum of cool stars: Line identifications, indices for spectral classification and the stellar content of the Seyfert galaxy NGC 1068', *A&A*, **280**, 536
11. Origlia, L., Ferraro, F.R., Fusi Pecci, F., Oliva E.: 1997, 'Infrared stellar absorption lines around 1.6 μm: a new metallicity scale for old stellar populations', *A&A*, **321**, 859
12. Ortolani, S., Barbuy, B., Bica, E.: 1991, 'Metal–rich globular clusters – Giant branch morphology as a metallicity criterion', *A&A*, **249**, L31
13. Ortolani, Bica, E., S., Barbuy, B.: 1993, 'Blanketing effects in the very metal–rich bulge globular cluster Terzan 1', *A&A*, **267**, 66
14. Ortolani, S., Barbuy, B., Bica, E., Renzini, A., Marconi, G., Gilmozzi, R.: 1999, 'HST Observations of Terzan 1: A Second Parameter Globular Cluster in the Galactic Bulge', *A&A*, in press
15. Peterson, C.J.: 1993, 'Integrated photometric properties of globular clusters', in *Structure and Dynamics of Globular Clusters*, G.S. Djorgovski & G. Meylan eds., ASP, S. Francisco (CA), ASP Conf. Ser., **50**, 337

Speaker: L. ORIGLIA

A. CHIEFFI: What is the effect of changing the abundance in the synthetic spectra of cool stars?

The continuum opacity in the spectra of cool stars is mainly due to the H^- ion while the contribution of He and metals is negligible.

L. ORIGLIA: The effect of changing He on the spectra of cool stars is practically only due to the change in H^- abundance, that is very small.

R. GALLINO:

1) Does the $^{12}C/^{13}C$ value you quoted refer to the average of the stars you considered, or does it refer to the AGB star only?

2) Together with an increase of [O/Fe] by, I say +0.3 to +0.5 dex, I would let [C/Fe]=0. What $^{12}C/^{13}C$ would be in this case?

L. ORIGLIA: Our abundances refer to the five stars on the RGB. There are some evidences that ^{12}C is slightly depleted along this sequence, while the $^{13}C/^{12}C$ is enhanced.

CAN MIRA VARIABLES TELL US THE CHEMICAL ABUNDANCES IN STELLAR SYSTEMS?

MICHAEL FEAST
Astronomy Department, University of Cape Town
(mwf@artemisia.ast.uct.ac.za)

AND

PATRICIA WHITELOCK
South African Astronomical Observatory
(paw@saao.ac.za)

Abstract. A period-metallicity relation for oxygen-rich Miras derived from globular clusters is applied to similar variables in the galactic Bulge. The metallicity distribution in the Bulge thus obtained is in good agreement with that derived from K giants. This suggests that the periods of Miras are useful indicators of metallicity distributions, at least in old systems such as galactic bulges. For systems such as the LMC the situation is not yet clear.

1. Introduction

Whilst detailed, element by element, abundance analyses can now be extended to relatively faint stars, the overall metallicity of a star remains an important characteristic. Even to determine this directly for very faint stars remains a formidable problem. This is especially so when we wish to measure the spread of metallicities within a given system, since this evidently requires data on a large number of stars. In the present paper we suggest that the periods of Mira variables when suitably calibrated can provide this type of abundance data for old stellar systems such as the Bulge region of our own Galaxy. This is of particular interest since the period determination of Mira variables in the bulges of nearby galaxies, as well as in highly obscured regions of the Bulge of our own Galaxy, is now within reach.

F. Matteucci and F. Giovannelli (eds.), The Evolution of the Milky Way, 229–237.

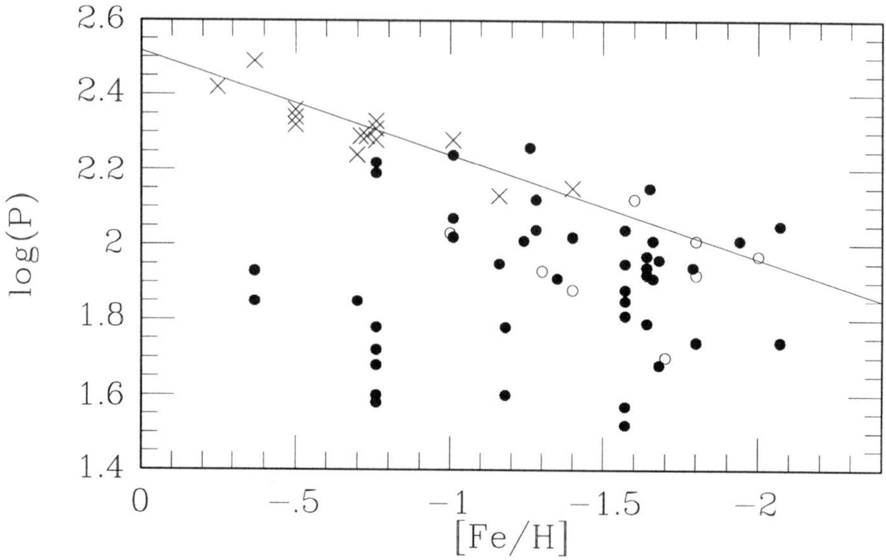

Figure 1. The pulsation period (in days) of variable stars is shown as a function of their [Fe/H]. Cluster Miras and SRs are shown as crosses and filled circles, respectively. Open circles represent field SRs with spectroscopically determined [Fe/H] values.

2. The Mira Period-Metallicity Relation

It has long been known that Miras (in the present paper we are only concerned with oxygen-rich Miras) are highly evolved, old, low mass stars and that their kinematics are a function of period (e.g. Feast 1963). This indicates that the period of a Mira in the general field is a function of metallicity or initial mass (or both). It is not possible to address this problem further in the general field since the complexity of Mira spectra has so far precluded direct spectroscopic measurements of metallicity for them. However, it is known (Feast 1981) that for those Miras which are members of globular clusters there is a relation between the Mira period and the metallicity of the cluster. This relation was quantified (Feast 1992) to [Fe/H] $\propto$ 3.6 $\log P$ (where P is the period).

The presently available data are shown in fig 1. Here the log of the period is plotted against metallicity, conventionally denoted as [Fe/H], for Miras and semiregular (SR) variables in globular clusters. The data used are primarily from Frogel and Whitelock (1998), details kindly made available by Dr J. Frogel. Some additional SR variables in globular clusters are also included in the plot as well as two additional cluster Miras, V16 in NGC 362 (Lloyd Evans 1983) and V1 in NGC 121 (an SMC cluster). The open circles represent SR variables in the field of the galactic halo for which there are spectroscopic determinations of [Fe/H] available (Preston and Wallerstein

1963, Luck and Bond 1985, Leep and Wallerstein 1981, Giridhar *et al.*, 1998, 1999), taking means where necessary. In a given globular cluster the SRs lie along an evolutionary track (or more precisely an isochrone), increasing period going with increasing luminosity (Whitelock 1986). For metal-rich clusters the AGB track terminates with the Miras (if any) in the cluster. Evidently an evolutionary track in fig 1 is vertical. The sloping line shown has been drawn by eye to characterize the longest period at each metallicity and hence may be taken as indicating the end-point of AGB evolution at each metallicity. The slope of the line was as determined in the earlier work on Miras (Feast 1992). Various linear fits to the Mira data are possible depending on how the observations are weighted and the adopted relative errors in the two co-ordinates. The line shown however, seems satisfactory for the present purpose as a fit to the Miras, and has the advantage of giving some weight to the upper limit for the low metallicity SR variables. In the lower metallicity clusters there are no Mira variables as normally defined, but the SR variables near the sloping line are presumably in an equivalent phase of stellar evolution to the Miras in the more metal-rich clusters. The line is given by the relation:

$$[\mathrm{Fe/H}] = 3.60 \log P - 9.06. \tag{1}$$

Figure 1 thus indicates that, at least in globular clusters, the period of a Mira is a good indicator of metallicity.

Frogel and Whitelock (1998) have shown recently that, within the inevitable uncertainty of small number statistics, the Miras in globular clusters have another useful property. This is that the ratio of the number of Miras in a cluster to the total luminosity of the cluster in the infrared (at K, 2.2μm) is independent of metallicity (over the range of metallicities for which cluster Miras are found). These results suggest that if the cluster $[\mathrm{Fe/H}] - \log P$ relation is applicable to a stellar system, not only will it yield the range of metallicities in the system, but also, the relative numbers of Miras of different periods will provide a useful estimate of the frequency distribution of these metallicities in the parent population.

3. Application to the Galactic Bulge

We can now test and discuss the cluster $[\mathrm{Fe/H}] - \log P$ relation by applying it to Miras in the galactic Bulge. To do this we have combined data on Miras in the two Baade windows, Sgr I and the NGC 6522 field. Combining the data from the two fields improves the statistics considerably.

In carrying out an analysis such as this it is essential that the results are not biased by selection effects. Although the infrared (K) and bolometric luminosities of Miras increase with increasing period, the visual luminosity

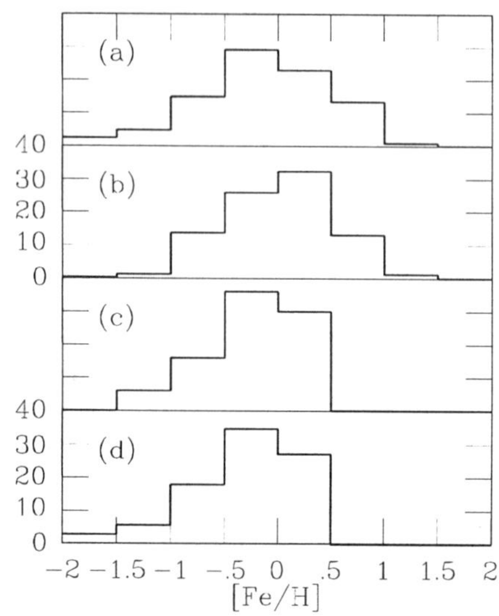

Figure 2. Histograms of the metallicity distribution of galactic Bulge stars: (a) Miras (as derived from periods); (b) K giants (Sadler *et al.*); (c) K giants (McWilliam and Rich); (d) Miras renormalized as described in the text. The samples have been normalized to the same effective total number of objects.

decreases because the longer the period, the cooler the star. The initial South African observations in the Sgr I field were on blue sensitive plates and the variables discovered on them (Oosterhoff and Ponsen 1968) were mainly of short period. The later work was on red sensitive plates and the proportion of longer period Miras found was larger (Lloyd Evans 1976). Later still Miras detected by IRAS extended further the long period coverage (Feast 1986, Glass 1986, Glass *et al.*, 1995), see Glass *et al.* fig 2. For the NGC 6522 field we have used the periods from Lloyd Evans. This samples seems essentially complete as it includes all except one of the IRAS Miras in the field. In the case of the Sgr I field we use the data from Glass *et al.* which includes the IRAS Miras together with periods from Lloyd Evans for Miras not observed by Glass *et al.*

Where necessary, in this first application, we simply extrapolate equation 1 linearly to higher metallicities, outside the range of the globular cluster Miras. The consequences of this extrapolation will be discussed below.

The frequency distribution of Mira metallicities derived in the above manner may be compared with the frequency distribution of metallicities of K giants in the NGC 6522 field. For this we have used the values of [Fe/H] derived by Rich (1988) for 88 stars (his solution 1 rescaled according to equation 6 of McWilliam and Rich (1994)), and the [Fe/H] values for the 262 Bulge K giants derived from low dispersion spectra by Sadler *et al.* (1996) and plotted in their fig 11. Figure 2 shows this comparison. The Mira distribution (total number of stars = 112) and the Sadler *et al.* K giant distribution have been scaled to match the total number of K giants in the Rich sample.

Figure 2 shows that the distributions of the Miras and the K giants from Sadler *et al.* agree very well. Sadler *et al.* quote a mean [Fe/H] for their sample of -0.11 ± 0.04 with a dispersion of $\sigma_{[Fe/H]} = 0.46$ whilst the Miras give a mean [Fe/H] of -0.16 ± 0.06 and $\sigma_{[Fe/H]} = 0.59$. These values are not significantly different for the two samples. The Mira metallicity distribution agrees well with that of the Rich sample except at the high metallicity end. The agreement of the Mira distribution with either the Rich or the Sadler *et al.* sample would be made considerably worse by moving the Miras one box in either direction.

There are at least three possible (alternative) reasons for the apparent lack of agreement of the Miras with the Rich sample at the high metallicity end.

(1) At the high metallicity end, both the Mira and the Sadler *et al.* distributions depend on extrapolations of the (entirely different) metallicity calibrations used. One can to some extent circumvent possible problems with the extrapolation by simply comparing the number of Rich K giants with [Fe/H] > 0.0 with the number of Miras in this same metallicity range. There are 30 Rich K giants with [Fe/H] > 0, whilst the scaled number of Miras is 37. These numbers are not statistically different. Thus the results do not preclude the possibility that the frequency distributions of Rich K giants and Miras are the same at all metallicities. A change in the slope of the [Fe/H] $- \log P$ relation at high metallicities would suffice to bring the two distributions into satisfactory detailed agreement.

(2) On the other hand there also seems some possibility that the Rich K giant distribution might be deficient at the metal-rich end. This would occur if all the very metal-rich K giants were below Rich's magnitude limit. In other words, in the magnitude range observed by Rich such metal-rich stars might be late M giants. In that case a better comparison between the Rich stars and the Miras is obtained by omitting the long period Miras which are predicted to have [Fe/H] > 0.5, and rescaling. The result of doing this is shown in fig 2(d). The agreement of the two distributions is improved in this way.

(3)Finally, it should be borne in mind that the apparent difference between the Rich K giant distribution and the other two distributions may be simply a result of small number statistics.

Our conclusion regarding the Mira metallicity distribution differs somewhat from that of Frogel and Whitelock (1998). They also obtained a metallicity distribution from the periods of Bulge Miras (their fig 8). This agrees less well than ours with the K giant distribution (either that of Rich or that of Sadler *et al.*) both at high and low metallicities. The main reason for this is that Frogel and Whitelock used a much steeper [Fe/H] – $\log P$ relation than is consistent with our fig 1. This is due to their relation being drawn to take into account some metal-rich SR variables which are below the limiting line that the Miras actually fit.

At first sight our conclusions seem at odds with another of the results derived by Frogel and Whitelock. They concluded that integrated over all metallicities, the number of Miras per giant brighter than $M_{bol} = -1.2$ in the NGC 6522 field was less than in globular clusters. However, at the present time this result must be considered quite uncertain. The result depends partly on a comparison of the total number of Miras in the NGC 6522 field with the number of M type stars brighter than a certain apparent magnitude in that field and the assumption that stars of the same apparent magnitude have the same absolute magnitude. Whilst this is true for a globular cluster it is not so for the galactic Bulge since in the Bulge the stars are significantly spread along the line of sight. The spread (~ 2 mag) in the Mira infrared-PL relation in the Sgr I field (Glass *et al.*, 1995) compared with the narrow PL relation in the LMC (Feast *et al.*, 1989) is evidence for such a spread. This will evidently complicate any analysis. In addition the relative numbers of Miras to giants in globular clusters is not calculated directly but derived from a theoretical relation between the total luminosity of a cluster and the number of giants brighter than a certain value. Whilst our own results are in satisfactory agreement with the hypothesis that the ratio of numbers of giants to total luminosity does not depend critically on metallicity, the absolute calibration of this ratio depends on theoretical giant branch isochrones.

4. Is the Mira Period-Metallicity Relation Universal?

The evidence just discussed suggests that Mira of a given period in the galactic Bulge have the same metallicity as Miras of that period in globular clusters. Is this result universal?

There is evidence of differences between Miras of the same period in different environments. Glass *et al.* (1995) have compared the period - infrared colour relations for Miras in the Sgr I field with those in the LMC (their fig 4). There is some uncertainty in these relations due to uncertainties in the correct interstellar absorption to adopt. However, no adopted relative absorption between the LMC and Bulge fields, with a standard reddening law, will bring all the period-colour relations into agreement. If we make the $P - (J - K)$ relations agree then $H - K$ is redder in Sgr I than the

LMC at a given period and $J - H$ is bluer. In view of this result a useful quantity is:

$$\phi = (J - H)_0 - (H - K)_0. \tag{2}$$

This quantity is rather insensitive to the adopted interstellar reddening. There is no significant difference in ϕ between the Bulge Miras and those in globular clusters ($\Delta\phi(Glob.Cl. - Bulge) = +0.03 \pm 0.03$). However, a difference, varying somewhat with period, does exist between the Miras in the LMC and those in the Bulge. Using the stellar models of Bessell *et al.* (1989), Feast (1996) used this difference in ϕ to deduce that the LMC Miras at a given period were metal deficient compared with those in the Bulge by $\sim$ 0.4 dex. Differences in ϕ reflect differences in the strength of H_2O bands at a given period. This will be affected not only by differences in the atmospheric [O/H] ratio but also by any difference in the [C/H] ratio (due for instance to dredge up processes) since oxygen is preferentially locked up in the CO molecule. Thus whilst it seems safe to assume that the LMC Miras are deficient in oxygen compared with those in the Bulge the exact amount is still uncertain. Whether an oxygen deficiency implies a deficiency in other elements (e.g. iron) is also uncertain. As summarized by Gilmore and Wyse (1991) young objects in the LMC seem to have a lower [O/Fe] than similar galactic objects but it is not known if this extends to older stars. It remains unclear therefore whether the LMC Miras would fit the period-metallicity relation of fig 1. However, it is noteworthy that the Mira in the SMC globular cluster NGC 121 which is plotted at $\log P = 2.15$ (Thackeray 1958) and [Fe/H] = –1.4 (Stryker *et al.*, 1985) in fig 1 fits the adopted relation closely. This suggests that at least at short periods the Magellanic Cloud Miras fit the adopted relation [1].

The period distribution of O-rich Miras in the LMC peaks at shorter periods than in the galactic Bulge. Thus the O-rich LMC Miras in Table II of Hughes and Wood (1989) together with equation 1 above yield a distribution of LMC O-Mira metallicities peaked between an [Fe/H] of –1.0 and –0.5. However, unlike the galactic Bulge there are also carbon Miras in the LMC and it is not entirely clear how the overall metallicity distribution would be affected by taking these into account, even if equation 1 applies to the LMC O-Miras.

[1](1) There is a also a difference in the $(K_0 - m_{bol}) - \log P$ relation between the LMC and the Bulge Miras which is probably a metallicity effect (see Feast and Whitelock 1999).

(2) The value of ϕ at a given period is also a function of the pulsation amplitude (Whitelock *et al.* in preparation). This is believed to be due to the strengthening of the H_2O bands as the atmospheric extension is increased by pulsation. It is unlikely that this is the cause of the LMC - Bulge difference.

5. Conclusions

Our main conclusion is that applying the metallicity scale for Miras as a function of period set by globular clusters, to Miras in Bulge fields, leads to a metallicity distribution in good agreement with that shown by Bulge K giants in the sample of Sadler *et al.* The agreement is also good with the metallicity distribution of the (revised) Rich K giant sample except at the very metal-rich (long period) end where a significant extrapolation of the derived cluster period-metallicity relation is required. A modification of this extrapolation could remove this discrepancy. Alternatively the K giant sample of Rich might be deficient in very metal-rich stars. Thus at least in the Bulge, the Miras provide a metallicity tracer for a significant population.

References

1. Bessell, M.S., Brett, J.M., Scholz, M. & Wood, P.R. (1989) *A&AS*, **77**, 1
2. Feast, M.W. (1963) *MNRAS*, **125**, 367
3. Feast, M.W. (1981) in: Iben, I. & Renzini, A., eds., *Physical Processes in Red Giants*, Reidel, Dordrecht, p. 193
4. Feast, M.W. (1986) in: Israel, F.P., ed., *Light on Dark Matter*, Reidel, Dordrecht, p. 339
5. Feast, M.W. (1992) in: Bergeron, J., ed., *Highlights of Astronomy*, vol. 9, Kluwer, Dordrecht, p. 613
6. Feast, M.W. (1996) *MNRAS*, **278**, 11
7. Feast, M.W., Glass, I.S., Whitelock, P.A. & Catchpole, R.M. (1989) *MNRAS*, **241**, 375
8. Feast, M.W. & Whitelock, P.A. (1999) in: Heck, A. & Caputo, F., ed., *Post-Hipparcos Cosmic Candles*, Kluwer, Dordrecht, p. 75
9. Frogel, J.A. & Whitelock, P.A. (1998) *ApJ*, **116**, 754
10. Gilmore, G. & Wyse, R.F.G. (1991) *ApJ*, **367**, L55
11. Giridhar, S., Lambert, D.L. & Gonzalez, G. (1998) *PASP*, **110**, 671
12. Giridhar, S., Lambert, D.L. & Gonzalez, G. (1999) *PASP*, **111**, 1269
13. Glass, I.S. (1986) *MNRAS*, **221**, 879
14. Glass, I.S., Whitelock, P.A., Catchpole, R.M. & Feast, M.W. (1995) *MNRAS*, **273**, 383
15. Hughes, S.M.G. & Wood, P.R. (1989) *AJ*, **99**, 784
16. Leep, E.M. & Wallerstein, G. (1981) *MNRAS*, **196**, 543
17. Lloyd Evans, T. (1976) *MNRAS*, **174**, 169
18. Lloyd Evans, T. (1983) *MNRAS*, **204**, 961
19. Luck, R.E. & Bond, H.E. (1985) *ApJ*, **292**, 559
20. McWilliam, A. & Rich, R.M. (1994) *ApJS*, **91**, 749
21. Oosterhoff, P. Th. & Ponsen, J. (1968) *Bull. Ast. Inst. Neth. Sup.*, **3**, 79
22. Preston, G. & Wallerstein, G. (1963) *ApJ*, **138**, 820
23. Rich, R.M. (1988) *AJ*, **95**, 828
24. Sadler, E.M., Rich, R.M. & Terndrup, D.M. (1996) *AJ*, **112**, 171
25. Stryker, L.L., Da Costa, G.S. & Mould, J.R. (1985) *ApJ*, **298**, 544
26. Thackeray, A.D. (1958) *MNRAS*, **118**, 117
27. Whitelock, P.A. (1986) *MNRAS*, **219**, 525

Speaker: M. FEAST

S. ORTOLANI: Is it possible that Miras interpreted as super-metal rich (more than McWilliam and Rich distribution) are coming from a younger population?

M. FEAST: It is possible that the longest period Miras are the youngest. However this is not certain. More metal rich clusters contain longer period Miras. So the problem comes down to asking if more metal rich globulars are younger than less metal rich ones. As you know there is no present agreement on this matter.

ABUNDANCES IN OPEN STAR CLUSTERS: WHAT WE HAVE TO WORK WITH

RANDY L. PHELPS
California State University, Sacramento
Department of Physics and Astronomy
6000 J Street
Sacramento, CA 95819 USA

Abstract. By virtue of the fact that ages and distances can be reliably be determined for them, open clusters provide important tools for investigating the chemical evolution of the Milky Way. In this review, our current understanding of Galactic evolution, as derived from open clusters will be summarized, as will attempts to improve upon the knowledge gained from them.

1. Introduction

Star clusters are laboratories with which a wide range of astrophysical problems may be investigated, including the formation and structure of the disk of our Galaxy. Old open clusters (here defined to be clusters older than the Hyades, with an age of approximately 800 Myr) provide unique probes of the formation and development of the Galactic disk, since they span a wide range in ages, they can be dated quite reliably, they can be seen to large distances, and the brightest members are strong-lined, luminous red giants which are well suited to metallicity (e.g., Geisler 1987; Geisler et al. 1992; Friel & Janes 1993) and radial velocity studies (Scott, Friel & Janes 1995). It is for these reasons that many open cluster studies have been undertaken to derive the Galactocentric radial abundance gradient of the disk, the age-metallicty gradient and age of the disk, along with other relations within the disk.

The purpose of this contribution is to 1) review the foundations of our current understanding of the chemical evolution of the Galactic disk using open clusters; 2) to highlight the strengths and weaknesses of the

F. Matteucci and F. Giovannelli (eds.), The Evolution of the Milky Way, 239–247.

foundations and 3) to indicate current and future work to shore up those foundations.

2. Foundations of Our Current Understanding of Open Cluster Abundances

2.1. THE GALACTOCENTRIC RADIAL ABUNDANCE GRADIENT

Open clusters have been used for several decades in an attempt to investigate the chemical evolution of the Galactic disk. Janes (1979) first used open clusters, by employing DDO and UBV photometry, to derive the Galactocentric radial abundance gradient. Using 41 open clusters he derived [Fe/H] = -0.075 $\pm$ 0.034 dex kpc^{-1}. Panagia & Tosi (1981), by matching theoretical isochrones to HR diagrams of 20 clusters with ages less than 1 Gyr old, derived [Fe/H] = -0.095 $\pm$ 0.034 dex kpc^{-1}, while Cameron (1985), using UBV photometry of clusters spanning a wide age range, derived [Fe/H] = -0.11 $\pm$ 0.02 dex kpc^{-1}. Janes et al. (1988), by using a weighted fit to the published parameters in the Lynga (1987) catalog of open cluster data, derived [Fe/H] = -0.113 dex kpc^{-1}. By separating the clusters into age groups, he found [Fe/H] = -0.07 dex kpc^{-1} for clusters with ages less than 200 Myr, and [Fe/H] = -0.14 dex kpc^{-1} for clusters older than 200 Myr.

Among the most widely referenced works on the radial abundance gradient is that of Friel & Janes (1993). In that study, moderate resolution (4 Å) spectra were obtained in order to derive [Fe/H] from indices measuring Fe and Fe-peak element blends, the CN λ4216 Åband and Mg b and Mg H features. The spectra were calibrated using standard stars with detailed abundance analyses (Cottrell & Sneden, 1986), along with several 'standard clusters'. Friel & Janes (1993) obtained spectra within 24 clusters, with abundances being derived by averaging 6-7 stars/cluster. Data for an additional 9 clusters were taken from the literature. Cluster parameters (e.g., age, distance) were taken from the Lynga (1987) catalog of open cluster data. From this study, Friel & Janes (1993) derived a Galactocentric radial abundance gradient of [Fe/H] = -0.095 $\pm$ 0.017 dex kpc^{-1} using all 33 clusters, or [Fe/H] = -0.088 $\pm$ 0.017 dex kpc^{-1} using only the 24 clusters for which they themselves had obtained spectra.

A subsequent revision of the Friel & Janes (1993) result was presented by Friel (1995), using additional spectroscopic results and a more uniform set of cluster properties. From this analysis, Friel (1995) derived [Fe/H] = -0.091 $\pm$ 0.014 dex kpc^{-1}. A further revision was presented by Friel (1998), using yet more spectroscopic data and yielded [Fe/H] = -0.06 $\pm$ 0.01 dex kpc^{-1}, a gradient that is shallower than previously derived. At about the same time, Carraro, Ng & Portinari (1998), using a sample of 37 clusters with age determinations based on calibration of a synthetic color magnitude technique, derived metallicities on the Friel & Janes (1993) and Friel (1995) abundance scale. They derived an overall abundance gradient in the Galactic disk of [Fe/H] = -0.09 dex kpc^{-1}, and found that it was most likely independent of cluster age.

An alternative position has been put forth by Twarog, Ashman & Anthony-Twarog (1997). Using 76 clusters with abundances based upon DDO photometry and/or moderate dispersion spectroscopy, they interpret the cluster abundance distribution to be that of two groupings, with a sharp discontinuity at a Galactocentric radial distance of 10 kpc. At distances less than 10 kpc, clusters have a mean metallicity of [Fe/H] $\sim$ 0, while those exterior to 10 kpc have a mean metallicity of [Fe/H] $\sim$ -0.3.

2.2. THE GALACTOCENTRIC AGE-METALLICITY RELATION

Geisler (1987), using Washington photometry of 7 outer disk clusters, concluded that the chemical enrichment of Galactic anti-center clusters is more like that of LMC than the solar neighborhood, based on the cluster age-metallicity relation (AMR). These results were based, in part, on the observation that four of the clusters had abundances as low as, or lower than, 47 Tuc. Among the most widely referenced works on the AMR within the disk is that of Friel & Janes (1993). From their analysis, it was determined that there is essentially no correlation between cluster age and metallicity over an age range of about 1-8 Gyr. There is a tendency for clusters with Galactocentric radii less than 10 kpc to have lower abundances, but no apparent age-metallicity relation was found.

This result was confirmed in the Friel (1995) study, which used a more robust and uniform age ranking of open clusters. Phelps, Janes & Montgomery (1994) undertook a comprehensive literature search for open cluster photometry, and over a several year period, undertook a CCD photometric survey of candidate old (ages greater than the Hyades) open clusters. Using published isochrone age estimates for a sample of old open clusters, Janes & Phelps (1994) calibrated color-magnitude diagram morphological age indicators and derived a Morphological Age Index (MAI) that provides a means to rank cluster ages on a relative scale. The technique used is similar to that developed by Anthony-Twarog & Twarog (1985), and later modified by various investigators. The resulting age ranking of $\sim$ 75 old open clusters, and 15-20 clusters older than M67 found by Janes & Phelps (1994), were used by Friel (1995) to derive her AMR.

Carraro et al. (1998) recently performed a comprehensive analysis of the open cluster AMR. In their study, they found that field stars and clusters trace a similar AMR, with an excess of metal-rich objects with ages between 5-9 Gyr. These are the objects that give the open cluster AMR, in particular, a 'flat' appearance in most studies

2.3. OTHER RELATIONS WITHIN THE DISK

Another metallicity-related relation within the disk is that of metallicity versus distance above or below the Galactic plane. Friel & Janes (1993) found no apparent correlation of chemical abundance with a cluster's vertical distance from the Galactic plane. This result was confirmed by Friel (1995) and Friel (1998). Carraro et al. (1998) concluded that the existence of a vertical gradient could not be established from their data, a result that is consistent with the lack of a detectable gradient by Twarog et al. (1997). The preponderance of evidence to-date, therefore, suggests that no abundance gradient perpendicular to the Galactic disk exists.

Other relations that are not specifically metallicity-related, but are nonetheless of interest for studies of Galactic evolution relate to the age distribution of clusters, and how clusters of various ages are spatially distributed. The most comprehensive study of these studies was undertaken by Janes & Phelps (1994), with subsequent revisions by Friel (1995). In these studies, the well-known result that old open clusters are found only for Galactocentric distances greater than about 7.5 kpc was reinforced by the inclusion of substantially more clusters. Additional results indicate that no correlation exists between old open cluster age and distance perpendicular to the Galactic plane (although the scale height of old clusters is about 375 pc compared to the 55 pc scale height of clusters with ages less than 600-800 Myr), and that there is no correlation between cluster age and Galactocentric distance (beyond 7.5 kpc).

Another noteworthy result from Phelps et al. (1994) and Janes & Phelps (1994) was the age distribution of old open clusters, which suggested that the oldest open cluster (Berkeley 17) had an age as old as the youngest globular clusters. Phelps (1997) derived an age of 10-13 Gyr for Berkeley 17. Kaluzny (1994) derived a lower age ($\sim$ 9 Gyr), as did Carraro et al. (1999), where an age $\leq$ 10 Gyr was derived. Berkeley 17, therefore, appears to among the oldest open clusters, and is perhaps *the* oldest open cluster, depending on the adopted age. A determination of its true absolute age would place important constraints on the age of the Galactic disk (see the contribution by Carraro in this proceedings).

3. How Strong Are the Foundations?

Recent studies have significantly improved our understanding of the evolution of the Galactic disk, both chemically and spatially. Nevertheless, differences in the published data, and occasional misinterpretations of their uncertainties, have led to important differences in interpreting the meaning of the data. Until these differences are addressed, controversy will remain about the formation and evolution of the Galactic disk.

For example, the Friel & Janes (1993) metallicity study, and its subsequent revision by Friel (1995), have formed the basis for many studies of the chemical evolution of the disk. Recent results, however, suggest that the metallicity zero-point used in these studies may be in error (e.g., Gratton 1999; Twarog et al. 1997). Friel (1998), in fact, based on additional cluster data and a recalibration of the spectroscopic indices, derived a Galactocentric radial abundance gradient of [Fe/H] = -0.06 ± 0.01 dex kpc^{-1}, which is significantly shallower than that found in her earlier studies. As discussed by Friel & Janes (1993), an error in the derived reddening of a cluster can result in a metallicity error corresponding to about +0.06 dex for every 0.05 mag increase in the derived E(B-V). These, and higher, levels of reddening uncertainties are well within the realm of possibility for many of these distant, highly reddened open clusters. Future revisions of cluster properties, and relations constructed based on them, are likely.

An important individual example is Berkeley 21. This cluster, with an adopted [Fe/H] of -0.97 (Friel & Janes 1993), has been considered to be the most metal poor open cluster. Recent results, however, suggest that this metallicity may be too low. Hill & Pasquini (1999), using spectra of 3 giants, derived an abundance of [Fe/H] = -0.54 ± 0.2 dex, a result that is in line with a recent photometric study of the cluster (Tosi et al. 1998) that suggested a metallicity closer to -0.5.

The Friel (1995) results may also be prone to revisions as new ages are derived. As previously discussed, the Friel (1995) 'ages' are based on the MAI of Janes & Phelps (1994). These 'ages' have been misused in the literature, since the MAI was intended only to provide a relative age ranking of clusters, and not as a true age. As emphasized by Janes & Phelps (1994), the MAI is based on correlations of the morphological age parameters for open clusters with any/all published isochrone ages at that time - the MAI is thus based on a 'hodgepodge' of stellar input physics and it should not be used for absolute ages!

Even without the uncertainties associated with the MAI, age estimates for clusters are often controversial. In large part this results from the fact that cluster ages are based on eyeball matches to cluster color-magnitude diagrams, and remain susceptible to an individual investigators 'opinion' as to how good the match is. An important example is the case of Berkeley 17. Phelps (1997), based on a VI photometric study, suggested that the cluster has an age of 10-13 Gyr, based solely on the photometry. No use of the Friel (1995) metallicity was made since, as pointed out in Phelps (1997), the Friel (1995) metallicity determination used a cluster reddening obtained from preliminary photometry that was found to be in error by 0.1

mag. Other studies of the cluster (e.g., Kaluzny 1994; Carraro et al. 1999) indicated a lower age of 9-10 Gyr. The Carraro et al. (1999) study, however, based on a reanalysis of the Phelps (1997) technique of age determination, concluded that an age of 11 $\pm$ 1 Gyr, was also not ruled out.

4. Current and Future Work to Shore Up the Foundations

Recent photometric work by various groups (e.g., Phelps, Janes and collaborators, Carraro and collaborators, Twarog, Anthony-Twarog and collaborators, Kaluzny and collaborators) and spectroscopic work (e.g., Friel and collaborators) have greatly improved our knowledge of open cluster properties, and hence our understanding of the chemical evolution of the disk of our Galaxy. Despite this improvement, however, numerous discrepancies for individual clusters remain (see above), as does the detailed interpretation of the meaning of the latest research results (see also above). Fortunately, several groups, undertaking complementary approaches, have begun work that promises to shore up the foundations of our current understanding.

4.1. HIGH-RESOLUTION SPECTROSCOPY

We are now at the point where further gains in metallicity studies require high-resolution spectroscopy. For example, determination of [O/Fe] and [α/Fe] are required to ensure that open cluster abundances are on the same scale as the globular clusters. Several groups have headed in the direction of high-resolution abundances studies, among them J. Brown and collaborators (c.f., Sneden et al. 1995), E. M. Green and collaborators (private communication) and Bragaglia & Tosi (e.g., this proceedings). The faintness of the stars in the most interesting old open clusters, coupled with the need for high dispersion, however, have limited these studies to scarcity of large aperture telescopes available for open cluster studies. The new generation of 6-10 meter telescopes should provide a revolution in open cluster abundance studies, if telescope time can be obtained.

4.2. HIGH-PRECISION PHOTOMETRY

Discrepancies in photometrically derived properties of clusters, and stars within them, also limit the interpretations made possible by open cluster studies. Several groups (e.g., Carraro and collaborators, Twarog and collaborators, Kaluzny and collaborators, Tosi and collaborators) have been instrumental in improving the available photometry for open clusters. For further progress to be made, additional high-precision, multiwavelength photometric studies must be undertaken, including the near-infrared (e.g., Carraro et al. 1999). Near-infrared studies will prove to be especially important for deriving V-K temperature calibrations of stars for high-resolution abundance studies.

4.3. A NEW PHOTOMETRIC SURVEY: THE PALOMAR-LAS CAMPANAS OBSERVATORY-NOAO OPEN CLUSTER SURVEY

Another approach for improving our understanding of the chemical evolution of the Galactic disk from open clusters involves the identification of yet more clusters for further study. For example, only two clusters (Be 20 and Be 29) are known with Galactocentric radii grater than 16 kpc. This severely limits our ability to probe the full evolutionary history of the Galactic disk.

In an attempt to improve the situation, a new survey has been initiated to increase the sample of old open clusters. The primary goal of the Palomar-Las Campanas Observatory-NOAO (PLCON) Open Cluster Survey (Phelps, in preparation) is the identification of new old open clusters to further studies of the evolution of the Galactic disk. For example, an increased number of known old open clusters will help to establish better the evolutionary history of the outer disk and help to confirm the age of the Galactic disk based on the age of the oldest open clusters (e.g., Phelps 1997).

Candidate old open clusters are selected based on several criteria. One criteria uses the fact that distant old open cluster will be small in angular size, and have a uniform distribution of the brightest stars rather than a few bright ones. Janes & Phelps (1994) found that the median size of old open clusters is 5 pc, so small angular-size clusters may be distant and the population of brightest stars (e.g, red giants) would be faint. Appearance, therefore, provides one means of identifying candidate old clusters. Another criteria uses the fact that old open clusters have a larger scale height (375 pc) above the Galactic plane than the 55 pc scale height of the younger open clusters (Janes & Phelps 1994). If one assumes a cluster has a diameter of 5 pc (Janes & Phelps 1994), then using the published diameters (e.g., Lynga 1987), a crude estimate of the cluster's distance from the Galactic plane can be obtained. Candidate old open clusters are taken to be those that have a 'significantly' higher distance from the Galactic plane (e.g., 3 scale heights) than the younger open cluster scale height.

Using a small number of telescopes (primarily the Palomar 1.5-meter and Las Campanas 1-meter telescopes), consistent sets of filters, and a small number of detectors, a uniform data set for open clusters will be obtained. The Survey strategy involved the observation of candidate old open clusters using short, medium and long exposures in each of the V and I passbands. To-date, some 500 open clusters have been observed.

A number of these clusters have been found to be old. Among the new old open clusters are Berkeley 52, with an age of 1-2 Gyr, ESO 098-SC03 which was independently found to be old (age 4-5 Gyr by Bica et al. 1999) and Berkeley 57. A complete accounting awaits the conclusion of the PLCON Survey, which is expected in several years.

5. Conclusion

The past decade has seen great progress in our understanding of the evolution of the Galactic disk, as inferred from studies of open clusters. A number of nagging questions remain, however. The path to improving our understanding involves higher-resolution spectroscopic studies, new high-precision photometric studies, and a sample of new objects to study. Several groups have risen to the challenge, and the future should lead to a clearer picture of what open clusters are telling us about the evolution of the Galactic disk.

References

1. Anthony-Twarog, B. J. & Twarog, B. A. (1985) *ApJ*, **291**, 595
2. Bica, E., Ortolani, S. & Barbuy, B. (1999) *A&AS*, **136**, 363
3. Cameron, L. M. (1985) *A&A*, **147**, 47
4. Carraro, G., Ng & Portinari (1998) *MNRAS*, **296**, 1045
5. Carraro, G., Vallenari, A., Girardi, L. & Richichi, A. (1999) *A&A*, **343**, 825
6. Carraro, G., Girardi, L. & Chiosi, C. (1999) *MNRAS*, **309**, 430
7. Cottrell, P. L. & Sneden, C. (1986) *A&A*, **161**, 314
8. Friel, E. D. (1995) *ARA&A*, **33**, 381
9. Friel, E. D. (1998) in *Galaxy Evolution: Connecting the Distant Universe with the Local Fossil Record* M. Spite, F. Crifo (eds.), p. 49. Kluwer: Dordrecht
10. Friel, E. D. & Janes, K. A. (1993) *A&A*, **267**, 75
11. Geisler, D. (1987) *AJ*, **94**, 84
12. Geisler, D., Claria, J. J. & Minniti, D. (1992) *AJ*, **104**, 1892
13. Gratton, R. (1999) in *Stellar Clusters and Associations: Convection, Rotation and Dynamos*, ASP Conference Series, in press
14. Hill, V. & Pasquini, L. (1999) *A&A*, **348**, 21
15. Janes, K. A. (1979) *ApJS*, **39**, 135
16. Janes, K. A., Tilley, C. & Lynga, G. (1988) *AJ*, **95**, 771
17. Janes, K.A. & Phelps, R. L. (1994) *AJ*, **108**, 1773
18. Kaluzny, J. (1994) *AcA*, **44**, 247
19. Lynga, G. (1987) *Lund Catalog of Open Cluster Data*
20. Panagia, N. & Tosi, M. (1981) *A&A*, **96**, 306
21. Phelps, R. L. (1997) *ApJ*, **483**, 826
22. Phelps, R. L., Janes, K. A. & Montgomery, K. A. (1994) *AJ*, **107**, 1097
23. Scott, J., Friel, E. D. & Janes, K. A. (1995) *AJ*, **109**, 1706
24. Sneden, C., Basri, G., Boesgaard, A. M., Brown, J. A., Carney, B. W. Kraft, R. P., Smith, V. V. & Suntzeff, N. B. (1995) *PASP*, **107**, 997
25. Tosi, M., Pulone, L. Marconi, G. & Bragaglia, A. (1998) *MNRAS*, **299**, 834
26. Twarog, B. A., Ashman, K. M. & Anthony-Twarog, B. J. (1997) *AJ*, **114**, 2556

Speaker: R. PHELPS

G.P. PIOTTO: If you are interested in finding an overlap between the ages of open and globular clusters, you might be interested in the case of Palomar 1 (Rosenberg et al., 1998), which has an age of 8 Gyr. Indeed, it is an open issue if this is the youngest globular cluster or one of the oldest open clusters.

R. PHELPS: Indeed I am, thank you.

GALACTIC GLOBULAR CLUSTER RELATIVE AGES: CLUES ON THE MILKY WAY EARLY EVOLUTION

G. PIOTTO
Dipartimento di Astronomia, Università di Padova
vicolo dell'Osservatorio, 5, I-35122, Padova

A. ROSENBERG
Instituto de Astrofísica de Canarias, Via Lactea s/n, Tenerife

I. SAVIANE, M. ZOCCALI
Dipartimento di Astronomia, Università di Padova
vicolo dell'Osservatorio, 5, I-35122, Padova

AND

A. APARICIO
Instituto de Astrofísica de Canarias, Via Lactea s/n, Tenerife

1. Abstract

Based on two new large, homogeneous photometric databases of 35 (from V and I groundbased data) and 15 (from B and V HST data) Galactic globular clusters, a set of distance and reddening independent relative age indicators has been measured. The observed $\delta(V-I)_{@2.5}$, $\delta(B-V)_{@2.5}$, and $\Delta V_{\rm TO}^{\rm HB}$ vs. metallicity relations have been compared to the relations predicted by two recent updated libraries of isochrones. Using these models and two independent methods, we have found that self-consistent relative ages can be estimated for our GGC samples. The main results are: (a) there is no age-metallicity trend; (b) there is no evidence of an age spread for clusters with [Fe/H]< -1.2, all the clusters of our sample in this range being old and coeval; (c) for the more metal rich clusters there are indications of a larger age dispersion, of the order of 10-15%; (d) there is no trend of the age with the distance from the Galactic center out to 25 Kpc.

A possible scenario for the Milky Way formation: The GC formation process started at the same zero age throughout the halo, at least out to ~ 25 kpc from the Galactic center. A fraction (40% ?) of the metal-rich ([Fe/H]> -1.2) globulars are formed at a later time ($\sim 15\%$ lower age).

F. Matteucci and F. Giovannelli (eds.), The Evolution of the Milky Way, 249–260.

2. Introduction

Galactic globular clusters (GGC) are the oldest components of the Galactic halo for which ages can be obtained. The determination of their relative ages and of any age correlation with metallicities, abundance patterns, positions and kinematics provides clues on the formation timescale of the halo and gives information on the early efficiency of the enrichment processes in the proto–Galactic material. The importance of these problems and the difficulty in answering these questions is at the basis of the huge efforts dedicated to gather the relative ages of GGCs in the last 30 years or so (VandenBerg et al. 1996, Sarajedini et al. 1997), and references therein].

The methods at use for the age determination of GGCs are based on the position of the turnoff (TO) in the color–magnitude diagram (CMD) of their stellar population. We can measure either the absolute magnitude or the de–reddened color of the TO. However, in order to overcome the uncertainties intrinsic to any method to get GGC distances and reddening, it is common to measure either the color or the magnitude (or both) of the TO, relative to some other point in the CMD whose position has a negligible dependence on age.

Observationally, as pointed out by Sarajedini & Demarque (1990) and VandenBerg et al. (1990), the most precise relative age indicator is based on the TO color relative to some fixed point on the red giant branch (RGB). This is usually called the "horizontal method". Unfortunately, the theoretical RGB temperature is very sensitive to the adopted mixing length parameter, whose dependence on the metallicity is not well established, yet. As a consequence, investigations on relative ages based on the horizontal method might be of difficult interpretation, and need a careful calibration of the relative TO color as a function of the relative age (Buonanno et al. 1998). The other age indicator, the "vertical method", is based on the TO luminosity relative to the horizontal branch (HB). Though this is usually considered a more robust relative age indicator, it is affected both by the uncertainty of the dependence of the HB luminosity on metallicity and the empirical difficulties to get the TO, and the HB magnitudes for clusters with only blue HBs.

Given these problems, it is still an open debate whether most GGCs are almost coeval (Stetson et al. 1996, Buonanno et al. 1998, VandenBerg 1999) or whether there was a protracted formation epoch of 5 Gyr (Sarajedini et al. 1997) or so (i.e. for 30-40% of the Galactic halo lifetime).

Indeed, there is a major limitation to the large scale GGC relative age investigations: the photometric inhomogeneity and the inhomogeneity in the analysis of the databases used in the various studies. Many previous studies frequently combine photographic and CCD data, different databases (obtained with different instruments with uncertain calibrations to standard systems and/or based on different sets of standards), or inappropriate CMDs were used. This inhomogeneity affects even many recent works, for which results can not yet been considered conclusive (see Stetson et al. 1996 for a discussion).

For this reason, since 1997 (Saviane et al. 1997), we started a long term project, to build up a large, photometrically homogeneus data base useful for relative age determination. We decided two approaches. First, we started a groundbased (GB) observational campaign in the V and I bands, using the smallest number of telescopes. Second, we also took advantage of our ongoing HST surveys (GO7470, GO8118, and GO8723) in B and V for a second, independent catalog of CMDs. The main results from the GB campaign are presented in the following, but see also Rosenberg et al. (1999, hereafter RSPA) for a more extensive and detailed description. In this paper we present also the first results from the HST database. It is worth mentioning that both investigations are still going on.

3. The data set and the methodology

The goal of our observational strategy was to obtain color differences near the TO region with an uncertainty ≤ 0.01mag, which allows a ≤ 1Gyr age resolution. For the GB investigation, we used 1-m class telescopes to build a large reference sample including all clusters within $(m-M)_V = 16$. We observed 52 of the 69 available clusters. Of the total sample, only 34 (plus Pal 12, Rosenberg et al. 1998b) were suitable for this study. The remaining objects were excluded due to several reasons: differential reddening, small number of member stars, large background contamination, bad definition of the RGB or HB. The huge reduction work has been part of Alfred Rosenberg Phd Thesis. A detailed description of the observation and reduction strategies are given in Rosenberg et al. (2000a,b), where the CMDs for the whole photometric sample are also presented.

As for the HST catalog, collecting all our snapshot and GO6095 data and the archive data, we have 32 CMDs. Of these, 15 are suitable for a relative age determination. The data reduction has been part of Manuela Zoccali PhD thesis. A detailed description of the observations and reduction and a presentation of all the CMDs will appear in Piotto et al. (2000).

In our attempt to be as homogeneous as possible, we have adopted the metallicities listed in Rutledge et al. (1997), calibrated on the Carretta & Gratton (1997) scale. The main results presented in the following sections would not change adopting the Zinn and West (1984) scale.

Recent discussions on the possible choices for the photometric parameters (which always measure the TO position with respect to some other

CMD feature with negligible dependence on age) can be found in Stetson et al. (1996), Sarajedini et al. (1997) and Buonanno et al. (1998). Our investigation is based on two "classical" reddening and distance independent parameters: the magnitude difference $\Delta V_{\rm TO}^{\rm HB}$ between the HB and the TO (vertical method), and the color difference $\delta(V-I)_{@2.5}$ [or $\delta(B-V)_{@2.5}$] between the TO and the RGB (horizontal method), where the RGB color is measured 2.5 magnitudes above the TO.

The details on the measurement of the observational parameters, and on the error estimate can be found in RSPA.

In order to interpret the results of our data samples, the theoretical isochrones computed by Straniero et al. (1997, hereafter SCL97), Cassisi et al. (1998, C98), and VandenBerg et al. (1999, V99) were used. These isochrones are the most recent ones which provide both $(B-V)$ and $(V-I)$ colors and use updated physics. It is important to notice that these theoretical models are completely independent. The trends of the theoretical quantities as functions of both age and metallicity were least-square interpolated by means of third-order polynomials, so that the observed parameters can be easily mapped into age and metallicity variations.

In order to calculate the theoretical values of $\Delta V_{\rm TO}^{\rm HB}$, we have to assume a relation for the absolute V magnitude of the HB as a function of the metal content. In particular, here we adopted $M_V(ZAHB) = 0.18 \cdot ([\rm Fe/H] + 1.5) + 0.65$, from the recent investigation of Carretta et al. (1999). The implications of this choice will be discussed in the following sections.

4. Clusters' relative ages

In this section, relative ages are obtained from the observed $\Delta V_{\rm TO}^{\rm HB}$ and $\delta(V-I)_{@2.5}$ parameters by comparison with the V99 and SCL97 models. As discussed in the Introduction, from the observational point of view, the horizontal method is a more precise relative age indicator than the vertical one (Sarajedini & Demarque 1990, VandenBerg et al. 1990), though potentially more difficult to interpret because of the uncertainty in obtaining the corresponding theoretical values. For this reason, we made an internal consistency check for the theoretical models, and selected those for which the relative age trend with the metallicity turned out to be same (within the errors) using both the horizontal and vertical method. While the V99 and SCL97 models satisfy this condition (cf. Figs. 1 and 2) for our sample of GGCs, C98 models do not. Further tests are required to identify the source of this problem (cf. RSPA for a possible explanation).

We want to note that the *absolute* ages obtained from the two methods are not the same. The age differences between the vertical and horizontal method are ~ 1.2 and ~ 1.5 Gyrs for the SCL and V99 models, respec-

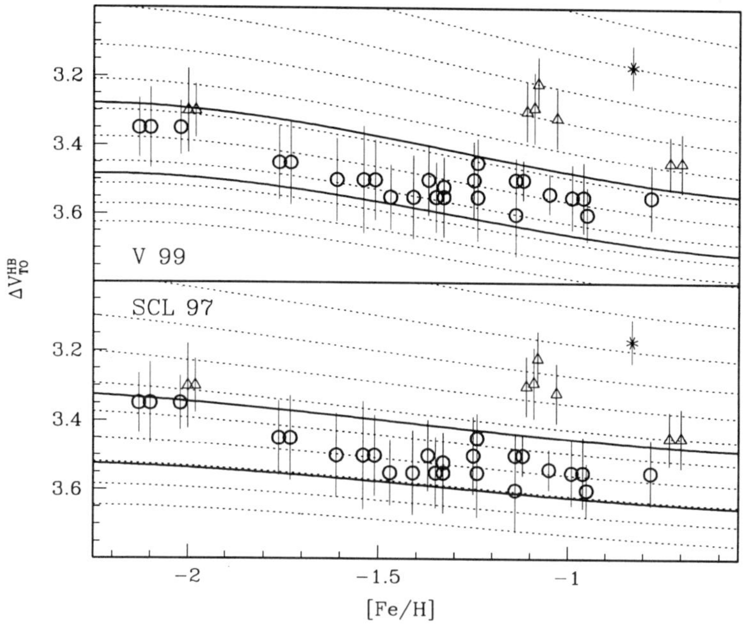

Figure 1. The measured $\Delta V_{\rm TO}^{\rm HB}$ parameter is plotted versus the metallicity. The dotted lines in the two panels show the theoretical trend for V99 (*top*) and SCL97 (*bottom*) models. The isochrones are spaced by 1 Gyr (starting from 18 Gyr at the bottom). The asterisk represents the cluster Pal 12. The two isochrones displayed as solid lines represent the ±1 standard deviation limits of the ΔV_{TO}^{HB} parameter for the entire sample (excluding Pal 12), and clusters falling within these (circles), are defined as "fiducial coeval". Note that the two independent models give the same fiducial coeval object selection.

tively. This discrepancy can be removed by adopting for the $V_{\rm HB}$ vs. [Fe/H] relation an appropriate constant. Far of being a problem for our purpose of measuring relative ages, these discrepancies can be a way to test the models and to fine-tune some still uncertain input parameters.

For reasons which will be further discussed below, the GB sample is photometrically more homogeneous than the HST one. Therefore, we will first present the results from the GB study.

The measured $\Delta V_{\rm TO}^{\rm HB}$ parameters are plotted versus the cluster metallicity in Fig. 1. The dotted lines are the isochrones from the V99 (top) and the SCL97 (bottom) models. Age is spaced by 1 Gyr steps, with the lowermost line corresponding to 18 Gyrs.

We notice in the figure that the clusters are distributed in a narrow band of ≤ 2 Gyrs width, apart from five clusters at [Fe/H] values between -1.1 and -0.9 (namely, NGC 2808, NGC 362, NGC 1261, NGC 1851 and Pal 12). Within the observational errors, the theoretical isochrones and the observed values show similar trends with metallicity for [Fe/H] ≤ -0.9. It must be stated that this result depends on the choice of the trend of the HB luminosity with [Fe/H], although the conclusions would be the same if the slope of the V_{HB} vs. [Fe/H] relation is changed by no more than $\pm 15\%$ (see also below).

The isochrones were used to tentatively select a sample of coeval clusters. We arbitrarely defined as coeval GGCs those clusters whose vertical parameter was within ± 1 standard deviation from the best-fitting isochrone. This interval is marked by thick lines in Fig. 1. Objects lying within this interval for both sets of theoretical models (that we will call fiducial coeval from here on) are marked by open circles in Fig. 1 and will be used later on to test the isochrones in the $\delta(V-I)_{@2.5}$ vs. [Fe/H] plane. Interestingly enough, the same set of coeval clusters is selected using both the SCL97 and the V99 isochrones, and using any slope α for the V_{HB} vs. [Fe/H] relation in the range $0.17 < \alpha < 0.23$ for the V99 isochrones and $0.15 < \alpha < 0.20$ for the SCL97 isochrones. The actual dispersion of the fiducial coeval clusters around the mean isochrone, is indeed consistent with a null age dispersion.

The age dispersions have been calculated assuming a mean absolute age of 13.2 Gyr (Carretta et al. 1999) for the mean age of the GGC bulk. The corresponding dispersions in terms of percent values are 9.2% and 9.8% (all clusters minus Pal 12), and 4.4% and 4.5% (coeval sample) age dispersion. These latter dispersions are fully compatible with the uncertainties in the ΔV_{HB}^{TO} values, strengthening the idea that the clusters selected as coeval must indeed have the same age.

The measured $\delta(V-I)_{@2.5}$ parameters are plotted versus the cluster metallicity in Fig. 2. The dotted lines in the figure represent the isochrones from V99 (upper panel) and SCL97 (lower panel), in 1 Gyr steps. The lowermost lines are the 18 Gyr and 17 Gyr isochrones, respectively.

Remarkably enough, Fig. 2 resembles Fig. 1: again, most clusters are located in a narrow sequence for $[Fe/H] \leq -0.9$, with the exception of the same previously identified five clusters, which result to have a younger age also in this case. Also, the trend with metallicity is conserved, with a similar uprise at the metal-rich end.

For the clusters at $[Fe/H] \leq -0.9$ dex, the run of $\delta(V-I)_{@2.5}$ is also reproduced by the isochrones. In this metallicity range, the clusters selected as fiducial coeval by the vertical method (open circles) still fall within a chronologically narrow band of ≤ 2 Gyr, showing a remarkable consistency between the two methods.

Apparently, the metal richer clusters are younger than the bulk of Galactic globulars. Once more, this result is totally model-dependent, and we must recall again that uncertainties in the color-temperature relations and mixing length calibration, as well as the run of the alpha elements con-

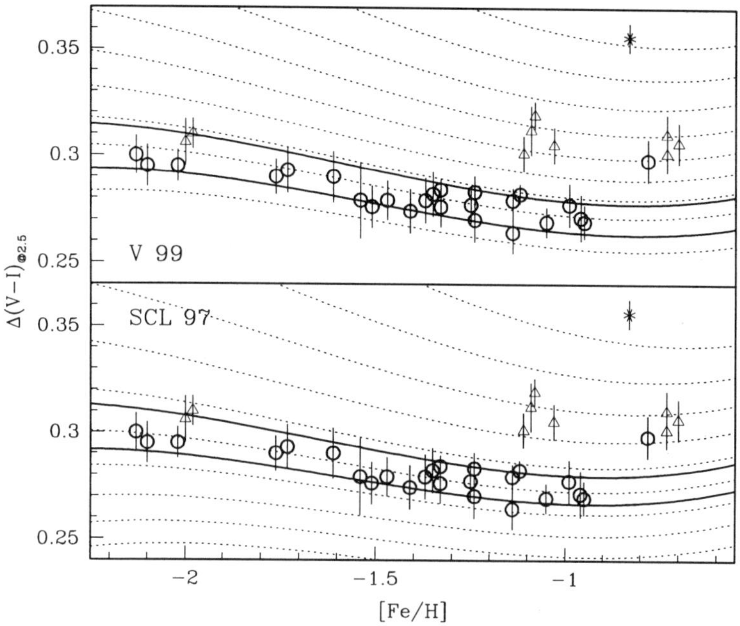

Figure 2. The measured $\delta(V-I)_{@2.5}$ parameter is plotted versus metallicity. The same two sets of theoretical models of Fig. 1 are shown (dotted lines). Age is spaced in 1 Gyr steps, the lowermost line corresponding to 18 Gyr and 17 Gyr isochrones, for V99 and SCL97, respectively. The fiducial coeval clusters selected in Fig. 1 are plotted as *open circles*. The two isochrones displayed as solid lines represent the ± 1 standard deviation limits of the $\delta(V-I)_{@2.5}$ parameter for the GGC sample (except Pal 12). Notice that the two clusters at [Fe/H]= −0.73 (NGC 6366 and NGC 6838) have the same $\Delta V_{\rm TO}^{HB}$, so they appear as a single point in Fig. 1.

tent and helium abundance with metallicity, could affect the relative ages obtained for the most metal rich objects (e.g. Stetson et al. 1996). The errors are the age intervals covered by the photometric error bars in the normalized age scale.

Using the SCL97 models, the percent age dispersions that we have from the horizontal method are 9.2% for the entire sample (with the exception of Pal 12) and 4.3% (for the fiducial coeval sample). Similar results would be obtained by comparison with V99.

Figure 3 shows the comparison of the observed vertical (*bottom panel*) and horizontal (*top panel*) parameters from 15 HST CMDs with the V99 models. Apparently, there is a larger dispersion in the data. This is in part due to the fact that the HST data have not been collected for this specific project. However, we cannot exclude that part of the problem could arise from the still unsolved difficulties in correcting the WFPC2 data for the CTE problems which might affect the different observations to different extents (Stetson 2000). Still, it is interesting to note that the V99 models follow the general trend of the observed parameters. And also the general trend of the ages with metallicity is mostly as discussed for the GB data: the clusters are essentially coeval for [Fe/H]< -1.2, while there is a larger age dispersion for the more metal rich clusters. The HST sample allows to extend the covered metallicity range. The HST data suggest some age dispersion ($\sim 10 \div 15\%$ for [Fe/H]> -1.2, up to [Fe/H]~ -0.4. A similar dispersion for $-1.2 <$[Fe/H]< -0.9 is present also in the GB data, and the HST sample simply strenghtens this result. Whether this age dispersion is present also at higher metallicities is less clear. There are too few points, and we have already discussed the large uncertainties associated to the calibration of the HST data.

Combining the HST and GB data, this age dispersion remains constant up to [Fe/H]~ -0.4, while it is not possible to confirm that the clusters with [Fe/H]> -0.9 are in general younger that the more metal poor ones, as found from the GB sample alone.

5. Discussion: Mean Age Distributions

In this section, the age vs. metallicity and Galactocentric distance trends will be discussed. We will use the relative ages obtained as discussed in the previous Section.

Fig 4 plots the relative percent ages vs. metallicity (*lower panel*) and Galactocentric distance (*upper panel*) for both the HST (*open circles*) and the GB (*filled circles*) data.

First of all, we note the very good agreement between the two data sets for [Fe/H]< -0.9, despite the problems with the HST data calibration. And it is also interesting to note how the HST sample complement the GB one, particularly at the highest metallicities.

For [Fe/H]< -0.9, no age-metallicity relation is visible. However, while for the metal poor clusters the age dispersion is very small, and compatible with zero, for $-1.2 <$[Fe/H]< -0.9 there is a larger dispersion, or, alternatively, two groups of clusters: a group of 8 objects, coeval with the 23 more metal poor ones, and a group of 6 clusters $\sim$25% younger. For [Fe/H]> -0.9, the GB sample would suggest some trend with metallicity which instead seems to mostly disappear considering also the HST GCs. Instead, there seems to be an age dispersion, similar to that already discussed for the intermediate metallicity clusters, at a level of $\pm 10 \div 15\%$. This result urgently requires further investigations, in particular based on more secure (in terms of calibration) GB observations.

In conclusion, our data do not reveal an age-metallicity relation in the usual sense of age decreasing (or increasing) with metallicity. What is found

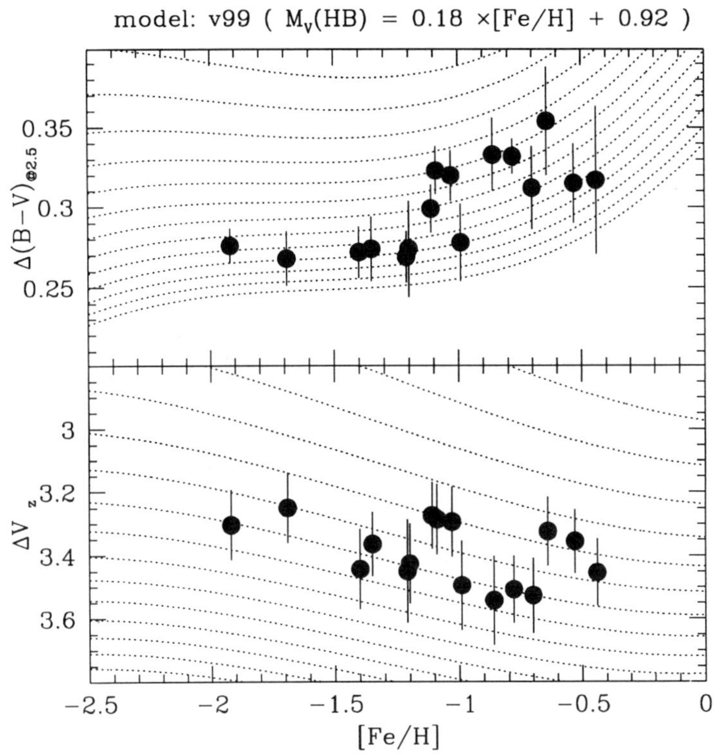

Figure 3. The measured $\delta(B-V)_{@2.5}$ (*upper panel*) and $\Delta V_{\rm TO}^{\rm HB}$ (*lower panel*) for the HST sample are compared with the V99 models. Age is spaced in 1 Gyr steps, the lowermost line corresponding to 18 Gyr. The general trend is similar to that described in Fig. 1 and 2

is an increase of the age dispersion (due to the presence of a few clusters with younger ages than the bulk of the GGCs) for the metal rich clusters, while the lower metallicity ones ([Fe/H]≤ -1.2) seem to be all coeval. This is in agreement with the results of Richer et al. (1996), Salaris & Weiss (1998), Buonanno et al. (1998). On the other side, Chaboyer et al. (1996) proposed an age-metallicity relation, of the order $\Delta t_9/\Delta$[Fe/H] $\simeq -4$ Gyr dex^{-1}, which is not present in our data set.

Some important clues on the Milky Way formation and early evolution can be obtained from the Galactocentric radial distribution of the GGC relative ages. It is represented in Fig. 4 (*upper panel*). There is no radial trend, apparently out to 25 Kpc from the Galactic center. This result is slightly different from what discussed in RSPA. But the difference arises only from the fact that including the HST sample we have more than twice as many metal rich clusters than in RSPA. Apparently, in RSPA all the clusters with [Fe/H]> -0.9 seemed younger, and we regarded this result as

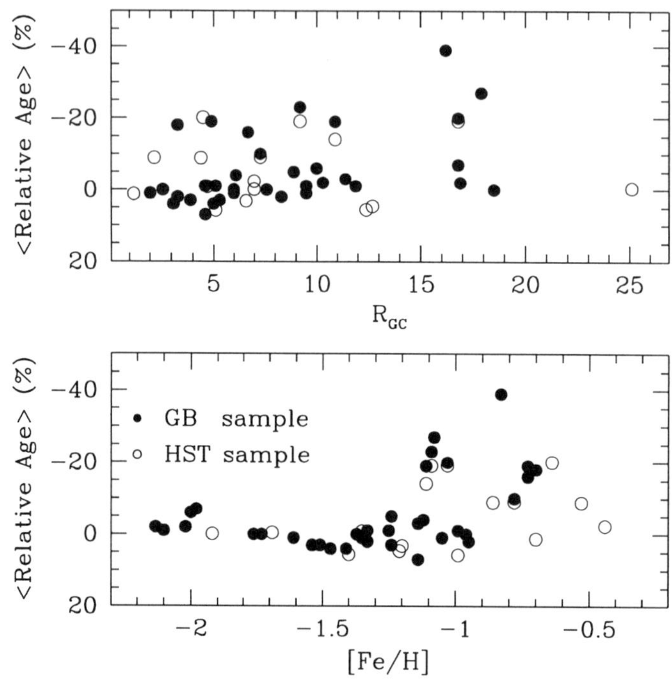

Figure 4. The mean relative ages (percentage values) for the HST and the GB samples are plotted against the metallicity and the distance from the Galactic center.

suspicious, as totally model dependent. Now, with a larger sample, we see that at the most metal rich end there are both younger clusters and objects as old as the most metal poor ones. While the relative age determination is still model dependent, we can now tentatively say that the indication of a larger age dispersion for more metal rich objects extends from [Fe/H]~ -1.2 to the most metal rich clusters. The result is roboust, and present in both data bases up to [Fe/H]~ -0.9, but it needs further investigation for more metal rich objects. In any case, at the moment it seems premature to say whether there is a real age dispersion or two formation episodes.

In summary, the following picture arises from our analysis:

- According to the current models, most of the clusters are coeval and old.
- A fraction of the intermediate and high metallicity clusters are significantly younger.
- The young clusters located at larger $R_{\rm GC}$ have typical halo kinematics.

Therefore, the GC formation process must have been started at the same zero age throughout the halo, at least out to $\sim$ 25 kpc from the center. A fraction (40% ?) of the more metal rich ([Fe/H]≥ -0.9) clusters formed at later times ($\sim 15 - 20\%$ of the halo age). Once again, we stress that this interpretation is model dependent, as it depends on the behavior of the isochrones at high metallicities, and, particularly for [Fe/H]> -0.9, it is based on a limited number of objects. These younger clusters do not identify a unique substructure of the Galaxy.

This research has been supported by the Ministero della Ricerca Scientifica e Tecnologica and by the Agenzia Spaziale Italiana. AR has been supported by the Italian Consorzio Nazionale Astronomia e Astrofisica.

References

1. Buonanno R., Corsi C.E., Pulone L., Fusi Pecci F., Bellazzini M., 1998, A&A, 333, 505
2. Carretta E., Gratton R., 1997, A&AS, 121, 95
3. Carretta E., Gratton R., Clementini G., Fusi Pecci F., 1999, astro-ph/9902086
4. Cassisi S., Castellani V., Degl'Innocenti S., Weiss A., 1998, A&AS, 129, 267 (C98)
5. Chaboyer B., Demarque P., Sarajedini A., 1996, ApJ, 459, 558 (C96)
6. Richer H.B. et al., 1996, ApJ, 463, 602 (R96)
7. Rosenberg A., Saviane I., Piotto G., Aparicio A., Zaggia S. R., 1998a, AJ, 115, 648
8. Rosenberg A., Saviane I., Piotto G., Held E. V., 1998b, A&A, 339, 61
9. Rosenberg A., Saviane I., Piotto G., Aparicio A. V., 1999, AJ, 118, 1306
10. Rosenberg A., Piotto G., Saviane I., Aparicio A., 2000a, A&AS, submitted
11. Rosenberg A., Aparicio A., Saviane I., Piotto G., 2000b, A&AS, submitted
12. Rutledge A.G., Hesser J.E., Stetson P.B., 1997, PASP, 109, 907
13. Salaris M., Weiss A., 1998, A&A, 335, 943 (SW98)
14. Sarajedini A., Chaboyer B., Demarque P., 1997, PASP, 109, 1321
15. Sarajedini A., Demarque P., 1990, ApJ, 365, 219 (SD90)
16. Saviane I., Rosenberg A., Piotto G., 1997. In: R.T. Rood & A.Renzini (eds.) Advances in Stellar Evolution, Cambridge University Press, Cambridge, p. 65 (SRP97)
17. Stetson P.B., 2000, in The Galactic Halo: From Globular Clusters to Field Stars, eds. Noels et al., in press
18. Stetson P.B., VandenBerg D.A., Bolte M., 1996, PASP, 108, 560 (S96)
19. Straniero O., Chieffi A., Limongi M., 1997, ApJ, 490, 425 (SCL97)
20. Sweigart A.V., 1997, ApJ, 474, 23
21. VandenBerg D.A., Swenson F.J., Rogers F.J., Iglesias C.A., Alexander D.R., 1999, ApJ, submitted (V99)
22. VandenBerg D.A., Bolte M., Stetson P.B., 1990, AJ, 100, 445 (V90)
23. VandenBerg D.A., Stetson P.B., Bolte M. 1996, ARA&A, 34, 461
24. VandenBerg D.A. 1999, ApJ, submitted
25. Zinn R., West M., 1984, ApJS, 55, 45

Speaker: G. PIOTTO

F. MATTEUCCI: How much younger than the others are the clusters at $R_{GC} \geq 8$ Kpc?

G.P. PIOTTO: We have to distinguish two groups of objects:
i) a group, namely N362, N1261, N1815, N 2808 (with -1/2< $[Fe/H]$ < -0.8) which are in our sample, and that is 15-20% younger than a counterpart of objects, at the same metallicity, which result co-eval with the metal poorest objects;
ii) a few other objects: PAL12 (which is in our sample), PAL1, Arp2, Rup106, Terzan7 which are more than 30% younger than the bulk of the galactic globular clusters. However, some or all of these objects might be clusters formed in other environments (dwarf galaxies?) and later on captured by our galaxy.

M. BELLAZZINI: At least two of the "young" clusters in your sample, namely Ter7 and Arp2, are not members of the MW GC system. In fact they belong to the SGR dSph galaxy. I think it would be safe to exclude these clusters from the sample if you want to measure the age dispersion of MW globulars.

G.P. PIOTTO: Indeed, these objects are not in our present sample, and we have not used them in calculating the age dispersion.

INFRARED OBSERVATIONS OF GLOBULAR CLUSTERS

L. PULONE
Osservatorio Astronomico di Roma
Via di Frascati 33, 00044 - Monteporzio Catone. Italy

Abstract. This communication will be focused upon two main topics: IR observations of Globular Clusters (GC) from the space and the reliability of the mass segregation corrections for Luminosity Functions (LF), observed far away from the half mass radius. The effects of the mass segregation corrections on the correlations among Global Mass Functions (GMF) of the GCs and their kinematical parameters will be finally discussed.

1. HST-NICMOS observations of Globulars

One of the most controversial aspects of the observations of the low Main Sequence (MS) in GCs is the behavior of the LF and therefore, of the MF approaching the substellar limit. As the H-burning limit is approached, the stars become cooler and redder and can be, therefore, better observed in Near IR where their highly non-blackbody spectra tend to peak (Allard and Hauschildt 1995). The observations of the NICMOS instrument, on board of HST are crucial in this context. During the life of the instrument we observed 3 clusters $\Omega\,Cen$ (Pulone *et al.*) 1998), $M\,4$ (Pulone, De Marchi, and Paresce 1999) and $NGC\,6397$ (De Marchi, Paresce, and Pulone 2000).

The main results can be summarized as follows:

a) a direct comparison of optical and IR observation has definitively clarified whether the WFPC2 is efficient at detecting faint, red dwarfs or whether it has missed a large amount of the very low–mass objects that could potentially account for a conspicuous fraction of the total mass of GCs. The same field of $NGC\,6397$ has been observed by NICMOS and by WFPC2 (De Marchi, Paresce, and Pulone 2000). All the objects detected with NICMOS in the H band with $170min$ exposure are also present in the same field with $90min$ WFPC2 image through the I filter. In particular, the faintest detected object is a star with $M_V = 13.8$, $M_I = 12.4$, $M_H = 10.7$.

F. Matteucci and F. Giovannelli (eds.), The Evolution of the Milky Way, 261–267.

By using the distance modulus of Reid (1998) and the Mass–Luminosity (ML) by Baraffe *et al.* (1997) for [M/H]=-1.5, this translates into a mass of $0.09 M_{\odot}$ star, just the one expected for an object close to the H-burning limit. Down to the bottom of the cluster's MS, NICMOS does not show any other faint, red object than those already detected by the WFPC2.

b) the validity of the physical assumptions in coding non-grey atmosphere models of low mass stars has been confirmed (Allard and Hauschildt 1995, Chabrier and Baraffe 1997). For example, the feature in the LFs corresponding to a 0.5 stellar mass, due to the dissociation of H2 molecules in the stellar atmosphere is well reproduced (Pulone *et al.* 1998). In the IR bands this feature is better visible than in the optical ones and, moreover, it could be used as a further distance indicator. This affects the reliability of the ML relationships, because we use them to obtain the MF of the observed clusters.

c) in all these cases, the mass equivalent to the maximum in the LFs (below $0.3 M_{\odot}$) has been reached and exceeded. All the LFs observed in all three clusters show the same consistent behavior: an increase with decreasing luminosity up to a peak at $M_H \sim 7$ a drop well before the incompleteness becomes significant. Within the errors, at $M_H \sim 10.5$ the LF are compatible with zero. By applying the best ML relation appropriate to the metallicities of the observed clusters, we obtain MFs showing a break in the slope at $0.3 M_{\odot}$. As regards $M\,4$ and $NGC\,6397$, no single exponent power-law is compatible with the data, regardless of the value of the exponent. Thus, the decrease in the LF below the maximum is an intrinsic feature of the MF and indicates an effective scarsity of stars, not exclusively due to the shape of the ML relationship. Drawing the first conclusion, we can say that, through the HST–IR NICMOS observations, the reliability of the WFPC2 optical exposures which reached the bottom of the GC's MS, is established. At the H–burning limit, in the optical band, what you see is what you get.

2. Taking into account the mass–segregation corrections

It is well known that, through the relaxation processes the stars inside a system tend to reach the equipartition of the energy. The heavier stars migrate toward the center of the clusters, while the smaller ones are pushed toward the external layers. If we observe stars located far away from the half mass radius, we have to correct the local LFs for the mass-segregation effect. Only properly corrected MFs have to be used in order to find possible correlations among structural, dynamical or kinematical parameters.

For LF observed at radial distances far from the half mass radius, we calculated the simplest dynamical models, like King–Michie, for $NGC\,6397$ and $M\,4$, by exploiting both optical and IR observations.

We faced the question whether the difference in inner and outer LFs (MFs) can be quantitatively accounted for by the two-body relaxation mechanism. This requires a simulation of the dynamical status of the clusters. We have used multi-mass King–Michie models, built up with an approach nearly identical to that of Gunn and Griffin (1979) and already described and used by Meylan (1987, 1988). Each model is characterized by a segmented power-law, where the exponent is allowed to vary with the mass, and by four structural parameters describing respectively the scale radius r_c, the scale velocity v_s, the central value of the dimensionless gravitational potential W_o, and the anisotropy radius r_a.

The mass stratification of the model allows us to correct the MFs calculated by LFs observed in regions far away from the half mass radius. It is well known that the difference between the local mass function and the global one is lowest at the r_{hm}. The aim of the simulation is to fit simultaneously the Surface Brightness Profile (SBP) and the Velocity Dispersion Profile (VDP). Fitting both, however, still allows a variety of shapes of the GMF. To break this degeneracy, we have complemented the Meylan's approach by imposing the condition that the model radial MFs, transformed into LFs, simultaneoulsy agree with the IR and optical observed LFs.

As regards *NGC* 6397, the SBP and VDP have been taken by Meylan and Mayor (1991). If the model GMF is allowed to flatten below $\sim 0.3 M_\odot$, both portion of the three LFs can be fitted while, at the same time, the SBP and VDP, are adequately reproduced. In particular we find that an exponential MF index $\alpha = 2 \pm 0.2$ for evolved stars ($M > 0.8 M_\odot$) and $\alpha = 1.6 \pm 0.2$ in the range $0.8 - 0.3 M_\odot$ which drops with $\alpha = -0.2 \pm 01$ at lower masses would not only produce a good fit in the three fields, but would also give a reasonable value for the cluster structural parameters: the concentration ratio ($c \sim 2.6$), the total cluster mass ($\sim 8.9x10^4 M_\odot$), the fraction of heavy remnants ($\sim 33\%$ in mass), and the anisotropy radius ($r_a/r_c \sim 20$) are all in good agreement with those of the 8 best models by Meylan and Mayor (1991).

This is an example of a positive situation, but there are other circumstances in which the problem is more complicated to solve.

Indeed, if we replay a similar game with *M* 4, we are forced to take into account non-canonical value of some structural parameter such as the core radius. In our simulations we have used the VDP by Peterson, Rees and Cudworth (1995). Concerning the SBP, we have compared the profile measured by Kron, Hewitt, and Wasserman (1984) with that by Trager, King and Djorgovski (1995), and found some discrepancies. Then, we have adopted a SBP, defined as the parametrized King-type profile with the canonical values of core radius $r_c = 50"$ and concentration $c = 1.6$ (Harris 1996). Let us note, however, that values as large as $r_c = 90"$ and $c = 2$ are found in the literature.

Although several models can reproduce the SBP and VDP, no choice of the GMF exponent gives local MFs, which converted in LFs via the ML relationship of the proper metallicity, agree with the observational data at $2'$ and $6'$.

We found that, by forcing the core radius to take the value of $\sim 90"$, both sets of LFs can be simultaneously fitted by an exponential GMF index $\alpha = 0.2$ in the range $0.8 - 0.25 M_\odot$, which drops with $\alpha = -0.4$. In this case, the concentration ($c = 1.5$), the total mass ($\sim 8.8x10^4 M_\odot$) and the index of the GMF above $0.8 M_\odot$ are also acceptable. Nevertheless, it would still be possible to find a model which uses the IMF by De Marchi and Paresce (1997) for *NGC* 6397 and simultaneoulsy fit the observed LFs. This would require an even higher value of the concentration ($c \sim 1.9$), still compatible with the literature data.

To summarize the results, we should be very cautious in considering the validity of the mass segregation corrections of LFs or MFs, because there are many sources of errors. On the other hand, the dynamical model too, could be inadequate or insufficient to describe the status of the clusters. The good fits we obtain do not necessarily guarantee that these models reflect the physical status of the cluster. Indeed, from the kinematics of its orbit (Dauphole *et al.* 1996), *M* 4 should have undergone an extensive tidal stripping and shocks, because it penetrates very deeply into the galactic bulge and its orbital plane is one of the closest to the galactic plane. These effects might have strongly modified the local mass distribution, particularly in the outer layers, and the hypothesis of perfect energy equipartition adopted could no longer be applied.

3. Playing with MF slopes and statistics

I shall now briefly discuss the relevance of all this in the context of correlations among GMFs and kinematical parameters of the Globulars. I will use the same data set adopted by Paresce in his talk on the mass functions of globular clusters (in this volume).

If we plot the slope of the LFs of the dozen of clusters observed with the HST down to the bottom of the main sequence versus, for example, the minimum distance from the galactic center, we wonder whether a correlation might be found (and we know it should).

By applying a statistical technique like the F-test, we could check the validity of the assumption of a flat distribution, and then the absence of a correlation, taking into account the errors. This is done by comparing the observed distribution to a simulated one obtained by a few thousands of points distributed supposing the errors are gaussian. A $F - value$ lower

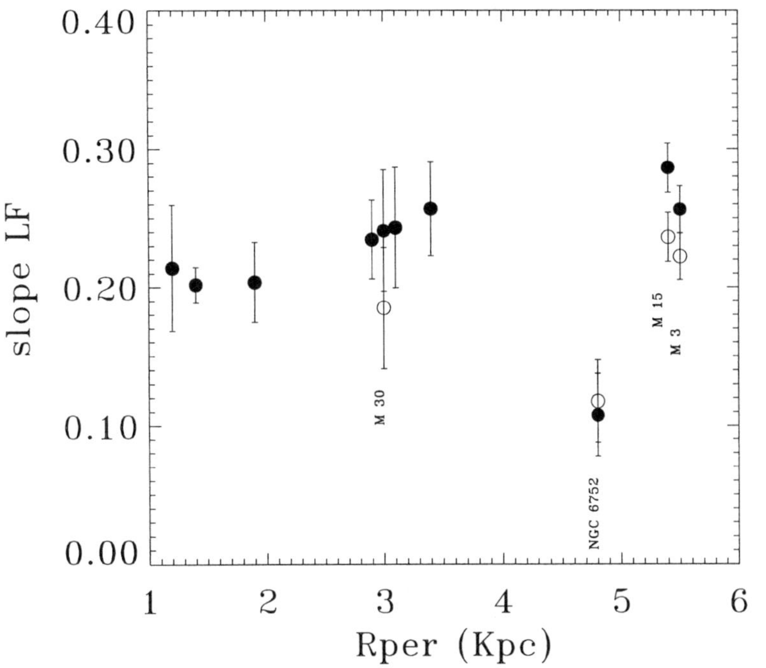

Figure 1. Slopes of the luminosity functions between the maxima at $M_I = 9$ and the dips $2mags$ brighter vs the perigalactic distances by Dinescu *et al.* (1999). The GC's whose LFs have been observed far away from the half–mass radius and then corrected for the mass segregation, are indicated by their name. The filled circles represent the uncorrected slopes. For the four clusters $M30$, $NGC6752$, $M15$, and $M3$, the corrected slopes have been indicated by empty circles

than 0.1 means that the two distributions are different and that there is a real correlation among the observed slopes and their perigalactic distances.

If we exclude $NGC\,104$, just to keep the sample as homegeneus as possible, looking at the graph of fig. 1, in which uncorrected slopes of the LFs (filled circles) are shown versus perigalactic distances as calculated by Dinescu *et al.* (1999), the hypotesis of a flat distribution has to be rejected. Indeed, the $F-value$ is 0.02. With the exclusion of $NGC\,6752$, this conclusion is reinforced. If we calculate and apply the mass segregation correction to the the LFs observed far away from the half mass radius, the F-test gives the surprising result that, on the basis of the errors, we cannot state the two distributions are different ($F - value = 0.17$). Even if we remove from the sample the outlier $NGC\,6752$, the hypotesis of a flat distribution cannot be rejected. This does not mean that there is no gradient. A correlation among slopes and perigalactic distances of clusters should be reasonably well founded. In other words, by taking into account the indeterminacies of the theoretical scenario and the observational errors, it is impossible at the moment to gauge the reality of gradients among LFs (MFs) and kinematical or dynamical parameters. While waiting for improvements in dynamical simulations, we must obtain GMF as precise as possible, by sampling as large a stellar content of the clusters as possible.

References

1. Allard, F. and Hauschildt, P. H. (1995) Model atmospheres for M (sub)dwarf stars. 1: The base model grid, *ApJ*, **445**, 433
2. Baraffe, I.,Chabrier, G., Allard, F. and Hauschildt, P. H. (1997) Evolutionary models for metal-poor low-mass stars. Lower main sequence of globular clusters and halo field stars, *A&A*, **327**, 1054
3. Chabrier, G. and Baraffe, I. (1997) Structure and evolution of low-mass stars, *A&A*, **327**, 1039
4. Dauphole, B., Geffert, M., Colin, J., Ducourant, C., Odenkirchen, M. and Tucholke, H. J. (1996) The kinematics of globular clusters, apocentric distances and a halo metallicity gradient, *A&A*, **313**, 119
5. De Marchi, G. and Paresce, F. (1997) The Initial Mass Function of Low-Mass Stars in Globular Clusters, *ApJ*, **476**, L19
6. De Marchi, G., Paresce, F. and Pulone, L. (2000) The mass function of Main Sequence stars in NGC 6397 from Near IR and Optical High Resolution HST observations, *ApJ*, in press
7. Dinescu, D. I., Girard, T. M., Van Altena, W. F. (1999) Space Velocities of Globular Clusters. III. Cluster Orbits and Halo Substructure, *AJ*, **117**, 1792
8. Gunn, J. E. and Griffin, R. F. (1979) Dynamical studies of globular clusters based on photoelectric radial velocities of individual stars. I - M3 (1979), *AJ*, **84**, 752
9. Harris, W. E. (1996) Catalog of Parameters for Globular Clusters in the Milky Way *AJ*, **112**, 1487
10. Kron, G. E., Hewitt, A. V. and Wasserman, L. H. (1984) Brightness profiles for 69 globular clusters determined by means of the electronic camera *PASP*, **96**, 198
11. Meylan, G. (1987) Studies of dynamical properties of globular clusters. III - Anisotropy in Omega Centauri, *A&A*, **184**, 144
12. Meylan, G. (1988) Studies of dynamical properties of globular clusters. IV - Detailed structure of 47 Tucanae *A&A*, **191**, 215
13. Meylan, G. and Mayor, M. (1991) Studies of dynamical properties of globular clusters. VI - The high-concentration cluster NGC 6397 *A&A*, **250**, 113
14. Peterson, R. C., Rees, R. F. and Cudworth, K. M. (1995) Radial velocities of stars in the globular cluster M4 and the cluster distance *ApJ*, **443**, 124
15. Pulone, L., De Marchi, G., Paresce, F. and Allard, F. (1998) The Lower Main Sequence of omega Centauri from Deep Hubble Space Telescope NICMOS Near-IR Observations,*ApJ*, **492**, L41
16. Pulone, L., De Marchi, G., and Paresce, F. (1999) The mass function of M 4 from near IR and optical HST observations, *A&A*, **342**, 440
17. Reid, I. N. and Gizis, J. E. (1998) The Distance to NGC 6397 by M-Subdwarf Main-Sequence Fitting, *AJ*, **116**, 2929
18. Trager, S. C., King, I. R. and Djorgovski, S. (1995) Catalogue of Galactic globular-cluster surface-brightness profiles, *AJ*, **109**, 218

Speaker: L. PULONE

G.P. PIOTTO: The fact that you cannot reproduce the mass segregation on M4 might be a consequence of the fact that the cluster is strongly modified by the interaction with the galactic potential. I believe that the case of M4 is very similar to the case of NGC6397, on this respect, as anticipated in Cool, Piotto and King (1996). This is a further indication that to speak of IMF for globular clusters might be somehow premature.

L. PULONE: I agree as regards the M4 case, but let us recall that we have to consider other sources of error such as indetermination in Surface Brightness profile in the core radius, in tidal radius and so on. As regards NGC6397; it has a LF slope not so different from that one of other clusters. Taking into account the orbits, as calculated by Dauphole (1996), NGC6397 does not penetrate deeply into the bulge. The perigalactic distance is around 3 Kpc and I do not expect a tidal stripping as severe as that one of M4.

Coming to your last comment about the IMF, I have just shown that within the observational error and the theoretical indeterminations, the global mass function of the globular clusters does not seem to be correlated in a simple way with dynamical or kinematical parameters. Moreover, we cannot reject at 1 σ level the hypothesis of a common global mass function.

ABSOLUTE MAGNITUDE OF RR LYRAE IN GLOBULAR CLUSTERS FROM PULSATIONAL PROPERTIES

V. RIPEPI
Osservatorio Astronomico di Capodimonte
Via Moiariello 16, Napoli, I-80131

F. CAPUTO
Osservatorio Astronomico di Roma
Via di Frascati, 43, Roma, I-00044

V. CASTELLANI
Dipartimento di Fisica, Univ. di Pisa
Piazza Torricelli, 2, Pisa, I-56100

AND

M. MARCONI
Osservatorio Astronomico di Capodimonte
Via Moiariello 16, Napoli, I-80131

1. Introduction

Globular Clusters (GC) probably formed within 1-2 Gyr from the Big Bang, so that their absolute ages provide a lower limit to the age of the universe. On the other hand, their age dispersion (i.e. relative ages) depends on the mechanism of galaxy formation and evolution. In this context accurate distances to GGC are fundamental because:

- they most heavily affects the final age estimate (Renzini 1991).
- along with dynamical and kinematical information, they are needed for a complete description of the galaxy formation and early evolution.
- the GCLF, once known and accurately calibrated, would make GC excellent standard candles for extragalactic distance determinations.

In what follows, we will show some preliminary results about the determination of GC distances through a pulsational method.

F. Matteucci and F. Giovannelli (eds.), The Evolution of the Milky Way, 269–274.

TABLE 1. Sources of photometry and metallicity for the selected clusters; asterisk means that the metallicities were translated in the Carretta & Gratton (1997) scale.

Cluster	Photometric Source	Metallicity Source
NGC 6362	Walker (1999)	Carretta & Gratton (1997)
M5	Caputo et al.(1999)	Carretta & Gratton (1997)
IC 4499	Walker & Nemec (1996)	Walker & Nemec (1996)*
M3	Carretta et al.(1998)	Carretta & Gratton (1997)
M2	Lee & Carney (1999)	Zinn & West (1984)*
M68	Walker (1994)	Carretta & Gratton (1997)
M15	Bingham et al.(1984)	Carretta & Gratton (1997)
NGC 5053	Nemec et al.(1995)	Harris (1996)*

2. Pulsational distances

As shown by Caputo (1997), pulsational distances to globular cluster can be derived by matching the observed and theoretical blue boundary of the instability strip in the $\log P - M_V$ plane.

The required ingredients are:

- Globular cluster with a good number of c type RR Lyrae with well measured periods, magnitudes and metallicities.
- Reliable theoretical boundaries of the instability strip as a function of chemical composition and stellar mass, such as those predicted by non linear, non local, time dependent convective models (see Bono & Stellingwerf (1994), Caputo et al. (1999)).
- Updated model atmospheres (Castelli et al. 1997a,b).

The main advantages of this pulsational method are:

- The blue boundary of the instability strip (FOBE) in the $\log P - M_V$ plane is a good theoretical prediction since it is little dependent on convection and metallicity.
- The $\log P - M_V$ diagram is reddening-free along the horizontal axis.
- If the observed blue boundary is well populated, the apparent cluster distance modulus (DM) can be derived with an internal uncertainty of about ± 0.03.
- The derived absolute magnitude of RR Lyrae does not depend on the cluster reddening

On the contrary, the main drawbacks of the method are:

- Only clusters with a rich population of c type RR Lyrae can be analysed.

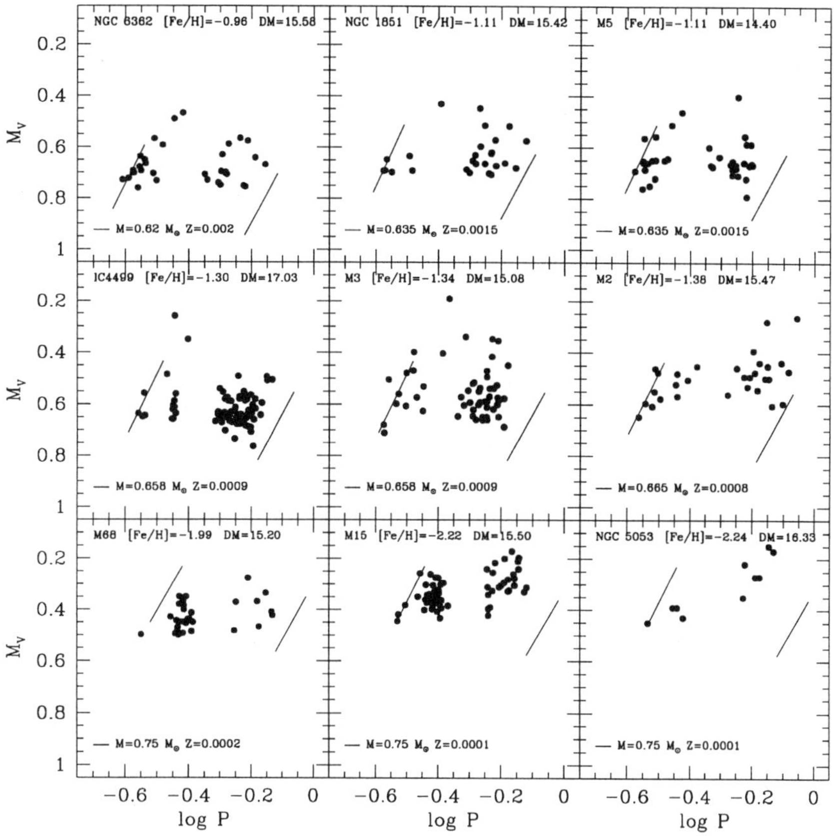

Figure 1. Determination of pulsational distances to the GCs in our sample. Filled circles are the observations; solid lines show the theoretical instability strip limits for the labeled values of mass and chemical composition. In each panel the estimated apparent distance modulus is also labeled.

- The theoretical blue boundary is dependent on the assumed evolutionary scenario for the mass of RR Lyrae as a function of the chemical composition.

3. Application of the method

As already pointed out two main ingredients are required to apply our method. As for the theoretical boundaries of the instability strip, we computed new pulsation models for $M = 0.75 M_{\odot}$, $Z = 0.0001$; $M = 0.70\ M_{\odot}$, $Z = 0.0004$ and $M = 0.62 M_{\odot}$, $Z = 0.002$ in addition to models by Caputo

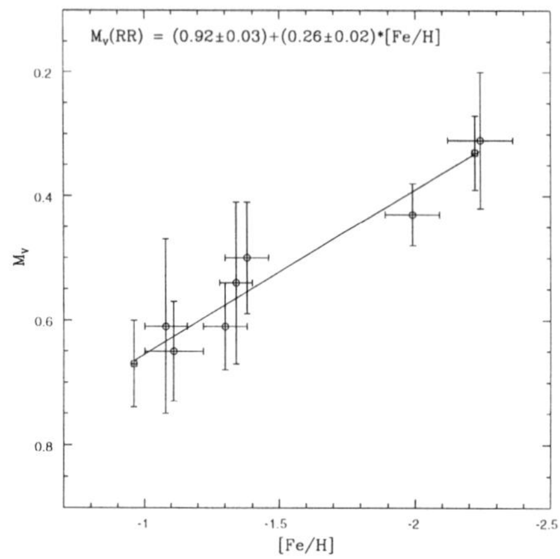

Figure 2. The $M_V(RR) - [Fe/H]$ relation the the selected sample of GC (see text).

et al. (1999) for $M = 0.65M_{\odot}$, $Z = 0.001$.
Note that for each chemical composition, the adopted mass follows current models of HB evolution. As for the observations, we selected a sample of 9 Galactic Globular clusters with data satisfying the requirements of our method. Details about the sources of photometry and metallicity are given in Table 1.

4. Results

The results of the application of our method to the GC sample described in the previous section is reported in Fig. 1

On the basis of Fig. 1 we can calculate the mean magnitude (and dispersion) of the RR Lyrae population of each individual cluster. We organized these data versus $[Fe/H]$ in Fig. 2; note that the reported errors on $[Fe/H]$ do not represent the total uncertainties on this quantity, but only a way to rank it.
In order to derive a mean relation $M_V(RR) - [Fe/H]$ we performed a least square fit to the data. As a result, we have found the following relation:

$$M_V(RR) = (0.92 \pm 0.03) + (0.26 \pm 0.02) * [Fe/H] \tag{1}$$

where the uncertainties on the parameters of the equation are the formal errors from the fit. Adding other possible error sources (photometry, model atmosphere etc.) we estimate that the real dispersion is at least twice the quoted value.

Even if no firm conclusion can be drawn from this preliminary analysis, we note that our results seem to support a value of the slope which is intermediate between the flatter ones estimated e.g. from horizontal branch models and BW analysis, and the steeper value given by Sandage (1993) (see Cacciari, 1999 for a comprehensive review on this subject).

As a final point, if we assume $V_0 = 18.94 \pm 0.04$ mag and $[Fe/H] = -1.9$ for the RR Lyrae variables in the LMC globular clusters (Walker, 1992), we obtain $DM_{LMC} \sim 18.51 \pm 0.07$ mag. This result is intermediate between the short and long distance scales (again we refer to Cacciari 1999).

As a conclusion we have to note that the results presented in this contribution have been obtained using a small selected sample of GC RR Lyrae observations, we expect that increasing the sample might change significantly these results.

References

1. Bingham, E. A., Cacciari, C., Dickens, R. J., Pecci, F. F. 1984, MNRAS, 209, 765
2. Bono, G., Stellingwerf, R.F. 1994, ApJS, 93, 233
3. Cacciari, C. 1999, in *Harmonizing Cosmic Distance Scales in a Post-HIPPARCOS Era*, ASP n. 167, ed. D. Egret & A. Heck, p.140
4. Caputo, F. 1997, MNRAS, 284, 994
5. Caputo, F., Castellani, V., Marconi, M., Ripepi, V. 1999, MNRAS, 306, 815
6. Carretta, E., Cacciari, C., Ferraro, F. R., Fusi Pecci, F., Tessicini, G. 1998, MNRAS, 298, 1005
7. Carretta, E., Gratton, R. G. 1997, A&AS, 121, 95
8. Castelli, F., Gratton, R. G., Kurucz, R. L. 1997a, A&A, 318, 841
9. Castelli, F., Gratton, R. G., Kurucz, R. L. 1997b, A&A, 324, 432
10. Harris, W.E. 1996, AJ, 112, 1487
11. Lee, J.W, Carney, B.W. 1999, AJ, 117, 2868
12. Nemec, J.M., Mateo, M., Schombert, J.M. 1995, AJ, 109, 618
13. Renzini, A., 1991, in *Observational Tests of Inflation*, Ed. T.Banday & T.Shanks (Dodrecht:Kluwer), p. 131
14. Sandage, A., 1993, AJ, 106, 719
15. Walker, A.R. 1992, ApJ, 390, L81
16. Walker, A.R. 1994, AJ, 108, 555
17. Walker, A.R. 1999, in preparation
18. Walker, A.R., Nemec, J.M. 1996, AJ, 112, 2026
19. Zinn, R., West, M. J. 1984, ApJS, 55, 45

Speaker: V. RIPEPI

G. CARRARO: How does the age of the metal richest GCs change due to your new distance estimates?

V. RIPEPI: We estimate that the ages of these clusters should increase by 1-2 Gyr.

GALACTIC GLOBULAR CLUSTERS AS CALIBRATORS OF SYNTHETIC LINE INDICES

CLAUDIA MARASTON
Universitäts-Sternwarte München
Scheinerstrasse 1, D- 81679 München, Germany

Abstract.
Models of Simple Stellar Populations (SSP) are commonly used to infer the ages and the metallicities of complex stellar systems (e.g. elliptical galaxies). In order to obtain reliable informations, these models have to be *calibrated* on globular clusters. In this contribution, we compare various synthetic optical colours and line indices to Galactic Globular Cluster (GGC) data. We examine different metallicity scales for GGCs, concluding that the classical Zinn (1985) scale leads to the best match between SSP models and data. The Hβ line strenghts in GGCs increase at metallicities [Fe/H] < -1.3, and the optical colours get bluer. These properties are due to the characteristics of the Horizontal Branch phase for large ages ($t >$ 12 Gyr) and low metallicities and are well reproduced by our models. As a consequence, composite stellar populations containing a small fraction of old and metal poor stars succeed in explaining the strong Hβ lines observed in elliptical galaxies (Maraston & Thomas 1999 [22]).

1. Introduction

The analysis of the stellar population features (e.g. colours, spectral indices, mass-to-light ratios) in galaxies set constraints on their ages and metallicities. The modeling can be performed taking into account the detailed chemical evolution and the star formation histories (see e.g. Guiderdoni & Rocca-Volmerange 1987 [17]; Chiosi, Bressan and Fagotto 1994 [3]; Matteucci 1996 [23]). Alternatively, Simple Stellar Population (SSP) models, defined as single age and single metallicity assemblies of single stars, can be used to analyse the light-averaged stellar populations in galaxies (see e.g. Faber et al. 1995 [3]; Buzzoni 1995 [6]; Greggio 1997 [16]). In both

F. Matteucci and F. Giovannelli (eds.), The Evolution of the Milky Way, 275–280.

cases SSP models are the fundamental ingredients. Hence the first step is to check if these models reproduce the basic observed properties of star clusters. Galactic globular clusters (GGC) allow the calibration for metallicities below ~ -0.5 dex and old ages (> 10 Gyr).

In this contribution we compare various synthetic optical colours and spectral indices with GGC data. We then discuss the impact of the calibration on the age determination of elliptical galaxies.

The SSP models presented here are a new set based on the population synthesis by Maraston (1998) [20]. The stellar input tracks are taken from Bono et al. (1997) [2] and S. Cassisi (1999, private communication). The calibration of the solar chemical composition models is shown in Maraston (1998) [20]. The synthetic line indices for SSPs are presented in Maraston, Greggio, & Thomas (1999) [21]. For more details on the modelling we refer the reader to Maraston (2000) (*in preparation*).

2. GGCs optical colours

Fig. 1 shows the $B-V$ colour (from Harris 1996 [18]), as a function of metallicity, taking different metallicity scales for GGCs into account. In the left-hand panel the iron abundance [Fe/H] values are taken from the Harris 1996 [18] catalogue, in which the Zinn (1985) [32] scale is adopted. In the other two panels the metallicity is taken from the recent compilation by Ferraro et al. (1999) [12]. In the central panel [Fe/H] is from Carretta & Gratton (1997) [9]. In the right-hand panel [M/H] represents the global metallicity, computed by Ferraro et al. (1999) [12] (see their Table 1) by adopting the equation from Salaris, Chieffi & Straniero (1993) [27]. These authors investigate the effect of the α-enhancement on theoretical evolutionary sequences. They conclude that α-enhanced isochrones are well simulated by standard solar-scaled isochrones with global metallicity $[M/H]=[Fe/H]+\log(0.638f_\alpha+0.362)$, in which f_α is the enhancement factor (see Ferraro et al. 1999 [12]).

Overplotted are SSP models of constant age: this work for t=15 Gyr (*solid lines*); Tantalo et al. (1996) [28] for t=15.85 Gyr (*dotted line*); Worthey (1994) [30] for t=17 Gyr (*long dashed line*); Buzzoni (1989) [5] for t=15 Gyr (*short dashed line*). The models of the various authors are formally consistent with respect to the description of the mass loss along the Red Giant Branch, parametrized à la Reimers (1975) [25] with an efficiency η=0.33 (Fusi Pecci & Renzini 1976 [13]). In the metallicity range considered here our computations agree with Tantalo et al. (1996) [28]. The $B-V$ colours of Worthey (1994) [30] and Buzzoni (1989) [5] models instead are too red. A similar conclusion is obtained by Barmby et al. 1999 [1] when comparing these SSP models to GGCs and M31 globular clusters.

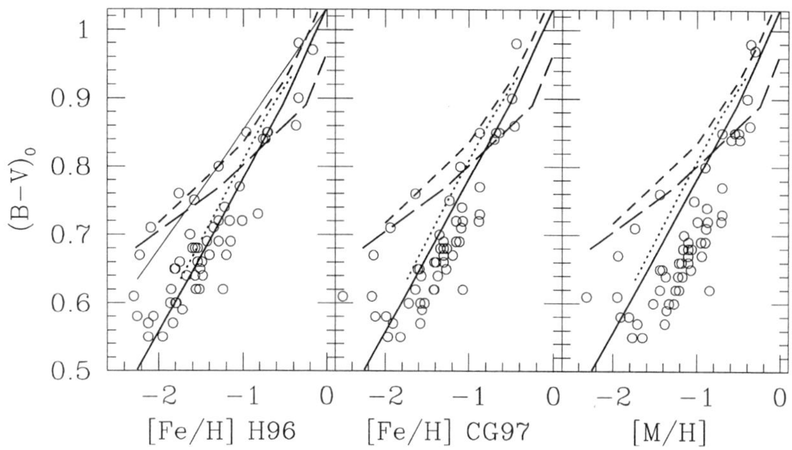

Figure 1. The $B-V$ colours of GGCs for different metallicity scales: the Harris 1996 (*left-hand panel*), based on the Zinn (1985) scale; the Carretta & Gratton (1997) (*central panel*); the global metallicity [M/H] scale (*right-hand panel*) from Ferraro et al. (1999). The SSP models are: this work (*solid lines*); Tantalo et al. (1996) (*dotted line*); Buzzoni (1989) (*short dashed line*); Worthey 1994 (*long dashed line*). The thin solid line in the left-hand panel are the models of this work computed assuming no mass loss along the Red Giant Branch phase (see the text).

2.1. THE MASS LOSS ON THE RED GIANT BRANCH

Fig. 1 shows that the matching of the optical colours of GGCs with [Fe/H]<-1.4 requires the inclusion of mass loss on the RGB in the evolutionary synthesis computations. The value η=0.33 is appropriate to describe the GGCs loci. For comparison, we compute a set of SSPs assuming no mass loss (thin solid line in the left-hand panel). The resulting $B-V$ is too red at a given metallicity. The match with the data requires an age $\sim$ 5 Gyr which is excluded from Colour-Magnitude diagram (CMD) studies (see e.g. Piotto, this volume).

2.2. THE METALLICITY SCALE

The Zinn (1985) [32] iron abundance [Fe/H] scale (Fig. 1, left-hand panel) yields the lowest metallicities for the GGCs at a given $B-V$, leading to the best agreement with the SSP models. The GGC global metallicities [M/H] (Fig. 1, right-hand panel) are in principle the closer representations of the metal content Z of the SSP models. Instead these are too high at a given $B-V$. In this case the matching of the data requires a SSP age of about 4 Gyr, in contradiction with the CMD ages. A possible explanation for this discrepancy is that the Zinn (1985) measurements, based on low dispersion spectroscopy, reflect more the global metallicity than the pure

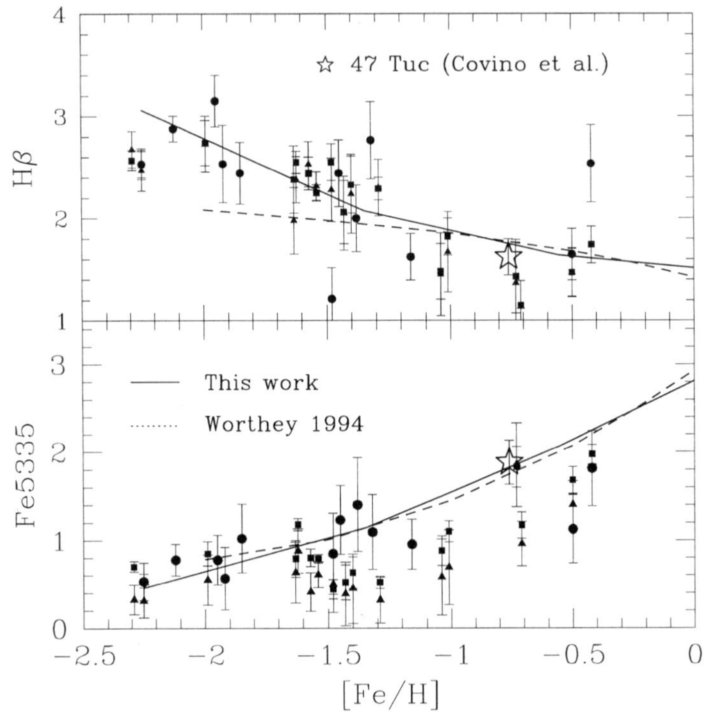

Figure 2. Calibration of the SSP models on GGCs. Data are from Burstein et al. 1984 (squares), Covino, Galletti, & Pasinetti 1995 (circles), and Trager 1998 (triangles). [Fe/H] (Zinn 1985 scale) is taken from Harris (1996). Solid lines are our SSPs for age $t = 15$ Gyr. Worthey (1994) SSPs for $t = 17$ Gyr are shown as dashed lines. The big star symbol represents the galactic globular cluster 47 Tuc.

iron abundance. The fact that SSP models based on solar-scaled tracks reproduce the data implies that the α-enhancement has no impact on optical colours like $B-V$. This is consistent with the conclusion by Salaris, Chieffi & Straniero 1993 ([27]), that the GGC ages determined by adopting the $\Delta(B-V)$ method are insensitive to the α-enhancement. Finally, the Carretta & Gratton (1997) iron abundance scale (Fig. 1, central panel) gives higher metallicities with respect to Zinn (1985) for the low-intermediate metallicity GGCs, yielding a worse match between models and data. The discrepancy may be the consequence of the small sample of giants stars on which the Carretta & Gratton scale is based.

3. GGC line strenghts

A key ingredient for the modelling of SSP synthetic line indices (Hβ, Mg_2, etc.) are the stellar line strenghts as a function of the basic stellar parameters, effective temperature, surface gravity and metallicity ($T_{\rm eff}$,g, [Fe/H]). GGC stars and field stars are used to construct these calibrations, called *fitting functions* (Worthey et al. 1994 [31]; Buzzoni, Gariboldi & Mantegazza 1992 [7]; Buzzoni, Mantegazza, & Gariboldi, 1994 [8]). Maraston et al. (1999) [21] show that synthetic indices are affected by uncertainties that cannot be neglected when using these models to derive the ages and the metallicities of stellar systems. These uncertainties originate from the population synthesis/stellar input and to a large extent from the use of different fitting functions. We focus here on Hβ and Fe5335, referring the reader to Maraston et al. [21] for the discussion of the other spectral indices. Fig. 2 shows Hβ and Fe5335 of GGCs as a function of [Fe/H] (Zinn (1985) [32]). Our SSPs (t=15 Gyr) computed adopting the Worthey et al. 1994 [31] fitting functions are shown as solid lines. Dashed lines display the Worthey (1994) models. As old, metal-poor ([Fe/H] <-1.3) stars spend part of their Horizontal Branch lifetimes at temperatures $T_{\rm eff}$ > 6000 K, Hβ is steeply rising with decreasing [Fe/H]. This feature is well reproduced by our models. The SSPs from Worthey (1994) [30] agree well with our computations above [Fe/H]~ -1.4; below this value instead they predict a more shallow trend. If the data are interpreted with the Worthey (1994) [30] models, the very metal-poor GGCs turn out to be younger than the more metal-rich ones. The synthetic Fe5335 indices are well consistent. Combining Hβ and Fe5335 provides a tool to break the age-metallicity degeneracy of stellar populations (Worthey 1994 [30]). For example, for 47 Tuc we obtain a 'spectroscopic age' of 15 Gyr, in excellent agreement with the age derived from the color-magnitude diagram (14 ± 1 Gyr, Richer et al. 1996, [26]). Note that the uncalibrated index $H\gamma_{\rm HR}$, instead, leads to a 'spectroscopic age' in excess of 20 Gyr (Gibson et al. 1999 [14]). Using the SSPs presented here, Maraston & Thomas (1999) show that strong Balmer lines can be produced in a globally old, metal rich population, contaminated by a small fraction of old, metal poor stars (see below).

4. Discussion

In this contribution we discuss the importance of calibrating Simple Stellar Population models. The galactic globular clusters are particularly suitable as calibrators in the low metallicity/old age regime. The Hβ indices of GGCs sharply rise below [Fe/H] < -1.3. This feature is well reproduced by our SSPs. The sharp increase of the Hβ index at low metallicities is the key ingredient for the conclusions presented in Maraston & Thomas (1999) [22]. The authors compute Composite Stellar Populations in which a small fraction of an old, metal poor population is added to an old, metal rich one. These composite models are able to reproduce the strong Hβ values observed in cluster elliptical galaxies (see e.g. Kuntschner & Davies [19]; Mehlert et al. 2000 [24]). This is alternative to the option of adding a young, metal rich population.

Acknowledgements

Santi Cassisi is acknowledge for kindly providing the stellar tracks and isochrones. A special thank to F. Matteucci and F. Giovannelli for the very lively and fruitful workshop. This work was supported by the "Sonderforschungsbereich 375-95 für Astro-Teilchenphysik" of the Deutsche Forschungsgemeinschaft.

References

1. Barmby, P., Huchra, J.P., Brodie, J.P., Forbes, D.A., Schroder, L.L. & Grillmair, C.J, 1999, AJ, in press, astro-ph 9911152
2. Bono, G., Caputo, F., Cassisi, S., Castellani, V. & Marconi, M. 1997, ApJ, 489, 822
3. Bressan, A., Chiosi, C., & Fagotto, F. 1994, ApJS, 94, 63
4. Burstein, D., Faber, S. M., Gaskell, C. M., & Krumm, N. 1984, ApJ, 287, 586
5. Buzzoni, A., 1989, ApJS, 71, 817
6. Buzzoni, A., 1995, ApJS, 98, 69
7. Buzzoni, A., Gariboldi, G., and Mantegazza, L., 1992, AJ, 103, 1814
8. Buzzoni, A., Mantegazza, L., and Gariboldi, G., 1994, AJ, 107, 513
9. Carretta, E., and Gratton, R.G., 1997, A&ASS, 121, 95
10. Covino, S., Galletti, S., & Pasinetti, L. E. 1995, A&A, 303, 79
11. Faber, S. M., Trager, S.C., González, J.J., Worthey, G., in Van der Kruit P. C., Gilmore G., eds, Proc. IAU Symp. 164, Stellar Populations. Kluwer, Dordrecht, p. 249.
12. Ferraro, F., , AJ, *in press*, astro-ph 9906248
13. Fusi-Pecci, F., Renzini, A., 1976, A&A, 46, 447
14. Gibson, B. K., Madgwick, D. S., Jones, L. A., Da Costa, G. S., & Norris, J. E. 1999, AJ, 118, 1268
15. González J.J., 1993, PhD thesis, Univ. of California, Santa Cruz
16. Greggio, L., 1997, MNRAS, 301, 166
17. Guiderdoni, B., Rocca-Volmerange, B., 1987, A&A, 186, 1
18. Harris, W. E. 1996, AJ, 112, 1487
19. Kuntschner, H., & Davies, R. L. 1998, MNRAS, 295, L29
20. Maraston, C., 1998, MNRAS, 300, 872
21. Maraston, C., Greggio, L., & Thomas, D. 1999, Ap&SS, in press, astro/ph-9906088
22. Maraston, C., Thomas, D., 1999, ApJL, submitted
23. Matteucci, F., 1996, Fund. Cosm. Physics, 17, 283
24. Mehlert, D., Saglia, R. P., Bender, R., & Wegner, G. 2000, A&AS, in press
25. Reimers, D., Mem. Soc. R. Sci. Liege, Ser. 6, vol. 8, 369
26. Richer, H. B., et al. 1996, ApJ, 463, 602
27. Salaris, M., Chieffi, A., and Straniero, O., 1993, ApJ, 414, 580
28. Tantalo, R., Chiosi, C., Bressan, A., and Fagotto, F., 1996, A&A, 311, 361
29. Trager, S.C., 1998, PhD thesis, Univ. of California, Santa Cruz
30. Worthey, G., 1994, ApJS, 95, 107
31. Worthey, G., Faber, S.M., González, J. and Burstein, D., 1994, ApJS, 94, 687
32. Zinn, R.J., 1985, ApJ, 293, 424

OLD OPEN CLUSTERS AS TRACERS OF GALACTIC EVOLUTION

A. BRAGAGLIA AND M. TOSI
Osservatorio Astronomico di Bologna, via Ranzani 1
I-40127 Bologna (Italy), e-mail: angela, tosi@bo.astro.it

G. MARCONI[1,2]
[1] *Osservatorio Astronomico di Roma, Via Osservatorio 5*
I-00040 Monte Porzio (Italy),
[2] *E.S.O., Santiago (Chile), e-mail: gmarconi@eso.org*

AND

E. CARRETTA
Osservatorio Astronomico di Padova, vicolo Osservatorio 5
I-35122 Padova (Italy), e-mail: carretta@pd.astro.it

1. Why (old) open clusters?

Open clusters (OC's) are important to study the properties of the Galactic disk, since they offer information on ages, both absolute and relative, and on the metallicity evolution, both in space and time. In fact, OC's are found in different regions of the disk, cover a large interval in age (from a few Myr to about 10 Gyr) and in metallicity ($Z=Z_{\odot}/20$ to supersolar). Moreover, their ages and distances are accurate, much more than for any other disk object like e.g., single field stars. Also, since OC's orbits do not generally take them too far away from their birthplaces, we may assume that their current position in the Galaxy is representative also of their original one, and that we are not smearing properties by mixing populations, always a problem when dealing with field stars.

To define at least a reliable ranking of the open clusters properties we need a large sample of objects whose age, distance and metallicity are accurately and homogenously known: see e.g., Janes & Phelps (1994), Carraro & Chiosi (1994), Friel (1995), Twarog et al. (1997). These authors have tried to define such samples, but had to sacrifice something in the quality of the data, often taken from the literature, and not as excellent as attain-

F. Matteucci and F. Giovannelli (eds.), The Evolution of the Milky Way, 281–287.

able with today's means, and/or on the homogeneity of the treatment to get a sample as large as possible.

We have started a few years ago a project, admittedly ambitious, to build a homogeneous and statistically significant sample of clusters of various ages, metallicities and positions in the disk. We work only on new photometry taken by our group, or on literature data of comparable quality. The clusters' properties are derived using the synthetic colour-magnitude diagram method (Tosi et al. 1991). We compare the observed diagram to a grid of synthetic ones, generated from a series of homogeneous sets of theoretical evolutionary tracks by several authors. The comparison is based both on morphology (e.g., main sequence shape, red clump position, gaps, etc.) and on population ratios (e.g., the luminosity function). We have already applied this method to nine open clusters, ranging in age from 0.1 to about 10 Gyr (see Bragaglia et al. 1999 for a recent review), and we have data for ten more.

We plan now to turn to spectroscopy, in particular to high resolution, in order to get precise information on the clusters' metallicity.

2. What do we know about open clusters' metallicities?

Of all the OC's we have studied photometrically, four have the metallicity measured by high dispersion spectroscopy, the best technique to get reliable results, with uncertainties of less than about 0.1 dex in [Fe/H] (see Gratton 1999 for a recent review). If we look at the whole sample of old OC's this is true for only 18 of the about 80 known. There are other ways to get the metallicity, like low resolution spectroscopy calibrated to high-res, as done by Janes and Friel (see e.g., Friel 1995; this allows uncertainties of about 0.15 dex), or Washington, UBV and DDO photometry (with uncertainties of about 0.2 dex). A little more than half of the old OC's sample has the metal abundance measured by at least one of the above methods. An interesting feature comes out of the comparison between metallicities measures by high-res spectroscopy and any of the other methods: they all underestimate [Fe/H] by at least 0.1 dex on average (even if the individual values may be off by much more).

Apart from a possible sistematic effect, does the technique used influence the derived general properties of the cluster sample? If we consider the old OC's we can study the existence of the radial metallicity gradient using metallicities derived by all methods. Figure 1 shows the different cases; at least part of the observed dispersion at all Galactocentric distances is real. If we determine the slope of the metallicity gradient we find values very similar to each other within the errors; they are indicated in Figure 1, and are of the order of –0.1 dex kpc^{-1}. Notice the larger slope obtained for

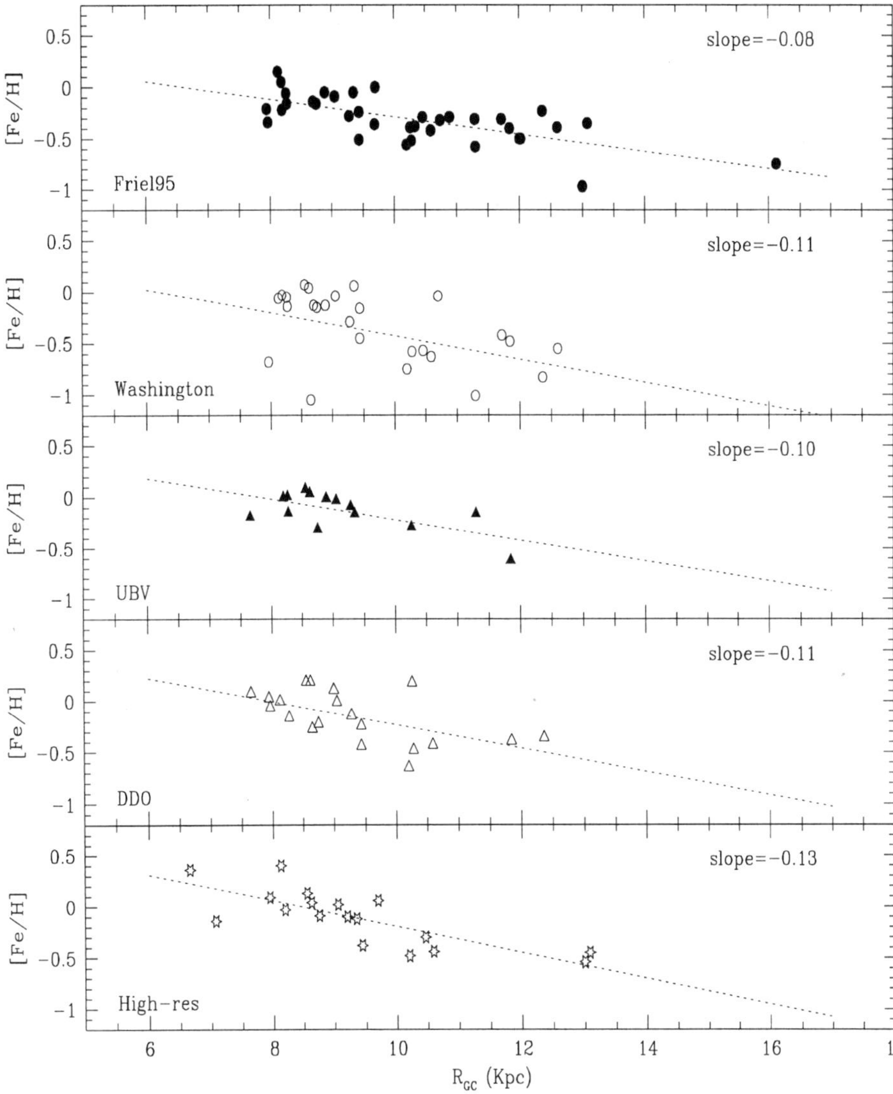

Figure 1. Old open clusters: run with Galactocentric distance of the observed metallicity, derived with different techniques (from top to bottom: by low resolution spectroscopy, by Washington, UBV, DDO photometry, and by high dispersion spectroscopy). The slope of the resulting metallicity gradient is indicated in each panel.

the high-res values; the sample is still too small to decide whether it is really significant, and it would be interesting to have many more high-res determinations to put on firmer grounds the absolute and relative values of OC's metallicities.

Regarding the slope of the radial gradient, it seems irrelevant which method is used to determine metallicities. On this basis, to study whether the metallicity gradient has changed with time, we use the Friel (1995) values, since they are the most numerous for the old OC's (36 OC's, compared to 26 with Washington, 14 with UBV and 20 with DDO photometry). If we divide the whole sample in four age intervals (age < 1, 1–3, 3–6, > 6 Gyr) we find that there is no variation of the gradient slope with age (see Figure 2). There may be a slight indication of a shallower slope for the younger OC's, but it's well within the errors.

Finally, using samples with metallicities obtained homogenously, we do not see any indication of the discontinuity at R_{GC} about 10 kpc assumed by Twarog et al. (1997) as an indication of the edge of the original thick disk.

3. The metallicity of NGC 6253 and future work

Our first direct contribution to the knowledge of OC's metallicities comes from NGC 6253, a cluster in the direction of the Galactic centre, with age $\simeq 3$ Gyr, and metallicity solar or (more probably) twice solar as derived from photometry alone (Bragaglia et al. 1997). We have high resolution spectra of four giant stars, taken with EMMI mounted on NTT as part of a backup program in a night not suitable for photometry. As a result, the signal-to-noise ratio (about 35) and the resolution (about 15,000) are not as good as one would desire to derive precise abundances from fine analysis of high resolution spectra. This is especially true for stars of metallity so high that continuum tracing is a very difficult task, and for this reason we compared the observed spectra with a field star of similar metallicity (ζ Cygni, [Fe/H]=+0.05), as shown in Figure 3.

On the basis of their radial velocities, all the four stars are cluster members. We have only studied the iron content of two red clump stars (see Carretta et al. 1999 for a more detailed description of the followed procedure), and derived for them the following values: #2971: [Fe/H] = +0.33, and #2508: [Fe/H]= +0.39. The overall uncertainty, resulting from the measured equivalent widths and the adopted atmospheric parameters, is 0.15–0.20 dex, so we may assume for NGC 6253: [Fe/H] = $+0.36 \pm 0.20$ dex. This is in very good agreement with the photometric study and with the value found by Piatti et al. (1998) from integrated spectra. NGC 6253 appears in Figure 1 (high resolution panel) as the innermost cluster of the whole sample, and its metallicity conforms to the Galactocentric gradient found for the other clusters.

We have recently acquired spectra of better quality of the same stars in NGC 6253, and we plan to analyze them to derive the abundances of

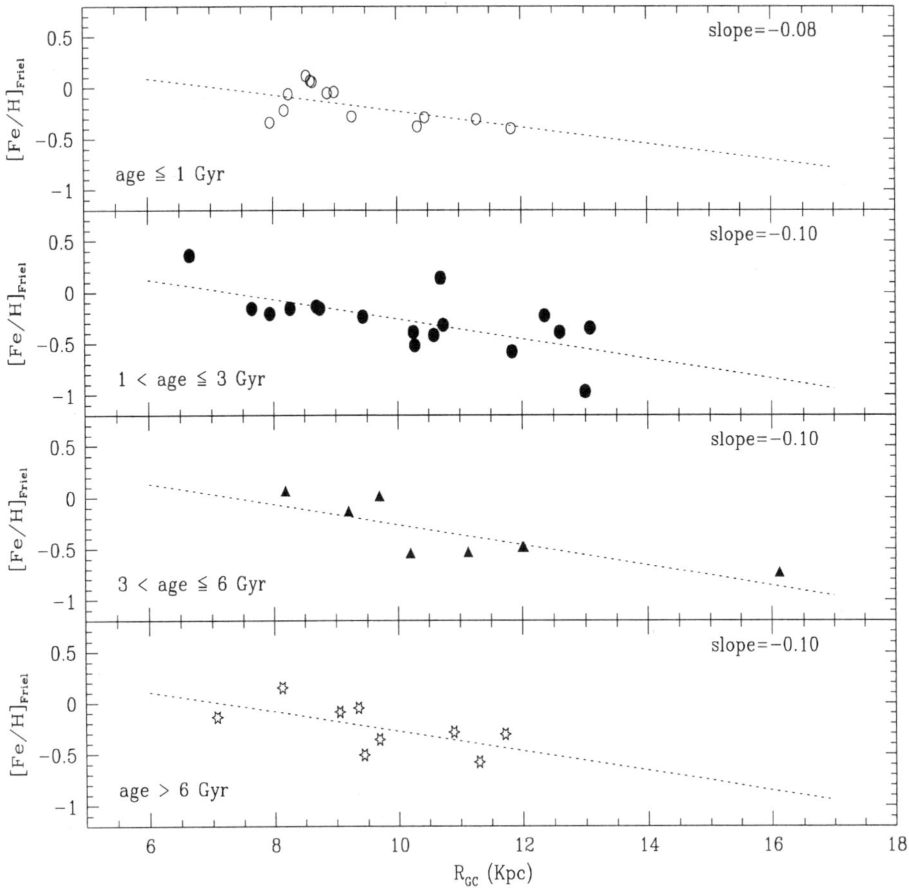

Figure 2. Run with Galactocentric distance of the observed metallicity (as derived by Friel 1995) for open clusters, divided into four age bins.

other elements. Our plan for the next future is to obtain high-resolution spectra in several open clusters in order to derive accurate and homogenous metallicities of a large sample of clusters. This is a prerequisite for a really meaningful study of the history of the chemical enrichment of the Galactic disk.

References

1. Bragaglia, A., Tessicini, G., Tosi, M., Marconi, Munari, U. (1997) UBVRI CCD photometry of the old open cluster NGC 6253, *Mon. Not. R. Astr. Soc.* **284**, 477
2. Bragaglia, A., Tosi, M., Marconi, G. Sandrelli, S. (1999) Old open clusters as trac-

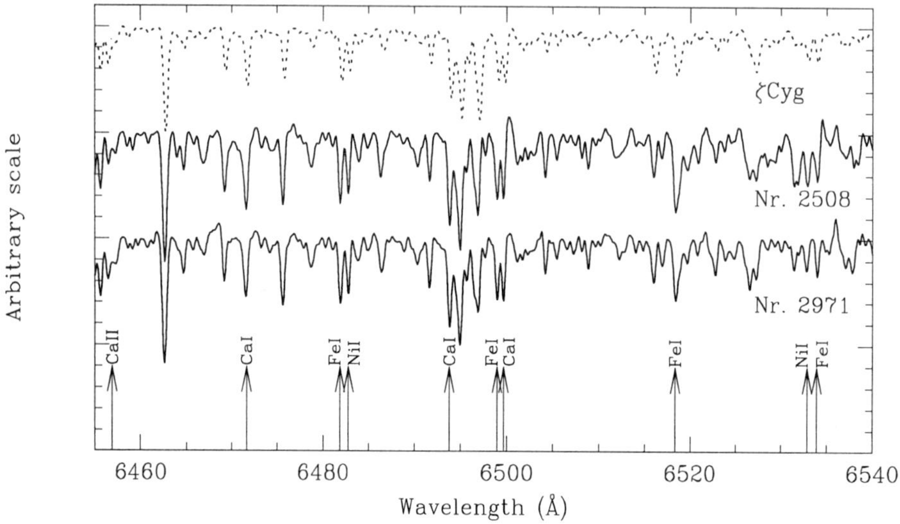

Figure 3. A region of the normalized spectra of the two NGC 6253 stars, compared to ζ Cyg: notice the shallower lines of the latter, indicative of a slightly lower metal content; the spectra have been arbitrarily shifted for clarity.

ers of galactic evolution: the cases of NGC2660, IC1311, and NGC4815, inn *Stellar clusters and associations : convection, rotation, and dynamos*, in press

3. Carraro, G., Chiosi, C. (1994) The Galactic system of old open clusters: age calibration and age-metallicity relation, *Astron. Astrophys.* **287**, 761
4. Carretta, E., Bragaglia, A., Tosi, M., Marconi, G. (1999), The metallicity of NGC 6253, in *Stellar clusters and associations : convection, rotation, and dynamos*, in press
5. Friel, E. (1995) The old open clusters of the Milky Way, *Ann. Rev. Astron. Astrophys.* **33**, 381
6. Gratton, R.G. (1999), Abundances in open clusters: results and concerns, in *Stellar clusters and associations : convection, rotation, and dynamos*, in press
7. Janes, K.A., Phelps, R.L. (1994) The galactic system of old star clusters: The development of the galactic disk, *Astron. J.* **108**, 1773
8. Piatti, A.E., Clarià, J.J., Bica, E., Geisler, D., Minniti, D. (1998) A Photometric and Spectroscopic Study of the Southern Open Clusters Pismis 18, Pismis 19, NGC 6005, and NGC 6253, *Astron. J.* **116**, 801
9. Tosi, M., Greggio, L., Marconi, G., Focardi, P. (1991) Star formation in dwarf irregular galaxies - Sextans B, *Astron. J.* **102**, 951
10. Twarog, B.A., Ashman, K.M., Anthony-Twarog, B. (1997) Some Revised Observational Constraints on the Formation and Evolution of the Galactic Disk, *Astron. J.* **114**, 2556

Speaker: A. BRAGAGLIA

M. GRENON's Comments:
1) For clusters inside the co-rotation the effect of the galactic bar may be important, for example for NGC6791 with u velocity component of 100km/sec. Old clusters may then be observed far from their birthplace. 2) The determination of distance of remote clusters depends on the local reddening law , dependent on the gas [M/H]. At r=15 Kpc the law should be that of LMC so how do you cope with the changing law with galactocentric distance?

G. CARRARO's Comment to M. GRENON's Comment: The epicyclical amplitude is large only for clusters inside the solar circle, due to the bar influence (for instance NGC6791). Clusters outside the solar ring should not feel too much the bar.

A. BRAGAGLIA:
1) see Carraro's comment/answer. 2) We have not studied clusters so far from the Galactic center; the only one we have in our sample near r=15 Kpc is Be21, which has complications from differential extinction. For the moment we stick to the standard extinction law.

IV-HALO, BULGE AND DISK

Excursion to the Volcano

ABUNDANCES OF GLOBULAR CLUSTERS IN THE GALACTIC BULGE

B. BARBUY
IAG/USP
São Paulo, Brazil

Abstract. Element ratios in two stars of NGC 6553 and one star in NGC 6528 reveal that α-elements are enhanced relative to Fe. The metallicity [Fe/H] $\approx$ -0.5 combined to overabundances of several elements gives an overall metallicity close to solar for both clusters.

1. Introduction

About 60 of the globular clusters in our Galaxy lie within 4 kpc of the Galactic bulge, within a radius of 20° around the Galactic center. A fraction of them would belong to the halo but probably a majority of these belong to the bulge population.

The bulk of the bulge stellar population is old, showing metallicities in the range -1.5 < [Fe/H] < +0.5. At the same time there is a large number of Miras, which seem to be a mixture of young/intermediate age Miras, together with luminous Miras belonging to the old population (Guarnieri et al. 1997).

It seems to us that the bulge stellar population is essentially old as predicted in models of galaxy formation "inside-out" (Larson 1990; Renzini 1999), but there seems to be as well a young, disk-like population, resulting from the dynamical evolution of a bar (Raha et al. 1992; Spergel et al. 1996; Pfenniger 1993).

If the bulge is formed rapidly in the beginning of the Galaxy life, a fast enrichment by SNs II at early times would result in a signature through [α-elements/Fe] enhanced. If the bulge is disk-like, then one would expect [α-elements/Fe] = 0.0.

In order to try to solve this question we are carrying out a program of spectroscopic determination of abundances in bulge stars.

F. Matteucci and F. Giovannelli (eds.), The Evolution of the Milky Way, 291–300.

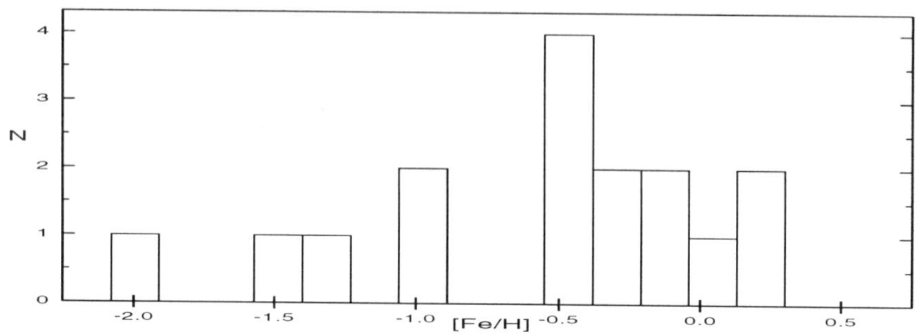

Figure 1. Metallicity distribution of 17 globular clusters within the inner 5° of the Galactic center.

2. The bulge stellar population

The metallicity distribution of low-mass field stars in Baade's Window (Rich 1988; McWilliam & Rich 1994) and that for bulge globular clusters within 5° of the Galactic center (Barbuy et al. 1998) are very similar. In Fig. 1 is shown the metallicity distribution of the 17 globular clusters within the inner 5° cf. Barbuy et al. (1998).

Detailed determinations of metallicity and abundance ratios for bulge stars are not many, due to the difficulties to observe such faint and reddened stellar populations. Low resolution spectra obtained by Rich (1988) indicated a mean abundance for bulge stars of [Fe/H] = +0.25. Sadler et al. (1996) derived [Mg/Fe] $\approx$ +0.3 from measurementes of the Mg_2 index in about 400 stars. Idiart et al. (1996) derived [Mg/Fe] = +0.45 from an integrated spectrum of Baade's Window.

McWilliam & Rich (1994, hereafter MR94) presented a high-resolution spectroscopic study of 12 bulge K giants of Baade's Window. Their main results are: [Fe/H] = -0.25, [Mg/Fe] = [Ti/Fe] = +0.3, [Ca/Fe] = [Si/Fe] = 0.0, [O/Fe] = 0.0 and [Eu/Fe] = slightly enhanced.

3. The bulge globular clusters NGC 6553 and NGC 6528

Ortolani et al. (1995) have shown that NGC 6528 and NGC 6553 should be nearly coeval with the halo clusters, and that the luminosity function of these clusters is very similar to that of field stars in Baade's Window, confirming that the clusters and bulge field form a same stellar population, making the clusters ideal laboratories for bulge studies.

Since 1990 we have been carrying out studies of globular clusters in the bulge, through imaging and spectroscopy. In Barbuy et al. (1998) we gathered the results obtained for all clusters within 5° of the Galactic center. In the present paper we present our first results on detailed abundances of these clusters. Note that Cohen et al. (1999) presented results for horizontal branch stars of NGC 6553. See also our previous discussions in Barbuy (1999a,b).

4. Derived metallicities

A detailed analysis was carried out for two stars in NGC 6553 (Barbuy et al. 1999) and one star in NGC 6528, using high resolution échelle spectra obtained at the ESO 3.6m telescope using CASPEC.

The effective temperatures were derived from HST VI and ESO 2.2m IRAC2 JK photometry. The reddening was corrected according to Guarnieri et al. (1998), where it was derived by comparing the colour of the subgiant branch of NGC 6553 to that of 47 Tuc.

MARCS model atmosphere grids by B. Plez (private communication), which are an extension of the grid by Plez et al. (1992) and Gustafsson et al. (1975) were employed - note that the use of different model atmospheres may have an important consequence on final results, as will be discussed below in a comparison with recent results by Cohen et al. (1999).

In order to derive the metallicity [Fe/H] we used FeI and FeII lines for which laboratory oscillator strenghts gfs by Fuhr et al. (1988) were available, and the measurement of equivalent widths was carried out by tracing a continuum close to the observed continuum.

In Fig. 2 we show the curve-of-growth of star 2 in NGC 6528: the spread of points is due to the blending of lines in this cool giant, and also to a S/N $\sim$ 30. It is possible to see however that it would be difficult to make this star much more metal-rich than [Fe/H] $\approx$ -0.4. We adopted [Fe/H] = -0.5$\pm$0.2 for this star.

5. Comments on results by Cohen et al. 1999

Reasons for a difference in metallicity by $\sim$ 0.3 dex relative to Cohen et al. (1999) for NGC 6553 are partly mentioned in Sect. 7.1 of Cohen et al.'s paper. Our comments are:

(i) Model atmosphere: we employed Plez et al. (1992, 1997) grid of atmospheric models, whereas Cohen et al. used Kurucz (1993) model atmosphere.

(ii) Barbuy et al. (1999) used laboratory oscillator strengths (gfs) by Fuhr et al. (1988), Martin et al. (1988), Wiese et al. (1969) and the compilation of laboratory gfs by MR94, whereas Cohen et al. (1999) used their own gfs - note that their gfs are similar to those obtained by us through fittings of the solar spectrum with the Holweger-Müller model atmosphere (Barbuy et al. 1999). We preferred however to use laboratory data, which are about 0.2 dex higher than Cohen et al.'s.

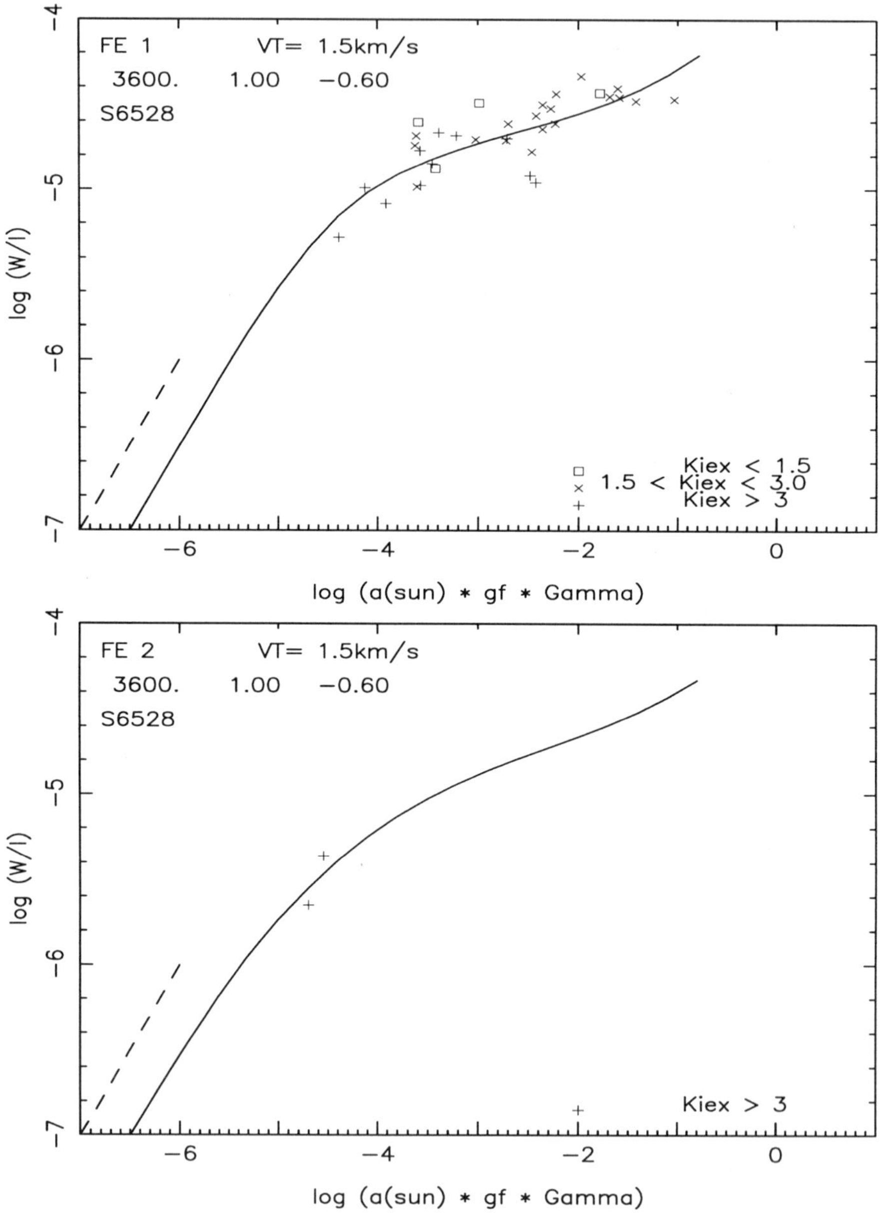

Figure 2. Curves-of-growth of FeI and FeII lines for star 2 (internal ring) in NGC 6528, of $T_{eff} = 3600$ K.

(iii) The major reason for the difference in metallicity between Barbuy et al. (1999) and Cohen et al. (1999) comes from the temperature scale adopted in Cohen et al. We have used their measurements of equivalent

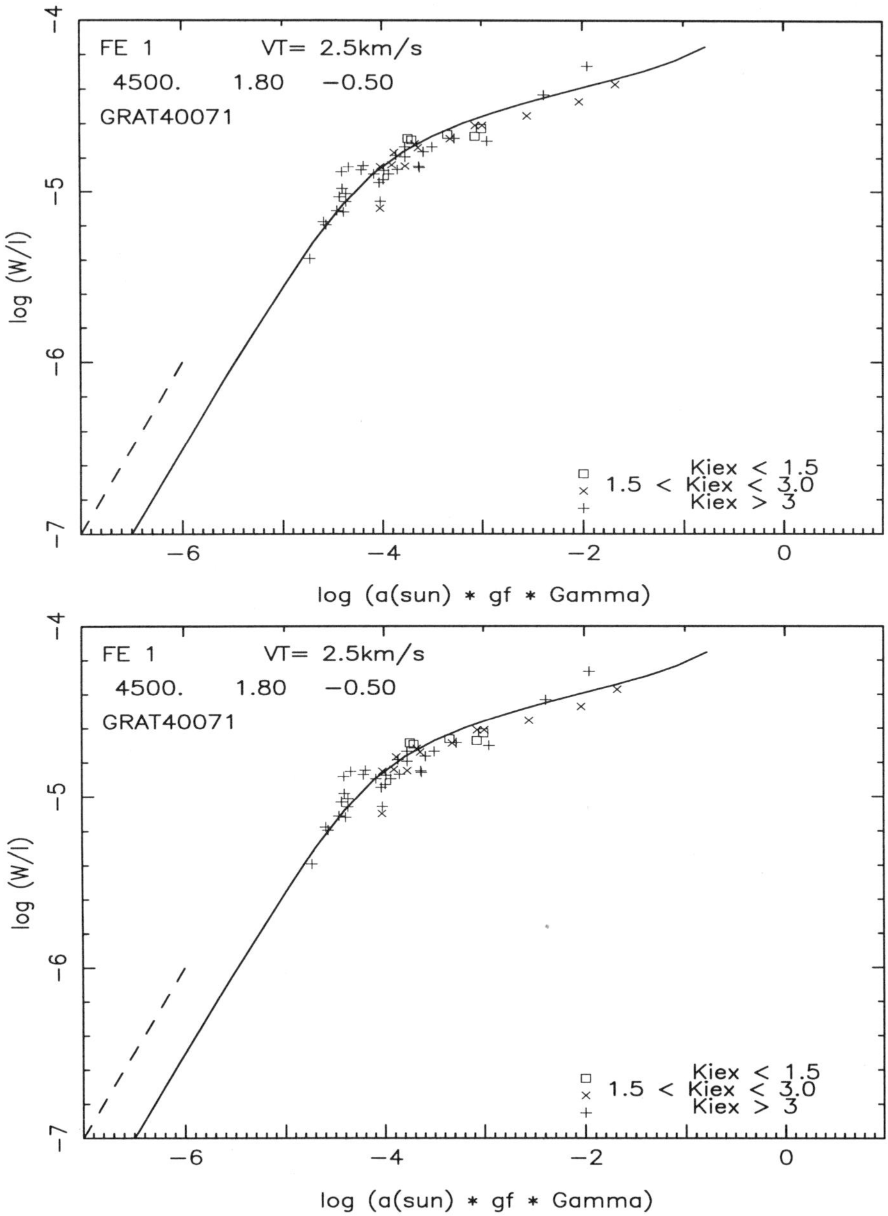

Figure 3. Curves-of-growth of FeI and FeII lines for star 40071 in NGC 6553, using data by Cohen et al. (1999), but replotting the curve-of-growth using MARCS model atmospheres.

widths for the star 40071, and recomputed a curve-of-growth, using however MARCS model atmospheres - this curve-of-growth is shown in Fig. 3, where the temperature is fitted to obtain a best excitation equilibrium, i.e. the least difference in the curves of growth of lines of different excitation potential. Using this criterion for this star we obtain (T_{eff}, log g, [Fe/H], v_t) = (4500, 1.8, -0.5, 2.5) instead of (4725, 2.3, -0.2, 2.5) given in Cohen et al.'s. In a general way, it seems to us that the new metallicity scale by Gratton et al. (1996) and Carretta & Gratton (1997) is a consequence of their hotter temperature scale since in their scale essentially all clusters are more metal-rich by $\sim$ 0.2 dex relative to other analyses.

6. Abundance ratios

Elemental abundances were derived by computig synthetic spectra line-by-line. In Table 1 are given the resulting abundance ratios for a mean of two stars in NGC 6553 (Barbuy et al. 1999), one star in NGC 6528 and a mean of results for 12 bulge K giants by MR94. Chemical evolution models for the bulge were presented by Matteucci & Brocato (1990), Matteucci et al. (1999), where a inside-out model for the formation of the Galaxy is adopted. An increased star formation efficiency, of a factor 10 relative to the solar neighbourhood is assumed, and a timescale for the bulge formation of about 10^7 yr is obtained. The main result is an enhancement of α elements, with the exception of calcium. Our results, including [Ca/Fe] < 0.0 (see Fig. 4) are in good agreement with Matteucci et al. (1999). Further observations of larger samples of stars with 8m class telescopes in order to achieve high S/N ratios are necessary in order to better establish the constraints on bulge formation history.

7. Super-metal-rich stars in the solar neighbourhood

Grenon (1998, 1999) presented a sample selected from the NLTT proper motion catalogue, for which Hipparcos proper motions, Geneva photometry and radial velocities with Coravel are available.

The selected sample consists of metal-rich stars with orbits typical of bulge stars presently in their apocenter, close to the solar neighbourhood. Hipparcos data allowed to derive an age of 10-11 Gyrs for this sample (Grenon 1998).

Subsamples of these stars were analysed in Barbuy & Grenon (1990) and Castro et al. (1997), where [Fe/H] > 0, [s-elements/Fe] < 0 (Castro et al. 1997), [α-elements/Fe] ≥ 0 were found. A series of spectra obtained with FEROS at ESO are under analysis.

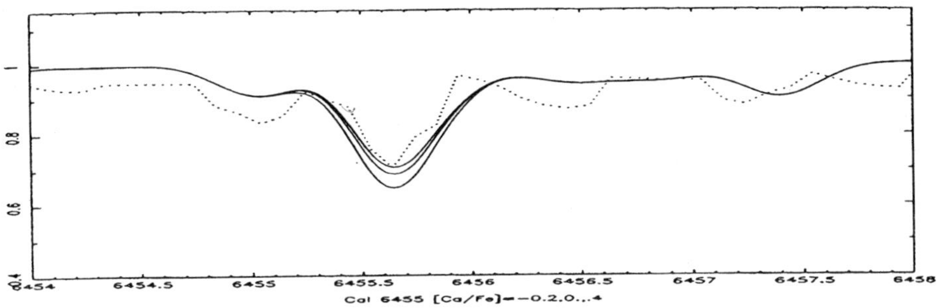

Figure 4. NGC 6528:Star 2 - CaIλ6455 Å line computed with [Ca/Fe] = -0.2, 0.0, +0.4.

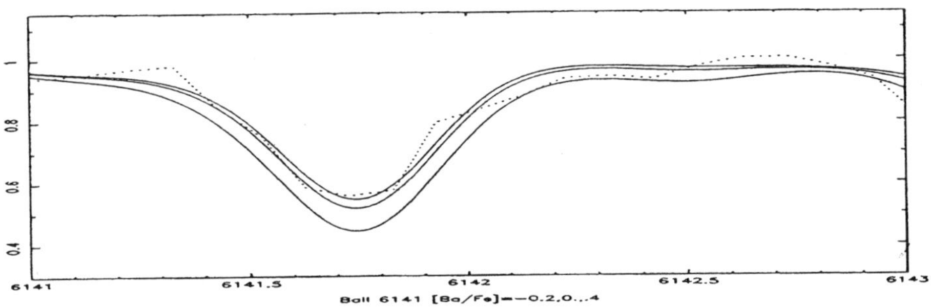

Figure 5. NGC 6528:Star 2 - BaIλ6141 Å line computed with [Ba/Fe] = -0.2, 0.0, +0.4.

8. Conclusions

A first detailed analysis of low-mass bulge stars was carried out by McWilliam & Rich (1994). We present high resolution detailed abundance analysis of individual stars in the bulge globular clusters NGC 6528 and NGC 6553 (Barbuy et al. 1999), which show [Fe/H] $\approx$ -0.5 combined to enhancementd of α-elements. For both clusters we obtain $[Z/Z_\odot] \approx$ -0.1. Note that 47 Tuc with $[Z/Z_\odot]$ = -0.32, would show a difference relative to these metal-rich clusters in the enhancement of α elements. Note that also ellipticals and bulges of spiral galaxies show enhancements of magnesium (Worthey et al. 1992).

A comparison of these results with those for field bulge stars by M-cWilliam & Rich (1994) show strong similarities, as for example, low [Ca/Fe] $\approx$ 0.

Our results show that the bulk of stellar populations in the Galactic bulge seems to be old, and their chemical enrichment proceeded in a fast timescale at early times (see discussions in Renzini 1998, 1999).

Further high resolution spectra to be obtained with 8m class telescopes are necessary to study larger samples of stars in these clusters. In order to understand the differences between Barbuy et al. (1999) and Cohen et al. (1999) it would be interesting to analyse stars from the cool giant branch down to the horizontal branch, covering a wide range of temperatures. Since the brightest stars in NGC 6528 (case of our presently studied star 2) have V = 15.8, 8m class telescopes will be necessary to carry out such work.

TABLE 1. Abundance ratios in NGC 6553, NGC 6528, and bulge field K giants (MR94)

[X/Fe]	NGC6553	NGC6528	MR94
[Fe/H]	-0.55	-0.5	–
Na	+0.65	+0.4	+0.20
O	–	+0.25	+0.03
Mg	+0.33	+0.40:	+0.35
Al	+0.50	+0.4	+0.58
Si	+0.35	+0.6	+0.14
Ca	+0.32	-0.2	+0.14
Ti	+0.60	+0.2	+0.37
Y	-0.20	–	+0.05
Zr	-0.40	-0.40	-0.53
Ba	-0.10	-0.2	+0.20
La	+0.13	–	+0.09
Eu	+0.0	0.0	+0.26

References

1. Barbuy, B. 1999a, *Galaxy evolution: connecting the distant universe with the local fossil record*, eds. M. Spite, F. Crifo, Kluwer Acad. Pub., in press

2. Barbuy, B. 1999b, *VLT Symposium*, eds. L. Pasquini, R. Kudritzki, Springer-Verlag, in press
3. Barbuy, B., Bica, E., Ortolani, S. 1998, *A&A* **333**, 117
4. Barbuy, B., Grenon, M. 1990, in *Bulges of Galaxies*, ESO Conf. and Workshop Proceedings 35, eds. B.J. Jarvis & D.M. Terndrup, p. 83
5. Barbuy, B., Renzini, A., Ortolani, S., Bica, E., Guarnieri, M.D. 1999, *A&A* **341**, 539
6. Carretta, E., Gratton, R.G. 1997, *A&AS* **121**, 95
7. Castro, S., Rich, R.M., Grenon, M. Barbuy, B., McCarthy, J. 1997, *AJ* **114**, 376
8. Cohen, J.G., Gratton, R.G., Behr, B.B., Carretta, E. 1999, *ApJ* **523**, 739
9. Fuhr, J.R., Martin, G.A., Wiese, W.L. 1988, *Atomic Transition Probabilities: Iron through Nickel*, Journal of Physical and Chemical Reference Data, vol. 17, suppl. no. 4
10. Gratton, R.G., Carretta, E., Castelli, F. 1996, *A&A* **314**, 191
11. Grenon, M. 1998, in Highlights of Astronomy, in press
12. Grenon, M. 1999, in *Galaxy evolution: connecting the distant universe with the local fossil record*, eds. M. Spite, F. Crifo, Springer-Verlag, in press
13. Guarnieri, M.D., Renzini, A., Ortolani, S. 1997, *ApJ* **477**, L21
14. Guarnieri, M.D., Ortolani, S., Montegriffo, P., Renzini, A., Barbuy, B., Bica, E., Moneti, A.: 1998, *A&A* **331**, 70
15. Gustafsson, B., Bell, R.A., Eriksson, K., Nordlund, Å: 1975, A&A 42, 407
16. Idiart, T., Freitas Pacheco, J.A., Costa, R.D.D. 1996, AJ 111, 1169
17. Kurucz, R. 1993, CD-ROM 18
18. Larson, R.B. 1990, *PASP* **102**, 709
19. Martin, G.A., Fuhr, J.R., Wiese, W.L. 1988, *Atomic Transition Probabilities: Scandium through Manganese*, Journal of Physical and Chemical Reference Data, vol. 17, suppl. no. 3
20. Matteucci, F., Brocato, E. 1990, *ApJ* **365**, 539
21. Matteucci, F., Romano, D., Molaro, P. 1999, *A&A* **341** , 458
22. McWilliam, A., Rich, R.M. 1994, *ApJS* **91**, 749 (MR94)
23. Ortolani, S., Renzini, A., Gilmozzi, R., Marconi, G., Barbuy, B., Bica, E., Rich, R.M.: 1995, *Nature* **377**, 701
24. Pfenniger, D. 1993, IAU Symp. 153, p. 387
25. Plez, B., Brett, J.M., Nordlund, A. 1992, A&A, 256, 551
26. Plez, B. 1997, private communication
27. Raha, N., Sellwood, J.A., James, R.A., Kahn, F. D. *Nature* **352**, 411
28. Renzini, A. 1999, astro-ph/9902108
29. Rich, R.M. 1988, *AJ* **95**, 828
30. Sadler, E., Terndrup, D.M., Rich, R.M. 1996, AJ 112, 117
31. Spergel, D.N., Malhotra, S., Blitz, L. 1996, in *Spiral Galaxies in the Near-IR*, eds. D. Minniti, H.-W. Rix, Springer-Verlag, p. 128
32. Wiese, W.L., Martin, G.A., Fuhr, J.R. 1969, *Atomic Transition Probabilities: Sodium through Calcium*, NSRDS-NBS 22
33. Worthey, G., Faber, S.M., Gonzalez, J.J. 1992, *ApJ* **398**, 69

Speaker: B. BARBUY

R. GALLINO: Concerning the heavy elements Ba, La, Eu, they are similar to disk stars except for a low Zr. Does this means that AGB stars had time to contribute (Be, La) as well as the r-sources (Eu)?

B. BARBUY: s-elements-to-iron are solar or slightly below solar; this probably means that AGB stars continue to contribute to the enrichment of these elements, along the Fe enrichment by SNIas.

E. CARRETTA: When you re-analyzed with curve-of-growth our results for a star in NGC6533, you found a lower metallicity, but also a lower gravity, log g = 1.8. Isn't it a too low gravity for a red AB star?

B. BARBUY: In any case, the lines show that the temperature you adopted is too high. Since the horizontal branch is overimposed on the red giant branch, maybe in fact you observed red giant branch stars. From your line data it is clear that the excitation equilibrium requires a lower temperature than what you have adopted.

S. CASSISI: How much are the α-elements abundance measurements affected by the use of model atmospheres computed for a solar-scaled mixture?

B. BARBUY: Kurucz has computed α-element enhanced models. Differences in derived abundances by using these models seem not to be very different. But you are right: the use of proper models in the analysis is important.

THE GLOBULAR CLUSTERS AND FIELD STARS IN THE GALACTIC BULGE

SERGIO ORTOLANI
Department of Astronomy, Padova University
Vicolo Osservatorio 5, I-35122, Padova, Italy

Abstract. We present the characteristics and the distribution of the clusters projected in the direction of the Galactic Bulge. The clusters are compared to the bulge stellar field population. It appears that these clusters are good tracers of the galactic bulge. The derived scenario is an old bulge stellar population, kinematically decoupled from the disk, with a peculiar chemical composition.

1. Overview

The galactic bulge is characterized by a high concentration of stars as well as of globular clusters in a rather small volume. Its kinematical mass, in a 1.5 kpc radius, is estimated between 2 and 3 $\times 10^{10}$ $M_{\odot}$ (Blum, 1995), which is nearly half of the mass of the galactic disk. In a radius of about 4 Kpc from the galactic center there are also about 50 known globular clusters (Barbuy et al., 1999a) which represent more than one third of all the known galactic globular clusters. The galactic bulge contains also a concentration of variable stars, mainly of RR Lyrae and Mira type. In a note published in 1946, Baade presented the results of his first survey in the region which will be called later the Baade Window, where 152 variables have been discovered in an area of 0.38 square degrees. He found that most of them were RR Lyrae type, with the long period next in frequency, as in the halo globular clusters. Comparing this result with the Milky Way fields previously investigated, Baade argued that the stellar population of the Sagittarius region is different from the disk population, which appeared instead dominated by eclipsing variables, followed by long period and irregular. He concluded then that *the faint corona of stars (the halo) surrounding our Galaxy represents the continuation of the nuclear bulge.*

F. Matteucci and F. Giovannelli (eds.), The Evolution of the Milky Way, 301–308.

It is not easy to explain the presence of RR Lyrae stars in a metal rich population such as the galactic bulge. In fact the metal rich globular clusters in the halo contain very few, if any, RR Lyrae. Lee (1992) pointed out that the presence of numerous RR Lyrae stars with a nearly solar abundance can be explained only if they are older than the galactic halo. Thanks to microlenses surveys, the number of known RR Lyrae stars recently increased by more than one order of magnitude. ¿From their pulsational properties only a small fraction of them results to be metal rich. These stars can be the metal poor halo component extending inside the bulge (Alcock et al., 1998) or the oldest component of the bulge.

The bulge includes also a large number of bright Miras and long period variables which have been interpreted in the past as evidence of a young population. This is still matter of controversy because of the degeneracy age-metallicity, but Guarnieri et al. (1997) presented evidence for the case of very bright long period variables in a globular cluster (NGC 6553) with high metallicity and an old age.

There is still an open debate in the literature whether the bulge is the inner extension of the halo, having formed in the same process, or whether it formed much later, perhaps as a consequence of the dynamical evolution of the disk (Larson 1990, Renzini 1999, Raha et al., 1992, Wyse et al., 1997).

New, interesting results are coming mainly from the the spectroscopy (see the review by Barbuy in these proceedings). There is today evidence that the bulk of the bulge stellar population is metal rich. The spectroscopic analysis recently obtained by Rich (1988), McWilliam and Rich (1994) and confirmed Sadler et al. (1996) and by Barbuy et al. (1999b) revealed (a) a broad metallicity distribution in the range $-1.5 < [Fe/H] < +0.5$, and (b) a strong [α-elements/Fe] enhancement. This signature is currently interpreted as a fast enrichment by SNs II at early times in the formation of the bulge before the iron contribution by ejecta of SNIa to the interstellar medium. Even if the time scale of SNIa contribution is not well known and model dependent, an indicative time of about 1 Gyr is suggested (Renzini, 1999). A scenario with an old bulge formed during a rapid enrichment process can be derived (see the models by Matteucci and Brocato, 1990).

The direct study of the bulge age via color-magnitude diagrams (CMDs) is an obvious task to constrain the formation process of our galaxy. However the measurements are affected by crowding, differential reddening across the field of view and by contamination due to the disk stellar population seen projected on the galactic bulge (Fig. 1). These effects produce an apparent selective upward shift (with an increasing factor at fainter magnitudes) of the turnoff, relatively to the HB, giving the impression of a younger age. Furthermore the color magnituide diagrams obtained in the field ap-

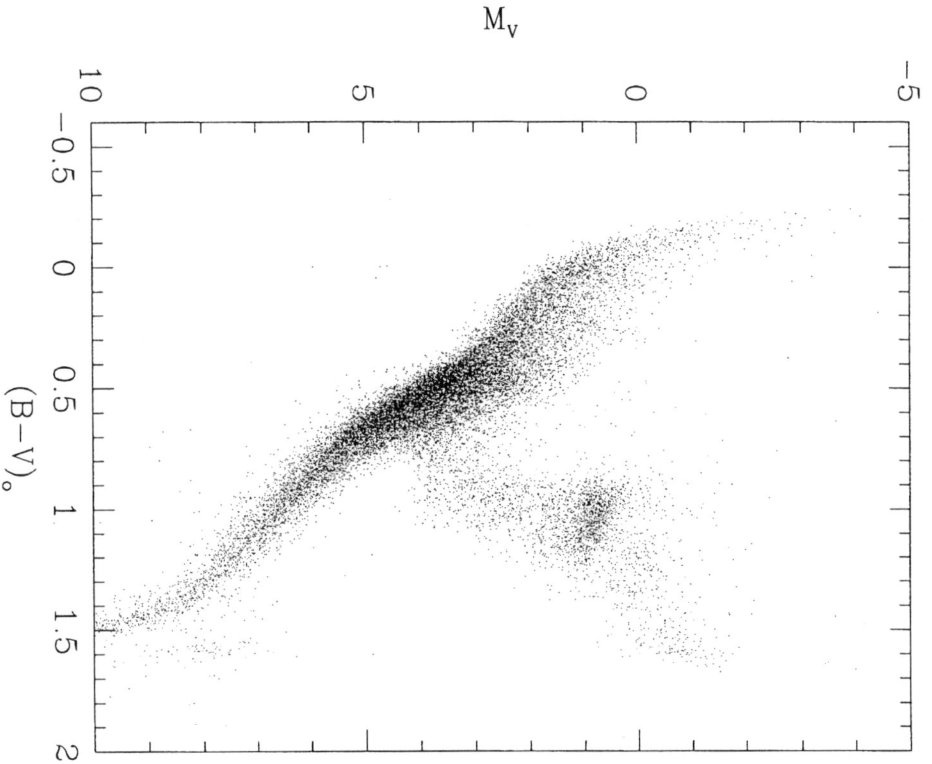

Figure 1. Schematic view of the vertical section of the Galaxy. The main windows through the disk, in the direction of the bulge are indicated.

pear broadened because they contains stars with different metallicity and because they are affected by depth effects.

Early data from HST WFPC1 observations in the Baade Window have been interpreted (Holtzman et al., 1993) as evidence for an intermediate age (around 10 Gyr) of the field population, while Ng et al. (1996), from the analysis of OGLE color-magnitude diagrams found a dominant old population (13-16 Gyr), with a minor, younger component (8-9 Gyr) associated to the bar.

2. The bulge globular clusters: bulge stellar population tracers

The globular clusters are simple stellar populations, so they are not affected by metallicity, age and distance spread. For this reason in the last 15 years we concentrated our studies on the globular clusters projected in the direction of the galactic bulge.

Most of the low galactic latitude clusters have been discovered by Terzan and collaborators more than 20 years ago (Terzan, 1968 and references therein) from wide field infrared plates. So far no other systematic surveys, aimed to discover globular cluters, have been undertaken. The available sample is still far to be complete. Barbuy et al. (1998) showed that there is still an empty wide strip extending about 1 degree above and 2 degrees below the galactic plane. A concentration of clusters, at least 20-30, hidden by heavy extinction, is expected there. These objects could be discovered in a wide field near infrared survey covering about 20 square degrees.

Until recently very few data were available in the literature for most of the clusters in the direction of the galactic bulge (Racine and Harris, 1989, Armandroff, 1989). In our systematic study we revised previous classifications and derived new metallicities, distances for about 50 clusters in the area included between $l = \pm 20^{o}$ and $b = \pm 15^{o}$, and we completed the analysis of all the known globular clusters within 5 ^{o} from the galactic center (Barbuy et al. 1998).

The method we used is based on the comparison of color-magnitude diagrams (CMDs) with templates. The best template is choosen according to the morphological comparison (which gives a first guess for the metallicity). The horizontal shift, derived from the best fit with the template, gives the reddening, while the vertical shift gives the apparent distance modulus. Integrated spectra have also been used as an additional guideline for the reddening and the metallicity. It is now possible to study the properties of this sample on a statistical base.

With a few exceptions the studied clusters are concentrated nearby the galactic center.

Most of them are metal-rich clusters, characterized by a red and very short HB and a very extended and tilted giant branch.

Some objects, (about10 % of our sample) turned out to be different objects (galaxies, HII regions, supernova remnants or open clusters, or simply non-existing, see e.g. Bica et al., 1995).

An important fraction of clusters classified in the previous literature as metal rich clusters resulted to be blue HB with very steep giant branches, resembling those of the most metal poor halo clusters such as M 30. The correct classification of a blue HB globular cluster in the bulge is more d-ifficult than a red HB because the blue tail is blended with the disk main sequence stars. This component contributes to the wide metallicity spread in the chemical composition of the bulge globular clusters, resemblig theis-persion of the field stars. Some blue HB clusters, however, could be metal rich but affected by dynamical effects (Rich et al., 1997). There is also one case of a relatively metal poor cluster showing a very short and red HB recalling the second parameter effect of some galactic halo clusters.

The total number of the red HB clusters is comparable to the blue HB ones, but the red HB clusters, are much more concentrated toward the galactic plane, with a ratio of red HB clusters over blue HB of two in the inner $\pm$ 5 ^{o}, while it decreases below one at a distance greater than 10 ^{o} from the galactic plane. This trend can be explained in a scenario of interloping metal rich bulge clusters and more metal poor halo clusters (Minniti, 1995). In spite of this different latitude trent, the two samples are well spatially mixed and their kinematics do not seem sharply different. This can be certainly due also to the low numbers, but we cannot exclude that at least some of the blue HB clusters formed in the bulge itself at early times, still mantaining orbits characteristic of the bulge.

These metal poor inner bulge clusters are of great interest because, in an inside-out formation model, they could be the oldest in the Galaxy. Combined deep VLT optical-near infrared observations below the turnoff and high resolution spectroscopy of the giant stars to get their chemical composition and kinematics, are needed to clarify the nature of these clusters.

3. Space distribution, kinematics and ages

The method used to find the distances is a key aspect in any space distribution analysis of globular clusters. In the case of low latitude clusters, affected by heavy absorption, the problem is complicated because of the possible dependence of the ratio R of the selective to absolute extinction (reddening to extinction) on several parameters such as temperature, spectral type, chemical composition of the observed stars and on the absolute value of the absorption itself (Grebel and Roberts, 1995). Since we derive the absorption from the reddening (obtained from the horizontal shift of the CMDs) the uncertainty on this coefficient is dominating all the other distance uncertainties when the reddening is higher than E(B-V) = 0.3-0.4

mag. Current estimates of the ratio R are from 2.7 to 3.6. Tests made on globular clusters observed in optical and in the near infrared (where the reddening law is more stable) such as Terzan 2 (Ortolani et al., 1997 and Kuchinski et al., 1995) and NGC 6553 (Guarnieri et al. 1998) show that the current adopted value R=3.1 is too low, and a value around 3.3 could be a more acceptable compromise. However higher statistics are needed to reveal any trend.

With this caution we found that all but two clusters projected within 5 degrees from the galactic center are located within 4 Kpc from the galactic center. This is also the radius of the Galactic Molecular Ring (Burton 1976). It is very well known that inside this ring there is a "hole" in the galactic disk, characterized by a pronounced lack of gas and population I tracers (Georgelin and Georgelin, 1976; Taylor and Cordes, 1993). This suggests a that the globular clusters projected in the direction of the inner galactic bulge belong to the bulge itself and are not connected to the disk.

The vertical distribution of the genuine inner bulge clusters within 5 degrees is strongly concentrated toward the galactic plane, following Kent's (1992) bulge exponential distribution, but it is quite different from that discussed by Armandroff (1989).

Our results from the kinematics confirm previous findings from Minniti (1995): the globular clusters in the galactic bulge show a slow rotation if compared to the disk, with a slope similar to that of the bulge field stars (see also Cote', 1999). Ibata and Gilmore (1995) reached the same conclusion from their data on the field stars.

Combining the spatial distribution and the results from kinematics, one can conclude that the clusters projected in the direction of the galactic bulge share the same properties of the bulge stellar population, while they are completely decoupled from the galactic disk.

As concerning the age, we detected the main sequence turnoff in three bulge clusters, NGC 6553, NGC 6528 (Ortolani et al., 1995) and NGC 6652 (Ortolani et al., 1994). Their HB - main sequence luminosity difference is comparable to the halo cluster 47 Tuc. In a first approximation this is an indication of a comparable age. NGC 6553 and 6528, in particular, have a chemical composition very similar to the Baade Window field giants (Barbuy et al., 1999b) and almost identical CMDs. Their luminosity functions, from the horizontal branch to the main sequence, are also very similar to that of the Baade Window (Ortolani et al., 1995, Fig. 3).

While the determination of a precise absolute age is very model dependent, this result implies that the age of the bulk stellar population in the bulge is comparable to the age of the relatively metal rich clusters in the halo. Preliminary results from another bulge window located nearby the globular cluster NGC 6558 confirm a luminosity function resembling that of NGC 6553. ¿From this point of view it appears that the original intuition by Baade was not too far from the results obtained with modern techniques.

The conclusion of our analysis is based only on a few clusters. The extention of this study to other clusters and the discovery of new clusters located within one degree from the gaalctic plane (from infrared surveys) are obvious targets for the next future.

Acknowledgements
We are very grateful to Giovanni Carraro and to Manuela Zoccali for their help in the preparation of the text.

References

1. Alcock, C. et al. 1998, *ApJ* **492**, 190
2. Armandroff, T. 1989 *AJ* **97**, 375
3. Baade, W. 1946, *PASP* **58**, 249
4. Barbuy, B., Ortolani, S., Bica, E., Desidera, S. 1999a, *A&A* **348**, 783
5. Barbuy, B., Renzini, A., Ortolani, S., Bica, E., Guarnieri, M.D. 1999b, *A&A* **341**, 539
6. Barbuy, B., Bica, E., Ortolani, S. 1998, *A&A* **333**, 117
7. Bica, E., Clariá, J.J., Bonatto, C., Piatti, A.E., Ortolani, S., Barbuy, B. 1995, *A&A* **303**, 747
8. Blum, R. D. 1995, *ApJ* **444**, L89
9. Burton, J.B. 1976, *Ann. Rev. A&A* **14**, 275
10. Cote', P. 1999, *AJ* **118**, 406
11. Geogelin, Y.M., Georgelin, Y.P. 1976, *A&A* **49**, 57
12. Grebel, E.K., Roberts W.J. 1995*A&A Suppl.* **109**, 293
13. Guarnieri, M.D., Ortolani, S., Montegriffo, P., Renzini, A., Barbuy, B., Bica, E., Moneti, A. 1998, *A&A* **331**, 70
14. Guarnieri, M.D., Renzini, A., Ortolani, S. 1997, *ApJ* **477**, L21
15. Holtzman, J.A. et al, 1993 *AJ* **106**, 1826
16. Ibata, R.A., Gilmore, G.F., 1995 *MNRAS* **275**, 605
17. Kent, S.M. 1992, *ApJ* **387**, 181
18. Kuchinski, L.E., Frogel J.A., Terndrup, D.M., Persson, S.E. 1995, *AJ* **109**, 1131
19. Larson, R.B. 1990, *PASP* **102**, 709
20. Lee, Y.W. 1992, *AJ* **104**, 1780
21. Matteucci, F., Brocato, E. 1990, *ApJ* **365**, 539
22. McWilliam, A., Rich, R.M. 1994, *ApJS* **91**, 749
23. Minniti, D. 1995, *AJ* **109**, 1663
24. NG, Y. K., Bertelli, G, Chiosi, C., Bressan, A., 1996 *A&A* **310**, 771
25. Ortolani, S., Bica, E., Barbuy, B., 1994 *A&A* **286**, 444
26. Ortolani, S., Renzini, A., Gilmozzi, R., Marconi, G., Barbuy, B., Bica, E., Rich, R.M.: 1995, *Nature* **377**, 701
27. Ortolani, S., Bica, E., Barbuy, B., 1997 *A&A* **326**, 614
28. Racine, R., Harris, W.E. 1989, *AJ* **98**, 1609
29. Raha, N., Sellwood, J.A., James, R.A., Kahn, F. D. *Nature* **352**, 411
30. Renzini, A. 1999, astro-ph/9902108

31. Rich, R.M. 1988, *AJ* **95**, 828
32. Rich, R.M., Sosin, C., Djorgovski, S. G., Piotto, G., King, I. R., Renzini, A., Phinney, E. S., Dorman, B., Liebert, J., Meylan, G. 1997, *ApJ* **484**, 25
33. Taylor, J.H., Cordes, J.M. 1993, *ApJ* **411**, 674
34. Terzan, A. 1968, *Comptes Rendus Acad. Sci. Paris* **267**, 1245

THE INITIAL MASS FUNCTION OF THE GALACTIC BULGE DOWN TO $\sim 0.15\,M_\odot$

M. ZOCCALI[1], S. ORTOLANI[1]
Dipartimento di Astronomia, Università di Padova
vicolo dell'Osservatorio, 5, I-35122, Padova

S. CASSISI[2]
Osservatorio Astronomico di Collurania
Via M. Maggini, I-64100 Teramo

J.A. FROGEL[3], A. GOULD[3], A.W. STEPHENS[3]
Department of Astronomy, Ohio State University
5040 Smith Lab., 174 West 18th Avenue, Columbus, OH 43210

A. RENZINI[4]
European Southern Observatory
K.Schwarzschild-Str. 2, D-85748, Garching bei München, Germany

AND

R.M. RICH[5]
Department of Physics and Astronomy
University of California, Los Angeles, CA 90095-1562

1. Abstract

We present a luminosity function (LF) for lower main sequence stars in the Galactic bulge near $(l, b) = (0°, -6°)$ to $M_J \sim 9.3$. This LF is derived from HST+NICMOS observations of a region of $22\farcs5 \times 22\farcs5$, with the F110W and F160W filters. Our derived mass function extends to 0.15 $M_\odot$ with a power law slope of $\alpha = -1.33 \pm 0.07$. Although shallower than the Salpeter one, this IMF is steeper than that recently found for the Galactic disk ($\alpha = -0.8$ and $\alpha = -0.54$ from the data of Reid & Gizis, 1997, and Gould et al. 1997, respectively, in the same mass interval), but is virtually identical to the disk IMF derived by Kroupa et al. (1993). The bulge IMF is also quite similar to the mass functions derived for those globular clusters which are believed to have experienced little or no dynamical evolution.

F. Matteucci and F. Giovannelli (eds.), The Evolution of the Milky Way, 309–315.

2. Introduction

The initial mass function (IMF) is a fundamental property of stellar populations, hence one of the most crucial ingredients in models of galaxy formation and evolution. It determines several key properties of stellar populations and galaxies. Yet, the IMF is usually taken as a free parameter, particularly at the low-mass end (for recent reviews on the IMF see Larson 1998; Scalo 1998, 1999). Observational constraints on the IMF are therefore of the greatest astrophysical importance.

Because faint, low-mass stars have such low temperatures, infrared observations give a crucial advantage over optical data. Moreover, in the near-IR the effects of extinction and differential reddening are considerably reduced. Finally, the relatively low IR background of HST, combined with diffraction limited resolution, gives a fundamental advantage when dealing with very faint sources in a crowded field. Therefore, the NICMOS near-IR cameras offer a unique opportunity to reach the faintest stars possible in the Galactic bulge, thus extending to lower masses the range over which the IMF is observationally constrained.

3. The Color–Magnitude Diagram

The selected field (l=0.277, b=−6.167) was observed with the NIC2 camera of NICMOS, on board HST, through the filters F110W and F160W, with a total integration times of 10240 s and 25600 s, respectively. The field of view has dimension of $22''.5 \times 22''.5$.

The images were pre-processed by the standard NICMOS pipeline CALNICA. The photometric reduction was carried out using the DAOPHOTII photometry package (Stetson 1987), in a way described in details in Piotto et al. (1999). Calibration to the standard CIT/CTIO system was performed according to the calibration equations determined by Stephens et al. (1999).

The observed color–magnitude diagram (CMD) for the 780 stars measured in our field is shown in Figure 1. The bulge main sequence (MS) is well defined from the turnoff ($J \sim 18$) down to magnitude $J \sim 24$ where the sequence starts to broaden and the density of stars falls abruptly. A prominent feature in this CMD is the sharp bend that is clearly visible at $J \approx 20.5$. Fainter than this point, the MS is almost vertical. As predicted by stellar models (Cassisi et al. 2000; Baraffe et al. 1997), this behavior is due to the competition between the tendency towards redder colors due to both the decreasing effective temperature and the increasing molecular absorption at optical wavelengths, and the increasing collision-induced absorption of molecular hydrogen at infrared wavelengths (CIA mechanism, Saumon et al. 1994).

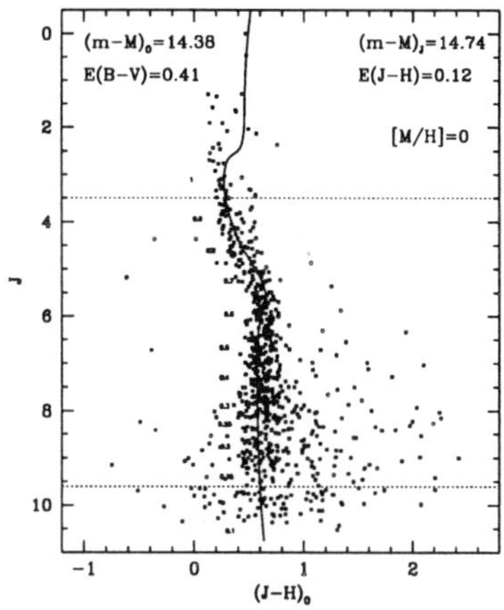

Figure 1. The observed color–magnitude diagram of the 780 stars in the NIC2 field. The solid line is the isochrone for a solar metallicity, 10 Gyr old population (Cassisi et al. 2000).

Also shown in Figure 1 is the theoretical isochrone by Cassisi et al. (2000). These models have been constructed by adopting the most updated input physics, such as stellar opacities, equation of state, and outer boundary conditions (see Cassisi et al. 2000, for more details). We adopt here the absolute distance modulus and reddening of this region of the Galactic bulge, as measured by Rich et al. (1998): $(m - M)_0 = 14.38$ and $E(B-V) = 0.41$. By assuming $R_V = 3.1$ the extinction is $A_V = 1.27$, which can be converted to the corresponding A_J and A_H by means of the relations given by Cardelli et al. (1989): $A_J = 0.282 A_V$ and $A_H = 0.190 A_V$. The isochrone shown in Figure 1 refers to solar metallicity ([Fe/H]=[α/Fe]=0) and an age of 10 Gyrs. The model is a satisfactory match to the general shape of the observed MS, in particular the position of the bend at $J \sim 20.5$ ($M_J \sim 5.75$) is well reproduced, even if its strength seems to be a little overestimated.

4. The Luminosity and Mass Functions

In order to obtain the stellar LF of our field, particular attention was devoted to estimating the completeness of our sample. Standard artificial-star tests were carried out on the NIC2 field, in the same way as described in detail by Piotto & Zoccali (1999). The resulting J-band LF is very smooth over the whole range from the bin at $J = 18$ (turnoff region) to the faint limit at $J = 24$. In the determination of the IMF, we did not use the first two bins that, according to our model, correspond to evolved stars, nor the very last bin ($J = 24.25$) because its completeness is $\sim 30\%$.

Since the field is located at low Galactic latitude, contamination by disk stars cannot be neglected. We evaluated this contamination using Kent's (1992) model for the K band luminosity distributions of the disk and bulge. Contamination by extragalactic objects is estimated using the NICMOS H-band galaxy counts (Yan et al. 1998).

The LF for low-mass stars ($M \lesssim 1 M_\odot$) can be converted into a MF which is the same as the IMF, since the stars are unevolved and their number is unaffected by dynamical processes. In order to transform the LF into the IMF we adopted a mass–luminosity relation (MLR) appropriate for a 10 Gyr, solar metallicity population, obtained from the models by Cassisi et al. (2000). This MLR is in very good agreement with the empirical MLR determined by Henry & McCarthy (1993) from a sample of visual and eclipsing binaries in the solar neighborhood.

The resulting IMF for the Galactic bulge is shown in Figure 2. Within the errors, the IMF can be represented over the entire mass range by a single power-law of the form $dN \propto M^\alpha dM$, having a slope $\alpha = -1.33 \pm 0.07$ (where Salpeter has $\alpha = -2.35$). It is worth noting that adopting the widely used distance modulus $(m-M)_0 = 14.5$ to the Galactic center (Reid 1993), instead of the value of 14.38 adopted for this paper, the resulting slope is $\alpha = -1.30$ over the whole range $1 < M/M_\odot < 0.15$. Hence the result is fairly insensitive to errors in the distance.

5. Comparison with other mass functions

Figure 2 shows the comparison between the IMF derived above and the IMF from Holtzman et al. (1998) for the stars in Baade's Window. For a more consistent comparison, we used the I-band LF from their Figure 5, and converted it into an IMF by means of the same theoretical MLR that we have used for our derivation (except for the color transformation to the I band instead of the J band), and without applying a correction for binaries. The IMF so derived turns out to be very similar to our bulge IMF for the $-6°$ field. Holtzman et al.'s Baade's Window IMF has a slope $\alpha = -2.2 \pm 0.2$ for $M > 0.5 M_\odot$, and $\alpha = -1.4 \pm 0.2$ for $M < 0.5 M_\odot$, virtually identical to our result over the wider mass range that was accessible for the $-6°$ field.

It is also interesting to compare the bulge IMF with the IMF of the Galactic disk. The comparison is shown in Figure 3, which displays our bulge IMF together with the disk IMF as recently derived by Reid & Gizis (1997) and by Gould et al. (1997). Reid & Gizis (1997) extract their LF from the study of a volume-complete sample of low mass stars with $\delta > -30°$ and within 8 pc from the Sun. The Reid & Gizis (1997) disk IMF is well represented by a power law, but its slope $\alpha = -0.80 \pm 0.20$ differs by about

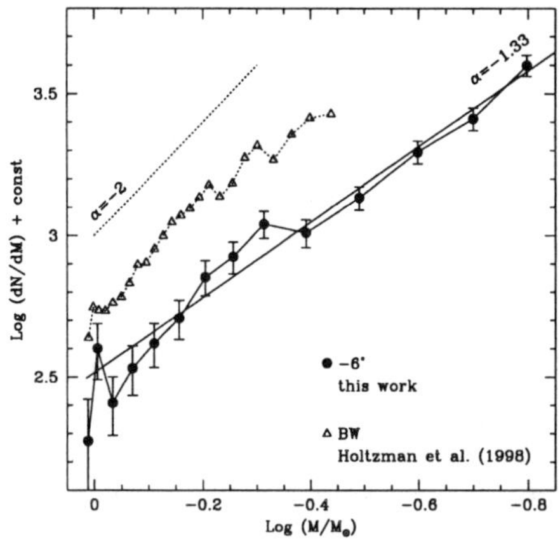

Figure 2. The IMF for the $-6°$ field (solid symbols). A single power-law with a slope of $\alpha = -1.33 \pm 0.07$ is able to fit the data in the whole mass range. Were the fit restricted to $M > 0.5 M_{\odot}$, a steeper slope $\alpha = -2.0 \pm 0.2$ would be obtained (dotted line). The quoted errors on the slopes are the formal errors on the fit. Also shown is the Baade's Window IMF from Holtzman et al. (1998)

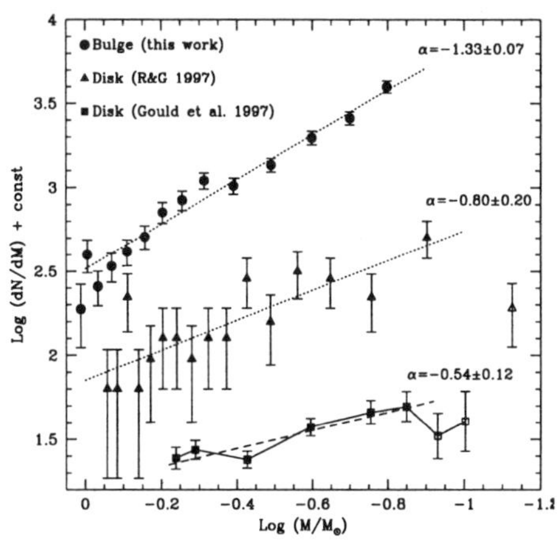

Figure 3. The bulge IMF shown in Fig. 2 is compared with two independent determinations of the IMF of the Galactic disk. The values shown in the figure, $\alpha = -0.8$ for Reid & Gizis (1997) and $\alpha = -0.54$ for Gould et al. (1997) are based on our fits to the data, based on the restricted range $M > 0.15\, M_{\odot}$ (solid symbols) and therefore differ slightly from the values reported by the original authors.

$2.4\,\sigma$ from that of the bulge. The disk IMF by Gould et al. (1997) is also shown in Figure 3. For their sample of disk M dwarfs they found an IMF with a slope $\alpha = -0.54 \pm 0.12$, in $1\,\sigma$ agreement with Reid & Gizis (1997) disk IMF, but definitely flatter than the bulge IMF. On the contrary, the

IMF for the -6° field is in very good agreement with the disk IMF obtained by Kroupa et al. (1993), for stars within ~ 5 pc from the Sun, which has a slope $\alpha = -2.2$ for $0.5 < M/M_\odot < 1$ and $\alpha = -1.3$ for $0.08 < M/M_\odot < 0.5$.

Finally, the bulge IMF is very similar to the MF measured in the Galactic globular cluster NGC 7078 (a high mass cluster with wide Galactic orbit) and steeper than the MF of less massive globulars, in closer orbits (Piotto & Zoccali 1999; Pulone et al. 1998; De Marchi et al. 1999). According to dynamical models (Vesperini & Heggie 1997) high mass clusters in wide orbits are less affected by tidal shocks, and therefore their MF is closer to the IMF. Our finding that the bulge IMF is similar to that of the less dynamically affected clusters suggests that the GCs and the bulge may have the same IMF. The similarity of the IMF of the solar metallicity bulge with that of NGC 7078 at [Fe/H] = −2 also suggests that the slope of the IMF is relatively independent of metallicity (see also Grillmair et al. 1998).

References

1. Baraffe, I., Chabrier, G., Allard, F., & Hauschildt, P. H. 1997, A&A, 327, 1054
2. Cardelli, J. A., Clayton, G. C., & Mathis, J. S. 1989, ApJ, 345, 245
3. Cassisi, S., et al. 2000, in preparation
4. De Marchi, G., Leibundgut, B., Paresce, F., & Pulone, L. 1999, A&A, 343, L9
5. Gould, A., Bahcall, J. N., & Flynn, C. 1997, ApJ, 482, 913
6. Grillmair, C. J., Forbes, D. A., Brodie, J. P., & Elson, R. A. W. 1998, AJ, 117, 167
7. Henry, T. J., & McCarthy, D. W.Jr. 1993, AJ, 106, 773
8. Holtzman, J. A., Watson, A. M., Baum, W. A., Grillmair, C. J., Groth, E. J., Light, R. M., Lynds, R., & O'Neil, E. J.Jr. 1998, AJ, 115, 1946
9. Kent, S. M. 1992, ApJ, 387, 181
10. Kroupa, P., Tout, C. A., & Gilmore, G. 1993, MNRAS, 262, 545
11. Larson, R. B. 1998, MNRAS, 301, 569
12. Piotto, G., & Zoccali, M. 1999, A&A, 345, 485
13. Piotto G., Zoccali M., King I.R., Djorgovski S.G., Sosin C., Dorman B., Rich R.M., Meylan G., 1999, AJ, 117, 264
14. Pulone, L., De Marchi, G., Paresce, F., & Allard, F. 1998, ApJ, 492, L41
15. Reid, I. N., & Gizis, J. E. 1997, AJ, 113, 2246
16. Reid, M. 1993, ARA&A, 31, 345
17. Rich, R. M., Ortolani, S., Bica, E., & Barbuy, B. 1998, AJ, 116, 1295
18. Saumon, D., Bergeron, P., Lunine, J. I., Hubbard, W. B., & Burrows, A. 1994, ApJ, 424, 333
19. Scalo, J. 1998, in: ASP Conf.Ser.142, The Stellar Initial Mass Function, Gilmore G., Howell D. (eds.), San Francisco, 201
20. Scalo, J. 1999, in: The Birth of Galaxies, Guiderdoni B., et al. (eds.), Gif-sur-Yvette, in press
21. Stetson, P. B. 1987, PASP 99, 191
22. Stephens, A. W., Frogel, J. A., Renzini, A., Ortolani, S., Jablonka, P., Rich, R. M., & Davies, R. 1999, AJ, submitted
23. Vesperini, E., & Heggie, D. C. 1997, MNRAS, 289, 898
24. Yan, L., Mc Carthy, P. J., Storrie-Lombardi, L. J., & Weymann, R. 1998, ApJ, 503, L19

Speaker: M. ZOCCALI

M. FEAST: In your analysis do you take into account the (significant) depth of the bulge in the line of sight in your field?

M. ZOCCALI: We did not take into account the effect of the depth of the bulge because the simulations we made revealed that it has some effect on the colour- magnitude diagrams, but not on the luminosity (and mass) function.

M. TOSI: 1) How does the fact that the adopted theoretical track doesn't fit well the observed TO region affect your IMF determination?

2) Wouldn't it be safer to limit the IMF determination to masses/magnitudes less affected by severe incompletness?

M. ZOCCALI: 1) We believe that this fact does not affect our result because we excluded the evolved (and turnoff) stars from our analysis. Moreover, our result for the slope of the IMF is not critically dependent on the first few points. 2) We truncated our LF where the completeness fell below 50%. Of course this limit is arbitrary, although generally applied, but we judge it safe enough for our purposes. Note that the errorbars of the IMF bins take into account the uncertainty in the completeness determination.

F. PARESCE: You used an unpublished M-L relation to obtain the IMF whose correctness is untested. What happen if you use a well known M-L relation such as the one from the Lyon group?

M. ZOCCALI: A direct comparison between the MLR we used and the one from the Lyon group demonstrates that the two sets of models are almost identical both in the colour-magnitude and in the mass-luminosity plane.

Therefore we believe that the derived IMF would be the same even adopting the MLR by the Lyon group.

RADIAL VARIABLES AS TRACERS OF STELLAR POPULATIONS

G. BONO
Osservatorio Astronomico di Roma,
Via Frascati 33, 00040 Monte Porzio Catone, Italy

AND

M. MARCONI
Osservatorio Astronomico di Capodimonte,
Via Moiariello 16, 80131 Napoli, Italy

1. Introduction

Since Baade's identification of two different stellar populations in the Milky Way, variable stars have been representing a fundamental tool for tracing the properties of both Pop. I and Pop. II stars. The prototypes of Pop. II and Pop. I variables are RR Lyrae and classical Cepheids respectively. As for RR Lyrae, the long term survey (more than ten year long) by Suntzeff et al. (1991) provided us with a complete mapping of the metallicity gradient from the outer halo down to the Galactic center, finding that the nature of the metallicity of the Galactic halo changes at the solar circle. In fact, according to these authors, the average metallicity outside the solar circle is $< [Fe/H] >= -1.65$ with a dispersion of 0.30 dex, whereas inside the solar circle the metallicity gradient is -0.06 dex Kpc^{-1}.

On the other hand, classical Cepheids are important tracers of the spiral structure (see e.g. Kraft & Schmidt 1963; Fernie 1968), playing a similar role in mapping the Galactic plane as OB associations and open clusters. Moreover Cepheids are used to constrain the disc rotation as recently discussed by Pont et al. (1997), Metzger et al. (1998) and Metzger & Schechter (1998).

The key feature for the use of RR Lyrae and Cepheids as tracers of stellar populations is the fact that we can estimate their distances with relatively high accurateness. In the following two sections, we discuss some predictions of theoretical pulsation models, showing the importance of observations in

F. Matteucci and F. Giovannelli (eds.), The Evolution of the Milky Way, 317–322.

the near infrared bands to optimize the role of RR Lyrae and Cepheids in fundamental problems such as the galactic and extraglactic distance scale.

2. RR Lyrae

The RR Lyrae stars are among the most popular old, low-mass standard candles. The main advantages of RR Lyrae stars are that they cover a wide metallicity range and can be easily identified in globular clusters and in the field of elliptical, and spiral galaxies. As a consequence, the RR Lyrae distance scale plays a fundamental role not only for estimating Galactic and Local Group distances, but also to estimate the intrinsic accuracy of distance determinations based on different distance indicators.

In this context, the main unsolved question remains the dependence of the luminosity of RR Lyrae on metallicity. A linear relation is usually suggested in the literature but both the zero point and the slope of such a relation are still very uncertain. As a consequence we are dealing with the so-called discrepancy between the "short" and the "long" distance scale (see e.g. Cacciari 1999 for dicussion). Moreover, both theoretical and empirical pieces of evidence suggest that RR Lyrae Luminosity-Metallicity relation -M_V vs $[Fe/H]$- is not linear when moving from metal-poor to metal-rich RR Lyrae stars. (see Bono et al. 1997, Cassisi et al. 1998, McNamara 1999).

The main uncertainties affecting empirical estimates of the $M_V(RR)$ vs $[Fe/H]$ relation are the following: 1) individual cluster distances are known with an accuracy of the order of 0.2 mag; 2) individual cluster reddenings are still affected by large uncertainties, and indeed literature values may also differ by more than 30-40%; 3) metallicity estimates based on the calibration of the ΔS parameter obtained by different authors present systematic differences of the order of 0.2 dex.

An interesting approach to overcome these uncertainties was suggested by Longomore et al. (1990, hereinafter L90). They found, on the basis of few K magnitude observations of RR Lyrae stars in 8 GGCs, that these objects follow, within the observational uncertainties, a Period-Luminosity relation in the K band (PL_K). On the basis of this evidence it goes without saying that the PL_K relation can supply accurate distance determinations of both cluster and field variables. In fact, K magnitudes in comparison with optical magnitudes present a mild dependence on metallicity and are roughly one order of magnitude less affected by uncertainties on reddening estimates. Moreover, and even more importantly, the luminosity amplitude of RR Lyrae stars in the K band is approximately a factor of 4 smaller than in the B band. As a consequence, the number of individual measurements necessary for providing an accurate estimate of the mean K magnitude are much smaller than in optical bands.

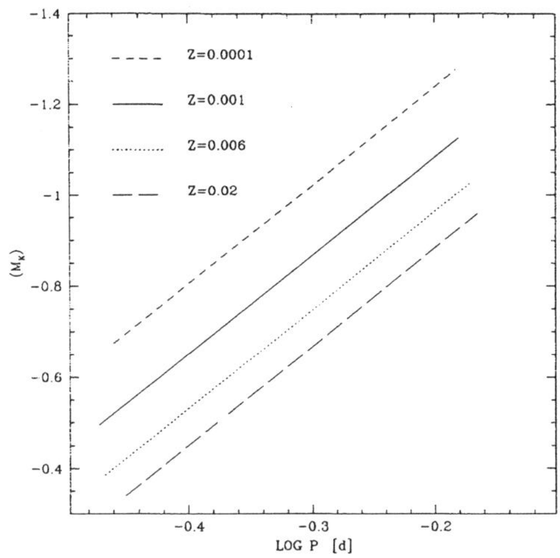

Figure 1. The predicted PL_K relation at various metal abundances (see labels).

In order to test whether pulsational models of RR Lyrae stars account for this firm empirical evidence we evaluated a PL_K relation by adopting several sequences of full amplitude, nonlinear convective models (e.g. Caputo et al. 1999) constructed by adopting a wide range of metallicities ($0.0001 \leq Z \leq 0.02$) and stellar masses ($0.75 \leq M/M_{\odot} \leq 0.53$). By adopting the atmosphere models provided by Castelli et al. (1997) to transform theoretical predictions into the observative plane we found the following metal-dependent relation:

$$(M_K) = -0.79(\pm0.02) - 2.17(\pm0.03)\log P + 0.154(\pm0.003)[Fe/H],$$

with a rms equal to 0.02 mag.

Fig. 1 shows the predicted PL_K relation at various metal abundances (see labels). Data plotted in this figure show quite clearly that the slope of the PL_K relation is almost unaffected by the adopted metallicity, whereas the zero-point significantly depends on this input parameter.

2.1. THEORY VALIDATION: THE CASE OF M3

In order to test the plausibility of the theoretical predictions, we applied the previous relation to the RR Lyrae sample in M3, observed by L90. By adopting the same reddening (E(B-V)=0.01 mag.) and metallicity ([Fe/H]=-1.66) suggested by these authors, we found a true distance modulus of 15.04±0.06 mag. On the other hand, if we adopt the metallicity scale suggested by Carretta & Gratton (1997), that is [Fe/H]=-1.34, we find a true distance modulus of 14.99±0.06 mag. These values have to be compared with the distance obtained by L90, namely 15.00 mag.

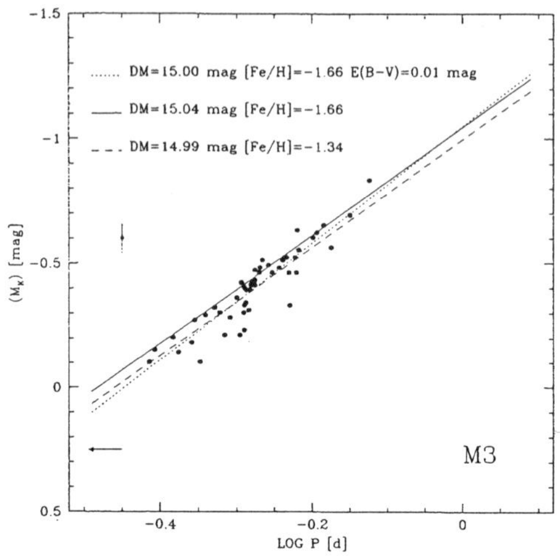

Figure 2. Comparison between the observed RR Lyrae in M3 in the M_K-log P plane, assuming the reddening and distance modulus values published in L90, and the theoretical M_K-log P-$[Fe/H]$ relation for $[Fe/H] = -1.66$ (solid line) and $[Fe/H] = -1.34$ (dashed line).

Fig. 2 shows the comparison between the observed RR Lyrae in M3 in the M_K-log P plane and the theoretical M_K-log P-$[Fe/H]$ relation for $[Fe/H] = -1.66$ (solid lines) and $[Fe/H] = -1.34$ (dashed line). The true moduli derived by fitting the observations with the theoretical relation for the two quoted $[Fe/H]$ values are also labeled.

The distance modulus used to transform the apparent magnitudes into the absolute scale is based on the absolute calibration adopted by L90 (15.00 mag.), by adopting the IR flux method version of the Baade-Wesselink analysis. The best fit solution through the data is represented with a dotted line and the error bar shows the magnitude dispersion around the regression line. The arrow marks the vertical shift (0.14 mag. toward a shorter modulus) of the best fit solution when adopting the absolute calibration by Jones et al. (1988) and Liu & Janes (1989) instead of the one based on the IR flux method version of the Baade-Wesselink analysis.

Classical Cepheids

Classical Cepheids, thanks to their characteristic Period-Luminosity (PL) and Period-Luminosity-Color (PLC) relations, are the most important primary distance indicators within the Local Group. Several secondary indicators, out to distances of the order of 100 Mpc, are calibrated through Cepheids, so that these variables are at the basis of the extragalactic distance scale, and in turn of the evaluation of the Hubble constant.

Unfortunately, apart from the intrinsic uncertainties of secondary indicators, the Cepheid distance scale is affected by systematic errors and the coefficients of PL and PLC relations are still debated.
Recent investigations, both from the theoretical and the observational point of view, have re-analyzed the problem of a possible metallicity dependence of Cepheid distance scale and the situation is still controversial (see e.g. Sasselov et al. 1997, Kennicutt et al. 1998, Saio & Gautschy 1998, Bono et al. 1999, Laney 2000) and the situation is still controversial (we refer the interested reader to the quoted papers).
In order to address the problem from a theoretical point of view, our group constructed an extensive grid of nonlinear, convective, Cepheid pulsation models, for different values of the stellar mass and three different chemical compositions, representative of Cepheids belonging to the Galaxy and the Magellanic Clouds. The results of these computations show that the topology of the instability strip, as well as the morphology and the amplitudes of light and velocity curves, do depend on the chemical composition. As a result PL and PLC relations (see Bono et al. 1999 for details) are affected by the metal abundance. Theoretical predictions support the evidence that, at fixed period, the mean luminosity decreases at larger metallicities. This effect, which is at variance with the quoted empirical trend, decreases toward longer wavelengths, thus confirming the validity of NIR photometry when using Cepheids as distance indicators (see McGonegal et al. 1982; Laney & Stobie 1994, hereinafter LS94).

Fig. 3 shows the location of fundamental pulsation models for the labeled values of stellar mass and chemical composition in the $< M_K >$-log P diagram. The plotted lines are the empirical PL relations in the K band by various authors (Madore & Freedman 1991 [MF91]; LS94; Gieren, Fouqué & Gomez 1997 [GFG97]).
As a conclusion, we notice that, due to the small dependence of the Cepheid PL_K on reddening and metallicity, new data in the NIR bands would allow us to derive more accurate distances and to perform better comparisons with theoretical predictions.

References

1. Bono, G., Caputo, F., Cassisi, S., Castellani, V., Marconi, M. 1997, ApJ, 479, 279
2. Bono, G., Caputo, F., Castellani, V., Marconi, M. 1999, ApJ, 512, 711
3. Cacciari, C. 1999, in *Harmonizing Cosmic Distance Scales in a Post-HIPPARCOS Era*, ASP n. 167, ed. D. Egret & A. Heck, p.140
4. Caputo, F., Castellani, V., Marconi, M., Ripepi, V. 1999, MNRAS, 306, 815
5. Carretta, E., gratton, R.G. 1997, A&AS, 121, 95
6. Cassisi, S., Castellani, V., Degl'Innocenti, S., Weiss, A. 1998, A&A 129, 267
7. Castelli, F., Gratton, R. G., & Kurucz, R. L. 1997a, A&A, 318, 841
8. Castelli, F., Gratton, R. G., & Kurucz, R. L. 1997b, A&A, 324, 432
9. Fernie, J. D. 1968, AJ, 73, 995

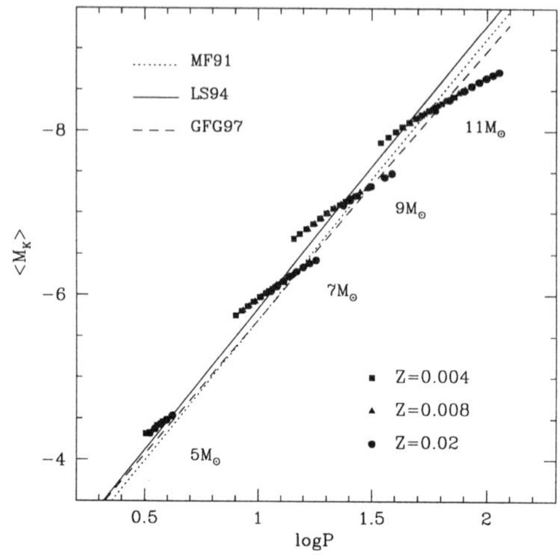

Figure 3. Location of fundamental pulsation models for the labeled values of stellar mass and chemical composition in the $< M_K >$-$\log P$ diagram, as compared with the empirical PL relations in the K band by various authors.

10. Gieren, W. P., Fouquè, P., & Gòmez, M. 1997, ApJ, 488, 74 [GFG97]
11. Jones, R. V., Carney, B. W., Latham, D. W. 1988, ApJ, 332, 206
12. Kennicutt, R. C., et al. 1998, ApJ, 498, 181
13. Kraft, R. P., & Schmidt, M. 1963, ApJ, 137, 249
14. Laney, D. 2000, in "The impact of large-scale surveys on pulsating star research", IAU Colloq. 176, in press
15. Laney, C. D., & Stobie, R. S. 1994, MNRAS, 266, 441 [LS94]
16. Liu, T., Janes, k. A. 1989, ApJS, 69, 593
17. Longmore, A. J., Dixon, R., Skillen, I., Jameson, R. F., Fernley, J. A. 1990, MNRAS, 247, 684 [L90]
18. Madore, B. F., & Freedman, W. L. 1991, PASP, 103, 933 [MF91]
19. McGonegal, R., McAlary, C. W., McLaren, R. A., & Madore, B. F., 1983, AJ, 269, 641
20. McNamara, D. H. 1999, PASP, 111, 489
21. Metzger, M. R., Caldwell, J. A. R., & Dchechter, P. L. 1998, AJ, 115, 635
22. Pont, F., Queloz, D., Bratschi, P., Mayor, M., 1997, A&A, 318, 416
23. Metzger, M. R. & Schechter, P. L. 1998, AJ, 116, 469
24. Saio, H. & Gautschy, A. 1998, ApJ, 1998, 498, 360
25. Sasselov, et al. 1997, A&A, 324, 471

ABUNDANCE GRADIENTS ALONG THE GALACTIC DISK

S. J. SMARTT
Institute of Astronomy,
University of Cambridge,
Madingley Road,
Cambridge, CB3 0HA,
United Kingdom

Abstract. The observtional evidence for abundance gradients along the Galactic disk is reviewed. Results from various methods and objects used to probe the metallicity of the disk are discussed and the quantatative results are compared.

1. Introduction

The chemical composition of the interstellar medium (ISM) varies between galaxies and is both position and time dependent within a galaxy. The metallicity sampled at any point in time and space is dependent on the history of chemical evolution of the system. Studies of abundances in nearby spiral galaxies and dwarf irregulars indicate the diversity of evolution scenarios occurring in the Local Universe (e.g. Skillman 1998 for a recent review). In particular, in non-barred spiral galaxies, the metallicity of the ISM decreases as galactocentric distance increases giving rise to decreasing abundance gradients. A wealth of observational examples exist in spirals other than the Milky Way (e.g Vila-Costas & Edmunds 1992, Garnett et al 1997, Henry & Howard 1995); with all these studies based on emission line analysis from H II regions. Measuring accurate abundance gradients in our own Galaxy can be done using similar H II region analyses, but alternative methods can also be used. These give valuable credibility to the quantitative results, and also allow other elements generally not seen in H II region data to be probed. However getting large, homogeneous data sets spanning the required radius to carry out a meaningful experiment and finding

F. Matteucci and F. Giovannelli (eds.), The Evolution of the Milky Way, 323–334.

consistent agreement between alternative methods has not proved straightforward in the recent past. We have progressed past the stage of correlating only a global metallicity (i.e. Z as mass fraction) along a radius to trying to determine the abundance of a wide range of individual elements and to look for differences and similarities between them. The data now allow us to do that, and also to compare individual elements across different methods. One must always be careful to pick similar elements in each comparison that one does between alternative methods. For example, one can get O abundances from H II region analyses (young objects with ages $\leq$20Myrs) but the 'metallicity' gradient is not *physically* comparable to the Fe abundances that one may derive from old open clusters (ages $\sim$6Gyrs), given the different processes that produce the elements and the epoch that they sample. This review will summarize work on abundance gradients along the Milky Way's disk and where the agreements and discrepancies now lie.

2. Methods to determine Galactic abundance

Since the mid-1970's there have been four main methods used to determine abundances in objects at different Galactocentric distances. These are; emission line analysis of ionized H II regions around massive star clusters; photospheric abundance analysis of early B-type stars; analysis of emission line regions of planetary nebulae (PNs); and a combination of photometric and spectroscopic index analysis of stellar populations in old open clusters. The main results are outlined in the following sections. There have been a few other methods employed (see Rolleston et al 2000, for a summary) but these have not enjoyed the same popularity or accuracy of measurement.

2.1. SUMMARY OF RESULTS FROM H II REGIONS

Elements which have diagnostic emission lines appearing in optical and infra-red spectra of H II regions are O, N, S, Ar and Ne; and in principle it is possible to derive an abundance for each element. However most individual studies produce measurements for 2 or 3 of these elements, depending on the method used and spectral coverage, with O, N and S being the most widely sampled. The largest uncertainties in these methods are the determination of a meaningful electron temperature and establishing the importance of temperature fluctuations within the gaseous nebulae. The latter has not yet been addressed in an extensive manner, but as Esteban & Peimbert (1995) report this may require 8-10m aperture telescopes for spatially resolved, high S/N data. Authors have dealt with the electron temperature problem in a variety of ways, from direct measurements using suitable diagnostic lines to photo-ionization models which assume some knowledge of the ionizing photon flux. The extensive study of abundances from H II regions by Shaver et al (1983) established a significant abundance gradient for O, using radio recombination lines as a measure of T_e. Afflerbach et al (1997) recently carried out a complimentary study of radio recombination and far-IR fine structure lines from a sample of Galactic H II regions. Their models simultaneously fit the observed features, yielding electron temperatures and densities, and abundances from the far-IR lines. The O gradient they derive is similar to that of Shaver et al (Table 1). Vílchez & Esteban (1996) have optically observed some distant objects, attempting to determine directly the electron temperature or alternatively (were the diagnostic line was not recovered) assuming an adopted temperature derived from photo-ionization models. They claim that the abundance gradient of O flattens in the *outer* Galaxy in the region $12 \leq R_g \leq 18$ kpc. However considering their data alongside that of the other H II region studies, there doesn't appear to be a strong argument in favour of a change in the slope in the outer Galaxy (see Fig. 1). It is certain that the abundances in their sample show consistently sub-solar values, but from these data alone one cannot conclude that a linear gradient is insufficient to describe the situation in the far outer Galaxy. For example fitting a straight line to all the data covering the 5–18 kpc region in Fig. 1 still results in a value of -0.07 dex kpc^{-1} and visually there is no strong evidence for the single linear relation to be inadequate.

Undoubtedly an abundance gradient for N and S also exists, however the exact value derived from the H II region methods is perhaps a little more uncertain than that of O. In Table 1 the values derived from the Shaver et al and the Afflerbach et al studies are listed along with that from Rudolph et al (1997). The latter paper used similar observations of the far-IR ionic lines of O, N and S as the Afflerbach et al work, but in many cases they were not able to determine a T_e directly and adopted a value suitable for the Galactocentric distance (from Shaver et al's relation). For N these three studies derive a value of between -0.07 and -0.11. It would appear that adopting the mean value of -0.09 ± 0.02 is well representative of the sample as a whole. Again Vílchez & Esteban suggest that between 12–18 kpc the gradient may be flat but Fig. 2 shows that a linear gradient is sufficient and that the evidence at present is not strong enough for one to suggest that a change in this simple relation should be declared necessary.

In Table 1 the values for sulphur abundance gradients show reasonable agreement, with only the Shaver et al results being discrepant. The reason for this is unclear. Again the distant data points derived by Vílchez & Esteban are *not clearly inconsistent* with a simple linear gradient.

TABLE 1. Summary of results from major studies of abundance gradients using H II regions as probes of the present day ISM metallicity

Element	Study	Gradient	Radial Baseline (kpc)
Oxygen	Shaver et al (1983)	-0.07 ± 0.02	$5 - 13$
	Vílchez & Esteban (1996)	-0.04 ± 0.01	$12 - 18$
	Afflerbach et al (1997)	-0.06 ± 0.01	$0 - 12$
Nitrogen	Shaver et al (1983)	-0.09 ± 0.02	$5 - 13$
	Vílchez & Esteban (1996)	-0.00 ± 0.01	$12 - 18$
	Afflerbach et al (1997)	-0.07 ± 0.01	$0 - 12$
	Rudolph et al (1997)	-0.11 ± 0.01	$0 - 17$
Sulphur	Shaver et al (1983)	-0.01 ± 0.02	$5 - 13$
	Vílchez & Esteban (1996)	-0.04 ± 0.03	$12 - 18$
	Afflerbach et al (1997)	-0.06 ± 0.01	$0 - 12$
	Rudolph et al (1997)	-0.08 ± 0.01	$0 - 17$
	Simpson et al (1995)	-0.07 ± 0.02	$0 - 11$

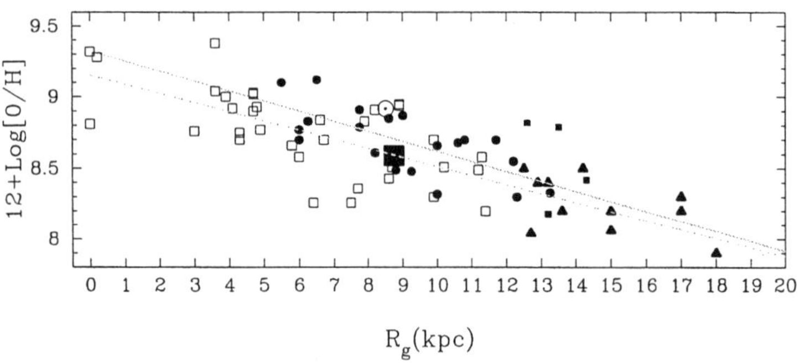

Figure 1. Oxygen abundances vs. galactocentric distances from the major galactic H II region studies. The solid symbols are: circles - Shaver et al (1983); squares - Fich & Silkey (1991); triangles - Vílchez & Esteban (1996). The open squares are from Afflerbach et al (1997). For references to the solar position and the orion abundance (large rectangle) see Rolleston et al (2000). The dashed line is a least-squares-fit to the Afflerbach et al data (-0.06 dex kpc^{-1}), and the solid line is a fit to *all* the other data points (-0.07 dex kpc^{-1}). One can see that the Vílchez & Esteban points are not inconsistent with a simple linear relation stretching to 18 kpc.

2.2. SUMMARY OF RESULTS FROM PLANETARY NEBULAE

Studies of the emission line regions in Planetary nebulae have also been a popular way of determining object abundances at large distances from the Sun, and Walter Maciel (these proceedings) has given a review on the meth-

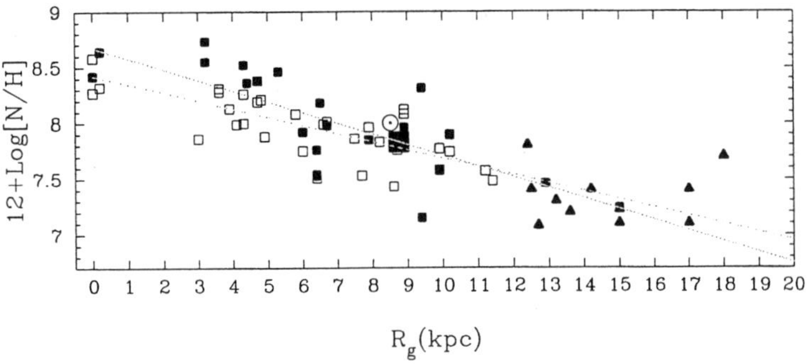

Figure 2. Nitrogen abundances vs. galactocentric distances from three major galactic H II region studies. The solid symbols are: triangles - Vílchez & Esteban (1996); squares - Rudolph et al (1997). The open squares are from Afflerbach et al (1997). For references to the solar position and the orion abundance (large rectangle) see Rolleston et al (2000). The dashed line is a least-squares-fit to the Afflerbach et al data (-0.07 dex kpc^{-1}), and the solid line is a fit to the Rudolph et al data (-0.11 dex kpc^{-1}). The Shaver et al data is not plotted (for sake of clarity) but is intermediate between the two. Again one can see that although Vílchez & Esteban claim that the gradient in the outer Galaxy may be flat, plotting their points in a Galaxy wide context would not convince one that any deviation from the simplest relation is required. The only way to really answer this is for more data points at very large distances.

ods and recent results in the field. Statistically larger samples of PN are generally available than either H II regions or early-type stars, with 200 being a typical sample size in recent studies (Maciel & Quireza 1992). However two complicating factors are that distance determinations are notoriously unreliable and PN sample a relatively large range in progenitor masses and hence ages (0-2 Gyrs for Type I and 8-10 Gyrs for Type III). Type I's have the most massive progenitors but may show some ON cycling which would strongly affect the O abundance in the nebulae (Maciel 1992). Type III's are also best avoided since they have come from quite low mass, old objects (in some cases Pop II stars) and hence are not directly comparable with the present day disk. Type II PN have been chosen by investigators best representing the young disk composition, and studies have attempted to work only with Type IIs. Although they should show pristine abundances (i.e. from the natal disk material) they are still 4-6 Gyrs old - which should be remembered in any comparisons. Attempts at comparing the abundance gradients of Types I, II and III individually have been done to try to uncover any time evolution, but no significant trend has been found (Maciel, these proceedings). The extensive studies of Maciel & Quireza (1999) and Maciel & Koppen (1994) have produced abundance gradients for O of -0.06 and -0.07 ± 0.01 dex kpc^{-1} respectively. Similar values are derived for the S gradient, hence excellent agreement exists between the gradients derived from PN and those outlined in Section 2.1.

2.3. SUMMARY OF RESULTS FROM EARLY-TYPE STARS

The spectral observation and analysis of massive, early-type stars has provided a valuable alternative and complimentary method of determining abundances in remote regions of the Galaxy. Generally, near main-sequence B-type stars (B0–B2 and V–III) have been used, and their photospheres are considered representative of the material from which they were born. The early B-types can be modelled satisfactorily in LTE (Rolleston et al 2000), and with absolute visual magnitudes of $-2 \leq M_v \leq -4$, they can be observed at high-spectral resolution at the edge of the Galaxy's optical disk. Their optical spectra contain absorption lines of C, N, O, Mg, Si (and depending on the quality of the spectra Al, S and Fe also). Some studies have indicated that fast rotating B-stars may mix some core material to the surface (Maeder 1987, Denissenkov 1994). This gas would be rich in N, and depleted in C and O, since it has been processed through the CNO cycle. In the worst case the modellers have predicted that N could be enhanced by upto 0.7 dex, but in such a scenario O will be depleted *by less than* 0.2 dex. Further there is no concrete observational evidence that a significant number of B-type stars (in the solar neighbourhood) do show such N enhancements. As O is relatively unaffected, even in this most extreme case, it is reasonably safe to assume that the photospheric abundance of O is representative of the present day ISM; certainly until there is strong observational data to tell us otherwise.

Three previous studies aimed at determining the abundances in B-stars across a relatively long radial baseline appeared to imply that that the gradient was much flatter than that determined from the PN and H II region data (Fitzsimmons et al 1992, Kaufer et al 1994, Kilian-Montenbruck et al 1994). Each of these studies is characterised by a relatively large sample of stars lying within about 2 kpc of the Sun and a few sparse points in the Perseus spiral arm (at ~12 kpc) or beyond. Hence the gradients are critically affected by the value of the furthest point. Smartt & Rolleston (1997) have discussed why each of these studies returned a flat gradient and the errors in the determination of distances and abundances that lead to these results. Smartt & Rolleston (1997) used a consistent analysis method for 50 stars in 21 young stellar clusters and associations with Galactocentric distances ranging from $6 \leq R_g \leq 18$ kpc to determine stellar O abundances. We derived an O abundance gradient of -0.07 ± 0.01 dex kpc^{-1} showing the first agreement with Galactic H II region and PN results (see Fig. 3)

We claimed that the crucial factor in determining gradients using stars is to analyse a homogeneous data set in a *consistent* manner. If one simply adopts the results from different authors and plots them together, one may mask the true relation by the scatter in the absolute abundances from

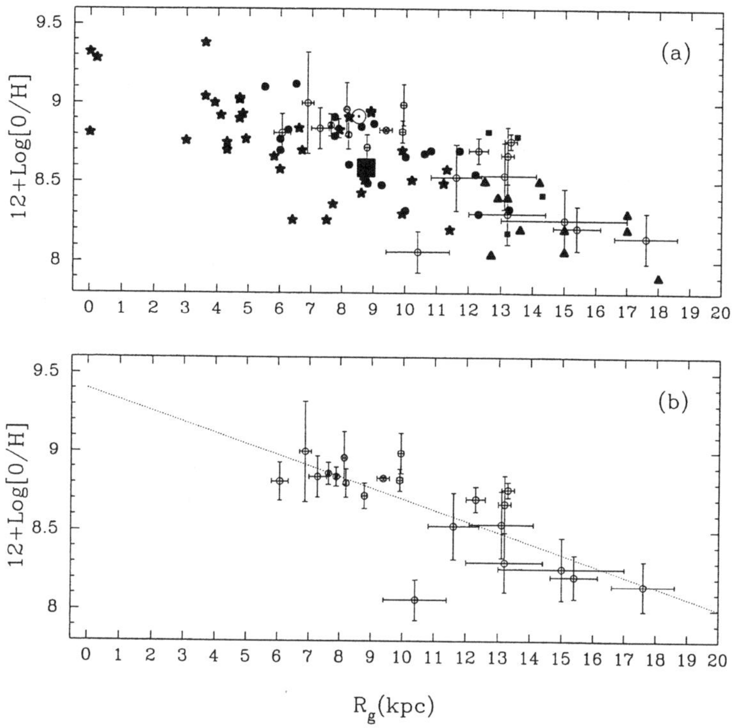

Figure 3. The B-star values from Smartt & Rolleston (1997) are plotted as open symbols with error bars. In (a) the H II region results are also shown (symbols as in Fig 1), and in (b) the fitted gradient in plotted.

one study to another. Different studies use various observational and reduction techniques, model atmospheres, atomic data, and line formation calculations so it is not surprising that comparisons between *absolute* values of abundances can vary. Gummersbach et al (1998), and Rolleston et al (2000) have reinforced this idea of analysing all available data consistently, and also determined an abundance gradient of O of -0.07 ± 0.01 dex kpc^{-1}.

2.4. SUMMARY OF RESULTS FROM OLD OPEN CLUSTERS

Spectroscopic and photometric indices of stellar populations in old open clusters have been used to determine an Fe/H stellar abundance and distance to the cluster, allowing a determination of an Fe abundance gradient (see review by Randy Phelps in these proceedings for details). Both Friel & Janes (1993) and Twarog et al (1997) have shown that the Fe abundance in the outer Galaxy appears lower than near the Sun, but have argued over the form of the gradient. Friel & Janes have derived a gradient of -0.09 ± 0.02 dex kpc^{-1} (in agreement with previous studies - references therein), while Twarog et al suggest that a step function may be more appropriate. We should be careful when comparing the results from these studies and our extreme Pop I objects - the massive stars and H II regions. First, Fe is produced mainly in SN Type Is and as such has had a distinctly different evolutionary history in the Galaxy (Edvardsson et al 1993) as compared to O which has its origin in SN Type IIs. So one should not compare the gradients of the two elements blindly as probes of the 'general metallicity' (i.e. Z by mass), as they may be somewhat different in nature. Also, the ages of the clusters used are up to 8 Gyr old hence they are not representative at all of the present disk as sampled by the young objects (e.g. Portinari & Chiosi 1999 predict a steeper O gradient at earlier epochs, although little work has been done on Fe). The fact that the magnitude of the Fe gradient as sampled by these old stars and the present day O gradient are so similar is quite a surprising (and, some might say coincidental!) result. How they are related theoretically (given the exclusive epoch and nucleosynthetic process they sample) is unclear, however future work on finding a present day Fe abundance gradient (e.g. through young stars) or an O gradient in old disk stars for more meaningful comparisons should be attempted.

2.5. ABUNDANCE RATIOS ALONG THE DISK

The B-type stars give us a large range in element abundances than are generally available through the nebular work. The elements Mg and Si are well observed in these objects and Rolleston et al (2000) has shown that the abundance of both of these elements are well correlated, supporting the idea that these elements are produced along with O from SN Type II.

In the solar neighbourhood and towards the low metallicity regions of the outer disk, a gradient in N/O is not apparent. This is inconsistent with N being a purely secondary element; if N has only a secondary origin (i.e. is produced from the C and O seed nuclei from the natal interstellar material through the CNO-cycle; Vila-Costas & Edmunds 1993) then the N/O ratio should vary linearly with O/H. Or if the secondary production was in any way important then N should at least have a noticeably steeper gradient than O. The results of Shaver et al, Vílchez & Esteban, Rolleston et al, and Gummersbach et al all indicate a virtually flat gradient of N/O. However Rudolph et al (1997) argue that while a flat N/O may be suitable for $R_{\rm g} > 6.2\,{\rm kpc}$, a step in the N/O ratio occurs at this radius and in the inner Galaxy N/O increases by 0.3 dex. This is the most extensive study of N abundances in the inner Galaxy, however preliminary results by Smartt et al (2000) on metal rich B-type stars towards the centre may support the far-IR nebular findings. This sample of main-sequence B-types and A-type supergiants within the inner Galaxy indicates that the gradient of O may flatten off, but that the gradients of Mg, Si and N in particular continue to rise. The flattening of the O gradient in the inner 6 kpc may provide the explanation for the Rudolph et al step function for N/O (as opposed to sharply increasing N). The fact that the Mg and Si abundances continue to rise as O levels off is surprising, given the close correlation of the elements at solar metallicities and below, and their expected common origins in massive SN Type II. One possible explanation is that at high metallicities the production of O in SN Type II is suppressed as more mass is lost through metallicity driven winds before the final explosion. This results in high *carbon* yields through the stellar wind ejecta, but relatively low O yields due to the smaller mass fraction left over which can be converted to O (Maeder 1992). Although Fig. 1 shows no strong evidence that the H II region results support this there are only few data points with O abundances *significantly* higher than the solar neighbourhood value. Further work on abundance ratios in metal rich disk stars is needed.

2.6. CONCLUSIONS AND FUTURE WORK

The main conclusions from this review on the observational evidence of abundance gradients in the disk are:

- Excellent quantitative agreement on the gradient of O now exists across different methods, conclusively proving its existence and magnitude. A value of -0.07 ± 0.01 dex kpc^{-1} is consistent with all major studies (within the region $6 \leq R_g \leq 18$ kpc). The historical discrepancy between the B-star and nebular results has been resolved.
- The studies reviewed above are also incredibly consistent in their conclusions on abundance gradients for the following elements; C: -0.07 ± 0.02, N: -0.08 ± 0.02, Mg: -0.07 ± 0.01, Si: -0.06 ± 0.01, S: -0.07 ± 0.02, Al: -0.05 ± 0.02,
- There is some evidence that the O gradient (and only O) levels off towards the Galactic centre, inwards of $R_g \simeq 6$ kpc. This may explain why the relatively flat N/O gradient from 6–18 kpc appears to jump in the inner regions.
- The claim by Vílchez & Esteban (1996) that the gradients of O, N and S flatten in the far outer Galaxy (at $R_g \geq 14$ kpc) does not appear consistent with current data. While we cannot definitively rule out such a scenario, there appears no observational evidence for introducing a change in the linear gradient when all the nebular work is considered together. The stellar data also does not require a change of slope for satisfactory fitting.

As mentioned in Section 2.4 a priority for the future is determining Fe abundance gradients in the present day disk as a means of comparing them to those found in older objects, and in constraining models of the evolution of both elements and of the Galaxy as a whole (Matteucci & Chiappini 1999). Such detailed work on the two other Local Group spirals (M31 and M33) is being carried out (Smartt 1999, Monteverde et al 1997, Venn 1997) and will improve when 8-10m telescopes are fully functional with their suite of high-resolution optical instruments. We will be able to determine accurate abundance gradients of (amongst others) C,N,O,Mg,Si,Fe,Cr,Ti from both A and B-type supergiants in these galaxies. This will allow comparison with the Milky Way to determine how mass, density and geometry of the galaxy as a whole affect the evolution of the observed gradients.

References

1. Afflerbach A., Churchwell E., Werner M. W., 1997, ApJ 478, 190
2. Denissenkov, P.A., 1994, A&A 287, 113
3. Edvardsson B., Andersen J., Gustafsson B., et al. 1993, A&A 275, 101
4. Esteban C., Peimbert M., 1995, Rev. Mex. Astron. & Astrof., SC3, 133
5. Fich M., Silkey M., 1991, ApJ 366, 107
6. Fitzsimmons A., Dufton P. L., Rolleston W. R. J., 1992, MNRAS 259, 489
7. Friel E. D., Janes K. A., 1993, A&A 267, 75
8. Garnett, D.R., Shields G.A. Skillman E.D., et al., 1997, ApJ 489, 36
9. Henry R. B. C., Howard J. W., 1995, ApJ 438, 170
10. Kaufer A., Szeibert Th., Krenzin R., Baschek B., Wolf B., 1994, A&A 289, 740
11. Kilian-Montenbruck J., Gehren T., Nissen P. E., 1994, A&A 291, 757
12. Maeder, A., 1992, A&A 264, 105
13. Maeder, A., 1987, A&A 178, 159
14. Maciel W. J., 1992, in Edmunds M.G., Terlevich R.J., Elements and the Cosmos, Cam. Univ. Press, p. 210.
15. Maciel W. J., Köppen J., 1994, A&A 282, 436
16. Maciel W. J., Quireza 1999, A&A 345, 629
17. Matteucci F., Chiappini C., 1999, in Proc. ESO Conference, Chemical evolution from Zero to High Redshift, Garching 1998, eds J. Walsh & M. Rosa, p. 83
18. Monteverde, M.I., Herrero, A., Lennon, D.J., Kudritzki, R.P., 1997, ApJ 474, L107
19. Portinari, L., Chiosi, C., 1999, A&A 350, 827
20. Rolleston W.R.J., Smartt S.J., Dufton P.L., Ryans R.S.I., 2000, A&A *submitted*
21. Rudolph A. L., Simpson J. P., Haas M. R., et al. 1997, ApJ 489, 94
22. Simpson J. P., Colgan S. W. J., Rubin R. H., et al, 1995, ApJ 444, 721
23. Shaver P. A., McGee R. X., Newton L. M., et al., 1983, MNRAS 204, 53
24. Skillman E. D., 1998 in Stellar Astrophysics for the Local Group, eds A. Aparicio, A. Herrero, F. Sanchez, Cambridge University Press, p. 457-518
25. Smartt S. J., Rolleston W. R. J., 1997, ApJ 481, L47
26. Smartt S. J., 1999, in Proc. ESO Conference, Chemical evolution from Zero to High Redshift, Garching 1998, eds J. Walsh & M. Rosa, p. 24
27. Smartt S. J., Venn K. A., et al., 2000 *A&A in prep*
28. Twarog B. A., Ashman K. M., Anthony-Twarog B. J., 1997, AJ 114, 2556
29. Vila-Costas M. B., Edmunds M. G., 1993, MNRAS 265, 199
30. Venn, K.A, 1997, Properties of Hot Luminous Stars ASP, V131, p177, ed. I. Howarth
31. Vílchez J. M., Esteban C., 1996, MNRAS 280, 720

Speaker: S. SMARTT

R. WYSE's Comment: I endorse the importance of the data for the outer disk - together with A. Ferguson and J. Gallagher we are finding, then obtaining, abundances for HII regions in the far outer disks of external galaxies.

R. WYSE: One of your B stars at intermediate distance looked significantly low in abundance. Could you comment further?

S. SMARTT: I agree that the very outer regions you have studied are important. The last data point I have is the furthest known HII region (and young cluster) in our galaxy. Undoubdetly more distinct ones exist but haven't yet been identified. The B-star at 10.5 Kpc lies within the local arm - Perseus arm gap and indeed has a very low metallicity. We think it may have formed within the spiral arm gap, with consequently anomalous abundances.

W. MACIEL: Concerning the "minimal difference" of the gradients from type II and type III planetary nebulae, I would like to comment that not only type III planetary nebulae show a flatter gradient, but their dispersion is higher. I think this is important to the question of the time evolution of the gradients.

S. SMARTT: I agree with this point also.

M. FEAST: If there are significant uncertainties in the derived distances (and this will be the case for many samples) the observed range in distances will generally be greater than the true range and the metallicity gradient in the disc will be underestimated. Do you think this could in particular explain the flat gradient shown by some distant objects at large galactocentric distances?

S. SMARTT: Concerning the B-stars first of all - our distances come from two methods: main-sequence fitting of the cluster CMDs (many of the stars analysed are members of young open clusters) and spectroscopic parallaxes. A robust error on the distance determinations is ± 1 Kpc, I don't think this uncertainty will seriously affect the B-star gradients. With regards to the HII regions (and to a lesser extent the PNs) I think there is not strong enough evidence to conclude that the gradient does in fact flatten (due to lack of data points). So your suggestion may indeed help explain why some authors have claimed flat behaviour.

G. ISRAELIAN: Do you have any He-weak or He-strong stars in your sample? Have you checked the abundances in your targets? Perhaps you could plot v *sini* (rotation) against [O/H], [C/H], etc. to check for any rotationally induced mixing effects.

S. SMARTT: All of the B-type spectra have between 4-6 HeI absorption lines visible. We derive a He abundance for each star during the atmospheric analysis. All stars have a "normal" He abundance of $\sim 11.0 \pm 0.3$ dex (the 0.3 dex is the scatter among the stars). We have tried to plot v *sini* vs. CNO ratios, but we conclude that our sample is not large enough to draw significant statistical conclusions. It would be desiderable to check such a relation with a large sample ($\sim$ 100) stars which have similar initial metallicity (e.g. amongst stars in clusters near the Sun).

THE AGE OF THE GALACTIC DISK

GIOVANNI CARRARO
Department of Astronomy, Padova University
Vicolo Osservatorio 5, I-35122, Padova, Italy

Abstract. I review different methods devised to derive the age of the Galactic Disk, namely the Radio-active Decay (RD), the Cool White Dwarfs Luminosity Function (CWDLF), old open clusters (OOC) and the Color Magnitude Diagram (CMD) of the stars in the solar vicinity. I argue that the disk is likely to be 8-10 Gyr old. Since the bulk of globulars has an age around 13 Gyrs, the possibility emerges that the Galaxy experienced a minimum of Star Formation at the end of the halo/bulge formation. This minimum might reflect the time at which the Galaxy started to acquire material to form the disk inside-out.

1. Introduction

Disks are quite common structures in the Universe. According to the classical theory of White & Rees (1978) spiral galaxies are considered the seeds of galaxies assembly due to gas cooling inside spinning dark matter halos. Apart from spirals, disks are seen also in the center of giant ellipticals; damped Lyman α clouds are commonly believed to be gaseous disks. Finally, S0 galaxies are expected to be disk galaxies depauperated of their gas.
However galaxian disks are fragile structures. N-body simulations convincingly show that the merging of two equal mass galaxies is fatal for disks. They are simply destroyed. The observational counterpart is represented -just to mention a source- in the IRAS database, which shows many disk on the way to be destroyed by major mergers.

Since our disk appears almost undamaged, this means that it did not suffer from strong mergers since its formation, and that its past life was relatively quiet. Of course we cannot forget that our disk has a warp, and that presently a dwarf galaxy - Sagittarius - is going to merge with our

F. Matteucci and F. Giovannelli (eds.), The Evolution of the Milky Way, 335–346.

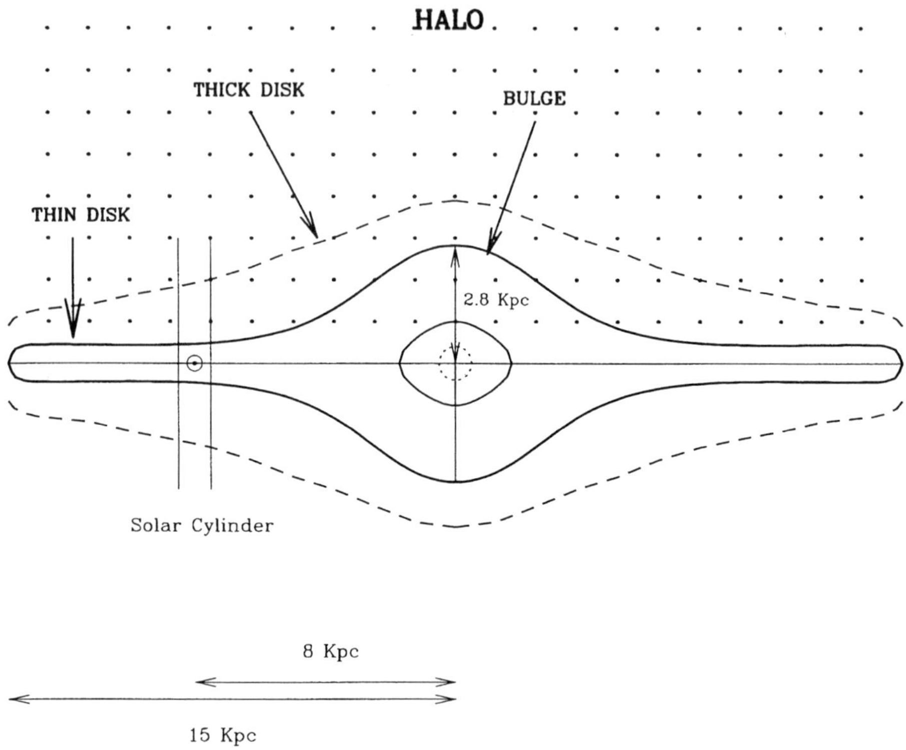

Figure 1. A sketch showing the vertical structure of the Galaxy. The solar vicinity is shown as a cylinder with a radius of 1 kpc.

Galaxy.

Just how the past life was quiet can be gauged by the ability of small galaxies to heat or thicken the disk. Quinn et al (1993) pointed out that even

satellites with masses of order 5% – 10% of the disk mass can thicken the disk by a factor of two or more if they dump all of their available orbital kinetic energy into the disk. Hence the presence of disks with scale heights less than a few hundred parsecs (*thin disks*) implies the absence of even minor encounters over the age of the *thin disk*. Nonetheless it may be possible for disks formed early in the lives of spirals to have been heated by satellites forming a thick disk within which a new *thin disk* was allowed to form. If this was the case, an estimate of the *thin disk* age can fix the time at which the last encounter occured.
The acute fragility of disks is a very important constraint on the evolution of the environment of protogalaxies that develop into current spirals. Either they were born in isolation (low density environment) or they managed to remove all potentially disk-disturbing debris early enough that a normal disk had time to form. It is clearly very interesting to look from our favorite observational point, the solar system, at the disk population to try to establish its age. This is important to constrain the zero point of chemical evolution models, the relationship between the disk and the other galaxy components, and to fix a rough chronology for the disk development and origin.

2. Age indicators

In the following I shall discuss five different methods devised to obtain an estimate of the age of the galactic disk, discussing their feasibility, robustness and limitations.
I shall start making a crucial point. Most methods to infer the age of the galactic disk pretend, or try, to give an age estimate for the entire disk (or even for the Galaxy as a whole), by using age indicators located in the near solar vicinity (see Fig. 1). Whether the solar neighborhood is really representative of all the disk is an open question, which does not hold only for this particular issue - the age of the disk -, but more generally for determining the global chemical evolution and the SF history of the galactic disk looking at nearby indicators (Carraro et al 1998). At present only old open clusters (although the sample is rather poor) can be used to derive an estimate for the age of a significant portion of the galactic disk. Moreover all the methods devised to address this topic make us of indicators located well inside the *thin disk*. Therefore I am going to discuss the age of the galactic *thin disk* (Gilmore et al 1989).

2.1. COOL WHITE DWARFS LUMINOSITY FUNCTION

The idea that the coolest white dwarfs (WD) can be used to determine the age of the local galactic disk dates back to Schmidt (1959). His idea

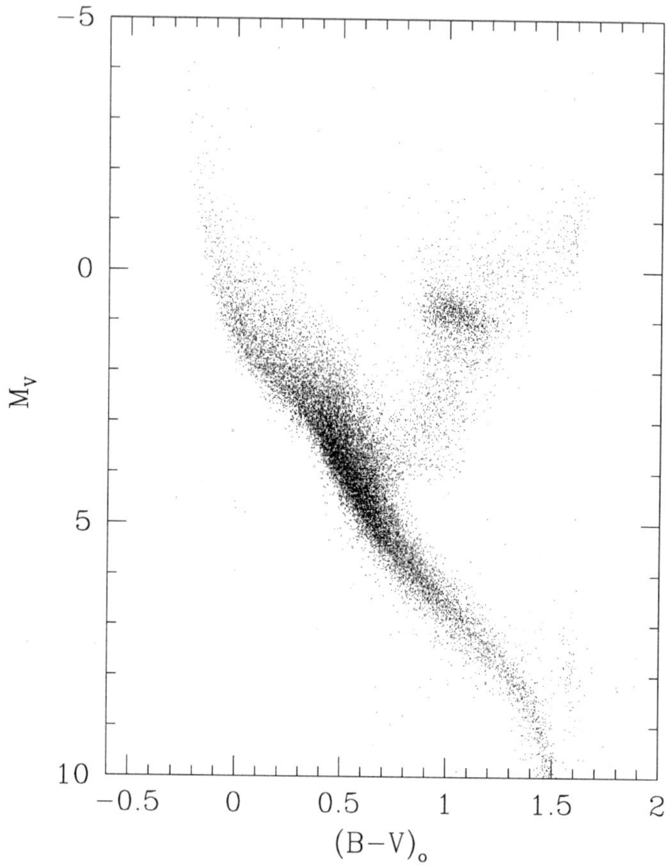

Figure 2. The CMD of nearby stars from Hypparcos. The red envelope of the subgiant region is used to set a minimum age of the galactic disc.

was the following: if we find the coolest white dwarf (just looking at the downturn of their luminosity function) and measure their cooling time, we can find the age of the disk just adding to this time the lifetime of the WD progenitor. Moreover WDs are easy to be modelled, much easier than Main Sequence (MS) stars (Mestel 1952). At that time it was not possible to detect the turn-down, and the idea failed. Moreover some uncertainties in the models arose, in particular the crystallization. In detail, at decreasing temperature ions start to crystallize, and the energy necessary to maintain the ions lattice causes the WD to cool to invisibility very rapidly.
More recently Winget et al (1987) succeeded to find the turn-down, and since then many authors built up CWDLF (Leggett et al 1998). In addition theorists now agree that the onset of rapid cooling is not reached for the

dominant stellar mass in the WD population ($0.6M_\odot$) until well below the luminosity of the observed shortfall. The most recent work is from Knox et al (1999). They find 52 CWDs using common proper motion binaries. Their sample is the only complete one both in luminosity and in proper motion. The resulting LF (see their Figs. 19,20 and 22) shows a clear short-fall at $logL/L_\odot = -4.2$, and predicts a larger number of WDs per unit volume, as compared with previous determination of the LF. Knox et al (1999) compared their LF with two sets of stellar models, by Wood (1992) and Garcia-Berro et al (1997). The best fit is obtained using Wood (1992) models, and provides a disk age of 9.0 ± 1.0 Gyr. Garcia-Berro et al (1997) models do not fit the LF maximum, and provide a somewhat greater age.

2.2. THE HYPPARCOS CMD

The CMD provided by Hipparcos satellite includes stars within about 150 parsecs from the Sun (see Fig. 2). It shows a well defined MS, a prominent clump of He-burning stars and a subgiant region. It represents of course a mix of stellar populations with spreads in age and in metallicity. The dating features used by Jimenez et al (1998) is the red envelope of the sub-giant region, and the methods adopted is the isochrone fitting one. Since the metallicity of the stars is not known - say from observations - the authors assume that the spread in color of the clump is a good indicator of a spread in metallicity. This is a rather crude approximation, as discussed by Girardi (this meeting). Nevertheless, it is possible to obtain a minimum age for this sample, assuming that the star populating the lower red envelope of the subgiant region has the maximum metallicity, since this metallicity provides the minimum age. So doing they obtain a minimum age for the disk of 8 Gyr.

2.3. NEARBY F & G STARS

Another method - although rather difficult - is the direct age estimate of single stars, like the 187 stars sample of Edvardsson et al (1993). To get an age estimate one needs star photometry, spectroscopy and distance, to put them in the $M_V - logT_e$ plane. In the case of the Edvardsson et al sample ages are inferred on the Vandenberg scale (1985). Removing from the sample the presumed thick disk stars, or stars whose orbital motions are not that of thin disk stars, a reasonable estimate for the age of the local galactic disk is $9 - 11$ Gyr. Recently the Edvardsson et al sample has been revised by Ng & Bertelli (1997), who re-computed the ages of those stars taking into account new distances from Hipparcos, correcting for the Lutz-Kelker effect, and putting them in the Bertelli et al (1994) scale. At older ages the new ages are slightly older, but the conclusion on the disk age is roughly the same.

2.4. THE RADIO-ACTIVE CLOCK

Butcher (1987) proposed to derive the age of the Galaxy by observing the radio-active nuclide ^{232}Th in stars of different ages, and relating the nucleosynthesis timescale to the stellar and galactic evolution. He considered the evolution of Nd, a stable nuclide and Th (half life 14 Gyr). The first point to stress is the extreme weakness of the spectral features in the measured stars spectra: in particular the Th line falls in a blend with Co. The errors related to the derived abundance are around 0.1 dex. The idea underlying this method is that after its formation, a stars does not modify its envelope abundances of Th and Nd but for radio-active decay. By measuring the ratio of the abundances of these elements, $[Th/Nd]$ in stars of different age, it is possible to reconstruct the decay evolution of Th. The basic assumption is that the growth rate of the two elements is the same, although Th is a r-process and Nd is partly a r- and partly a s-process. So doing, they conclude that no reliable chemical evolution model can account for this distribution without assuming an age less than 9 Gyr. The same result has been obtained by Morell et al (1992) who made a new abundance analysis on the same sample.

Clayton (1988) criticized these results, stressing that although the precise nature of r- and s- process is not clear enough, in principle one has to take into account their different evolution. In particular assuming that the contribution to the Nd abundance is about half from r- and half from s-processes, he showed that a simple model of chemical evolution can account for the Butcher distribution assuming ages greater than 12 Gyr, and concluding: "*An unbiased look at all methods together favours an age greater than 12 Gyr, although no single method is reliable. I point out that each nuclear method is still amenable to further improvements, but they alone will not be able to determine the Galaxy's age. Only a detailed and specific and correct model for the growth and chemical evolution of the solar neighbourhood can enable the galactic age to be inferred from radioactivity.*

2.5. OLD OPEN CLUSTERS

Old open cluster are well suited to address many issues concerning our disk (Friel 1995, Carraro et al 1998). For the present topic, they are in principle more suitable than the other indicators to derive a lower limit for the age of the galactic disk. In fact they are distributed in a larger portion of the disk. Good data have been obtained recently for clusters older than M 67, so it is actually feasible to use this sample to determine the age of the disk.

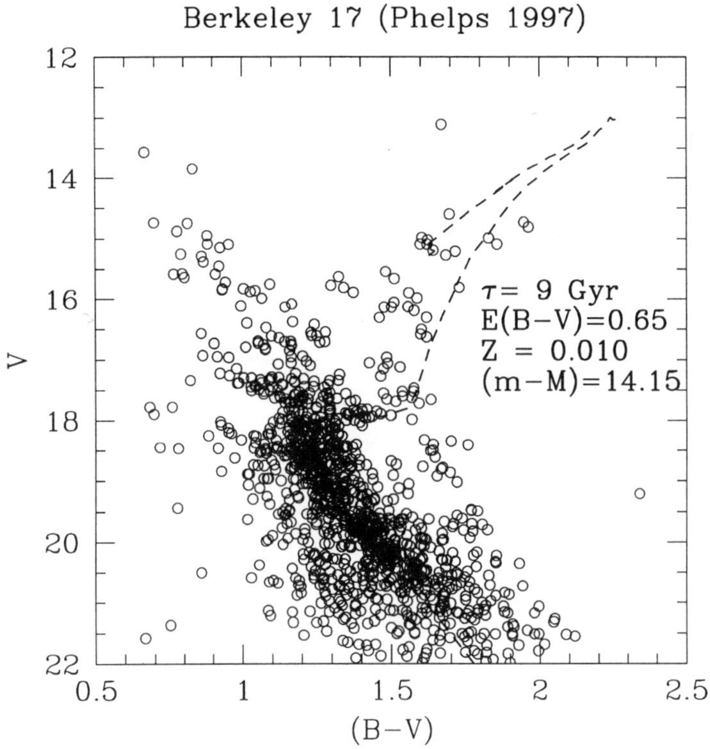

Figure 3. The CMD of Berkeley 17 from the V,I photometry of Phelps (1997). Overimposed is an isochrone from Girardi et al 1999 for the parameters listed in the plot.

However there is still debate on the role of the oldest open cluster. NGC 6791, often quoted as the oldest cluster (8-9 Gyr, Carraro et al 1999c), is a rather special object. Its nature is not completely clear, and somebody is suggesting that it could be a bulge globular pushed away by the bar, or the core of a dwarf spheroidal tidally stripped by our Galaxy (see Carraro et al 1999a for a detailed analysis on this cluster). Recently another cluster came out to be very old, Berkeley 17 (Fig. 3). It is actually quite old, with an age around 9 Gyr (Carraro et al 1999b), although optical photometry is not very good yet, and should be improved, being this cluster so important. If Berkeley 17 marks the age of the disk, its minimum age is around 9 Gyr. However the main drawback of open clusters is that their average lifetime is of the order of some 10^8 yr (Grenon 1990), so many old clusters might

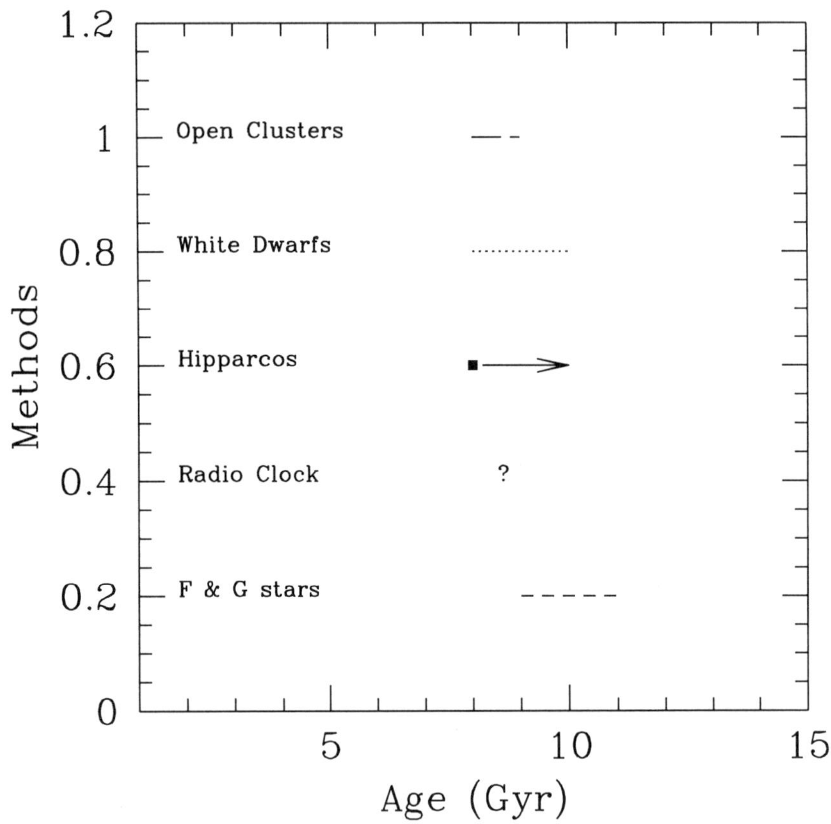

Figure 4. Summary of disk ages from different methods.

have been destroyed. Therefore the statistics of old open clusters is rather poor, and in principle they can provide only a lower limit for the age of the thin disk (Carraro et al 1999c). Anyhow their dating is rather simple and robust.

3. Conclusions

In the past several reviews were dedicated to the disk age issue. I would like to remind the reviews from Sandage (1990), van den Bergh (1990) and Grenon (1990).
Sandage at the 1989 Kingston Conference in Canada concluded that the disk overlaps in age the halo, whereas van den Bergh, at the same meeting, suggested that the disk is somewhat younger than the halo.

Obviously the question of a possible star formation delay at the end of the halo assembly depends on the age of the halo, and on the age of the disk. The age of the halo comes from the mean globular clusters age (12 – 15 Gyr at that time), whereas the age of the disk comes from a variety of methods, going from old open cluster to white dwarfs, from radioactive decay to stars in the solar neighbourhood. At that time the oldest open cluster was NGC 6791, with an age around 12 Gyr according to Sandage, who quoted Janes (1988), and of 7 Gyr according to van den Bergh, who quoted a preliminary work by Demarque et al (1992).
The conclusion of van den Bergh is supported also by Grenon (1990), who in addition stressed that in the solar vicinity there is a group of metal rich stars which seem to be older than open clusters, but whose birthplaces might be inside the bulge.

Ten years after the situation is not much different.
Gratton et al (1997) reported a mean age of the bulk of the halo globulars around 13 Gyr, which holds for all the halo clusters. There is indeed the evidence of a population of very young globulars (Pal 12 for instance), whose belonging to the halo is controversial.
Summarizing all the data I discussed above (see also Fig. 4) a plausible age for the disk is in the range 8 – 10 Gyr. Note that the age scale in the case of F & G stars and open clusters is the same as for globulars.
This seems to suggest the occurence of a hiatus, or minimun in the star formation history of the Galaxy, which might reflect the end of the halo/bulge formation. Afterwards the Galaxy started to acquire material to form the disk in an inside-out scenario.

Acknowledgements
I thank L. Girardi and M. Grenon for useful discussions, and Francesca Matteucci for her invitation.

References

1. Bertelli G., Bressan A., Chiosi C., Fagotto F., Nasi, 1994, A&AS 106, 275
2. Butcher H.R., 1987, Nature 328, 127
3. Carraro G., Fux R., Girardi L., 1999a, A&A in preparation
4. Carraro G., Ng Y.K., Portinari L., 1998, MNRAS 296, 1045
5. Carraro G., Vallenari A., Girardi L., Richichi A., 1999b, A&A 343, 825
6. Carraro G., Girardi L., Chiosi C., 1999c, MNRAS 309, 430
7. Clayton D.D., 1988, MNRAS 234, 1
8. Demarque P., Green E.M., Guenther D.B., 1992, AJ 103, 151
9. Edvardsson B., Andersen J., Gustafsson B., Lambert D.L., Nissen P.E., Tomkin J., 1993, A&A 275, 101
10. Friel E.D., 1995, ARA&A 33, 381
11. Garcia-Berro E., Isern J., Hernanz M., 1997, MNRAS 389, 973

12. Gilmore G., Wyse R.F.G., Kuijken K., 1989, ARA&A 27, 555
13. Girardi L., Bressan A., Bertelli G., Chiosi C., 1999, A&AS, in press
14. Girardi L., 1999, this meeting
15. Gratton R.G., Fusi Pecci F., Carretta E., Clementini G., Corsi C.E., Lattanzi M., 1997, ApJ 491, 749
16. Grenon M., 1990, in "Astrophysical Ages and Dating Methods", E. Vangioni Flamm, M. Casse and J. Adouze eds., Edition Frontieres, p. 153
17. Janes K.A., 1988, in "Calibration of stellar ages", A.G. Philip ed., p. 59
18. Jimenez R., Flynn C., Kotoneva E., 1998, MNRAS 299, 515
19. Knox R.A., Hawkins M.R.S., Hambly N.C., 1999, MNRAS 306, 736
20. Leggett S.K., Ruiz M.T., Bergeron P., 1998, ApJ 497, 294
21. Mestel L., 1952, MNRAS 112, 583
22. Morell O., Källander D., Butcher H.R., 1992, A&A 259, 543
23. Ng Y.K, Bertelli G., 1998, A&A 329, 943
24. Phelps R.L., 1997, ApJ 483, 826
25. Quinn P.J., Hernquist L., Fullagar D.P., 1993, ApJ 403, 74
26. Sandage A.R., 1990, J. Roy. Astron. Soc. Can., Vol. 84, No. 2, p.70
27. Schmidt M., 1959, ApJ129, 243
28. Van Den Bergh S., 1990, J. Roy. Astron. Soc. Can., Vol. 84, No. 2, p.60
29. White S.D.M., Rees M.J., 1978, MNRAS 183, 341
30. Winget D.E., Hansen C.J., Liebert J., Van Horn H.M., Fontaine G., Nather R.E., Kepler S.O., Lamb D.Q.,1987, ApJ 315, L81
31. Wood M.A., 1992, ApJ 386, 539

Speaker: G. CARRARO

M. GRENON's Comment: The open clusters may provide only lower limits for the disk age. Dynamical simulations show that none is expected to survive for 10 Gyr. When comparing the HR-diagram of NGC6791 with field stars with M_v from Hipparcos and [M/H] from photometry, the disk stars show a limiting isochrones 2-3 Gyr older than that of NGC6791, i.e. 10-11 Gyr for solar to metal rich stars.

G. CARRARO: Yup! Indeed the Edvardsson sample points to your direction. Open clusters must of course be used (why not), but I agree that they provide only a lower limit.

R. WYSE: You concluded that the halo gas did not form the disk, but I think a stroger argument supporting this conclusion is the very different angular momentum distributions of disk and halo $\longrightarrow$ the ejecta/leftover halo gas goes to the center of the Galaxy.

G. CARRARO: I agree with you. It is of course a strong argument, since also N-body-hydro simulations seem to favour the same picture, say the formation of the discs with material coming from outside.

R. PHELPS: A comment regarding Berkeley 17: the isochrone match for 9 billion years that you showed assumed a particular metallicity and reddening which, in my opinion, are quite uncertain. Additionally, the location of the match within the main sequence appears inconsistent with that chosen for NGC6791. I am willing to concede that Berkeley 17 may be as young as 9 Gyr, but I do not believe an age as great as 12 Gyr can be ruled out.

G. CARRARO: In Carraro et al. (1999) we show that, changing the metallicity from Z=0.007 to 0.017, which actually represents a large range, has no effect on the age determination but for the increase of the reddening at increasing metallicity. As for the MS fit, when no membership is available, it is common to fit the average MS, whereas when membership is available, the blue side should be the choice.

M. BELLAZZINI: I think that to draw a conclusion as "the age distributions of the halo and the disk do not overlap" you have to compare the two distributions measured on the same scale (i.e. same isochrones, etc.). Given the large errorbars of the actual age estimates there is no convincing evidence for this claim.

G. CARRARO: I reviewed several methods which lead to a reasonable age of 10 ± 1 Gyr. Since there is no WD LF for the halo, and dating of single stars is rather difficult, your argument holds only for clusters. The age of the disk in this case is 9 ± 1 Gyr. Gratton (1997) reports a mean age of 13 Gyr for the bulk of globulars. In my scale (Bertelli et al., 1996) the gap is even larger.

COMPACT HIGH–VELOCITY CLOUDS AT HIGH RESOLUTION

W.B. BURTON
Sterrewacht Leiden
P.O. Box 9513, 2300 RA Leiden, The Netherlands

AND

R. BRAUN
Netherlands Foundation for Research in Astronomy
P.O. Box 2, 7990 AA Dwingeloo, The Netherlands

Abstract. Six examples of the compact, isolated high–velocity clouds catalogued by Braun & Burton (1999) and identified with a dynamically cold ensemble of primitive objects falling towards the barycenter of the Local Group have been imaged with the Westerbork Synthesis Radio Telescope; an additional ten have been imaged with the Arecibo telescope. The imaging reveals a characteristic core/halo morphology: one or several cores of cool, relatively high–column–density material, are embedded in an extended halo of warmer, lower–density material. Several of the cores show kinematic gradients consistent with rotation; these CHVCs are evidently rotationally supported and dark–matter dominated. The imaging data allows several independent estimates of the distances to these objects, which lie in the range 0.3 to 1.0 Mpc. The CHVC properties resemble what might be expected from very dark dwarf irregular galaxies.

1. Introduction

Hierarchical structure formation scenarios suggest that galaxies form by continuous accretion of small, dark–matter dominated satellites. The possibility of an extragalactic deployment of high–velocity clouds has long been considered, and in various contexts, by Oort (1966, 1970, 1981), Verschuur (1975), Eichler (1976), Einasto et al. (1976), Bajaja et al. (1987), Burton (1997), Wakker & van Woerden (1997), Braun & Burton (1999; 2000, astro-ph/9912417), Blitz et al. (1999), and López–Corredoira et al. (1999). The discussion of Blitz et al. ties several HVC properties to the hierarchical structure formation and evolution of galaxies. In this context the extended

F. Matteucci and F. Giovannelli (eds.), The Evolution of the Milky Way, 347–359.

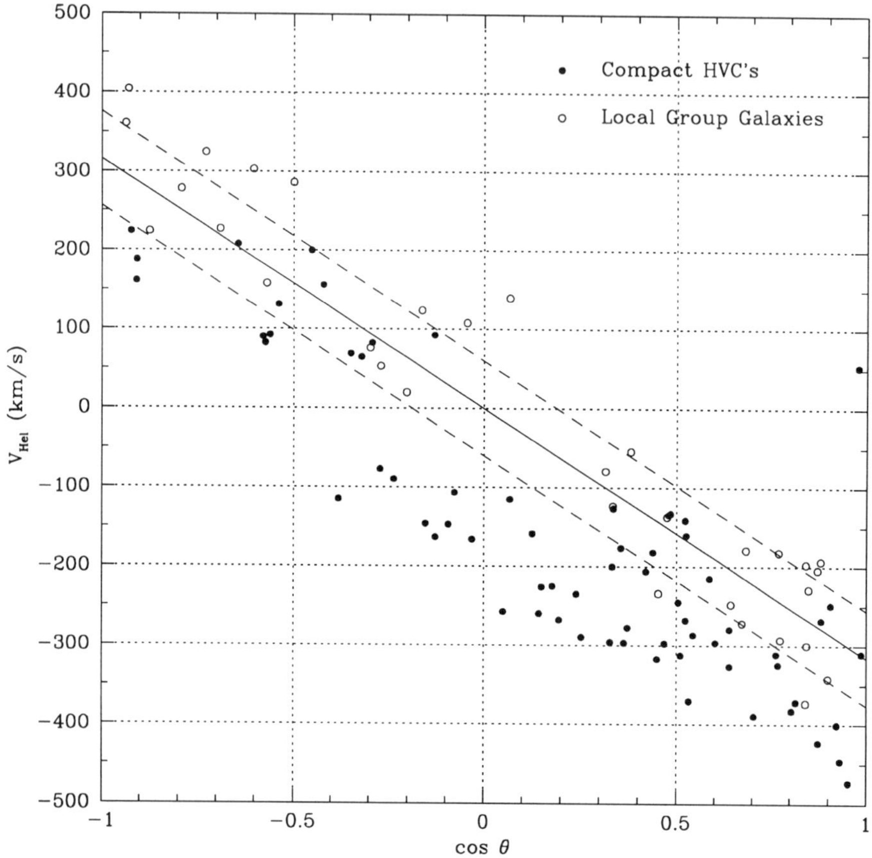

Figure 1. Variation of heliocentric velocity versus the cosine of the angular distance between the solar apex and the (l, b) direction of the compact, isolated high-velocity clouds catalogued by Braun & Burton (1999). The CHVCs are represented by filled circles; the galaxies constituting the Local Group, by open circles. The solid line represents the solar motion of $v_\odot = 316 \text{ km s}^{-1}$ toward $l = 90°$, $b = -4°$ (Karachentsev & Makarov 1996). The dashed lines represent the envelope one standard deviation ($\pm 60 \text{ km s}^{-1}$; Sandage 1986) about the velocity/angular–distance relation, pertaining for galaxies believed to be members of the Local Group. The kinematic and spatial deployment of the CHVCs is consistent with that of a dynamically cold ensemble spread throughout the Local Group, with a net infall velocity towards the barycenter of the Local Group of some 100 km s^{-1}.

HVC complexes would be nearby and currently being accreted onto the Galaxy, while the compact, isolated objects would be the primitive building blocks at larger distances, scattered throughout the Local Group.

The class of compact, isolated high–velocity clouds catalogued by Braun & Burton (1999) represent objects which plausibly originated under common circumstances, have shared a common evolutionary history, have not (yet) been strongly influenced by the radiative or tidal fields of the Milky Way or M31, and are falling towards the Local Group barycenter. The CHVC catalogue was based on survey data made with telescopes of modest resolution. The principal source was the Leiden/Dwingeloo Survey (LDS) of Hartman & Burton (1997), characterized by the angular resolution of $36'$ provided by the Dwingeloo 25–meter antenna; important additional data came from the more coarsely–sampled surveys of Hulsbosch & Wakker (1988) and of Bajaja et al. (1985) as analyzed by Wakker & van Woerden (1991), as well as from some new H I material observed at $21'$ resolution using the NRAO 140–foot telescope. The CHVCs are largely unresolved in angle in the single–dish catalogue, and therefore the large range in observed velocity widths can not be directly interpreted in terms of the intrinsic properties of individual gaseous entities.

Of the sample of 65 CHVCs catalogued by Braun & Burton (1999) only two had been subject earlier to interferometric imaging. Wakker & Schwarz had used the WSRT to show that both CHVC 114−10−430 and CHVC 114 − 06 − 466 exhibit a core/halo structure (with only some 40% of the single–dish flux recovered) that the linewidths of the resolved cores were substantially narrower than when the individual cores were blended at low resolution, and that several of the components displayed systematic velocity gradients. We have now imaged six additional CHVC fields using the Westerbork Synthesis Radio Telescope, and a further ten using the 304–meter Arecibo telescope. Selected properties of several of these fields are shown here. A complete discussion of the WSRT imaging is given by Braun & Burton (2000, astro-ph/9912417); a discussion of the Arecibo material is in preparation.

2. WSRT and Arecibo Data

Observations of the six CHVC fields imaged with the WSRT involved twelve–hour integrations in the standard array configuration having a shortest baseline of 36 meters. The effective velocity resolution was 1.2 times the channel spacing of 2.06 $\mathrm{km\,s^{-1}}$, over 256 spectral channels centered on the v_{LSR} of each source as catalogued by Braun & Burton (1999) on the basis of the single–dish spectra in the LDS. The angular and kinematic resolution afforded by the WSRT makes it well suited in important regards to detailed studies of the CHVC class of objects. Diffuse structures extending over more than about 10 arcmin are, however, not adequately imaged by the interferometer unless precautions are taken to eliminate the short–spacing bowl surrounding regions of bright extended emission. In a straightforward attempt to identify the column densities and overal extent likely to characterize any diffuse structures, we made use of the LDS data to determine

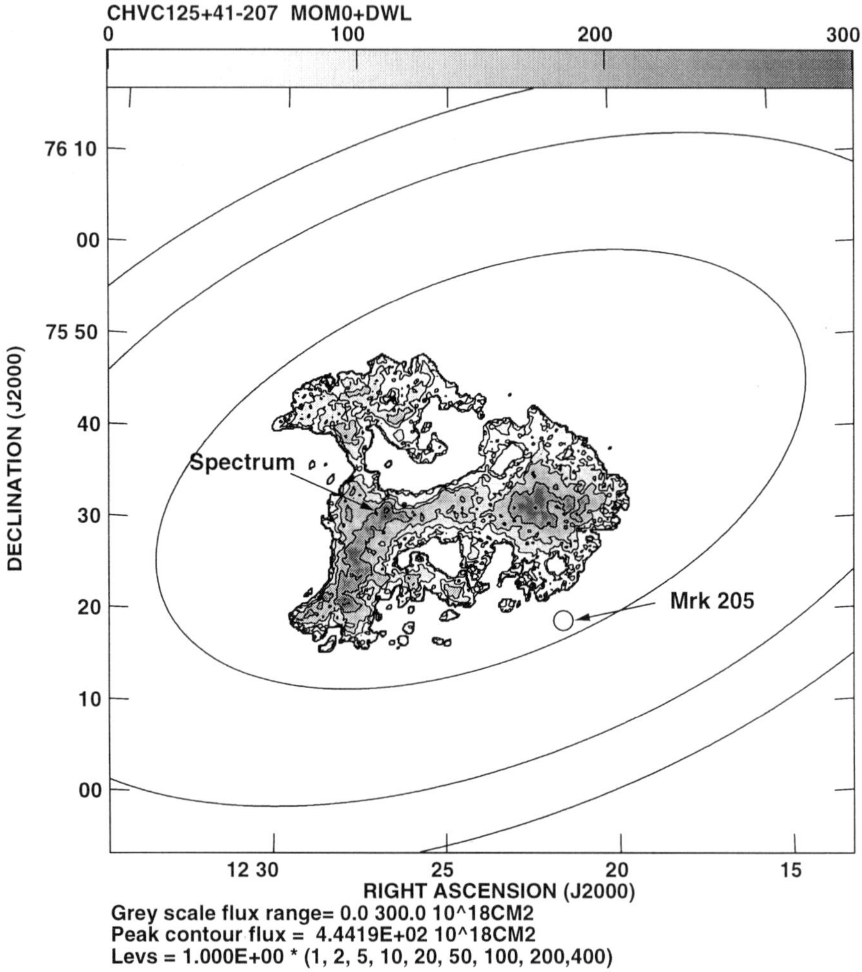

Figure 2. WSRT image of CHVC 125+41–207 displaying H I column densities at 28 arcsec angular resolution. $N_{\rm HI}$ was calculated assuming negligible opacity, and is displayed by contours at the levels indicated in units of 10^{18} cm^{-2} and a linear grey-scale running from 0 to 300×10^{18} cm^{-2}. The location of the Seyfert galaxy Mrk 205 is marked. This background source lies on a line of sight which penetrates the halo of the CHVC, where reconstruction of the integral flux using the composite WSRT and LDS single-dish data reveals a moderate diffuse-emission column depth. Bowen & Blades (1993) have measured Mg II absorption towards Mrk 205, and a metallicity substantially subsolar.

the emission from an elliptical Gaussian with dimensions and orientation as measured in the LDS, and with a total flux sufficient to recover the LDS integrated emission.

We have also recently observed ten CHVCs with the Arecibo telescope. The Arecibo facility is well-suited to provide sensitivity to the total column

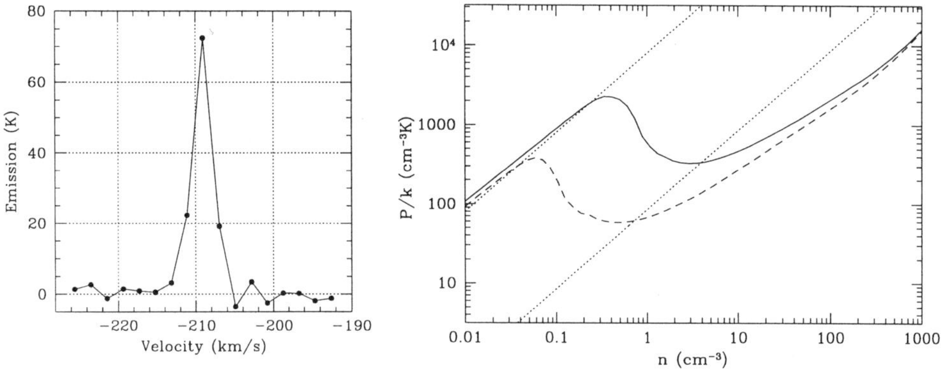

Figure 3. *left:* H I spectrum observed in the direction indicated in Fig. 2 of one of the bright emission knots in CHVC 125+41−207 . The spectrum is unresolved at a channel separation of 2 $\mathrm{km\,s^{-1}}$, indicating a core temperature of less than 85 K and quiescent turbulence. *right:* Equilibrium temperature curves for H I in an intergalactic environment characterized by a metallicity of 10% of the solar value and a dust–to–gas ratio of 10% of that in the solar neighborhood, calculated for two values of the neutral shielding column depth, namely 10^{19} $\mathrm{cm^{-2}}$ (solid line) and 10^{20} $\mathrm{cm^{-2}}$ (dashed line). The upper dotted line indicates the 8000 K temperature characteristic of the WNM; the lower one, the kinetic temperature of 85 K observed in the opaque core of CHVC 125+41−207. The volume density is tightly constrained for this temperature; a distance then follows from the measured column density and angular size.

density at relatively high resolution. This is especially important for CHVC targets which are of a size comparable to the field of view of most synthesis intruments. The Arecibo targets were observed with the new Gregorian feed and the narrow L–band receiver with two bandpass settings, namely 6.25 MHz and 1.56 MHz (yielding $\Delta v = 1.3$ and 0.32 $\mathrm{km\,s^{-1}}$, respectively) each centered on the $v_{\rm LSR}$ of the CHVC as determined by Braun & Burton (1999). The Arecibo targets were first mapped on a grid of $1° \times 1°$ size on a fully–sampled 90 arcsec lattice, in short integrations, in order to determine the locations of the peak flux concentrations; then at one or more of these principal components long–integration spectra were accumulated in a cut made at constant declination by repeating drift scans over the same 2° in right ascension. Typical column density sensitivies of some $10^{17.5}$ $\mathrm{cm^{-2}}$ over 20 $\mathrm{km\,s^{-1}}$ were reached, an $N_{\rm HI}$ regime largely unexplored (cf. Zwaan et al. 1997).

3. The Exceptionally Narrow Core in CHVC 125+41−207

The compact high–velocity cloud CHVC 125+41−207 is typical of the class of objects in several regards. Figure 2, adopted from Braun & Burton (2000, astro-ph/9912417; see also Burton & Braun 1999) shows several cool, quiescent cores embedded in a diffuse, warmer halo. The spectrum plotted in

Fig. 3, observed towards the brightest of these cores, has a linewidth which is completely unresolved by the effective resolution of the WSRT imaging. The velocity channels adjacent to the line peak have intensities below 20% of the maximum value. Such a width is one of the narrowest measured in H I emission, and contrains both the kinetic temperature and the amount of turbulence. An upper limit to the thermal–broadening FWHM of 2 km s^{-1} corresponds to an upper limit to the kinetic temperature of 85 K. The physical situation is yet more tightly constrained, because the brightness temperature in this core is observed to be 75 K; thus a lower limit to the opacity follows from $T_b = T_k(1 - e^{-1})$, yielding $\tau \geq 2$. Any broadening which might be due to macroscopic turbulence is less than 1 km s^{-1}.

The tightly–constrained temperature found for CHVC 125+41−207 allows an estimate of the distance to this object. Wolfire et al. (1995a, 1995b) show that a cool H I phase is stable under extragalactic conditions if a sufficient column of shielding gas is present and if the thermal pressure is high. Calculations of equilibrium conditions which would pertain in the Local Group environment have been communicated to us by Wolfire, Sternberg, Hollenbach, and McKee, and are shown in Fig. 3 for two bracketing values of the shielding column density, namely 10^{19} cm^{-2} and 10^{20} cm^{-2}. The figure shows that the equilibrium volume densities corresponding to the observed value of $T_k = 85$ K lie in the range 0.65 to 3.5 cm^{-3}. Thus provided with this range of volume densities, and having measured both the column depth of the cool core and its angular size, the distance to CHVC 125+41−207 follows directly from $D = N_{HI}/(n_{HI}/\theta)$, yielding a value in the range 210 to 1100 kpc. Further considerations of several opaque, cool cores made by Braun & Burton (2000, astro-ph/9912417) suggest that a distances between 0.5 and 1 Mpc are most plausibly representative of these objects.

Measurements of metallicity of high–velocity clouds are important to discussions of the nature of the phenomenon. If the clouds are primitive objects scattered throughout the Local Group, the gas would not be substantially enriched in heavy elements produced by stellar evolution. On the other hand, if the anomalous velocities had been generated by supernova explosions or some other energetic occurance in the Galactic disk, for example according to the precepts of the galactic fountain scenario (Shapiro & Field, 1976; Bregman, 1980) then the gas would be substantially enriched, with the already moderately high metallicities characteristic of the Galactic disk further enhanced by the circumstances of the ejection event. The small angular size of the CHVCs, and the amount of substructure being revealed at high resolution, will make it generally difficult to find suitable background sources. But the diffuse halo of CHVC 125+41–207 overlaps the Seyfert galaxy Mrk 205, and in this direction Bowen & Blades (1993) de-

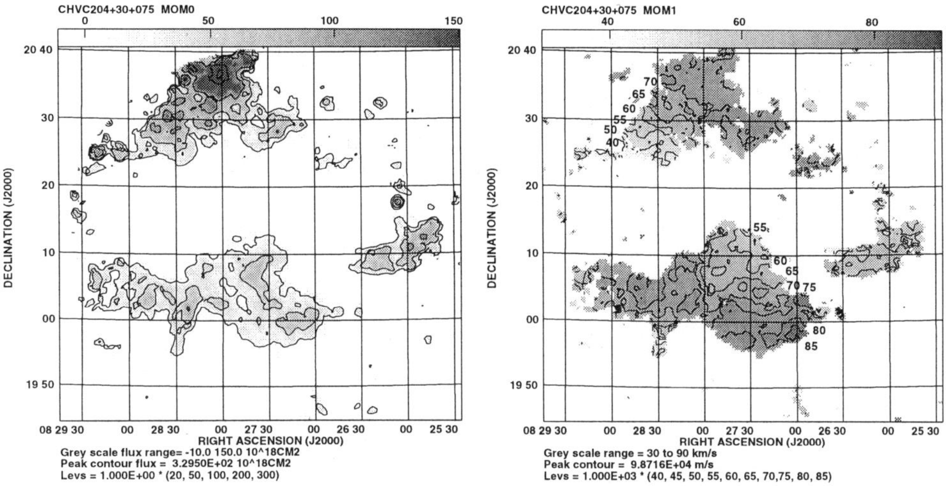

Figure 4. *left:* Westerbork image of CHVC 204+30+075 showing $N_{\rm HI}$ (calculated assuming negligible opacity) at an angular resolution of 1 arcmin; contours are drawn at levels of 20, 50, 100, 200, and 300 $\times$ 10^{18} $\rm cm^{-2}$. *right:* Intensity–weighted line–of–sight velocity, with contours of $v_{\rm LSR}$ showing systematic kinematic gradients across the two principal components of the CHVC object, consistent with rotation; contours are drawn in steps of 5 $\rm km\,s^{-1}$ from 40 to 85 $\rm km\,s^{-1}$.

tected Mg II absorption at $v_{\rm LSR} = -209$ $\rm km\,s^{-1}$. We determine a metallicity of this object in the range 0.04 to 0.07 solar.

4. Rotation in the Cores of CHVC 204+30+075

The narrowest FWHM of the CHVCs cataloged by Braun & Burton had a value of 5.9 $\rm km\,s^{-1}$; the broadest, a value of 95 $\rm km\,s^{-1}$. Under higher resolution, the characteristic width narrows as objects are resolved into several principal components, moving relative to each other. If the objects with multiple cores are to be stable, distances of order several hundred kpc are required. Some of the compact cores imaged owe their large total width in the LDS single–dish data to velocity gradients. The resolved WSRT image of CHVC 204+30+075 is shown in Fig. 4. The object shows two principal components each of which is elongated; furthermore, each of the elongated structures shows a systematic velocity gradient along the major axis.

The velocity gradients exhibited by the two principal components of CHVC 204+30+075 can be modelled in terms of circular rotation in a flattened disk system. Fig. 5 shows the results of fitting a standard tilted–ring model to the data. The fits display velocity rising slowly but continuously

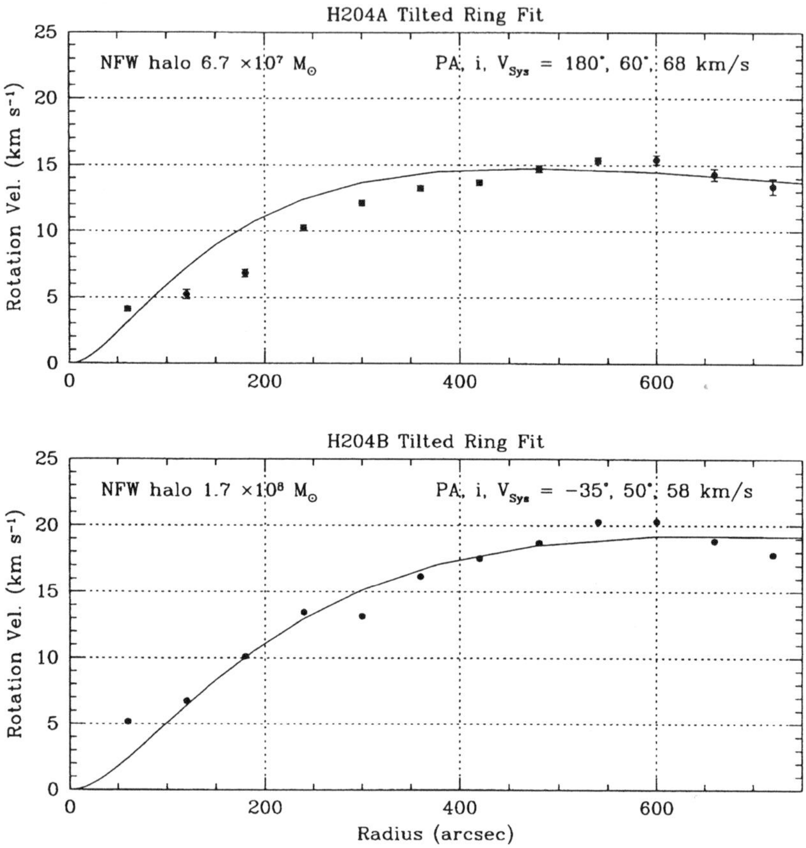

Figure 5. Rotation velocities fit by a standard application of the tilted–ring method to kinematic gradients revealed in the two principal components of CHVC 204+30+075 shown in Fig. 4. (The upper panel pertains to the component at higher declination.) The best–fit position angle, inclination, and systemic velocity are indicated. The solid lines show the rotation curves of Navarro, Frenk, & White (1997) cold–dark–matter halos of the indicated masses.

with radius to an amplitude of some 15 $\mathrm{km\,s^{-1}}$ in the one case and to some 20 $\mathrm{km\,s^{-1}}$ in the other, and then flattening to a constant value beyond about 500 or 600 arcsec. Estimates of the contained dynamical mass follow from the rotation curves if the distance is assumed, and the total gas mass follows from the integrated H I fluxes. At an assumed distance of 0.7 Mpc, the two principal clumps of CHVC 204+30+075 have $M_{\rm dyn} = 10^{8.1}$ and $10^{8.3}$ $\mathrm{M_\odot}$, and gas masses (including H I and helium of 40% by mass) of $10^{6.5}$ and $10^{6.9}$ $\mathrm{M_\odot}$, respectively for the upper and lower concentrations shown in Figures 4 and 5. The dark–to–visible mass ratios for these concentrations are 36 and 29, respectively.

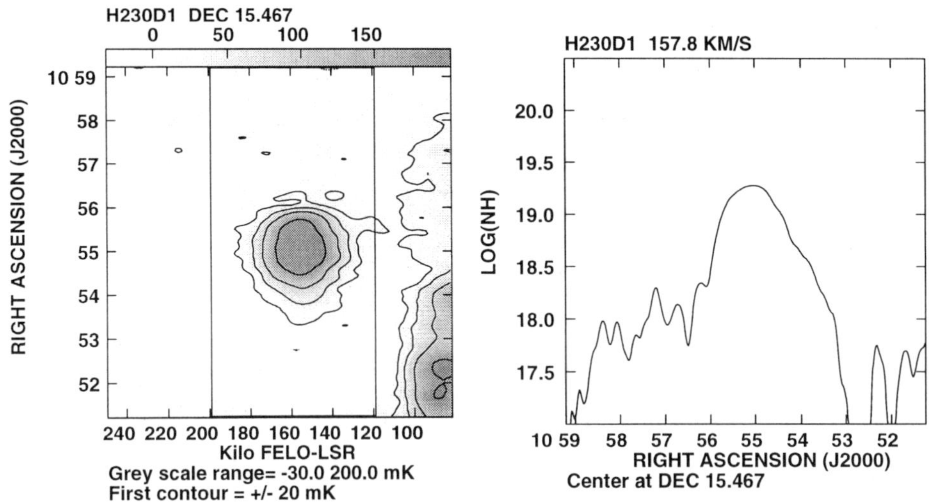

Figure 6. *left:* Position, velocity cut through CHVC 230+61+165 observed with the Arecibo telescope, at an angular resolution of 3.′3. The cut samples right ascension along the fixed declination 15.°467. This compact object shows no kinematic gradient along the cut sampled. *right:* Variation of $N_{\rm HI}$ with position, sampled at the velocity (158 km s^{-1}) of the peak of the column-density distribution. The $N_{\rm HI}$ values plotted here and in Fig. 7 are based on calibrated intensities in units of $T_{\rm b}$.

The shape of the modelled rotation curves for both of the CHVC 204+30+075 components is reproduced by the standard cold–dark–matter halo as presented by Navarro et al. (1997). At the assumed distance of 0.7 Mpc, the Navarro et al. halos fit to the two components have masses of $10^{7.8}$ M$_\odot$ (within 9.3 kpc) and $10^{8.2}$ M$_\odot$ (within 12.6 kpc), respectively.

5. The Objects CHVC 230+61+165 and CHVC 158−39−285

The WSRT imaging of CHVC 230+61+165 revealed a simple, faint structure. The Arecibo telescope is particularly well-suited to such targets, because of its sensitivity to low H I brightnesses. A position, velocity cut through this object at the location of the peak $N_{\rm HI}$ is shown in the lefthand panel of Fig. 6. No kinematic gradient is revealed. However the cut does show an interesting characteristic which several other of the CHVCs observed at Arecibo also show, namely a tendency to be more sharply bounded on one side of the cut than on the other. In the case of CHVC 230+61+165, the higher right–ascension boundary is sharper than the lower one down to $N_{\rm HI} = 10^{18.5}$. (We will consider this property further in our full discussion of the Arecibo observations.)

The Arecibo data on CHVC 158 − 39 − 285 are shown in Fig. 7. This

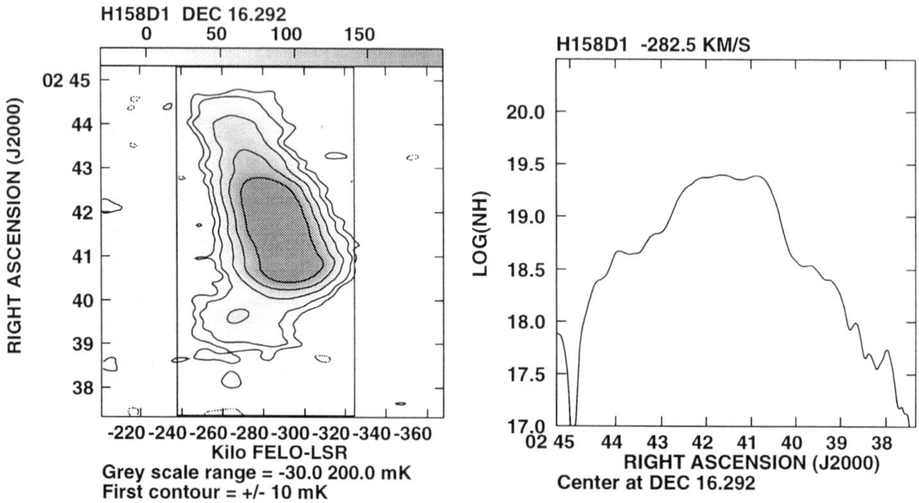

Figure 7. *left:* Position, velocity cut through CHVC 158−39−285 observed with the Arecibo telescope at an angular resolution of 3$'\!.$3. The cut samples right ascension along the fixed declination 16$^\circ\!.$292. This CHVC shows a kinematic gradient consistent with rotation, spanning some 40 km s^{-1}, as well as the characteristic core/halo morphology. *right:* Variation of $N_{\rm HI}$ with position, sampled at the velocity (−282.5 km s^{-1}) of the peak of the column–density distribution.

object is also a simple one, with only one component revealed. The position, velocity cut through the location of the peak $N_{\rm HI}$ value also shows that one side of the object is more sharply bounded than the other. The prominent kinematic gradient is consistent with simple rotation within the high $N_{\rm HI}$ core. The plots on the righthand panels of Fig. 6 and Fig. 7 show the variation of $N_{\rm HI}$ with position across the two CHVCs. The cores are embedded in diffuse material with characteristic $N_{\rm HI}$ values of order $10^{18.5}$ cm^{-2}.

6. Discussion

The WSRT and Arecibo high–resolution imaging shows that the morphology of the compact high–velocity clouds is characterized by one or more quiescent, low–dispersion compact cores embedded in a diffuse, warmer halo. The gas in the cores is identified with the cool neutral medium (CNM) of condensed H I at temperatures in the range 50 – 200 K; the halo gas is identified with the warm neutral medium. Such a nested geometry is expected if the CNM is to be stable in the presence of an ionizing radiation field of the sort expected in the Local Group environment. The cores contribute typically about 40% of the H I flux, while covering about 15% of the surface in the CHVC.

The high–resolution imaging data allows two independent distance estimates to be made. A distance for CHVC 125+41–207 follows from assuming rough spherical symmetry and equating the well–constrained volume and column densities of the compact cores. Another distance constraint (coupled with a dark–to–visible mass ratio) follows from consideration of the stability of CHVCs having multiple cores in a common envelope but having large relative velocities. The evidence available indicates that CHVCs have characteristic sizes of about 10 kpc, H I masses of about 10^7 $M_\odot$, and are at distances of 0.4 – 1.0 Mpc.

The imaging has also shown that the compact cores of CHVCs are commonly rotating. The observed kinematic gradients can be fit by rotating flattened disks, yielding dynamical masses for assumed distances. Dark–to–visible mass ratios of order 10 – 40 at $D = 0.7$ Mpc are indicated. The rotation curves shown here agree well with the cold–dark–matter halo predicted by Navarro et al. (1997).

The gaseous properties characteristic of CHVCs bear similarities to those of many dwarf irregular galaxies populating the Local Group (e.g. Young & Lo 1996, 1997a, 1997b). It is an important astrophysical challenge to establish the details of these similarities. It is particularly important to establish whether any of the CHVCs contain stars as well; the stellar density might be very low, but detection of any stellar component would lead to improved distances and would constrain the evolutionary history, and would augment the information on the faint end of the Local Group luminosity function. Failure to detect stars would imply that CHVCs are very primitive proto–galactic objects dominated by dark–matter halos. In either case, it seems plausible that the CHVCs are the missing Local Group satellite systems predicted by Klypin et al. (1999) and Moore et al. (1999).

Acknowledgements

We are grateful to M.G. Wolfire, A. Sternberg, D. Hollenbach, and C.F. McKee for providing the equilibrium temperature curves shown in Fig. 3, and to P. Perillat for assistance during our Arecibo observing session. The WSRT is operated by the Netherlands Foundation for Research in Astronomy, under contract with the Netherlands Organization for Scientific Research. The Arecibo Observatory is part of the National Astronomy and Ionosphere Center, which is operated by Cornell University under a cooperative agreement with the National Science Foundation.

References

1. Bajaja, E., Cappa de Nicolau, C.E., Cersosimo, J.C., Loiseau, N., Martin, M.C., et al., 1985, *ApJS*, **58**, 143
2. Bajaja, E., Morras, R., and Pöppel, W.G.L., 1987, *Pub. Astr. Inst. Czech. Ac. Sci.*, **69**, 237
3. Blitz, L., Spergel, D.N., Teuben, P.J., Hartmann, D., and Burton, W.B., 1999, *ApJ*, **514**, 818
4. Bowen, D.V., and Blades, J.C., 1993, *ApJ*, **403**, L55
5. Braun, R., and Burton, W.B., 1999, *A&A*, **341**, 437
6. Braun, R., and Burton, W.B., 2000, *A&A* (submitted); astro-ph/9912417
7. Bregman, J.N., 1980, *ApJ*, **236**, 577
8. Burton, W.B., 1997, in *The Physics of the Galactic Halo*, eds. Lesch H., Dettmann R.-J., Mebold U., Schlickeiser R. (Berlin: Akademie Verlag), 15
9. Burton, W.B., and Braun, R. 1999, *Nature*, **402**, 359
10. Eichler, D., 1976, *ApJ*, **208**, 694
11. Einasto, J., Haud, U., Jôeveer, M., and Kaasik, A., 1976, *MNRAS*, **177**, 357
12. Giovanelli, R., 1981, *AJ*, **86**, 1468
13. Hartmann, D., and Burton, W.B., 1997, *Atlas of Galactic Neutral Hydrogen*. Cambridge University Press, Cambridge
14. Karachentsev, I.D., and Makarov, D.A., 1996, *AJ*, **111**, 794
15. Klypin, A., Kravtsov, A.V., Valenzuela, O., and Prada, F., 1999, *ApJ*, **522**, 82
16. López-Corredoira, M., Beckman, J.E., and Casuso, E., 1999, *A&A*, **351**, 920
17. Moore, B., Ghigna, S., Governato, G., Lake, G., Quinn, T., Stadel, J., and Tozzi, P., 1999, *ApJ*, **524**, 19
18. Navarro, J.F., Frenk, C.S., and White, S.D.M., 1997, *ApJ*, **490**, 493
19. Oort, J.H., 1966, *Bull. Astr. Inst. Netherlands*, **18**, 421
20. Oort, J.H., 1970, *A&A*, **7**, 381
21. Oort, J.H., 1981, *A&A*, **94**, 359
22. Sandage, A., 1986, *ApJ*, **307**, 1
23. Shapiro, P.R., and Field, G.B., 1976, *ApJ*, **205**, 762
24. Verschuur, G.L., 1975, *ARA&A*, **13**, 257
25. Wakker, B.P., and Schwarz, U., 1991, *A&A*, **250**, 484
26. Wakker, B.P., and van Woerden, H., 1991, *A&A*, **250**, 509
27. Wakker, B.P., and van Woerden, H., 1997, it ARA&A, **35**, 217
28. Wolfire, M.G., Hollenbach, D., McKee, C.F., Tielens, A.G.G.M., and Bakes, E.L.O., 1995a, *ApJ*, **443**, 152
29. Wolfire, M.G., McKee, C.F., Hollenbach, D., Tielens, A.G.G.M., 1995b, *ApJ*, **453**, 673
30. Young, L.M., and Lo, K.Y., 1996, *ApJ*, **462**, 203
31. Young, L.M., and Lo, K.Y., 1997a, *ApJ*, **476**, 127
32. Young, L.M., and Lo, K.Y., 1997b, *ApJ*, **490**, 710
33. Zwaan, M.A., Briggs, F.H., Sprayberry, D., and Sorar, E., 1997, *ApJ*, **490**, 173

Speaker: B. BURTON:

R. WYSE's Comment: The latest high-resolution N-body simulations of Cold Dark Matter-dominated Universe are finding a factor of 10 more dwarf-galaxy-mass satellites around galaxies like the Milky Way than are observed in actual dwarf galaxies, so the possibility of many more dark-dwarfs is intriguing. However, it is not clear that the existence of these dense dark halos is compatible with the persistence of a **thin** disk.

THE PRESENT RATE OF SUPERNOVAE

M. TURATTO[1]
Osservatorio Astronomico di Padova
vicolo dell'Osservatorio 5, 35122 Padova, Italia

Abstract. We present and discuss the most recent determination of the rate of Supernovae in the local Universe. A comparison with other results shows a general agreement on the gross values but still significant differences on the values of the rates of various SN rates in different kinds of galaxies. The rate of SNe, used as a probe of Star Formation, confirms the young progenitor scenario for SNII+Ib/c. The increasing diversity of SNe reflects also in the SN yields which may affect the chemical evolution of the Galaxy but, because of the limited statistics, we cannot estimate the contributions of the new subtypes yet. It is also expected that in a few years observational determinations of the SN rates at various look-back times will be available.

1. Introduction

In the context of the chemical evolution of the galaxies the supernova (SN) rate and its history play a fundamental role. But, while there are several theoretical models predicting the time evolution of the SN rates with time, observational determinations at different look-back times are not yet available. An estimate of the rate of SNIa at z$\sim$ 0.4 has only be published so far (Pain et al. 1997). The efforts of the high-z searches have been focused more on the detection and follow-up of the SNe for the determination of the geometry of the Universe than on the careful analysis of their databases for the SN rates.

Different is the case of the estimates of the rate in the local Universe. Searches of bright SNe are active since several decades and their archives have been exploited to this purpose since the sixties. The determinations of the rates have been carried out on different samples and with different techniques. In some case the differences among various estimates are larger

F. Matteucci and F. Giovannelli (eds.), The Evolution of the Milky Way, 361–369.

than 100 %. In this paper we discuss the reasons of these uncertainties and their relation to the chemical evolution of the Galaxy.

2. Definition of SN rate

The SN rate is observationally defined in its general form as the number of SNe exploding in a given sample of galaxies during a certain time interval

$$\nu(yr^{-1}) = N_{SN} N_{gal}^{-1} \Delta t^{-1}$$

and is measured in yr^{-1}. However, since the galaxies can have very different stellar content, one expects that the average value depends strongly on the composition of the galaxy sample. Indeed, with the growing number of discoveries it appeared that critical parameters for the SN rates were the galaxy type, with late spirals more prolific than early spirals, and the galaxy luminosity, which in first approximation measures the number of available progenitor stars (Tammann 1970, Cappellaro & Turatto 1988).

The SN rate can therefore be computed for each galaxy type as

$$\nu^{gtype}(SNu) = N_{SN}^{gtype} (\sum_{j}^{Ngal} \sum_{i}^{obs} k_i L_j^{gtype} \Delta t_i^{gtype})^{-1}$$

where the index j refers to the sample galaxies of that type and i refers to the individual observations of each galaxy. Convenient units for the rate are the SNu, where 1 SNu corresponds to 1 SN per 10^{10} solar luminosities per century. Because of the normalization to galaxy luminosity, SNu scale as H_o^2. The parameter k_i accounts for various observational selection effects.

3. Efforts for better SN rates

It is evident from the last equation that in order to have better estimates of the SN rates one can: 1) enlarge the statistical samples (N_{SN}, N_{gal}), 2) estimate with higher precision the surveillance times, or 3) determine more reliable correction terms for the selection effects.

For the sample selection there are in practice two alternatives. We can, for instance, select all galaxies within a given volume, making reasonable assumptions on the surveillance times (cfr. Tammann 1994). Consequently the SN rate is computed using all the SNe discovered inside the volume during that period. The main advantage of this approach is in the large samples directly available form the galaxy and SN catalogues (e.g. Barbon et al. 1999). The drawback is in the critical assumption of the surveillance time which can hardly be the same for each galaxy.

TABLE 1. The pooled SN searches

SN search	type(*)	SN discoveries(**)	reference
Asiago	pg	51	Cappellaro et al. 1993 A&A 268, 472
Crimea	pg	33	Tsvetkov 1983 SA 27, 22
O.C.A.	pg	16	Pollas 1994, in Les Houches 1990, Bludman et al. p.769
Calan/Tololo	pg	12	Hamuy et al. 1993 AJ 106, 2392
Evans	vis	54	Evans 1997 PASA 14, 204

(*) pg=photographic; vis=visual
(**) some of which earlier reported by other searchers

The alternative is the survey approach which limits the galaxies to those surveyed by a specific SN search and the SNe to those discovered during the same search. The method allows for each galaxy a precise determination of the *control time*, i.e. the time during which a possible SN of a given type remains above the detection limit. This determination takes into account the distribution of the observations during the survey, the distance of the galaxy, the properties of the SN types and the observational limitations of the survey. The method, introduced already by Zwicky (1938), found several applications (e.g. Cappellaro & Turatto 1988, Evans et al. 1989). In this case the drawback is the small size of the sample.

In order to overcome such limitation we developed a strategy to treat simultaneously the databases of several of SN searches (Cappellaro et al. 1993, 1997a). The use of the five SN searches listed in Tab. 1 led us (Cappellaro et al. 1999) to build the largest sample ever used for the determination of SN rates, 137 SNe in 9346 RC3 galaxies, for a total control time of the sample of 24954 yr.

The determination of the correction terms for the biases in the discovery of SNe is crucial and controversial. Two biases are by far the most important: the effect of inclination in spirals and the deficiency of discoveries in the central regions of distant galaxies.

Several authors claimed that the bias due inclination was not affecting visual and CCD searches but only photographic ones (Evans et al. 1989, Muller et al. 1992). The composite sample above allows us to perform a direct check handling separately photographic and the Evans' visual searches. The rate of SNe computed for the Evans' sample with the control–time technique described above is 2.6 times greater in face–on spirals than in inclined ones, showing that this bias is almost as severe as in photographic searches for which the factor is 2.9. The fact that different detectors provide similar results, is an indication that the effect is intrinsic to the galaxies rather than to the observational technique.

The natural explanation is that SNe in inclined spirals suffer in average heavier dust extinction than those in face–on galaxies, thus are more difficult to discover. If this interpretation is correct we expect (1) that searches working at longer wavelengths are less affected, (2) that the effect is stronger in late type spirals and (3) that it is more important for core–collapse SNe which are associated to a younger population, than for SNIa associated to an older population.

So far contradictory answers have been given to the points above. The effect seems stronger in late than in early type galaxies (Cappellaro et al. 1997a), only marginally smaller in the visual that in the blue bands, as said above, (probably significantly smaller for red CCD searches) but comparable for core–collapse and thermonuclear SNe (Cappellaro et al. 1997a). Larger statistics is necessary for unambiguous answers.

In the attempt to correct for the bias due to inclination we have tried two different approaches (Cappellaro et al. 1999). First we applied the $\sec\alpha$ term corresponding to a plane parallel distribution of the dust in the disks of spirals. This results in an apparent overcorrection of the SN rate in edge–on galaxies. Then we explored the possibility to apply a model for the dust and the SN distributions consistent with our knowledge of the galaxies and SN progenitors. Even this more physical treatment showed some limitations indicating that the dust and the progenitor distributions in real galaxies are more complex than in simple models.

The study of the deficiency of SN discovery in the central regions of the parent galaxies has been discussed in detail in Cappellaro & Turatto (1997). The effect is critical in distant galaxies with about 50% of SNe exploding in galaxies with recession velocities larger than 7500 km s^{-1} lost in the central parts, in particular, the innermost 10 arcsecs. Interestingly the effect seems correlated to the inclination in spirals with more inclined galaxies more affected. The effect for the SN searches of Tab.1 was empirically determined in Cappellaro et al. (1997a) and accounted for also in Cappellaro et al. (1999).

4. The SN rates per unit B luminosity

The SN rates derived with the procedure and the sample described above are reported in Table 2. The errors include those due to the statistics and those due to the input parameters and the selection effects.

A comparison of these rates with other recent determinations is shown in Figure 1. There is a reassuring overall agreement among rates based on different samples and with different recipes. The largest differences are with Tammann et al. (1994) and for SNIa in ellipticals and for SNII. We believe

TABLE 2. SN rate(in SNu(B)) for the combined sample.

galaxy type	rate [SNu] Ia	Ib/c	II	All
E-S0	0.18 ± 0.06	< 0.01	< 0.02	0.18 ± 0.06
S0a-Sb	0.18 ± 0.07	0.11 ± 0.06	0.42 ± 0.19	0.72 ± 0.21
Sbc-Sd	0.21 ± 0.08	0.14 ± 0.07	0.86 ± 0.35	1.21 ± 0.37
Sm,Irr,Pec	0.40 ± 0.16	0.22 ± 0.16	0.65 ± 0.39	1.26 ± 0.45

$1\,\mathrm{SNu(B)} = 1\,\mathrm{SN}\,(100\mathrm{yr})^{-1}\,(10^{10}L_{\odot}^{\mathrm{B}})^{-1}$, $\mathrm{H}_o = 75$ km s^{-1} Mpc^{-1}

that such differences are mainly due to biases in the SN discoveries in the fiducial sample. Our new rates of SNIa show a shallower dependence on the galaxy types than our previous works of 1997 and 1993, which were based on the same approach but on smaller samples and with different bias corrections. At that time the increase of the rate of SNIa per unit luminosity moving from early to late type galaxies was considered as an argument in favor of different ages of the precursors of SNIa, with progenitors in spirals younger than those in ellipticals. This new result might be considered in favor of the conventional scenario of similarly old progenitors in all galaxy types. We warn that the high rate of SNIa observed in the heterogeneous class of Sm+Irregulars+Peculiars is uncertain to ±40%.

As in the previous works, the rates of SNII increase from early to late spirals but now the peak values are smaller. This is mainly due to the smaller correction terms used for the inclination effect. Similar trends are shown also by SNIb/c which however remain definitely less frequent than type Ia.

5. SN rates vs. Star Formation Rate

The normalization of the rates to the B luminosity of the galaxies has been a natural choice. In fact, the B magnitude is the magnitude most frequently available in the literature and it scales, in first approximation, with the galaxy mass since stars of different population have significant emission in this band.

In principle, the normalization of the rates to photometric bands sensitive to specific stellar populations might give important insights onto the SN progenitor scenario. If in all galaxies the flux at a given wavelength were due to a single stellar population, then the rate of the SNe associated to this population, normalized to the luminosity in this band, should be constant. For instance, infrared H or K bands would be particularly suit-

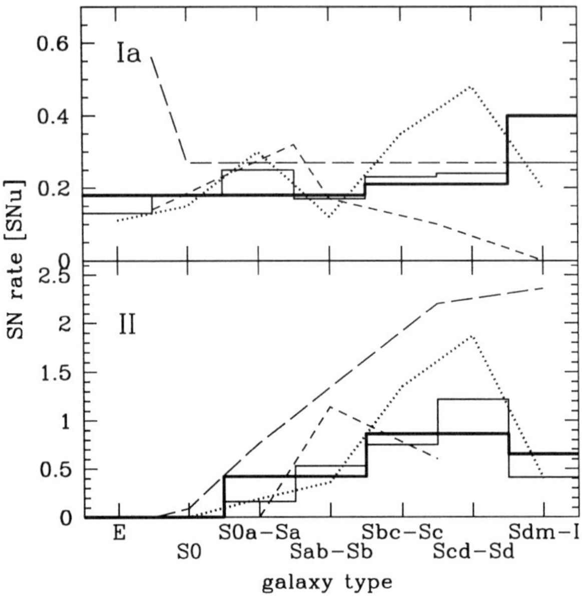

Figure 1. Comparison between different recent estimates of the SN rates scaled to H−o = 75 km s^{-1} Mpc^{-1} (Cappellaro et al. 1999, thick continuos line; Cappellaro et al. 1997a, thin continuous line; Tammann et al. 1994, long dashed line; van den Bergh & McClure 1994, short dashed line; Cappellaro et al. 1993, dotted line).

able for the study of the type Ia progenitors since they sample well the old stellar population. Unfortunately, the sample of galaxies with published infrared photometry is still too small for any direct analysis. Different is the situation for the integrated far infrared luminosities which are available for a large sample of galaxies thanks to the IRAS survey. L_{FIR} has been used often as an indicator of SFR and therefore we may try to gain information on the progenitors of core–collapse SNe.

Contrary to the expectations, the rate of core–collapse SNe (cfr. column 3 of Table 3) varies of almost a factor two moving from early and late spirals. Taken to face value this means that SNII and Ib/c are not associated to massive progenitors. Actually we think that this inconsistency is due to the contamination of the FIR emission associated with dust around massive stars by the cool component associated to the general stellar radiation field. In other words we consider this a further evidence that in normal galaxies L_{FIR} is not a reliable probe of SFR.

TABLE 3. SN rates per unit FIR luminosity.

galaxy type	SN rate [SNuIR] Ia	II+Ib/c	All
E-S0	1.8 ± 0.8		1.8 ± 0.8
S0a-Sb	0.6 ± 0.2	2.0 ± 0.5	2.7 ± 0.5
Sbc-Sd	0.6 ± 0.1	3.5 ± 0.6	4.1 ± 0.6

$1\,\text{SNuIR} = 1\,\text{SN}(100yr)^{-1}(10^{10}L_{\text{FIR},\odot})^{-1}$.

This finding is supported by the following argument. The effect of the general radiation field, due also to old stars, can be partially removed normalizing the FIR to the B luminosity to get the so–called infrared excess. The SN rates (in SNu(B)) computed independently for the three classes of IRAS not detected, low and high infrared excess galaxies are similar for type Ia. Instead for type II+Ib/c SNe the rate of IRAS not detected galaxies is significantly lower than in galaxies with high FIR excess (Table 6 of Cappellaro et al. 1999) in agreement with the young progenitor scenario.

Also the colors of galaxies can be used as indicators of the stellar populations with bluer galaxies hosting younger, more massive stars. We computed the SN rates (in SNu(B)) in galaxies with different optical integrated colors (Cappellaro et al. 1999). We found that the observed rate of SNIa is independent on the galaxy colors while the rate of core–collapse SNe is higher in bluer spirals than in galaxies of the same morphological type but with redder colors.

It is also interesting to note that the number of massive progenitors predicted by the star formation rate, derived by evolutionary synthesis models of galaxies with different colors, closely matches the observed number of core–collapse SNe.

6. Frequency and Nucleosynthesis

Supernovae are primary contributors to the chemical enrichment of the galaxies because they return to the ISM the elements synthesized by the hydrostatic and explosive burning processes, with the exception of the material trapped in the condensed remnants of core–collapse SNe. From a general point of view core–collapse SNe, descending from young massive progenitors determine the enrichment in the early galactic evolution while supernovae of type Ia are responsible of the heavy elements enrichment in the later phases of the galactic evolution on time scales which depend on the binary configuration and mass ratios. Current efforts are devoted to determine the masses of enriched material released by different kinds of explosions given their relative frequency.

For type II it is expected that the yields vary with the stellar mass, mass cut, explosion energy and neutron excess (Nakamura et al. 1999). The effect of these parameters have been modeled theoretically but observational confirmations are still needed. The well studied explosions of SN 1987A, SN 1993J and SN 1994I with progenitor masses between 13 and 20 $M_\odot$ produced about 0.07 $M_\odot$ of ^{56}Ni. Early works on the late light curves indicated that most SNII produced similar amounts of radioactive material (Turatto et al. 1990). In the last years data have been accumulated which show a different scenario (Danziger, this meeting). In particular there seems to be an increase of the ^{56}Ni production for stars up to about $25M_\odot$ (e.g. SN 1997ef, $0.15M_\odot$ of ^{56}Ni, and 1998bw, 0.7 $M_\odot$) after which the gravitational potential wins and a small ^{56}Ni mass is ejected due to fallback (e.g. SN 1997D, $0.002M_\odot$of ^{56}Ni; Turatto et al. 1998). The variation of the mass cut as a function of the progenitor mass seems to account for the observed trends of iron peak element abundance ratios in stars of low metallicity (Nakamura et al. 1999). The study of a second object with an expected low ^{56}Ni production, SN 1999eu a twin of SN 1997D at early time, will help to strengthen this hypothesis. Unfortunately the objects for which both the progenitor mass and the ejected ^{56}Ni mass are available is so limited that no kind of statistics is still possible.

Also for type Ia SNe, once considered a very homogeneous class, there is now evidence that the yields can be different. Cappellaro et al. (1997b) showed that the photometric properties of SNIa require a range of a factor 10 of the radioactive material synthesized during the explosion (0.1 to 1 $M_\odot$) and a factor 2 in the total mass of the ejecta. From a theoretical point of view such variations can be accounted for with different precursor masses (Chandrasekhar and sub-Chandrasekhar), and flame speeds. An attempt to determine the relative frequency of different Type Ia subtypes has been performed by Cappellaro et al. (1997a). Again going to analyze very particular SN subtypes we faced the problem of poor statistics. For faint SNIa, producing small amounts of ^{56}Ni, we found an intrinsic rate about 1/4 of that of all SNIa although they constitute only 5% of all SNIa discoveries. Also the percentage of SNIa producing large amounts of ^{56}Ni is expected to be small. In fact the observed number is relatively small although they are brighter than the average, hence easier to discover.

7. Conclusions

The determinations by various authors of the rates of SNe in the local Universe are in substantial agreement although large differences are still present on the rates of individual SN types in specific galaxy types. Improvements are possible in the treatment of the databases and in the analyses of the selection effects. However the slow increase of the statistics at low redshifts is the limiting factor.

On the contrary, huge is the number of discoveries at high (z= 0.5 – 1.0) redshifts thanks to the projects aiming to determine the geometry of the Universe with the use of SNe as distance indicators. The analysis of these databases can already allow the determination of the rates at such early epochs. Data at intermediate redshifts are still missing. In order to cover this gap we have started a SN search in the southern hemisphere which is starting to produce the first interesting results (SNe 1999ey, 1999gt, 1999gu). In few years observational estimates of the SN rates at various look-back times will provide strong constraints to the Galaxy evolutionary models.

References

1. Barbon,R., Buondí, V., Cappellaro, E., Turatto, M., 1999, A& AS 139, 531
2. Cappellaro, E., Turatto, M., 1988, A&A 190, 10
3. Cappellaro, E., Turatto, M. 1997, in: Nato – ASI on Thermonuclear Supernovae, eds. R. Canal, P. Ruiz-Lapuente, J. Isern, Kluwer Academic Publisher, Dordrecht p. 77
4. Cappellaro, E., Turatto, Benetti,S., Tsvetkov, D.Yu., Bartunov, O.S., Makarova, I.N., 1993, A&A 273, 383
5. Cappellaro, E., Turatto, M., Tsvetkov, D.Yu., Bartunov, O.S., Pollas, C., Evans, R., Hamuy, M., 1997a, A&A 322, 431
6. Cappellaro, E., Mazzali,P., Benetti,S., Danziger, I.J., Turatto, M., Della Valle,M., Patat,F., 1997b, A&A 328, 203
7. Cappellaro, E., Evans,R., Turatto, M., 1999, A&A 351, 459
8. Evans,R., van den Bergh,S., McClure,R.D., 1989, ApJ 345, 752
9. Muller, R.A., Marvin, H.J., Pennypacker, C.R., Perlmutter, S., Sasseen, T.P., Smith, C.K. 1992, ApJ 384, L9
10. Nakamura,T., Umeda,H., Nomoto,K., Thielemann,F., Burrows,A., 1999, ApJ 517, 193
11. Pain, R., et al. 1996, ApJ, 473, 356
12. Tammann, G.A., 1970, A&A 8, 458
13. Tammann, G.A., Löffler,W., Schröder,A., 1994, ApJS 92, 487
14. Turatto,M., Cappellaro,E., Barbon,R., Della Valle,M., Ortolani,S., Rosino,L., 1990, AJ 100, 771
15. Turatto,M., et al. 1998, ApJL 498, 129
16. van den Bergh, S., McClure, R.D. 1994, ApJ 425, 205
17. Zwicky,F., 1938, ApJ 96, 28

THE ROLE OF CLASSICAL NOVAE IN THE GALACTIC NUCLEOSYNTHESIS

M. DELLA VALLE[1]
Arcetri Astrophysical Observatory
Largo E. Fermi, 5, 50125, Firenze

1. Introduction

The outbursts of classical novae result from thermonuclear runaways occurring in the H-rich envelopes accreted onto the white dwarf components of nova systems. The *empirical* assessment of the contribution of novae to the chemical evolution of the Milky Way needs three *quantities*: 1) the galactic nova rate (average number of nova outbursts per year); 2) the mass of the ejected envelopes; 3) the abundances of heavy elements in the nova ejecta.

2. The Galactic nova rate

Nova rate estimates obtained by several authors in the last 50 years are rather discrepant. Despite their high luminosity at maximum (fast novae can achieve $M_V \sim -9$, e.g. Della Valle and Livio 1995), several factors prevent us from discovery most events. Our position within the Galaxy coupled with: *a)* the interstellar extinction in the galactic disk, *b)* the location of novae within the Galaxy, mainly at low galactic latitudes, *c)* seasonal effects, bad weather and other interruptions (illness, vacation, fishing...), makes the current galactic nova survey largely uncompleted. An analysis of nova discoveries of the last decades, indeed, points out an *observed* rate of about 3-4 events per year. On the other hand, the estimates based on extrapolating this *local* rate to the whole Galaxy are quite higher: 100 novae/yr (Allen 1954), 50 novae/yr, (Kopylov 1955), 270 novae/yr (Sharov 1972), 73±24 novae/yr (Liller and Mayer 1987). The dramatic discrepancy existing between the above reported estimates is due to the difficulties to properly account for the incompleteness factors. Aware of these problems, Della Valle (1988a, his Fig. 1) and Della Valle and Livio (1994) have attempted a different approach (see also Ciardullo et al. 1990). These authors

F. Matteucci and F. Giovannelli (eds.), The Evolution of the Milky Way, 371–377.

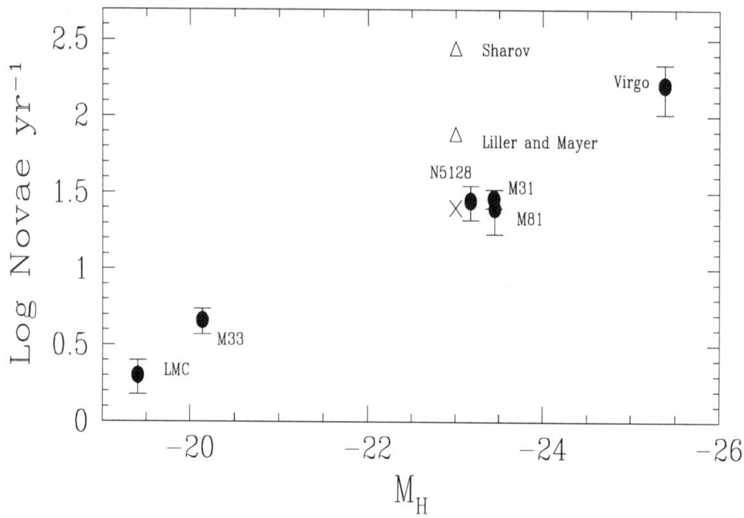

Figure 1. The nova rate as a function of the H-Luminosity of the parent galaxies. The 'cross' marks the extrapolated value for the nova rate of the Milky Way.

have shown that the nova rates in extragalactic systems, such as M31 (Arp 1956, Rosino 1964, 1973, 1989, Capaccioli et al. 1989), M33 (Della Valle et al. 1994), NGC 5128 (Ciardullo et al. 1990), LMC (Graham 1979) correlates with the luminosity of the respective parent galaxies (Fig. 1).

By extrapolating these data to the H (or K) luminosity of the Milky Way, Della Valle and Livio (1994) found an average number of nova outbursts of 24 novae/yr. However this figure still brings a large uncertainty. The previous authors estimated a rate of $\sim$ 20 novae per year (or even less) if the nova production in the Galaxy is mainly maintained by the *bulge*, up to 50 novae per year if the novae are mainly produced in the *disk*. One point seems clear: even upper limits of the *extragalactic-based* nova rate estimates are definitely smaller than the rates based on *galactic* observations. Shafter (1997) has recently reviewed the topic of galactic-based nova rates and was able to derive R=35±11 year^{-1}, after making proper allowance for the incompleteness factors. In the following we assume a rate of novae for our galaxy of: **R=30±10 year^{-1}**

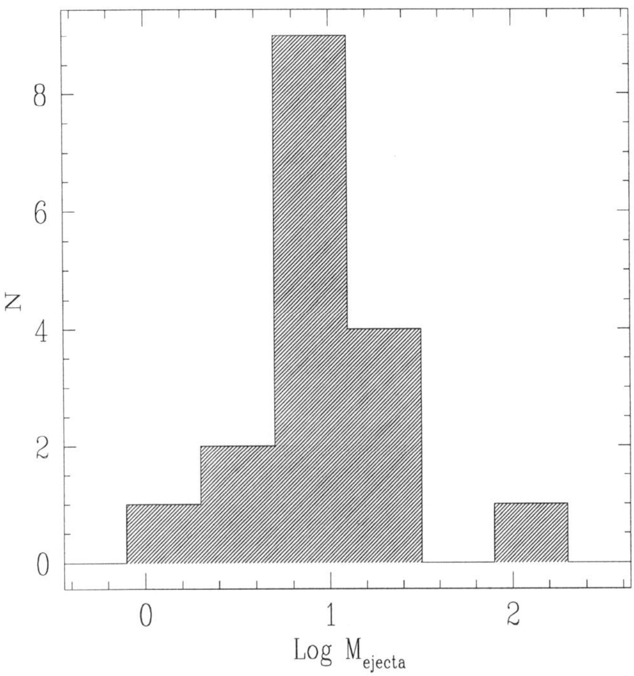

Figure 2. Frequency distribution of the masses of the nova envelopes. Masses in units of Log $10^{-5} M_{\odot}$.

3. The mass of the ejecta

About 20 mass measurements of nova shells are currently available in literature (out of more than 200 galactic novae to date discovered). In Fig. 2, we present the mass distribution of the ejecta (data from Warner 1995 and references therein, and Della Valle et al. 1997). The sample is small, nevertheless, it is apparent that the data peak close to $\sim$ $a\ few \times 10^{-4}$ solar masses.

Interestingly enough, Fig. 3, shows that the ejected mass marginally correlates (90% confidence) with the class of speed of the nova, according to:

$$\text{Log } M_{env} = 0.274(\pm 0.197) \times \text{Log } t_2 + 0.645(\pm 0.283)$$

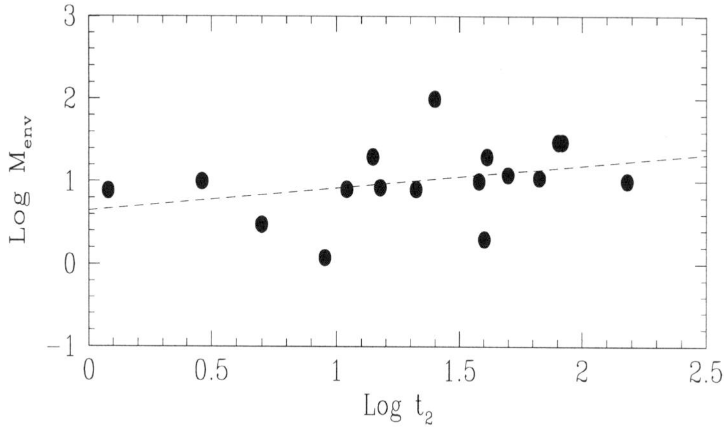

Figure 3. Masses of the nova envelopes as a function of the rate of decline. t_2 is the time which the nova takes to drop 2 magnitudes from maximum light.

This result is not unexpected. Since the pressure at the bottom of the accreted layer is proportional to $M_{WD} \times \Delta M_{acc}$, the massive white dwarfs need to accrete less material than light ones before starting thermonuclear runaways. Thus *fast* novae, which originate from massive white dwarfs (e.g. Livio 1992), are expected to eject less massive envelopes than *slow* ones. By combining the previous value of the galactic nova rate (Fig. 1) with the average mass of nova envelopes (Fig. 2), and assuming that the nova rate is a constant over a Hubble time, we derive that about $3 \times 10^7 M_\odot$ are processed through the ejecta over the lifetime of the Galaxy. Since the mass of the interstellar gas/dust content of the Galaxy is $\sim 3 \times 10^9 M_\odot$ (e.g. Allen 1972), it follows that only nuclei enriched, in the nova ejecta, by a factor at least ~ 100 can contribute significantly to the galactic nucleosynthesis.

4. The abundances in the nova ejecta

Recognition of unusual chemical abundances in the nova ejecta dates back many years (e.g. Payne-Gaposchkin 1957, Williams 1977, Gallagher and S-tarrfield 1978). This topic has been recently reviewed by Livio and Truran (1994). One point stands clearly: all novae, for which reasonable data are

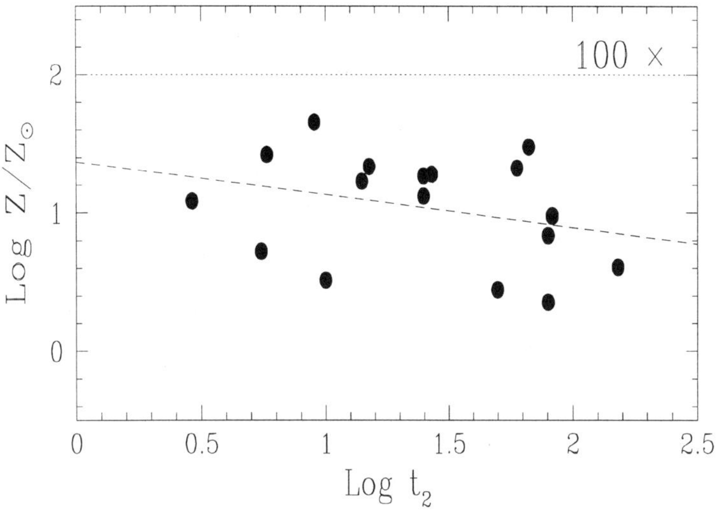

Figure 4. Heavy elements abundances in the nova envelopes as a function of the rate of decline

available (even admittedly characterized by large uncertainties), appear to be enriched in either He or/and heavy elements. This result is not unexpected because TNR proceed in the accreted H-rich envelope by means of the C-N-O burning chain.

Fig. 4 (abundances data from Livio & Truran and references therein; rates of decline from Della Valle 1988b) illustrates the average trend of the heavy elements abundance in the nova ejecta as a function of the rate of decline. Two facts emerges: 1) the available data suggest that the enrichment into the nova envelopes is below the $\sim$ 100 time threshold; 2) the heavy elements abundance in the ejecta marginally correlates (90% confidence) with the speed class of the nova according to:

$$\text{Log } Z/Z_{\odot} = -0.237(\pm 0.193) \times \text{Log } t_2 + 1.367(\pm 0.285)$$

Since fast novae eject the less massive envelopes (see Fig. 3), the correlation presented in Fig. 4 goes in the direction to further soften the contribution of classical novae to the chemical evolution of the Galaxy.

5. Concluding Remarks

On the basis of the data reviewed in this paper a number of conclusions can be sketched out.

1) Novae contribute to the galactic nucleosyntesis, but it is unlikely that they can dramatically affect the chemical evolution of the Milky Way (see also Truran 1990);

2) Novae can produce interesting concentrations only for rare isotopes (which are enhanced in the ejecta by more than a factor $\sim$ 100 to the solar abundances) such as: ^{13}C; ^{15}N (Sparks, Starrfield and Truran 1978; Williams 1985, Peimbert and Sarmiento 1984); ^{22}Na and ^{26}Al (Hillebrandt and Thielemann 1982, Kolb and Politano 1997); ^{22}Ne (Livio and Truran 1994) and ^{7}Li (Norgaard and Arnould 1975, Starrfield et al. 1978, D'Antona and Matteucci 1991, Romano et al. 1999, Travaglio et al. 2000);

3) Mass measurements of the nova envelopes based on the assumption that the shells are almost homogeneously filled in (which implies filling factors in the range 0.1–1) could be misleading. If the filling factor is close to $10^{-2/-5}$ rather than $\sim$ 1, as the recent observations of T Pyx might suggest (Shara et al. 1997), dramatic changes will be inevitable in the mass measurements of the material processed through the nova ejecta.

Acknowledgments. It is my pleasure to thank N. Panagia, P. Marziani and A. Bianchini for helpful discussions.

References

1. Allen, C. W. 1954, MNRAS, 114, 387
2. Allen, C. W. 1976, *Astrophysical Quantities*
3. Arp, H. C. 1956, AJ, 61, 15
4. Ciardullo, R., Ford, H. C., Williams, R. E., Tamblyn, P., & Jacoby, G. H. 1990, AJ, 99, 1079
5. Capaccioli, M., Della Valle, M., D'Onofrio, M., Rosino, L. 1989, AJ, 97, 1622
6. Della Valle, M. 1988a, in The extragalactic distance scale; ASP 100th Anniversary Symposium, Victoria, Canada, June 29-July 1, 1988. San Francisco, CA, Astronomical Society of the Pacific, 1988, p. 73, 74.
7. Della Valle, M. 1988b, Ph.D. thesis, Padova University
8. Della Valle, M., Rosino, L., Bianchini, A., & Livio, M. 1994, A&A, 287, 403
9. Della Valle, M., Livio, M. 1994, A&A, 286, 786
10. Della Valle, M., Livio, M. 1995, ApJ, 452, 704
11. Della Valle, M., Gilmozzi, R., Bianchini, A., Esenoglu, H. 1997, A&A, 325, 1151
12. D'Antona, F., Matteucci, F. 1991, A&A, 248, 62
13. Gallagher, J.S., Starrfield, S. 1978, ARA&A, 16, 171
14. Graham, J.A. 1979, PASP, 91, 79
15. Hillebrandt, W.; Thielemann, F.K 1982, ApJ, 255, 617
16. Kolb, U.; Politano, M. 1997, A&A, 319, 909
17. Kopylov, I. M. 1955, Izv. Krymskoi Astrofiz. Obs., 13, 23
18. Liller, W., & Mayer, B. 1987, PASP, 99, 606

19. Livio, M. 1992, ApJ, 393, 516
20. Livio, M., Truran, J. 1994, ApJ, 425, 797
21. Norgaard, H., Arnould, M. 1975, A&A, 40, 331
22. Payne-Gaposchkin, C. 1957, *The Galactic Novae*
23. Peimbert, M., Sarmiento, A. 1984, Astronomy Expr., 1, 97
24. Romano, D., Matteucci, F., Molaro, P., Bonifacio, P. 1999
25. Rosino, L. 1964, Ann. d'Astrophys., 27, 498
26. Rosino, L. 1973, A&AS, 9, 347
27. Rosino, L., Capaccioli, M., D'Onofrio, M., Della Valle, M. 1989, AJ, 97, 83
28. Shafter, A. 1997, ApJ, 487, 226
29. Shara, M.M., Zurek, D.R., Williams, R.E., Prialnik, D., Gilmozzi, R., Moffat, A.F.J. 1997, AJ, 114, 258
30. Sharov, A. S. 1972, Soviet Astron., 16, 41
31. Sparks, W. M.; Starrfield, S.; Truran, J. W. 1978, ApJ, 220, 1063 Starrfield, S.; Truran, J. W.; Sparks, W. M.; Arnould, M. 1978, ApJ, 222, 600
32. Travaglio et al. 2000, in preparation
33. Truran, J.W. 1990, in Physics of Classical Novae. IAU Colloquium No.122, held in Madrid, Spain, on June 27-30, 1989. Editors, A. Cassatella, R. Viotti; Publisher, Springer-Verlag, Berlin, Germany, p. 373
34. Warner, B. 1995, Cataclysmic Variable Stars (Cambridge:Cambridge Univ. Press)
35. Williams, R.E., 1977, in Interaction of Variable Stars with their Environment, IAU Colloq. 42, held in Bamberg, September 6-9, 1977, Bamberg: Remeis-Sternwarte, 1977, edited by Rudolf Kippenhahn, J. Rahe, and W. Strohmeier., p.242
36. Williams, 1985, in ESO Workshop on Production and Distribution of C, N, O Elements, Garching, West Germany, May 13-15, 1985, p. 225-232

FAINT STARS IN THE URSA MINOR DWARF SPHEROIDAL GALAXY: IMPLICATIONS FOR THE STELLAR IMF

ROSEMARY F.G. WYSE
Physics and Astronomy Department, The Johns Hopkins University, Baltimore, MD 21218, USA

GERARD GILMORE
Institute of Astronomy, Cambridge CB3 0HA, England

SOFIA FELTZING
Lund Observatory, Box 43, 221 00 Lund, Sweden

AND

MARK HOUDASHELT
Physics and Astronomy Department, The Johns Hopkins University, Baltimore, MD 21218, USA

Abstract.

The stellar Initial Mass Function (IMF) is a crucial ingredient in models of star formation and chemical evolution. Star counts in systems with simple star-formation histories are particularly straightforward to interpret. We present the faint stellar luminosity function in an external galaxy, the Ursa Minor dwarf Spheroidal (dSph). This relatively-nearby companion galaxy to the Milky Way has a stellar population with narrow distributions of age and of metallicity, remarkably similar to that of a classical halo globular cluster, i.e. old and metal-poor. The integrated luminosity of the Ursa Minor dSph is also similar to that of a globular cluster, but its central surface brightness is many orders of magnitude lower. Further, again in contrast to globular clusters, the internal dynamics of the Ursa Minor dSph are apparently dominated by dark matter, with $(M/L)_V \sim 80(M/L)_\odot$. Perhaps surprisingly, we determine for the Ursa Minor dSph a stellar luminosity function, and inferred IMF, that is identical, down to our completeness limit of $\sim 0.4M_\odot$, to that of the globular clusters of similar age and metallicity. Thus the stellar IMF long ago and far away, in a dark-matter-dominated galaxy, was the same as that in star clusters in the Milky Way.

F. Matteucci and F. Giovannelli (eds.), The Evolution of the Milky Way, 379–386.

1. Introduction

The stellar Initial Mass Function (IMF) is a crucial ingredient in models of star formation and chemical evolution, controlling aspects such as rate of return of enriched (and unenriched) gas to the interstellar medium. There are many physical parameters that one might have expected to affect the star-formation process and (hence?) the IMF, such as metallicity, gas density, pressure, specific angular momentum of the gas, magnetic field strength, etc. However, the available data are consistent with the null hypothesis of an invariant IMF, when averaged over regions larger than one open cluster (e.g. papers in Gilmore & Howell 1998).

The derivation of an IMF from star counts is simplified if the system under study consists of coeval stars, all with the same metallicity. The most populous star clusters in the Milky Way are the old, metal-poor, globular clusters and their low-mass IMFs have been much-studied (e.g. Piotto's and Paresce's contributions to this volume). The dwarf spheroidal (dSph) companion galaxies to the Milky Way were once thought also to consist of old, metal-poor populations, but newer and better data have revealed that a typical dSph satellite has a composite stellar population, with a range of ages and metallicities (e.g. Mateo 1998; Hernandez, Gilmore & Valls-Gabaud 1999). The Ursa Minor dSph (distance $\sim$ 65kpc) is rather unusual, having within the observational uncertainties essentially a mono-age ($\sim$ 15 Gyr) and mono-metallicity ([Fe/H]~ -2.2) stellar population (Olszewski & Aaronson 1985; Feltzing, Gilmore & Wyse 1999; Mighell & Burke 1999) similar to that of the old metal-poor halo globular clusters, in particular to M92 and M15. The total luminosity of the UMi dSph is $L_V \sim 2 \times 10^5 L_\odot$, comparable to globular clusters, but its central surface brightness is only 25.5 mag/sq$''$, some 10 magnitudes fainter than that of a typical globular cluster.

Further, most of the dSph companions of the Milky Way have stellar velocity dispersions that are in excess of those expected if stars with a normal mass function dominate their internal gravitational potentials (e.g. Mateo 1998). Neither orbital motions within stellar binaries (Hargreaves, Gilmore & Annan 1996) nor Galactic tidal effects (e.g. Piatek & Pryor 1995) can inflate the internal velocity dispersions enough to explain these measurements. Judicious viewing angles and rather eccentric orbits are required for the observed dSph to be actually disrupted remnants, with no dark matter (Klessen & Kroupa 1998). Rather, the stellar kinematics imply the presence of gravitationally-dominant dark matter; the internal kinematics of the Ursa Minor dSph imply a mass-to-light ratio of $\sim$ 80 (Hargreaves *et al.* 1994; Mateo 1998).

The nature of the dark matter is of obvious interest. This material should be cold, to be dominant on these small scales (e.g. Tremaine & Gunn 1978; Gerhard & Spergel 1992). Dark haloes composed of non-baryonic cold dark matter have a density profile that is incompatible with the observed rotation curves of gas-rich dwarf galaxies, and of low surface-brightness disk galaxies (e.g. Moore *et al.* 1999). Could the dark matter be dissipated baryons (e.g. Burkert & Silk 1997)? Stars of mass less than about half that of the Sun have $(M/L)_V > 10(M/L)_\odot$, and are thus viable candidates, provided that the stellar initial mass function (IMF) in these systems is very different from that of the solar neighborhood, or in globular clusters.

Thus we obtained deep images, with the Hubble Space Telescope, of the Ursa Minor dSph, to determine the faint stellar luminosity function. Data were obtained with WFPC2 (V-606 & I-814), STIS (LP optical filter) and NICMOS (H-band), in a field close to the center of the dSph (program GO 7419: PI Wyse, Co-Is Gilmore, Tanvir, Gallagher & Smecker-Hane; due to successive failures of HST the data acquisition phase of this project has only just been completed). The estimation of the contamination by foreground stars and background galaxies required acquisition of similarly-exposed data for an offset field ~ 2 tidal radii away from the Ursa Minor dSph, at similar Galactic coordinates to the UMi field ($\ell = 105^o$, $b = 45^o$); this field shows no evidence for Ursa Minor member stars, itself an interesting result, constraining tidal effects.

2. Faint Luminosity Functions

2.1. GLOBULAR CLUSTERS

We will utilise comparisons between the data for the Ursa Minor dSph and globular clusters of the same metallicity and age (M92 and M15). With age and metallicity constant, a direct comparison between the luminosity functions corresponds to a comparison between the underlying stellar mass functions. This one need not introduce uncertainties due to the transformation between mass and luminosity, allowing robust conclusions as far as *differences* in mass function are concerned. In making this direct comparison between the faint luminosity functions of globular clusters and dwarf galaxy we implicitly assume that the binary-star population is the same in each system. Note that we do detect a marginally significant redward excess in the color distribution across the main sequence of the Ursa Minor dSph (Feltzing *et al.* 1999); this is currently under further investigation.

However, the stellar mass function in globular clusters may be modified by both internal and external effects and thus in not all cases is the present-day mass function a good estimate of the initial mass function. The Piotto, Cool & King (1997) data we use were obtained at intermediate radius within M15 and M92, where the effects of mass segregation are

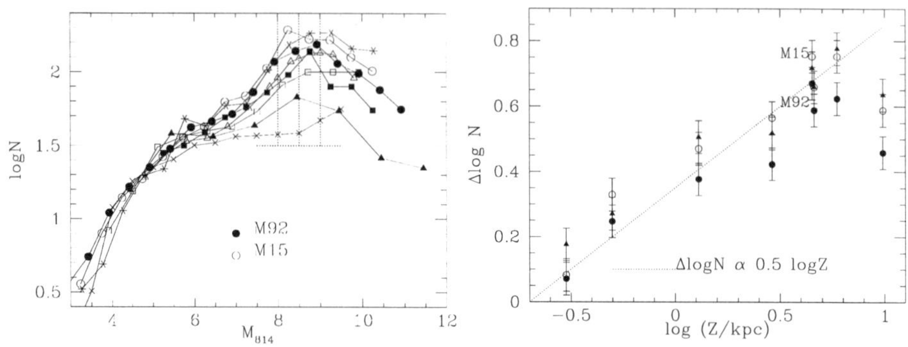

Figure 1. Based on Elson *et al.* (1999). The left panel shows the available luminosity functions for Galactic globular clusters, with the dotted lines at $M_{814} = 8.0, 8.5, 9.0$ illustrating the measurement of the quantity $\Delta \log N(M_{814}) = \log N(M_{814}) - 1.5$. The right panel shows this quantity, which is a measure of the flattening at the faint end, evaluated for $M_{814} = 8.0, 8.5, 9.0$ (filled circles, open circles and triangles), plotted against the present distance of the globular cluster from the Galactic Plane. Both M92 and M15 have steep luminosity functions, arguably the primordial shape.

expected to be small and the local faint luminosity function should be a good estimate of the global faint luminosity function. Further, as discussed by Elson *et al.* (1999; see Figure 1 here) and by Piotto & Zoccali (1999), M92 and M15 have fairly steep main-sequence luminosity functions, while other globular clusters, particularly those for which external dynamical effects such as tidal shocking by the disk or bulge of the Milky Way may be important, have flatter faint luminosity functions (these differences occur fainter than the limits of our present Ursa Minor dSph data). Indeed, analyses of the available estimates of the true space motions of globular clusters suggest that both M92 and M15 are fairly robust against both tidal shocking and evaporation (e.g. Dinescu, Girard & van Altena 1999). Thus we will assume that the present-day faint luminosity functions of M92 and M15, at the locations observed, may be taken as reasonable estimates of the initial luminosity functions, and hence IMF.

The data for all three systems (UMi dSph, M92 and M15) were obtained in the same HST bandpasses, thereby also avoiding the complications of photometric transformations.

2.2. URSA MINOR FAINT LUMINOSITY FUNCTION

The WFPC2 data presented here are discussed in more detail in Feltzing *et al.* (1999). They consist of 8 × 1200s in each of F606W and F814W filters. Standard HST data reduction techniques were followed. The scatter in the zero points and photometric calibrations is $\sim 6\%$, which is small compared to the 0.5 magnitude binning we adopt for the luminosity functions be-

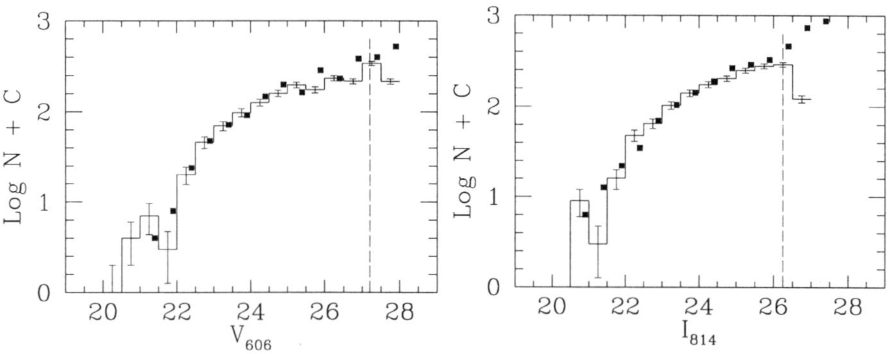

Figure 2. Comparison between the completeness-corrected Ursa Minor luminosity functions (histograms; 50% completeness indicated by the vertical dashed line) in the V-band (left panel) and the I-band (right panel) and the same for M92 (points; from Piotto *et al.* 1997), suitably normalised.

low. Identical procedures were applied to both the UMi and the offset-field datasets. The completeness of the data was determined by adding artificial stars to the original images and then re-processing them. The luminosity functions discussed here include only stars detected in both V and I.

By mid-1999 we had obtained seven 2900s exposures with STIS (LP optical filter) of each of the Ursa Minor field and the offset field. We also obtained a single STIS LP exposure (1200s) of the globular cluster M15 in a field with extant deep WFPC2 V and I data (Piotto *et al.* 1997). The STIS LP magnitudes for the Ursa Minor dSph may be transformed to I-814 using the M15 data. Standard data reduction procedures as above were applied to our data.

The derived, completeness-corrected, Color-Magnitude Diagram-based, WFPC2 V-band and I-band luminosity functions for the Ursa Minor dSph are compared in Figure 2 to those for the globular cluster M92 (courtesy of G. Piotto). We find a remarkable similarity between them, down to our 50% completeness limit, which corresponds to $\sim 0.4M_{\odot}$ (using the Baraffe *et al.* 1997 models).

A direct comparison between our STIS optical luminosity functions for an adjacent field in the Ursa Minor dSph and for the globular cluster M15 is shown in Figure 3 (left panel; only those bins at least 50% complete are shown). Again, there is very good agreement in slopes of luminosity functions, and again our 50% completeness limit for the Ursa Minor data is around $0.4M_{\odot}$.

Note that these results will be improved soon; the data aquisition was completed in November (1999) and the complete dataset is now being analysed.

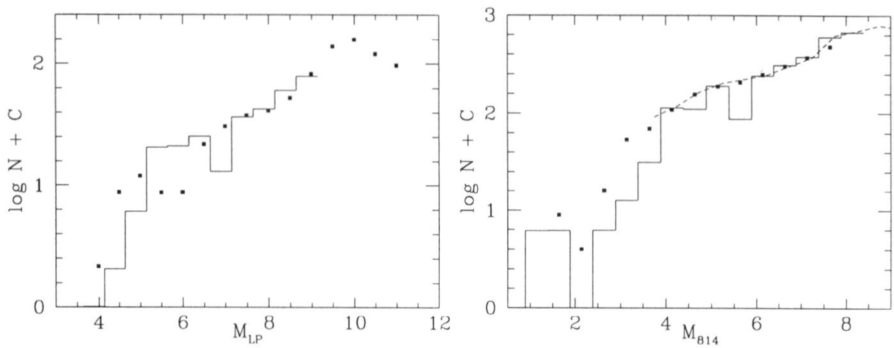

Figure 3. Left panel: comparison between the completeness-corrected STIS LP luminosity functions, suitably normalised, for the Ursa Minor dSph (solid line) and the globular cluster M15 (points). Right panel: comparison between the STIS-based I-band luminosity function of the Ursa Minor dSph (solid line), the observed I-band luminosity function (points), and the luminosity function corresponding to a power-law mass function with slope -1.35 (dashed line).

3. Mass Functions

The luminosity function corresponding to a mass function of slope -1.35 (where the Salpeter slope is -2.35) may be calculated using the Baraffe *et al.* (1997) models. This provides a reasonable fit to the data for M92 and for M15 (Piotto & Zoccali 1999). Further, Zoccali *et al.* (1999) found this mass function to fit their data for the faint stellar luminosity function of the bulge of the Milky Way in Baade's window. Figure 3 (right panel) shows our WFPC2 I-814 luminosity function for Ursa Minor, together with that derived from our transformed STIS LP data and the predictions from the Baraffe *et al.* models. This mass function appears to be an excellent description of the underlying IMF in the Ursa Minor dSph too.

Chemical elemental abundances for stars in dwarf spheroidal galaxies can be used to place constraints on the past high-mass IMF (cf. Wyse 1998), and are within range of the present generation of large telescopes (see Smecker-Hane & McWilliam 1999 for a review of the present, rather sparse, data).

4. Conclusions

The main-sequence stellar luminosity function of the Ursa Minor dSph, and implied IMF, down to $\sim 0.4 M_{\odot}$, is indistinguishable from that of the halo globular clusters M92 and M15, systems with the same old age and low metallicity as the stars in the Ursa Minor dSph. However, the globular clusters show no evidence for dark matter, while the Ursa Minor dSph is apparently very dark-matter-dominated. Thus the low mass stellar IMF for stars that formed at high redshift (reflecting their old age) is invariant in going from a low-surface-brightness, dark-matter-dominated external galaxy, to a globular cluster within the Milky Way. This encompasses a wide range of physical parameters. The effects of metallicity on the IMF are apparently also small (e.g. von Hippel *et al.* 1996).

Thus when calculating chemical evolution models, we are justified in holding the IMF constant. And we are spurred to answer the intriguing questions: Why? How do dSph evolve? What is the dark matter?

5. Acknowledgments

Based on observations with the NASA/ESA Hubble Space Telescope, obtained at STScI, operated by AURA Inc, under NASA contract NAS5-26555. Support for this work was provided by NASA grant number GO-7419 from STScI. We thank J. Gallagher, N. Tanvir & T. Smecker-Hane for discussions.

References

1. Baraffe, I., Chabrier, G., Allard, F. & Hauschildt, P. (1997) *A&A*, **327**, 1054
2. Burkert, A. & Silk, J. (1997) *ApJL*, **488**, L55
3. Dinescu, D.I., Girard, T. & van Altena, W.F. (1999), *AJ*, **117**, 1792
4. Elson, R., Tanvir, N., Gilmore, G., Johnson, R.A. & Beaulieu, S. (1999) in: Y.-H. Chu, N. Suntzeff, J. Hesser & D. Bohlender eds., *New Views of the Magellanic Clouds*, IAU Symposium 190, ASP, San Francisco, in press
5. Feltzing, S., Gilmore, G. & Wyse, R.F.G. (1999) *ApJ*, **516**, L17
6. Gerhard, O. & Spergel, D. (1992) *ApJL*, **389**, L9
7. Gilmore, G. & Howell, D. eds., *The Stellar Initial Mass Function*, ASP conf. series vol 142, ASP, San Francisco
8. Hargreaves, J., Gilmore, G. & Annan, C. (1996) *MNRAS*, **279**, 108
9. Hargreaves, J., Gilmore, G., Irwin, M. & Carter, D. (1994) *MNRAS*, **271**, 693
10. Hernandez, X., Gilmore, G. & Valls-Gabaud, D. (1999) *MNRAS*, in press
11. Klessen, R. & Kroupa, P. (1998) *ApJ*, **498**, 143
12. Mateo, M. (1998) *ARAA*, **36**, 435
13. Mighell, K. & Burke, C. (1999) *Ap*, **118**, 366
14. Moore, B., Quinn, T., Governato, M., Stadel, T. & Lake, G. (1999) *MNRAS*, submitted
15. Olszewski, E. & Aaronson, M. (1995) *AJ*, **90**, 2221
16. Piatek, P. & Pryor, C. (1995) *AJ*, **109**, 1071
17. Piotto, G. & Zoccali, M. (1999) *A&A*, **345**, 845
18. Piotto, G., Cool, A. & King, I.R. (1997) *AJ*, **115**, 1888
19. Smecker-Hane, T.A. & McWilliam, A. (1999) in : I. Hubeny, S. Heap & R. Cornett eds., *Spectro-Photometric Dating of Stars and Galaxies*, ASP, San Francisco, in press
20. Tremaine, S. & Gunn, J.E. (1978), *Phys Rev Lett*, **286**, 27
21. von Hippel, T., Gilmore, G., Jones, D. & Tanvir, N. (1996) *AJ*, **112**, 192
22. Wyse, R.F.G. (1998) in: G. Gilmore G. & D. Howell, eds., *The Stellar Initial Mass Function*, ASP conf. series vol 142, ASP, San Francisco, p. 89
23. Zoccali, M., Cassisi, S., Frogel, J., Gould, A. *et al.* (1999) preprint

Speaker: R. WYSE

I.J. DANZIGER: What is the difference between Ursa Minor Dwarf and Fornax Dwarf, because Fornax does have a range in metallicity?

R. WYSE: We lack good models of the chemical evolution of dSph. Not only does the range of metallicity vary from dSph to dSph, the range of stellar ages varies also, and thus the star formation history.

G. CARRARO: DM should influence the SF process, and hence the IMF. How can a DM system and a DM free system have the same IMF?

R: WYSE: Presumably stars form in over-dense self-gravitating clouds, so that the larger-scale background dark matter potential is not important in determining the details of the star formation process.

M. BELLAZZINI: A. Delgado et al. recently published a ground based wide field CMD of Ursa Minor, showing that a significant blue plume population is present also in this dSph. So it seems that after all Ursa Minor is not a pure old population system.

R. WYSE: There are indeed a few stars bluewards of the dominant turnoff, as could be seen in the colour-magnitude diagram of our HST data, that I showed in my talk. However, these stars cannot be understood as simply younger stars of the same (low) metallicity as the dominant population; they are too blue. Younger isochrones of even lower metallicity could perhaps fit them, but these are not (yet) available. Of course this would make the chemical evolution even more interesting.

B. PAGEL: It is not surprising that there is such a narrow range in metallicity? The broader range found in some other dSphs is much more natural. This seems to me to be yet another manifestation of the G-dwarf problem!

R. WYSE: Indeed we do not have a good understanding of how dSph evolve, or what determines the variations in stellar populations. The most naive models would predict an early type II-supernovae-driven wind, which would truncate the chemical evolution.

EVOLUTION OF HEAVY ELEMENTS IN OMEGA CENTAURI: THE PUZZLE OF S-ENRICHMENT

R. GALLINO
Dipartimento di Fisica Generale, Universita' di Torino, 10125 Torino, ITALY

V.V. SMITH
Department of Physics, University of Texas El Paso, TX 79968 USA

M. BUSSO
Osservatorio Astronomico di Torino, Strada Osservatorio 20, I-10025 Pino Torinese, ITALY

AND

K. CUNHA
Department of Physics, University of Texas El Paso, TX 79968 USA

1. Introduction

Stars in the globular cluster ω Cen span a range of metallicities, from [Fe/H] from -1.8 to -0.8 (Norris et al. 1996, Suntzeff & Kraft 1996). High-resolution spectroscopic analyses of a number of red giant stars (Smith et al. 2000, Norris & Da Costa 1995) has allowed the abundance determination of many elements that can be classified under different groups. The first includes O, Na, Mg, and Al, which show anticorrelations of [Na/Fe] and [Al/Fe] versus [O/Fe] and [Mg/Fe] similar to what is observed in many globular clusters (Kraft 1994; Kraft et al. 1997). A complete interpretation of the behavior of these elements is still lacking. They might indicate deep internal mixing in the stars now evolving along the giant branch by the so-called cool bottom process (CBP) operating in low mass stars (Charbonnel 1994, 1995, Wasserburg, Boothroyd, & Sackmann 1995, Charbonnel, Brown, & Wallerstein 1998). The CPB phenomenon is possibly driven by rotation (Sweigart & Mengel 1979, Langer et al. 1998).

F. Matteucci and F. Giovannelli (eds.), The Evolution of the Milky Way, 387–400.

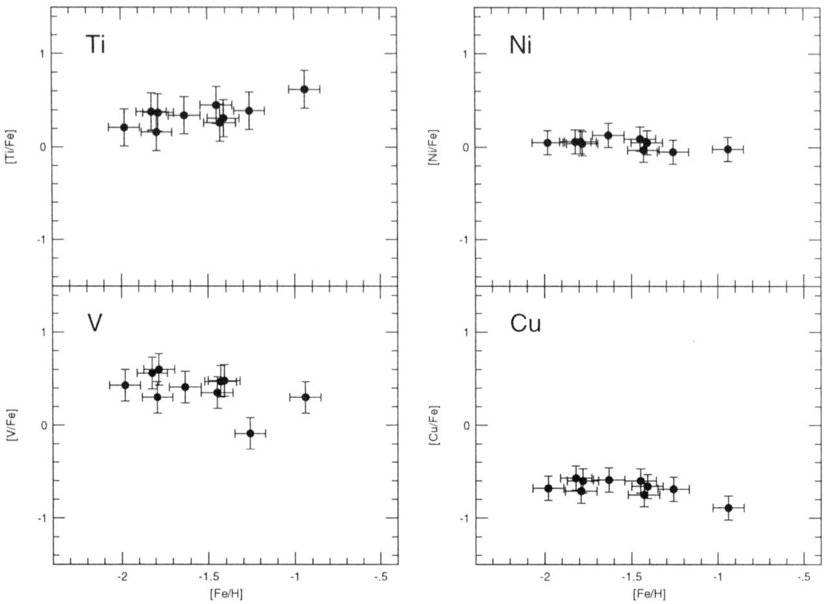

Figure 1. Abundances for [Ti/Fe], [V/Fe], [Ni/Fe], and [Cu/Fe] versus [Fe/H] from Smith et al. (2000). No apparent trends with [Fe/H] are noted: any increase of [Ti/Fe] with increasing [Fe/H] is influenced heavily by the most metal-rich program star. The value of [V/Fe]$\sim +0.4$ is influenced by the neglect of hfs in the V I lines. [Ni/Fe] is ~ 0.0, with very little scatter. The low (and constant) value of [Cu/Fe] ~ -0.6 is real and is different from the field halo stars over this metallicity range: the field stars show a similarly low [Cu/Fe] ratio at [Fe/H] $= -2.0$, but this then increases as [Fe/H] increases.

The second group includes Si, Ca, K, Sc, Ti. Examination of these elements show nearly constant abundance ratios [X/Fe] with [Fe/H], at values that are quite typical of halo stars, with [Si/Fe] $\sim +0.5$, [Ca/Fe] $\sim +0.3$, [Sc/Fe] $\sim +0.1$, [Ti/Fe] $\sim +0.4$. The results for [Ti/Fe] versus [Fe/H] are shown in the first panel of Fig. 1 as an example, whose abundance in ω Cen have certainly been produced by Type II SNe. No decreasing trend of [X/Fe] with [Fe/H] is apparent for the stars of the highest metallicity observed in ω Cen, indicating no apport of Fe to the gas by Type Ia SNe. The third set includes the iron group elements V, Cr, Mn, Fe, Ni, Co. In Fig. 1 the vanadium and nickel abundances observed by Smith et al. (2000) are shown in the second and third panels, respectively. The average [V/Fe] $\sim +0.4$ might be overestimated due to the neglect of hyperfine splitting in the lines used in the analysis.

The abundances of copper presented in Smith et al. (2000) are the first ones obtained in ω Cen stars. Fig. 1 (fourth panel) shows a nearly constant [Cu/Fe] ratio of about -0.6. This constancy differs from the increasing trend in field halo stars (Sneden, Gratton, & Crocker 1991), which goes from [Cu/Fe] $= -0.7$ to -0.3 over the range [Fe/H] $= -2$ to -1. They suggested that Cu is contributed by both Type II SNe and the weak *s*-process occurring in massive stars during hydrostatic burning. However, in the light of the secondary nature of the weak *s*-process, this contribution appears unlikely (Raiteri et al. 1993). In fact, Matteucci et al. (1993) suggested that, besides a moderate contribution by Type II SNe, there must be a substantial contribution to Cu from Type Ia SNe to explain the galactic trend of Cu. Since ω Cen shows clear evidence of enrichment from Type II SNe, and no evidence of enrichment by Type Ia SNe, the constant and low [Cu/Fe] must probably reflect only the contribution from Type II SNe. Note that field stars at [Fe/H] $= -1$ show [Cu/Fe] ≤ -0.2, about $+0.8$ dex higher that in ω Cen. Further studies of Cu are certainly desired.

2. The abundances of the heavy elements

The most interesting group includes the the neutron-rich heavy elements Y, Zr, Ba, La, Ce, Nd, mostly of *s*-process origin (Arlandini et al. 1999), Mo and Rb, of 50% to 20% of *s*-process origin, and Eu, a typical *r*-process element. The abundances of [Rb/Fe], [Y/Fe], [Zr/Fe], and [Mo/Fe] are shown in Fig. 2, and those of [Ba/Fe], [La/Fe], [Ce/Fe], and [Nd/Fe] are shown in Fig. 3. One peculiar aspect of the chemical enrichment in ω Cen is that the *s*-process neutron-rich elements increase to a much larger degree than the other elements (such as Ca, Fe, or Ni). The second peculiar aspect is that the heavy *s* elements in ω Cen stars at [Fe/H] ≥ -1.5 appear up to one order of magnitude higher than in the corresponding field halo stars. The heavy *s*-process elements are believed to be synthesized in asymptotic giant branch (AGB) stars, while most of the other elements studied in ω Cen almost certainly are created in massive, core-collapse supernovae (Type II SN). The large [*s*-process/Fe] abundance ratios found in the most metal-

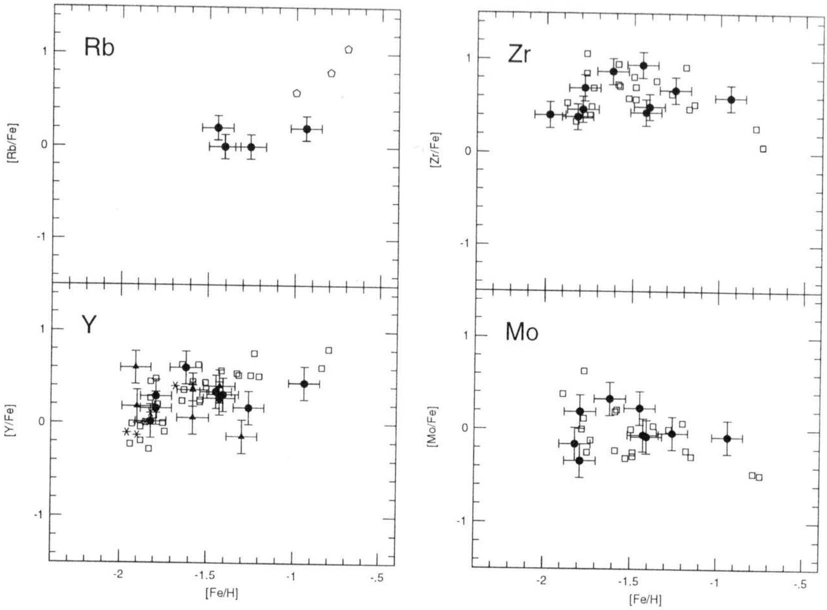

Figure 2. Abundances for [Rb/Fe], [Y/Fe], [Zr/Fe], and [Mo/Fe] versus [Fe/H]. Rubidium is detected in four of the giants from this study (filled circles) and also shown are three giants from Vanture, Wallerstein, & Brown (1994–open pentagons). Results for Y, Zr, and Mo also include results from François, Spite, & Spite (1988–six-pointed stars), Smith, Cunha, & Lambert (1995– filled triangles), and Norris & Da Costa (1995–open squares). The general trends for Y, Zr, and Mo are for small increases ($\sim +0.3$ dex) in [X/Fe] as [Fe/H] increases from -1.85 to -1.7, followed by almost constant values. As noted by Norris & Da Costa, the low [Zr/Fe] ratios found for their two most metal-rich stars are an artifact due to lack of including ZrO in their analysis.

rich ω Cen stars approach the values found in single AGB stars undergoing the third dredge up (TDU) and such extreme s-process abundances are not found in any other stellar population (the barium and CH stars result from

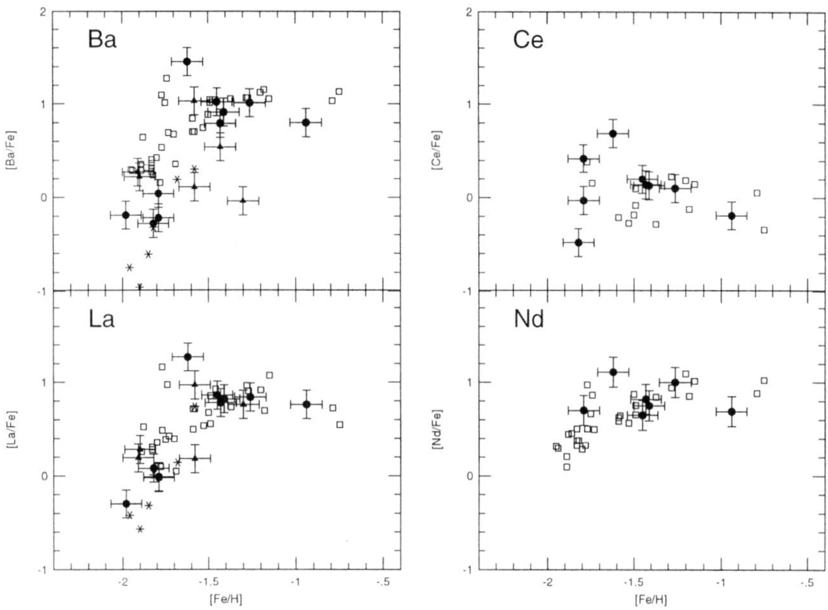

Figure 3. Abundances for [Ba/Fe], [La/Fe], [Ce/Fe], and [Nd/Fe]. Results are from Smith et al. (2000) (filled circles), from from François et al. (1988–six-pointed stars), Smith et al. (1995–filled triangles), and Norris & Da Costa (1995–open squares). The general trends exhibited for these elements are for large increases in [X/Fe] as [Fe/H] increases from -1.85 to -1.7, followed by almost constant values.

direct mass transfer of AGB-polluted material in a binary system). Lloyd Evans (1983) first suggested that the *s*-enriched stars in ω Cen formed directly out of gas that had been substantially enriched by AGB stars by the chemical self-enrichment of this globular cluster.

In general, the heavy *s*-process elements display increasing [X/Fe] ratios, which tend to flatten at $[Fe/H] \geq -1.5$. The increase in the heavier elements

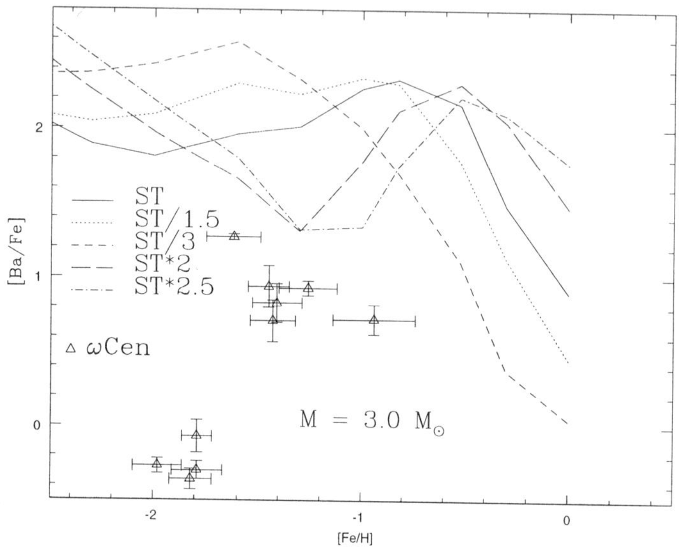

Figure 4. Abundances for [Ba/Fe] versus [Fe/H] from Smith et al. (2000) as compared with predicted abundances for AGB stellar models of $M = 3\ M_{\odot}$ with different choices for the ^{13}C pocket.

Ba, La, Ce, and Nd is larger than found for Y, Zr, or Mo. The increasingly dominant *s*-process enrichment presumably reflects the addition of AGB star ejecta to the chemical evolution of the gas in ω Cen. This *s*-process rich material can be compared to the heavy-element predictions from models of the AGB nucleosynthesis run at the appropriate metallicities. They are computed as post-processing on the results of complete stellar evolutionary models, and have been described by Straniero et al. (1997), Gallino et al. (1998a), Vaglio et al. (1998), Busso et al. (1999). For each mass and metallicity, 5 different models are computed according to a different choice of the abundance profile of ^{13}C, the main neutron source. Neutrons are released in radiative conditions in the interpulse period in a tiny layer placed at the top of the intershell zone (the ^{13}C *pocket*): this pocket drives a significant part of the *s*-process. The ^{13}C pocket formation derives from proton downflows occurring in the He- and C-rich intershell at TDU. We still need to parameterize it because the process is still far from being settled (Iben & Renzini 1982, Hollowell & Iben 1988, Herwig et al. 1997, 1998, Langer et al. 1998). We made the calculations using the ^{13}C-pocket prescriptions indicated as standard (ST) by Gallino et al. (1998a), and scaling it upwards (by factors of 2 and 2.5) and downwards (by factors of 1.5 and 3).

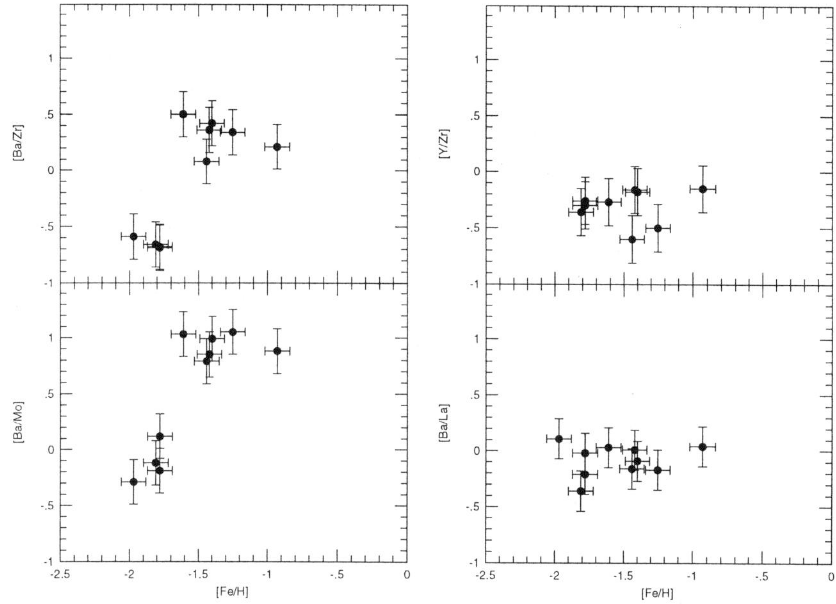

Figure 5. Abundances for [Ba/Zr], [Y/Zr], [Ba/Mo], and [Ba/La] versus [Fe/H].

In Fig. 4 the predicted [Ba/Fe] ratios versus [Fe/H] are compared with the observed ones from Smith et al. (2000). Predictions are for the envelope of AGB stars of 3 $M_\odot$ at the last TDU episode when the remaining envelope mass is 0.8 $M_\odot$ (Gallino et al. 1998a). Cases with different choices of the ^{13}C pocket are plotted. The ST case can be taken as representative of the average value among the different choices (see also Travaglio et al. 1999). The comparison with observations in red giants of ω Cen indicates that a dilution factor of 10 between the extant gas and material from AGB winds if sufficient to explain the inherited Ba enhancement of the newborn stars at [Fe/H] ≥ -1.5.

In Fig. 5 the abundance ratios [Ba/Zr], [Y/Zr], [Ba/Mo], and [Ba/La] versus [Fe/H] are shown. The [Ba/La] trend reflects the classical expectation of the s-process main component for these two elements in the solar system, (81% and and 62% s-process contribution to solar, respectively, see Arlandini et al. 1999). Of major importance for the study of the chemical evolution of the Zr-peak elements are the very large sudden increase of [Ba/Zr] and [Ba/Mo] between [Fe/H] $= -1.8$ and -1.6, by one order of magnitude. This result cannot be understood in the framework of the classical analysis of the main component, where the s-process contributions to solar Zr and Mo are 83% and 50%, respectively (with no room for any r-process contribution to Y!). Instead, it can be interpreted in the framework of the chemical evolution of the Galaxy of the s-process elements (Gallino et al. 1998b, Travaglio et al. 1999). With increasing metallicity the AGB stars first mainly contribute to Pb, then mostly to the Ba-peak elements, eventually to the Zr-peak elements. The [Ba/Zr] predicted abundances versus [Fe/H] for AGB stellar models of $M = 3\ M_\odot$ and different choices for the ^{13}C pocket are shown in Fig. 6. This ratio can be taken as representative of the gas from which new generations of stars come out. The comparison with the observed abundance ratio [Ba/Zr] plotted in the first panel of Fig. 5 shows a general agreement with predicted ratios. Similar agreement exist for the other ratios shown in Fig. 5, including the spectacularly high [Ba/Mo] ~ 1 at [Fe/H] ≥ -1.5.

In contrast to the other heavy elements studied, which are dominated by s-process contributions, Eu, an r-process product, does not increase with Fe. The average [Eu/Fe] abundance is about 0, but with a large scatter, a number of giants showing rather low [Eu/Fe] ratios, as first noted by Smith, Cunha, & Lambert (1995). The behavior of Eu makes clear that the strong increase of other heavy nuclei must be due to the s-process in the chemical evolution of ω Cen, arising from nucleosynthesis and evolution of AGB stars.

Gilroy et al. (1988) first showed that in the very metal-poor stars of the field halo ([Fe/H] ≤ -2.0), the r-process component dominates amongst the heavy, neutron-rich elements, as earlier suggested on theoretical grounds by Truran (1981). More recent studies have largely confirmed this conclusion: e.g. Gratton & Sneden (1994), McWilliam et al. (1995), and Ryan, Norris, & Beers (1996), with a summary by Sneden in the review by Wallerstein et al. (1997). Note that if the low initial [Ba/Fe] $= -0.6$ in ω Cen is due to the r-process, this requires an initial average value [Eu/Fe] $= 0.1$ (from the relative contributions of the r-process to these elements, see Travaglio et al. 1999, Raiteri et al. 1999). This is about 0.4 dex lower than in halo stars of similar metallicity and raises the possibility that the initial gas content of

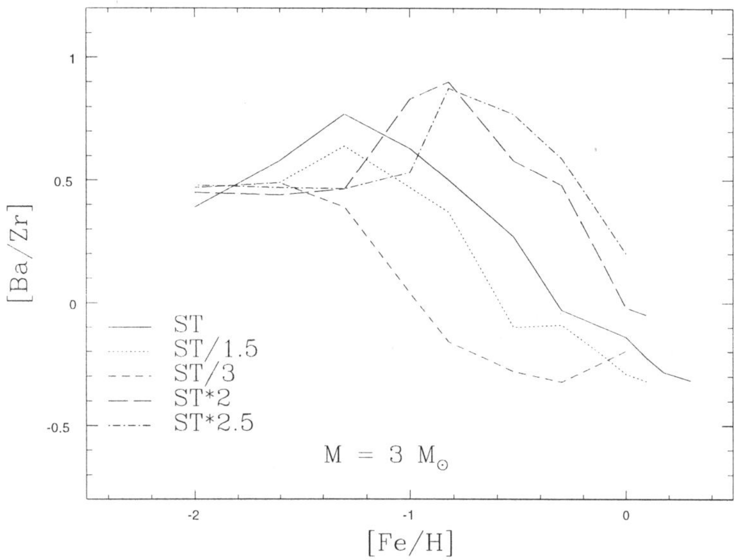

Figure 6. Predicted abundances for [Ba/Zr] versus [Fe/H] for AGB stellar models of $M = 3\ M_\odot$ with different choices for the ^{13}C pocket.

ω Cen was separated from a parcel of the Halo having lower than average r-process contributions (see also Smith et al. 1995). In fact, halo stars do show a large intrinsic scatter in heavy elements (McWilliam et al. 1995).

3. Rubidium abundances and low-mass AGB stars

Additional insights into the nature of the AGB stars responsible for the majority of the s-process enrichment in ω Cen are provided by the Rb abundances. The abundance of Rb, relative to a nearby s-process element such as Sr, Y, or Zr, is controlled by the s-process neutron density, over densities of $\sim 10^7 - 10^{10}$ cm^{-3}. Rubidium in s-enriched stars has been studied previously by Tomkin & Lambert (1983, 1999), Smith & Lambert (1984), Malaney & Lambert (1988), and Lambert et al. (1995). The neutron density sensitivity arises from the s-process branch point at ^{85}Kr (ground state with a half-life of 10.7 yr), which at low neutron densities ($N_n \leq 1 \times 10^8$ cm^{-3}) β-decays to ^{85}Rb, while at a higher neutron density preferentially suffers neutron capture producing ^{86}Kr and then ^{87}Kr, which quickly β-decays to ^{87}Rb. The two Rb isotopes have different neutron capture cross-sections (the one of ^{85}Rb being a factor of ten higher), with ^{85}Rb not significantly fed by the s-process. Thus low neutron densities can result in Rb abundances roughly 10 times less than at high neutron densities.

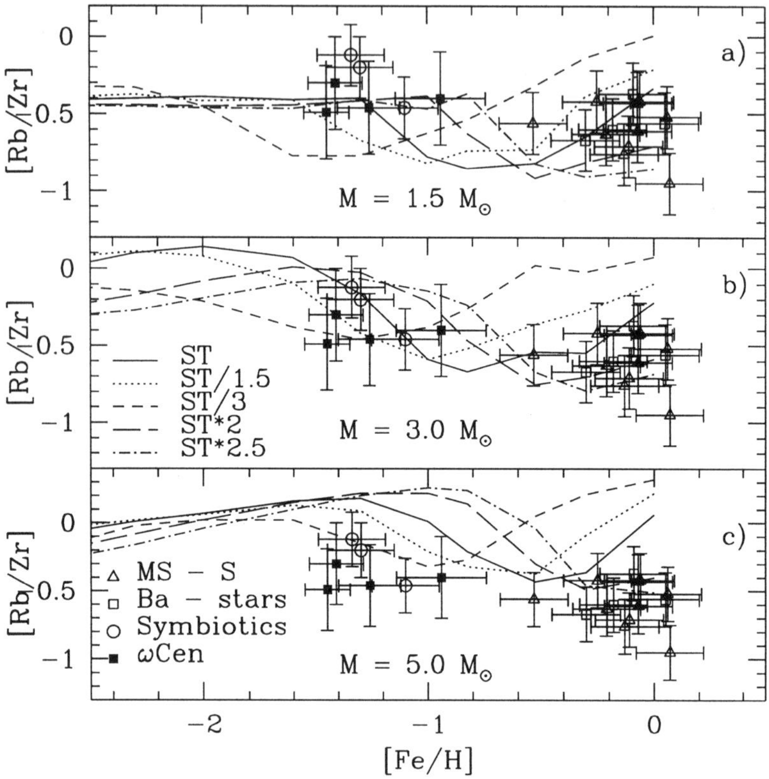

Figure 7. Abundances for [Rb/Zr] versus [Fe/H] as compared with AGB stellar models for $M = 1.5$, 3, and 5 $M_\odot$. Observational data for ω Cen (filled squares) are from Smith et al. (2000), for galactic halo symbiotic stars (open circles) from Smith et al. (1996, 1997), Pereira et (1998), for galactic disk MS, S and Ba stars (open triangles) from Lambert et al. (1995), Busso et al. (1995) and references therein.

The *s*-process neutron density is controlled by both the neutron source and the mass of the parent AGB star. Two possible neutron sources exist for the *s*-process: ^{22}Ne(α,n)^{25}Mg or ^{13}C(α,n)^{16}O. As discussed by Clayton (1988), these two sources will produce quite different signatures of neutron density and neutron exposure at low metallicities. The primary ^{13}C source produces typically lower neutron densities, which however increase with the neutron exposure as the metallicity decreases (up to a maximum of 6×10^8 n/cm^3 at the lowest metallicities of ω Cen stars and for the most efficient *s*-process models adopted). The ^{22}Ne does not show the same behavior, being mainly a secondary product of CNO nuclei, whose abundance declines with declining metallicity. Also, the ^{13}C source is the dominant neutron source for lower-mass AGB stars ($M \leq 3$ $M_\odot$), while ^{22}Ne is predicted to be the major neutron source for the highest mass AGB stars (Iben & Renzini 1983, Busso et al. 1999).

A comparison of the observed [Rb/Zr] ratios for the ω Cen stars with AGB models reveals that, in general, the various 5 $M_{\odot}$ models produce [Rb/Zr] ratios that are too large to fit the observed stellar ratios. The 3 $M_{\odot}$ models provide better overlap with the observations in some cases, but it is the 1.5 $M_{\odot}$ models that yield the best overall fits. A combination of 1.5 $M_{\odot}$ and 3.0 $M_{\odot}$ models provides complete overlap with the observations. This is in good agreement with the overall heavy-element abundance distributions discussed in §3. The heavy-element abundances in the ω Cen stars point to low-mass AGB stars as dominating the enrichment of the s-process in this cluster. The observed spread may correspond, as different model tracks suggest, to an intrinsic spread in s-process efficiencies. Note the lifetimes of these lower-mass stars are $1 - 3$ Gyr, pointing to a protracted period of star formation, evolution, and chemical enrichment in ω Cen.

4. Conclusions

According to the results presented and discussed in the previous sections, the chemical evolution that describes the abundance patterns observed within ω Cen seems understandable as a combination of Type II SNe plus AGB enrichment, occuring over a time period well in excess of 10^9 years.

Enrichment in elements through the iron-peak is presumed to derive from mixing of Type II SN ejecta with the cluster's residual gas from which the stars with $[Fe/H] > -1.8$ formed. The α-rich elements show enhancement proportional to Fe, so that their $[X_i/Fe]$ are constant. This is predicted in galactic chemical evolution models where Type II SNe control the nucleosynthesis (with little, or no contribution from Type Ia SNe).

For elements heavier than iron, there is evolution of the relative abundances. This is most noticeable for the heavy elements Ba through to Nd, but is seen also for the lighter counterparts Rb, Y and Zr. In sharp contrast to Ba, the mean Eu abundance (relative to Fe) may even decline rather than increase with increasing [Fe/H]. At the low-metallicity end ($[Fe/H] = -1.8$) the ω Cen heavy-element abundance distribution is characterized by an r process, while the contrasting behavior of Ba and Eu means that the heavy elements become dominated by the s process as [Fe/H] increases.

Quantitative interpretation of the observed heavy-element s-process abundances is possible using predicted yields for s-process products from AGB models. Heavy element abundances at $[Fe/H] = -1.0$ are best explained as the result of AGB nucleosynthesis for masses of $1.5 - 3.0$ $M_{\odot}$. Rubidium abundances provide additional clues to the stellar masses of the contributing AGB stars and confirm the conclusion from the heavy-element distributions that low-mass stars were significant contributors to the s process in ω Cen.

Why ω Cen displays such a large s-process component is possibly related to the relatively high mass of the cluster. The AGB ejecta, because of their low velocity winds, are more efficiently retained in the cluster relative to the much faster moving Type II supernova ejecta. The main-sequence lifetimes of low-mass AGB stars require that chemical enrichment occurred over timescales of $2 - 3 \times 10^9$ yr, if ω Cen evolved in a simple way in which the more metal-rich stars formed from gas contaminated by debris from earlier generations of stars. This timescale agrees with recent results from Wallerstein & Hughes (1999) and Hilker & Richtler (1999), who find age spreads of $3 - 4$ Gyr based upon Strömgren photometry and main-sequence isochrones. These timescales constrain the possible origins and history of ω Cen.

Since ω Cen is the most massive of globular clusters, it may not be surprising that the earliest supernovae were unable to expel all the residual gas. Other lower mass globular clusters do not show such strong signatures of chemical evolution, except for perhaps some primordial variations in the light elements C, N, O, Na, or Al, and in the r-process element Eu in the cluster M15 (Sneden et al. 1997). As these lower-mass clusters certainly were born with higher-mass stars which ejected processed material, it is apparent that such processed ejecta were not retained with any great efficiency. The larger mass cluster ω Cen, on the other hand, was apparently able to efficiently retain AGB ejecta, which was incorporated into a later generation (or generations) of stars. However, the large s-process to Fe enhancements would indicate that ω Cen was not massive enough to retain large fractions of ejecta from Type II SNe, resulting in a larger retention of AGB ejecta (s-process elements) than Type II SN ejecta (such as Fe) relative to other globular-cluster populations studied to date.

References

1. Arlandini, C., Käppeler, F., Wisshak, K., Gallino, R., Lugaro, M., Busso, M., Straniero, O. 1999, ApJ, 525, 886
2. Busso, M., Gallino, R., & Wasserburg, G.J. 1999, ARA&A, 37, 239
3. Charbonnel, C. 1994, A&A, 283, 811
4. Charbonnel, C. 1995, ApJ, 453, L51
5. Charbonnel, C., Brown, J.A., & Wallerstein, G., 1998, A&A, 332, 204
6. Clayton, D.D. 1988, MNRAS, 234, 1
7. François, P., Spite, M., & Spite, F. 1988, A&A, 191, 267
8. Gallino, R., Arlandini, C., Busso, M., Lugaro, M., Travaglio, C., Straniero, O., Chieffi, A., & Limongi, M. 1998a, ApJ, 497, 388
9. Gallino, R., Arlandini, C., Busso, M., Lugaro, M., Travaglio, C., Straniero, O., Chieffi, A., & Limongi, M. 1998b, in Nuclei in the Cosmos V, ed. N. Prantzos & S. Harissopulos, (Paris: Editions Frontières), 216
10. Gilroy, K., Sneden, C., Pilachowski, C.A., Cowan, J.J. 1988, ApJ, 327, 298

11. Gratton, R.G., & Sneden, C. 1987, A&A, 178, 179
12. Gratton, R.G., Sneden, C. 1994, A&A, 287, 927
13. Herwig, F., Blöcker, T., Schönberner, D., & El Eid, M. 1997, A&A, 324, L81
14. Herwig, F., Schönberner, D. & Blöcker, T. 1998, A&A, 340, L43
15. Hilker, M., & Richtler, T. 1999, to appear in The Galactic Halo: From Globular Clusters to Field Stars (Proceedings of the 35th Liège International Astrophysical Colloquium), eds. A. Noels et al. (preprint astro-ph/9910370)
16. Hollowell, D., & Iben, I. Jr. 1988, ApJ, 333, L25
17. Iben, I. Jr., & Renzini, A. 1982, ApJ, 263, L23
18. Iben, I. Jr., & Renzini, A. 1983, ARA&A, 21, 271
19. Kraft, R.P. 1994, PASP, 106, 553
20. Kraft, R.P., Sneden, C., Smith, G.H., Shetrone, M.D., Langer, G.E., Pilachowski, C.A. 1997, AJ, 113, 279
21. Lambert, D.L., Smith, V.V., Busso, M., Gallino, R., & Straniero, O. 1995, ApJ, 450, 302
22. Langer, N., Heger, A., Woosley, S.E., & Herwig, F. 1998, in Nuclei in the Cosmos V, ed. N. Prantzos & S. Harissopulos, (Paris: Editions Frontières), 129
23. Lloyd Evans, T. 1983, MNRAS, 204, 975
24. Malaney, R.A., & Lambert, D.L. 1988, MNRAS, 235, 695
25. Matteucci, F., Raiteri, C.M., Busso, M., Gallino, R., & Gratton, R.G. 1993, A&A, 272, 421
26. McWilliam, A., Preston, G.W., Sneden, C., & Searle, L. 1995, AJ, 109, 2757
27. Norris, J.E., & Da Costa, G.S. 1995, ApJ, 447, 680
28. Norris, J.E., Freeman, K.C., & Mighell, K.J. 1996, ApJ, 462, 241
29. Raiteri, C.M., Gallino, R., Busso, M., Neuberger, D., & Käppeler, F. 1993, ApJ, 419, 207
30. Raiteri, C.M., Villata, M., Gallino, R., Busso, M., & Cravanzola, A. 1999, ApJ, 518, L91
31. Ryan, S.G., Norris, J.E., & Beers, T.C. 1996, ApJ, 471, 254
32. Smith, V.V., Cunha, K., & Lambert, D.L. 1995, AJ, 110, 2827
33. Smith, V.V., & Lambert, D.L. 1984, PASP, 96, 226
34. Smith, V.V., Suntzeff, N.B., Cunha, K., Gallino, R., Busso, M., Lambert, D.L. & Straniero, O. 2000, AJ, in press
35. Sneden, C., Gratton, R.G., & Crocker, J.H. 1991, A&A, 246, 354
36. Sneden, C., Kraft, R.P., Shetrone, M.D., Smith, G.H., Langer, G.E., Prosser, C.F. 1997, AJ, 114, 1964
37. Straniero, O., Chieffi, A., Limongi, M., Busso, M., Gallino, R., & Arlandini, C. 1997, ApJ, 478, 332
38. Suntzeff, N.B., & Kraft, R.P. 1996, AJ, 111, 1913
39. Sweigart, A.V., & Mengel, J.G. 1979, ApJ, 229, 624
40. Tomkin, J., & Lambert, D.L. 1983, ApJ, 273, 722
41. Tomkin, J., & Lambert, D.L. 1999, ApJ, 523, 234
42. Travaglio, C., Galli, D., Gallino, R., Busso, M., Ferrini, F., & Straniero, O. 1999, ApJ, 521, 691
43. Truran, J.W., 1981, A&A, 97, 391
44. Vaglio, P., Gallino, R., Busso, M., Travaglio, C., Straniero, O., Chieffi, A., Limongi, M., Arlandini, C., & Lugaro, M. 1998, in Nuclei in the Cosmos V, ed. N. Prantzos & S. Harissopulos, (Paris: Editions Frontières), 223
45. Vanture, A.D., Wallerstein, G., & Brown, J.A. 1994, PASP, 106, 835
46. Wallerstein, G., et al. 1997, Rev. Mod. Phys., 69, 995
47. Wallerstein, G., & Hughes, J. 1999, in The Galactic Halo: From Globular Clusters to Field Stars (Proceedings of the 35th Liège International Astrophysical Colloquium), eds. A. Noels et al., in press
48. Wasserburg, G.J., Boothroyd, A., Sackmann, J.-I. 1995, ApJ, 447, L37

Speaker: R. GALLINO

F. MATTEUCCI: It seems to me that the [α-/Fe] vs. [Fe/H] pattern found in ω Cen resembles to the pattern found in bulge stars. As a consequence of this, one could infer that ω Cen suffered a very strong star formation rate like the bulge.

R. GALLINO: This is certainly an important ingredient, to which one has to add the peculiar s-process enhancement. A rapid consumption of the initial gas allows AGB winds to be more easily polluting the central region (and being easily retained because of low velocity winds). Maybe not all SNII ejecta are retained, only 10-20% of it, allowing a typical dilution factor 1/10 of AGB gas winds/cluster gas. A time scale of $\sim$ 1 Gyr is required by low mass AGB evolutionary time scales. The first SNIa are maybe blowing off all the remaining gas at the time. Further AGB winds are possibly of too low amount to allow further stellar condensations (and maybe stripped during the transit of the cluster within the galactic disk).

V-STELLAR EVOLUTION AND NUCLEOSYNTHESIS

An open air session

STELLAR EVOLUTION WITH ROTATION

A. MAEDER
Geneva Observatory
CH-1290 SAUVERNY, Switzerland

1. Introduction

The existing grids of star models without rotation allow us to already explain a number of properties of massive and WR stars (cf. Maeder and Conti,1994). In particular, the sequences of star clusters are well reproduced, as well as the relations of filiation between the various kinds of massive stars: O–type stars, blue– and red–supergiants, Wolf–Rayet (WR) stars. The observed abundances of CNO elements in WN stars and of He, C, O and Ne in WC stars are nicely accounted for, as well as the luminosities of WR stars, the number ratios WR/O and WC/WN in galaxies of various metallicities. The consequences of mass loss for the chemical yields are quite important and allow us in particular to understand the evolution of the CNO elements during the galactic history.

Despite these many successes, a number of serious discrepancies between current models and observations have been noticed over recent years. They concern particularly the helium and nitrogen abundances in massive O– and B–type stars and in giants and supergiants. The presence of of a large number of red supergiants at low metallicities was also a feature difficult to account for (Langer and Maeder, 1995). The observations show that the role of rotation has been largely overlooked. Indeed, all the model outputs (tracks in the HR diagram, lifetimes, actual masses, surface abundances, nucleosynthetic yields, supernova precursors, etc...) are greatly influenced by rotation. It turns out that stellar evolution is basically a function of mass M, metallicity Z and angular velocity Ω.

2. Basic input physics for models with rotation

Let us mention the main physical processes now included in our models with rotation.

F. Matteucci and F. Giovannelli (eds.), The Evolution of the Milky Way, 403–416.

- *"Shellular" rotation:* in a rotating star, the turbulence is highly anisotropic with a very strong horizontal component, which limits the degree of horizontal differential rotation. Thus rotation is essentially constant on isobars (Zahn, 1992). A law of the form $\Omega = \Omega(\mathrm{r})$ can be used, at least at not too fast rotation.
- *Basic equations of structure:* the classical scheme of equations for uniformly rotating stars, or for stars with a rotation constant on cylinders, has been written by Kippenhahn and Thomas (1970). However, this scheme does not apply for shellular rotation and the equations have to be reformulated (Meynet and Maeder 1997).
- *Shears:* they are the main instability sources in rotating stars. We take as an hypothesis that the excess of energy in the shear is injected in the spectrum of turbulence at the thermal timescale. This is leading to a new diffusion coefficient (Maeder 1997), identical to that by Zahn (1992) in the case of very high rotation and to that for semiconvection by Langer et al. (1983) in the case of no rotation.
- *Meridional circulation:* it is accounted for with the recent improvements by Zahn (1992) and Maeder and Zahn (1998). This provides a new expression which properly includes the reduction of the velocity of meridional circulation due to the μ–gradients. This is a major difference with previous treatments, in which it was considered that μ–gradients oppose their effects to that of the meridional circulation. Indeed, μ–gradients do not exist by themselves, they only appear when the matter starts rising due to to the meridional circulation and thus they just act as to reduce the effects of the circulation.
- *Surface effects:* the global increases of the mass loss rates due to rotation are accounted for (cf. Friend and Abbott 1986). Further developments will allow us to account for the anisotropic mass loss over the stellar surface, which selectively removes the angular momentum (Maeder 1999a).

Inside a rotating star the angular momentum is transported by convection, turbulent diffusion and meridional circulation. For shellular rotation, the equation of transport of angular momentum in the vertical direction is in lagrangian coordinates (Zahn 1992).

$$\rho \frac{d}{dt}\left(r^2\Omega\right)_{M_r} = \frac{1}{5r^2}\frac{\partial}{\partial r}\left(\rho r^4 \Omega U(r)\right) + \frac{1}{r^2}\frac{\partial}{\partial r}\left(\rho D r^4 \frac{\partial \Omega}{\partial r}\right) . \tag{1}$$

$\Omega(r)$ is the mean angular velocity at level r, $U(r)$ the vertical component of the meridional circulation velocity and D the diffusion coefficient due to the sum of the various turbulent diffusion processes, in particular the shears. If both $U(r)$ and D are zero, we just have the local conservation of the angular momentum $r^2\Omega = const.$ for a fluid element in case of contraction or expansion.

The solution of eqn. 2 gives the *non–stationary solution* of the problem. The rotation law is not arbitrary chosen, but is allowed to evolve with time as a result of transport by meridional circulation, diffusion processes and contraction or expansion. In turn, the differential rotation built–up by these processes generates some turbulence and meridional circulation, which are themselves functions of the rotation law. This coupling provides a feedback, and the self–consistent solution for the evolution of $\Omega(r)$ has to be found. The characteristic times associated, to both the processes of meridional circulation and diffusion are

$$t_{\text{circ}} \simeq \frac{R}{U} \quad , \qquad t_{\text{diff}} \simeq \frac{R^2}{D} \, . \tag{2}$$

One has to be careful that the characteristic time for the mixing of the elements is different (see below). The characteristic time t_{circ} is the time necessary for the meridional circulation to establish an equilibrium rotation law and t_{diff} is the same for diffusion.

The transport of angular momentum by circulation is often treated as a diffusion process (Endal and Sofia 1978; Pinsonneault et al 1989; Fliegner and Langer 1995; Chaboyer et al 1995ab; Heger et al 1999). We see from eqn. 1 that the term with U (advection) is functionally not the same as the term with D (diffusion). Physically advection and diffusion are quite different: diffusion brings a quantity from where there is a lot to other places where there is little. This is not necessarily the case for advection. A circulation with a positive value of $U(r)$, i.e. rising along the polar axis and descending at the equator, is as a matter of fact making an *inward* transport of angular momentum. If this process were treated as a diffusive function of $\frac{\partial\Omega}{\partial r}$, even the sign of the effect may be wrong.

The transport of chemical elements is also governed by a diffusion–advection equation like eqn. 1 (Endal and Sofia 1978; Schatzman et al 1981; Langer 1991, 1992; Heger et al 1999). If the horizontal component of the turbulent diffusion is large, the vertical advection of the elements can be treated as a simple diffusion (Chaboyer and Zahn 1992) with a diffusion coefficient D_{eff},

$$D_{\text{eff}} = \frac{\mid rU(r) \mid^2}{30 D_h} \, , \tag{3}$$

where D_h is the coefficient of horizontal turbulence, for which the estimate is $D_h = |rU(r)|$ (Zahn 1992). Eqn. 3 expresses that the vertical advection of chemical elements is severely inhibited by the strong horizontal turbulence characterized by D_h. Thus, the change of the mass fraction X_i of the chemical species i is simply

$$\left(\frac{dX_i}{dt}\right)_{M_r} = \left(\frac{\partial}{\partial M_r}\right)_t \left[(4\pi r^2\rho)^2 D_{\text{mix}} \left(\frac{\partial X_i}{\partial M_r}\right)_t\right] + \left(\frac{dX_i}{dt}\right)_{\text{nucl}} . \quad (4)$$

The second term on the right accounts for composition changes due to nuclear reactions. The coefficient D_{mix} is the sum $D_{\text{mix}} = D + D_{\text{eff}}$, where D is the term appearing in expression 1 and D_{eff} accounts for the combined effect of advection and horizontal turbulence. The characteristic time for the mixing of chemical elements is therefore

$$t_{\text{mix}} \simeq \frac{R^2}{D_{\text{mix}}} . \quad (5)$$

The characteristic time for chemical mixing is not t_{circ} given by eqn. 3, as has been generally considered (Schwarzschild 1958). This makes the mixing of the chemical elements much slower, since D_{eff} is very much reduced. In this context, we recall that several authors have reduced by arbitrary factors, up to 30 or 100, the effect of the transport of chemicals in order to better fit the observed surface compositions (Pinsonneault et al 1989; Chaboyer et al 1995ab; Heger et al 1999). This is no longer necessary with the more appropriate expressions given above.

3. The evolution of the rotational velocities

Detailed calculations (Meynet and Maeder 2000) show that the Ω–profile in a rotating star rapidly reaches an equilibrium (Denissenkov et al 1999), in a time t_{circ} inferior to $3 \cdot 10^4$ yr at 20 $M_\odot$, as a result of a compensation between the outwards transport of angular momentum by shears and the inwards transport by meridional circulation. Then the evolution of the internal $\Omega(r)$ is slower, keeping about the same profile during the MS phase. From the end of the MS phase onwards, the central contraction leads to a steep increase of the angular velocity in the central regions. At the end of the He–burning phase, the ratio of the central to the surface Ω may reach a factor 10^5, so that we may wonder whether new instabilities set in.

The main results on the evolution of the surface rotation velocity v_{rot} are illustrated in Fig.1. Firstly, we notice that without mass loss, the rotation velocities would easily reach the break–up limit during the MS phase. With a small mass loss as in the cases of the model with 12 $M_\odot$, or that with 20 $M_\odot$ at Z=0.004, the surface velocity keeps about constant during the MS phase. For more massive stars with higher mass loss rates, v_{rot} decreases very much during most of the MS phase. This occurs because mass loss,

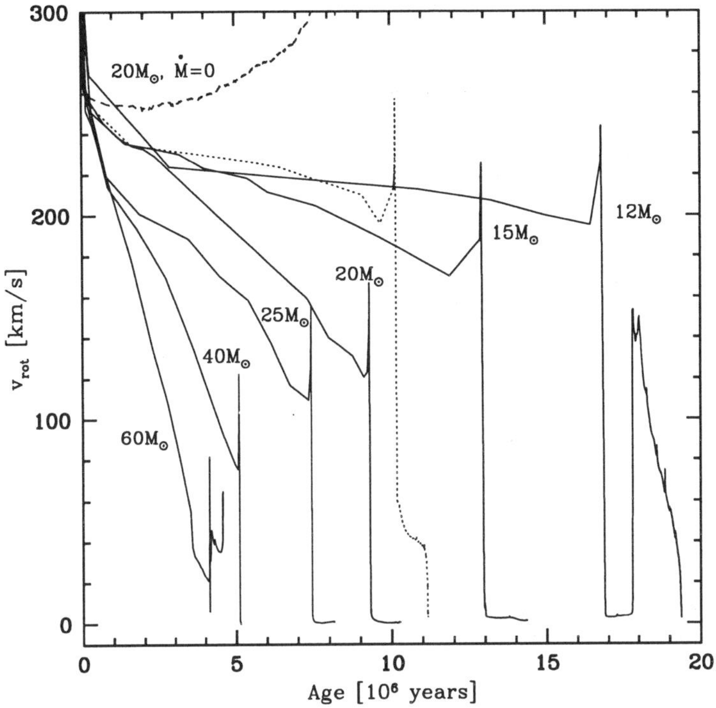

Figure 1. Examples of the evolution of the surface equatorial velocity as a function of time for stars of various initial stellar masses (Meynet and Maeder 2000). The continuous lines refer to solar metallicity models, the dotted line corresponds to a 20 $M_\odot$ star with $Z = 0.004$. The dashed line corresponds to a 20 $M_\odot$ star without mass loss.

even if isotropic, removes more angular momentum than the average in the star. This decrease of $v_{\rm rot}$ tends to blur the possible relation between fast rotation and He– or N– excesses.

It is interesting to examine where break–up velocities can be reached. There are two interesting cases. 1) Fast rotators below 12 $M_\odot$ tend to reach the critical velocity near the end of the Main-Sequence. This can be put in relation with the occurence of Be stars. Also, more stars should reach the break–up velocity when the metallicity is lower, since then the mass loss rates are also lower. This may explain the recent finding (see Fig. 2) of a higher fraction of Be stars at lower metallicities (Maeder et al 1999). 2) The second situation where critical velocities are reached is typically for very massive stars like the 60 $M_\odot$ star at the end of the MS. There the star starts making some bluewards horizontal excursions in the HR diagram. It is tempting to identify this situation with the LBV and the problem of the Ω–limit (Langer 1997). However, the finding by Langer was for stars

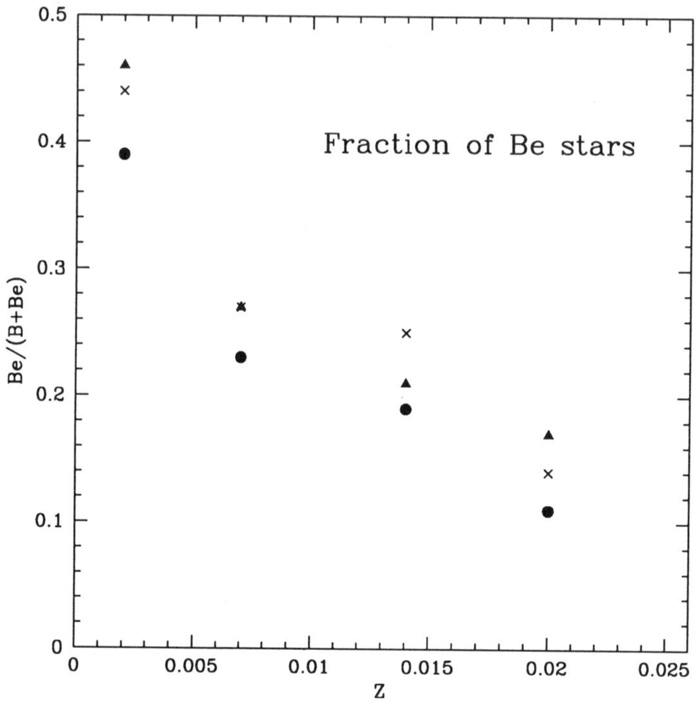

Figure 2. Relation between the number ratio Be/(B+Be) and the local metallicity for groups of clusters in the SMC, LMC, outer of the Milky Way and in the interior of the Milky Way. A sample of 19 clusters with a total of 979 stars in the range of types O9 to B3 was analysed (Maeder et al 1999). To test the validity of the results, the number counts were made in different magnitude intervals, the dots refer to counts made in the magnitude interval M_V = -5, -1.4, the crosses to the interval -5, -2 and the triangles to the interval -4, -2.

with solid body rotation, which usually keep $v_{\rm rot}$ constant during the MS evolution. Here, we show that the break–up rotation also occur for stars with a more realistic internal rotation law.

4. The HR diagram, lifetimes and age determinations

The models show that the convective cores are slightly increased by rotation. The height of the μ–discontinuity at the edge of the core is reduced and there is a mild composition gradient built up from the core to the surface, which may then be slightly enriched in helium and nitrogen. For low or moderate rotation, the convective core shrinks as usual during MS evolution, while for high masses ($M \geq 40$ M$_{\odot}$) and large initial rotations ($\frac{\Omega}{\Omega_{\rm crit}} \geq 0.5$), the convective core grows in mass during evolution. If this occurs, the star will bifurcate to the blue (cf. Maeder 1987; Langer 1992) towards the classical tracks of homogenous evolution and likely produce W–R stars.

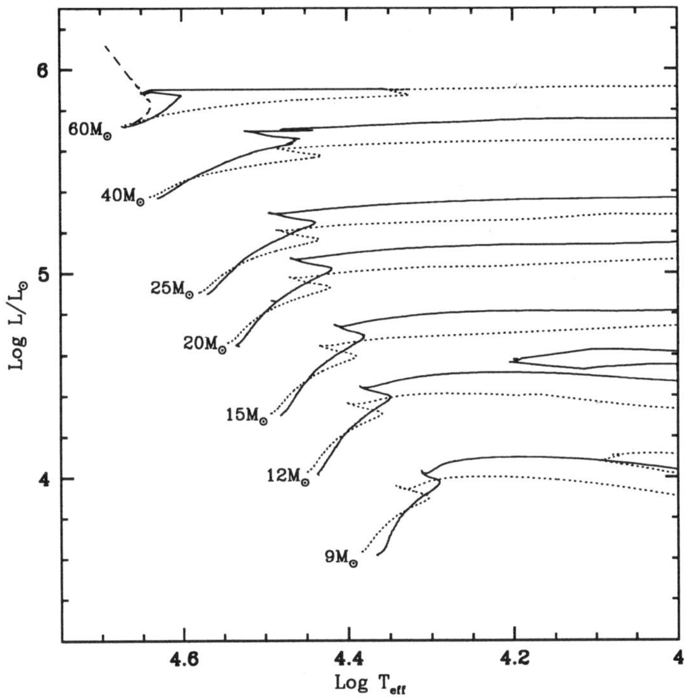

Figure 3. Part of the evolutionary tracks for non-rotating (dotted lines) and rotating (continuous lines) models with solar metallicity (Meynet and Maeder 2000). The rotating models have an initial velocity $v_{\rm rot}$ of 300 km/s. The dashed track corresponds to a very fast rotating star ($v_{\rm rot}$ ~ 400 km/s) of 60 $M_\odot$, which follows a nearly homogeneous evolution.

Fig. 3 shows the overall HR diagram for rotating and non–rotating stars during the MS phase and slightly beyond. The atmospheric distortions produce a small shift to the red in the HR diagram by several 0.01 in (B-V), with on the average only a small change of luminosity. During MS evolution, the luminosity of the rotating stars grows faster and the tracks extend farther away from the ZAMS, as in the case of a moderate overshooting (Maeder 1987; Langer 1992; Talon et al 1997). This effect introduces a significant scatter in the mass–luminosity relation (Meynet 1998), in the sense that fast rotators are overluminous with respect to their actual masses. This may explain some of the discrepancies between the evolutionary masses and the direct mass estimates in some binaries (Penny et al

1999). In this context, we recall that for a decade a severe mass discrepancy beween spectroscopic and evolutionary masses was claimed by some authors (Groenewegen et al 1989; Kudritzki et al 1992; Herrero et al 1992). Most of the problem has nowadays collapsed and was shown to result from insufficiencies in the atmospheric models.

The lifetimes t_H in the H–burning phase grow only moderately because more nuclear fuel is available, but at the same time the luminosity is larger. The net result is an increase by about 20 to 30 % for an initial velocity of 200 km/s (Talon et al 1997; Meynet and Maeder 2000). This influences the isochrones and the age determinations. As an example, for $v_{rot} = 200$ km/s, the isochrone of log age = 7.0 is the same as that of log age = 6.90 without rotation. Thus, accounting for rotation could lead to ages larger by about 25 % for O– and early B–type stars. If a bluewards track occurs, the larger core and mixing lead to much longer lifetimes in the H–burning phase. In this case, the fitting of time–lines becomes hazardous.

5. Rotation and surface abundances, effects on the chemical yields

There are numbers of evidences of He– and N– excesses in massive stars at all stages of their evolution. For MS O–type stars, all fast rotators show some He–excesses (Herrero et al 1998; Lyubimkov 1996). For the early B–type stars, there seems to be an increase of He– and N–abundances as the star is moving away from the zero–age sequence (Lyubimkov 1996). An increase of the N–abundance by a factor 2 – 3 for an O–star with the average v_{rot} of 200 km/s is the order of magnitude typically considered as a constraint for recent stellar models (Heger et al 1999). There is a boron depletion (Venn et al 1996) by a bout a factor of 4 in stars which also show N–excess. This is supporting the idea that rotational mixing occurs throughout the star (Fliegner et al 1996).

For OB–supergiants, He– and N–excesses seem to be the rule (Walborn 1988). As shown by McEarlan et al (1999) the determination of the helium abundance also depends on the adopted value for microturbulence. Villamariz and Herrero (1999) and Herrero et al (1999) point out however that the helium discrepancy is only reduced, but not solved when microturbulence is accounted for. Values of [N/H] amounting to 0.6 dex have been found for B–supergiants around 20 $M_\odot$ (Venn 1995ab). The values of [N/H] for galactic A–type supergiants around 12 $M_\odot$ lie between 0 and 0.4 dex (Venn 1995ab, 1999). All these values are globally consistent with the above results of Lyubimkov (1996). A striking point is that for the A–type supergiants in the SMC, the N/H excesses are much larger spanning a range [N/H] = 0 to 1.2 dex (Venn et al 1998; Venn 1999). Barbuy et al (1996) found both N–enrichments and normal compositions among the slow rotating F–G supergiants. Na–excesses have been found in yellow supergiants (Boyarchuk and Lyubimkov 1983). Two reactions could be at work: $^{22}Ne(p,\gamma)^{23}Na$ and $^{20}Ne(p,\gamma)^{21}Na$. The interpretation based on the first one seems preferable (Denissenkov 1994), because the second one may have a too low rate.

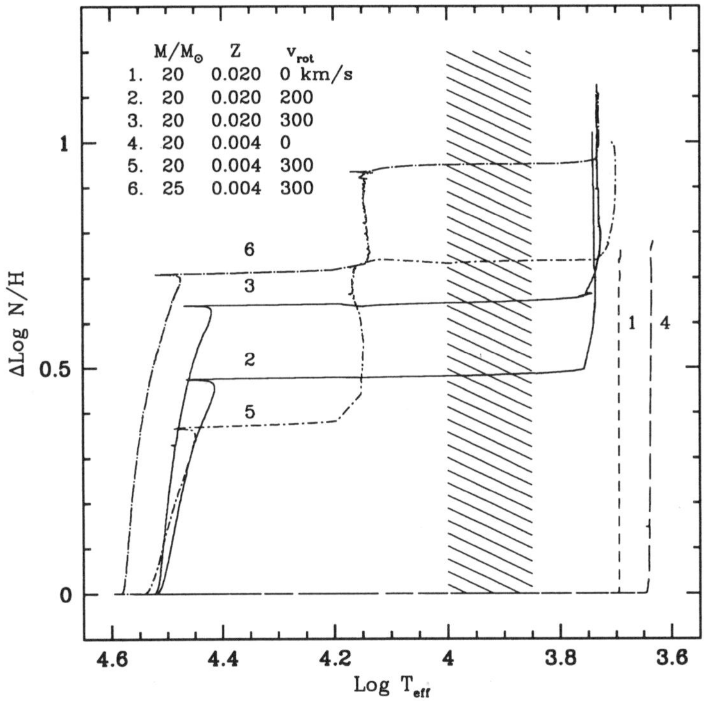

Figure 4. Evolution as a function of log $T_{\rm eff}$ of $\Delta \log \frac{N}{H} = \log(N/H) - \log(N/H)_i$, where N and H are the surface abundances (in number) of nitrogen and hydrogen respectively, the index i indicates initial values. The initial masses, metallicities and rotational velocities are indicated. The shaded area corresponds to the range of observed values (Venn 1999) for A-type supergiants in the SMC.

Fig. 4 illustrates the changes of the N/H ratios from the ZAMS to the red supergiant stage for 20 and 25 $M_\odot$ stars (Meynet and Maeder 1999). For non–rotating stars, the surface enrichment in nitrogen only occurs when the star reaches the red supergiant phase; there, CNO elements are dredged–up by deep convection. For rotating stars, N–excesses occur already during the MS phase and they are larger for high rotation and initial stellar masses. At the end of the MS phase, for solar metallicity Z = 0.02, the predicted excesses amount to factors 3 and 4 for initial $v_{\rm rot}$ = 200 and 300 km/s respectively. At lower metallicity, the N–enrichment during the MS phase is smaller, likely due to the lower mass loss; however, there is a very large increase (up to a factor of ~ 10) for late B–type supergiants, because the star spends a lot of time in the blue phase and mixing processes have time to work. The predictions of Fig. 4 are in agreement with the observed excesses for galactic B– and A–type supergiants (Venn 1995ab). Also the very large excesses observed for A–type supergiants in the SMC (Venn 1999) are remarkably well accounted for.

The chemical yields are strongly modified by rotation. The larger He–cores obtained in rotating models at core collapse imply larger production of helium and other α-nuclei elements (Heger 1998). This is by far the most important effect of rotation on the chemical yields. In addition, by enhancing the mass loss rates and by making the formation of W–R stars easier, rotation favours the enrichment of the interstellar medium by stellar winds. Indeed, the stronger the winds, the richer the ejecta in helium and carbon and the lower in oxygen. This effect is responsible for the growth of the [C/O] ratio during galactic evolution.

The rotational diffusion during the H–burning phase enriches the outer layers in CNO processed elements. Some ^{14}N is extracted from the core and saved from further destruction. The same can be said for ^{17}O and ^{26}Al, a radio–isotope with a half–life of 0.72 Myr. The mixing in the envelope of rotating stars also leads to a faster depletion of the temperature sensitive light isotopes, for instance lithium and boron (Fliegner et al 1996).

The early production of N in the evolution of galaxies seems to imply that some N is produced in massive stars (Matteucci 1986; Thuan et al 1995). The problem is that the usual stellar models do not show such a production without adhoc assumptions. Models of rotating stars allow us to clearly identify the conditions for the production of primary N: the star must have an He–burning core and a thick and long–lived H–burning shell, then diffusion and transport of new ^{12}C from the core to the shell may generate some primary ^{14}N. W–R stars do not seem favourable, since the H–shell does not live long enough, being quickly extinguished and removed by mass loss. Low metallicity supergiants that have not suffered large mass loss are a very favourable site able to produce some primary nitrogen, especially if rotation is faster at low metallicities.

6. The formation of WR–stars

Rotation plays a important role in the formation of WR stars in several respects (Maeder 1987; Fliegner and Langer 1995; Maeder 1999b; Meynet 1999). There is no need any more to enhance the mass loss rates to reproduce the observed W–R/O number ratios (Maeder and Meynet 1994). A fast rotating star may enter the W–R phase, still burning hydrogen in its core (Maeder 1987; Fliegner and Langer 1995; Meynet 1999), thus skipping the LBV phase and spending more time in the W–R phase.

Rotation favours the formation of W–R stars from lower initial stellar masses both through its effects on the mass loss rates and on the mixing. Typically for the non–rotating models, the minimum mass for W–R star formation is betwen 35–40 $M_\odot$. It decreases to about 25 $M_\odot$ for initial $v_{rot} = 300$ km/s. This effect may help to explain the low luminous WN stars often reported. Higher rotational velocities lead to longer W–R lifetimes. As an example for a 60 $M_\odot$ model, the W–R lifetime is increased by more than a factor 3 when rotation is included. The durations of the WN and of the transition WN/WC phases are also very much increased. The ratio of the lifetimes of the WC to the WN phase is reduced (Meynet and Maeder 2000). High rotations lead to less luminous WC stars. This is because a fast rotating star enters the W–R stage earlier in its evolution and thus begins to loose large amounts of mass early. Therefore, fast rotators enter the WC phase with a small mass and a low luminosity; the final masses are also smaller.

Surface abundances characteristic of the WR stars may appear in a rotating star, not only as a result of the mass loss which uncovers the inner layers, but also as a result of mixing in radiative zones. During the WN phase, the surface abundances show some differences with respect to the case of no rotation. Indeed the N/C, N/O ratios obtained at the surface of the rotating WN models may have not yet reached the full nuclear equilibrium, in contrast with the non–rotating case where nuclear equilibrium is reached as soon as the star enters the WN phase. The CNO ratios are however close to the equilibrium values. During the transition WN/WC phase, nitrogen enhancements can be observed simultaneously with carbon and neon enhancements. After this transition phase, the ^{22}Ne enhancement reaches more or less the same high equilibrium level whatever the initial angular velocity, in agreement with the determinations of the neon abundance at the surface of WC stars (Willis et al 1997).

ACKNOWLEDGEMENTS: I express my best thanks to Dr. Georges Meynet for a beautiful cooperation on stellar evolution over many years.

References

1. Barbuy B., De Medeiros J.R. , Maeder A. 1996, *Astron. Astrophys.* 305, 911
2. Boyarchuk A.A., Lyubimkov L.S. 1983. *Izv. Krym. Astrofiz. Obs.* 66, 130
3. Chaboyer B., Demarque P., Pinsonneault M.H. 1995a, *Ap. J.* 441, 865
4. Chaboyer B., Demarque P., Pinsonneault M.H. 1995b, *Ap. J.* 441, 876
5. Chaboyer B., Zahn J.P. 1992. *Astron. Astrophys.* 253, 173
6. Denissenkov P.A. 1994. *Astron. Astrophys.* 287, 113
7. Denissenkov P.A., Ivanova N.S., Weiss A. 1999. *Astron. Astrophys.* 341, 181
8. Endal A.S., Sofia S. 1978, *Ap. J.* 220, 279

9. Fliegner J., Langer N. 1995. In *Wolf–Rayet Stars: Binaries, Colliding Winds, Evolution*, IAU Symp. 163, Eds. K.A. van der Hucht and P.M. Williams, Kluwer, 326
10. Fliegner J., Langer N., Venn K.A. 1996. *Astron. Astrophys.* 308, L13
11. Friend D.B., Abbott D.C. 1986. *Ap. J.* 311, 701
12. Groenewegen M.A.T., Lamers H.J.G.L.M., Pauldrach A. 1989. *Astron. Astrophys.* 221, 78
13. Heger A. 1998. Ph.D. thesis, Max-Planck-Institut für Astrophysik, *MPA* 1120
14. Heger A., Langer N., Woosley S.E. 1999. *Ap. J.* in press
15. Herrero A., Corral L.J., Villamariz M.R,. Martin E.L. 1999. *Astron. Astrophys.* 348, 542
16. Herrero A., Kudritzki R.P., Vilchez J.M., Kunze D., Butler K., Haser S. 1992. *Astron. Astrophys.* 261, 209
17. Herrero A., Villamariz M.R., Martin E.L. 1998. In *Boulder–Munich II: Properties of Hot, Luminous Stars*, ASP Conf. Ser., Ed. I. Howarth, 131, 159
18. Kippenhahn R., Thomas H.C. 1970. In *Stellar Rotation*, IAU Coll. 4, Ed. A. Slettebak, pp. 20.
19. Kudritzki R.P., Hummer D.G., Pauldrach A., Puls J., Najarro F., Imhoff J. 1992. *Astron. Astrophys.* 257, 655
20. Langer N. 1991. *Astron. Astrophys.* 243, 155
21. Langer N. 1992. *Astron. Astrophys.* 265, L17
22. Langer N. 1997. In *Luminous Blue Variables: Massive Stars in Transition*, ASP. Conf. Ser., Eds. A. Nota and H.J.G.L.M. Lamers, 120, 83
23. Langer N., Fricke K.J., Sugimoto D. 1983. *Astron. Astrophys.* 126, 207
24. Langer N., Maeder A. 1995. *Astron. Astrophys.* 295, 685
25. Lyubimkov L.S. 1996. *Astroph. Sp. Sci.* 243, 329
26. Maeder A. 1987. *Astron. Astrophys.* 178, 159
27. Maeder A. 1997. *Astron. Astrophys.* 321, 134
28. Maeder A. 1999a. *Astron. Astrophys.* 347, 185
29. Maeder A. 1999b. In *Wolf–Rayet Phenomena in Massive Stars and Starburst Galaxies*, IAU Symp. 193, ASP Conf. Ser. Eds. K.A. van der Hucht et al. pp. 177.
30. Maeder A., Conti P.. 1994. *Ann. Rev. Astron. Astrophys.* 32, 227
31. Maeder A., Grebel E., Mermilliod J.C. 1999. *Astron. Astrophys.* 346, 459
32. Maeder A., Meynet G. 1994. *Astron. Astrophys.* 287, 803
33. Maeder A., Zahn J.P. 1998. *Astron. Astrophys.* 334, 1000
34. Matteucci F. 1986. *MNRAS* 221, 911
35. McErlean N.D., Lennon D.J., Dufton P.L. 1999. *Astron. Astrophys.* 349, 553
36. Meynet G. 1998. In *Boulder–Munich II: Properties of Hot, Luminous Stars*, ASP Conf. Ser., Ed. I. Howarth, 131, 96
37. Meynet G. 1999. In *Variable and Non–Spherical Stellar Winds in Luminous Hot Stars*, IAU Coll. 169, Ed. B. Wolf et al. pp.377
38. Meynet G., Maeder A. 1997. *Astron. Astrophys.* 321:465
39. Meynet G., Maeder A. 2000. *Astron. Astrophys.* in press
40. Penny L.R., Gies D.R., Bagnuolo W.G. 1999. In *Wolf–Rayet Phenomena in Massive Stars and Starburst Galaxies*, IAU Symp. 193, ASP Conf. Ser. Eds. KA van der Hucht et al. pp.86
41. Pinsonneault M.H., Kawaler S.D., Sofia S., Demarque P. 1989. *Ap. J. Suppl.* 338, 424
42. Schatzman E., Maeder A., Angrand F., Glowinsky R. 1981. *Astron. Astrophys.* 96, 1
43. Schwarzschild M. 1958. In *Structure and Evolution of the Stars*, pp. 183
44. Talon S., Zahn J.P., Maeder A., Meynet G. 1997. *Astron. Astrophys.* 322, 209
45. Thuan T.X., Izotov Y.J, Lipovetsky V.A. 1995. *Ap. J.* 445, 108
46. Venn K.A. 1995a. *Ap. J.* 449, 839
47. Venn K.A. 1995b. *Ap. J. Suppl.* 99, 659

48. Venn K.A. 1999. *Ap. J.* 518, 405
49. Venn K.A, Lambert D.L., Lemke M. 1996. *Astron. Astrophys.* 307, 849
50. Villamariz M.R., Herrero A. 1999. *Astron. Astrophys.* in press.
51. Walborn N.R. 1988. In *Atmospheric Diagnostics of Stellar Evolution*, IAU Coll. 108, ed. K. Nomoto, Springer, pp.70.
52. Willis A.J., Dessart L., Crowther P.A., Morris P.W., Maeder A., Conti P.S., van der Hucht KA. 1997. *MNRAS* 290, 371
53. Zahn J.P. 1992. *Astron. Astrophys.* 265, 115

Speaker: A. MAEDER

F. MATTEUCCI's Comment: Following Chieffi's comment: primary N from AGB stars is not enough to explain, for example, the high N observed in IZW18 if this galaxy is suffering its first starburst. The same is true for the N measured in very metal poor stars in the solar neighborhood. Therefore, primary N from massive stars seems to be required.

P. VENTURA: Did you follow the tracks presented here starting from the pre-main-sequence phase? In case, which are the assumptions you made about the starting total angular momentum of the star?

A. MAEDER: Here, we start on the ZAMS with a constant ω in the star. However in a few 10^4 yr, the internal distribution of $\omega(r)$ converges towards an asymptotic stage, which will then slowly evolve during the MS phase. The higher the initial rotational velocity, the faster the convergence.

I.J. DANZIGER: There are 3 or 4 examples of young oxygen-rich SN remnants which appear to have a ring structure. Could these be showing the imprint of the late evaluation of rotating stars such as you have described?

A. MAEDER: Over the surface of the rotating star, from the pole to the equator, the effective gravity and T_{eff} are decreasing according to von Zeipel theorem. Thus, if for some T_{eff} there is an opacity peak, this may produce a mass ejection at the corresponding latitude. I suggest this is the way symmetrical rings are formed. The fact that strong variations of the force multiplices have been found recently by Lamers tends to support this possibility.

THE EXPLOSIVE YIELDS OF MASSIVE STARS

A. CHIEFFI
Istituto di Astrofisica Spaziale
via Fosso del Cavaliere, I-00133 Roma, Italy

M. LIMONGI
Osservatorio Astronomico di Roma
via Frascati 33, I-00044 Monteporzio Catone, Italy

AND

O. STRANIERO
Osservatorio Astronomico di Collurania
I-64100 Teramo, Italy

Abstract.

In this review we will briefly discuss how the basic properties of an expanding shock wave, coupled to the behavior of the matter at various high temperatures and to the presupernova evolution of massive stars, contribute to produce their final explosive yields. An extended and detailed database of explosive yields recently computed by our group may be found in our evolutionary database ORFEO (Online Repository of Franec Evolutionary Output) at the address *www.mporzio.astro.it/~mandrake/orfeo.html*

1. Introduction

Since at least the pioneering work of Hoyle (1946) it has been recognized that massive stars are strong contributors to the chemical enrichment of the interstellar gas. These stars firstly evolve through all the various hydrostatic burning stages (H, He, C, Ne, O and Si) producing the well known onion-like chemical stratification and then they eject part of this synthetized material in the interstellar medium as a consequence of the explosion which ends their life as stars. Such an explosion is due to the rebounce of the inner core of the star after it has collapsed up to nuclear densities. The shock wave which forms after this rebounce unfortunately contains only a tiny fraction

F. Matteucci and F. Giovannelli (eds.), The Evolution of the Milky Way, 417–427.

of the $\sim 10^{53}$ ergs of gravitational energy liberated by the collapse, since most of this energy is released as neutrinos. A discussion of the properties (and problems) of the collapse and of the rebounce goes well beyond the purposes of this short review and hence we refer the reader interested in this topic to the papers by, e.g., Bethe (1990), Mezzacappa et al. (1998). Here it suffices to remind that the first shock wave produced during the rebounce does not seem able to escape the iron core (i.e. the highly neutronized region affected by the Si burning) since most of its energy is absorbed by the layers within which the shock wave moves. Hence this (prompt explosion) scenario has been superseded by the so called "delayed explosion" scenario in which it is suggested that part of the energy trapped within the neutrinosphere can be used to rejuvenate the shock wave, i.e. to provide it the amount of energy necessary to exit the iron core. Once the shock successfully escapes the iron core, it can freely move towards the surface of the star because it does not loose any more energy on its way out. Since at present a real autoconsistent and reproducible successful explosion has not been obtained yet, a number of fundamental questions remain opened; among the others: how much is the energy which can escapes the iron core?, how much is the delay (if it exists) between the first rebounce and the rejuvenation of the shock front? is there a fall back and at what extent?

The lack of a definite answer to these questions induced the people involved in the computation of the explosive yields coming from massive stars to simulate a successful explosion by depositing a certain amount of energy at the border of the iron core mass by means of a thermal or kinetic bomb (Thielemann, Hashimoto and Nomoto 1990, Thielemann, Nomoto and Hashimoto 1996), or by means of a piston (Woosley and Weaver 1982, Woosley and Weaver 1995): this extra energy immediately generates a shock wave which propagates outward. The amount of extra energy deposited at the border of the iron core mass is tuned so to match a specific final total kinetic energy of the mantle, usually of the order of $1 \div 1.2$ foes (1 foe $= 10^{51}$ ergs). A comparison among the yields produced by initiating the explosion by one or the other of these methods has been done by Aufderheide, Baron and Thielemann (1991): their main conclusion was that the artifact by which the shock is initiated (piston, thermal and/or kinetic bomb) does not alter the final yields by more than, say, 30%, and hence that they may be considered relatively affordable at least in this respect. The uncertainties connected to our lack of knowledge of the real energy escaping the iron core, the duration of the time delay and the possible existence of a fall back certainly constitute a much larger source of global uncertainty in the determination of the final yields.

Quite obviously also the presupernova structure plays a fundamental role in determining both the total amount and the chemical composition of the ejecta. In the following discussion we will combine the basic properties of an expanding shock wave to the behavior of the matter at very high temperature in order to outline the role played by the presupernova models in the production of the final yields. Before closing this section we want remind the reader the excellent reviews by Arnett (1973) and by Thielemann et al. (1999) on the explosive yields from massive stars.

2. Discussion

It has long been recognized (Weaver and Woosley, 1980) that the basic properties of a shock wave adiabatically expanding in a medium (which is essentially the present case, since the passage of the shock through the mantle occurs without loosing any significant amount of energy) may be understood without the necessity of solving the proper hydrodynamic equations but by simply making use of the "radiation approximation". In this approximation one assumes that the energy "E" of the shock is dominated by the radiation and that both the temperature and density profile behind the shock are flat. In this case one has that:

$$E = 4/3\pi r_s^3 a T_s^4$$

where "r_s" and "T_s" refer, respectively, to the radial location and the peak temperature of the shock front. The main informations included in this relation (and which have been confirmed by the integration of the proper hydrodynamic equations) are that a) the medium within which the shock front moves does not enter in the determination of the peak temperature: this means that its progressive reduction as the shock front moves outside is a simple consequence of its adiabatic expansion (a "geometrical" effect), independent on the properties (physical and chemical) of the presupernova model; b) the peak temperature left by the shock at the various radii "r" has a very mild dependence on the shock energy "E" since it scales as $E^{0.25}$: this is a welcome property because it alleviates our ignorance regarding the real net energy of the shock escaping the iron core.

Before proceeding further, it is necessary at this point to briefly summarize some basic properties of the matter at high temperature. Since the end of the sixties (Truran, Cameron and Gilbert, 1966) it is known that if the temperature raises above 5 10^9 K, the chemical composition of the matter does not depend any more on specific, efficient reaction rates but, on the contrary, all the direct and reverse strong processes come to an equilibrium and the relative abundances among the various isotopes simply follow a statistical law which is determined by the condition that each isotope is in equilibrium with the sea of free protons, neutrons and α particles (Clifford and Tayler, 1965). The weak processes, on the contrary, are not yet at the equilibrium in these conditions, so that their net efficiency determines the changing in the global neutronization, η. This physical condition in which the strong processes are in equilibrium while the weak ones are not, is called Nuclear Statistical Equilibrium (NSE): under these circumstances the chemical composition is not only a function of T and ρ but also depends (largely) on the degree of neutronization. How we have mentioned above, the NSE distribution has been largely analyzed since the 60's in a number of fundamental papers (Truran, Cameron and Gilbert 1966, Truran, Arnett and Cameron 1967, Bodansky, Clayton and Fowler 1968a,b, Clayton and Woosley 1969, Arnett, Truran and Woosley 1971) and since then it is known that the most abundant elements which form in these conditions are V, Cr, Mn, Fe, and Ni. The relative proportions among these elements and, what it is more important, the relative abundances among the various isotopes is a function of ρ, T and Y_e ($\equiv (1-\eta)/2$).

If the peak temperature reached by the shocked matter does not exceed 5 10^9 K, not all the processes can reach the full equilibrium with their reverse and in particular the first ones which go out of the equilibrium are those which bridge isotopes having atomic masses $A \simeq 44$. In other words a bottleneck begins to form around $A \simeq 44$ which does not allow the free redistribution of the matter in a typical NSE distribution; this condition is usually called Quasi Statistical Equilibrium (QSE). By the way, the existence of such bottleneck was firstly inferred by Truran (1968a,b), Michaud and Fowler (1972) and Woosley, Arnett and Clayton (1973) and extensively reanalized recently by Hix and Thielemann (1996). Since the matter exposed to the explosion is essentially made of elements having $A \ll 44$, the main effect of this barrier is that the matter which quickly accumulates as ^{28}Si cannot redistribute freely upward but it will remain partially locked as ^{28}Si. Since the strength of this barrier increases as the temperature lowers, the abundance of ^{28}Si increases as the temperature decreases. In this temperature range the relative abundances of all the isotopes depend on T, ρ, Y_e and the amount of ^{28}Si itself. The main elements which form in this region are V, Cr, Mn, Fe and Si. This QSE condition persists down to a temperature of $\simeq 4$ 10^9 K.

A further reduction of the peak temperature below 4 10^9 K almost completely inhibits the flux of the matter through the bottleneck; we are facing the situation in which two separate QSE clusters form, one above and one below the bottleneck at $A \simeq 44$. Within each cluster the various isotopes readjust on relative equilibrium abundances which depend on T, ρ, Y_e and $\sum^Q n_i$ (i.e. the sum by number of all the isotopes pertaining to the cluster). This condition of double QSE clusters persists down to a temperature of the order of 3.3 10^9 K and the main elements which form under these circumstances are Si, S, Ar, K and Ca.

If the peak temperature remains confined within 3.3 10^9 and 1.9 10^9 K the direct and reverse strong processes are essentially far from the equilibrium but both the temperature and the timescale of the expansion (cooling) of the matter are such that a significant amount of nuclear processing may occur through efficient reaction rates: it is clear that in this case the final chemical composition will depend on the detailed preexplosive chemical composition, the reaction rates and the timescale of the expansion and not only on few parameters as T, ρ, Y_e and the like. However, it is possible to recognize that the elements which are preferentially synthetized in this condition are Ne, Na, Al, Mg, P, Cl and Sc.

The outer part of the star, in which the peak temperature remains below 1.9 10^9 K does not experience any significant nuclear reprocessing during the explosion because the timescale of the expansion is too much shorter than the timescale on which reaction rates may operate. Hence it will be ejected by preserving the chemical composition determined by the previous hydrostatic burnings. Since these mass zones have been affected mainly by the H, He and C burnings plus the changings in the envelope due to dredge-up episodes, the main elements which are present in these layers are H, He, C, N, O, F, Ne, Na, Al and Mg (of course we are neglecting the possible production of light elements by neutrino's capture).

We explicitely mentioned the importance of the mantle expansion timescale in determining the final yields coming from the regions which experience a peak temperature lower than 3.3 10^9 K, since in this case the nuclear processing depends explicitely on the time through the reaction rates of the various processes. However, it must be reminded that also the final chemical composition of the matter exposed at much higher temperatures, i.e. matter whose chemical composition is the result of an equilibrium condition (i.e. apparently independent on the time), do depend on the expansion timescale. The reason is that the amount of free α particles increases very much with the temperature and for $T > 6 \div 7$ 10^9 K most of the matter (.9 by mass and even more) is locked as α particles. During the cooling phase which follows the passage of the shock front, the matter cannot (almost) istantaneously readjust on the NSE distribution of progressively lower temperatures since it exists another bottleneck which adds to the one already mentioned above, i.e. the 3α reaction. This process is out of equilibrium and hence the build up of C occurs on timescales which may (and do) enter in competition with the cooling timescales whose main effect is, of course, that of freezing the chemical composition. In other words it exists a competition between the timescale of build up of the NSE population (which means essentially Ni for a $Y_e \simeq 0.5$), which is controlled by the efficiency of the 3α rections, and the timescale of expansion which tends to freeze the chemical composition. This phenomenon, usually called "freeze out", is well known and it has been firstly discussed by Truran, Arnett and Cameron

(1967) and Arnett, Truran and Woosley (1971) and recently by Thielemann et al. (1999). It is possible to recognize two cases: the normal "freeze out" is the one in which most of the α particles may form C and hence the matter can (almost) instantaneously readjust on an equilibrium distribution which preferentially populates Ni (if $Y_e \simeq 0.5$); the α rich "freeze out", on the contrary, occurs when the timescale of the expansion is much shorter than that of build up of C and in this case the final chemical composition is characterized by a large amount of free α particles and a low Ni abundance. Also the regions which do not reach a temperature high enough to form preferentially α particles may experience a more or less faster "freeze out" but, in this case, the 3α process does not constitues a bottleneck any more (simply because there are few α particles) and hence the matter can readjust (almost) instantaneously on the equilibrium distribution proper of the current value of the temperature untill the "freeze out" temperature is reached. It goes without saying that a proper computation of the explosion of a massive stars is of overwhelming importance in order to establish the timescales of expansion of the various layers (especially the ones closest to the "Fe" core) and hence, in turn, to obtain the kind of "freeze out" they will experience during the explosion.

If we join the existence of these critical temperatures (and related chemical composition) to the fact that the shock wave cools simply as a consequence of the self expansion and does not depend on the structure within which it moves, it follows that we can define rather carefully and independently on the stellar structure the "volumes" within which the various nuclear conditions occur: a) full NSE (usually called complete Si burning), b) QSE (incomplete Si burning), c) double QSE clusters (explosive O burning), d) standard burning (explosive C and Ne burnings). To be more quantitative we can say that, for an initial energy escaping the iron core of, e.g., 10^{51} ergs (1 foe), the radii which mark the various zones a), b), c) and d) are, respectively:

$a)$ 3700 Km – $T = 5\ 10^9\ K$ – NSE (*complete Si burning*)

$b)$ 5000 Km – $T = 4\ 10^9\ K$ – QSE (*incomplete Si burning*)

$c)$ 6400 Km – $T = 3.3\ 10^9\ K$ – 2 QSE *clusters* (*expl. O burning*)

$d)$ 13400 Km – $T = 1.9\ 10^9\ K$ – *nucl. burn.* (*expl. C and Ne burnings*)

The final chemical composition of a star layer exposed to the passage of the shock front will depend only on the volume in which it will be located at the moment of the passage of the shock. Here becomes evident why the possible existence of a time delay between the collapse and the rejuvenation of the shock may play a pivotal role in determining the final explosive yields: in fact, during this time delay (which may range between 0 up to more than 1 sec.) the matter closer to the "Fe" core may significantly fall towards the center changing, even largely, the distribution of the mass (and, in turn, of the neutronization η) in the volumes defined above.

In addition to the critical radii a), b), c) and d), there is another critical quantity which must be mentioned: the mass cut. This is the limiting mass which separates the part of the star which will succesfully escape in the interstellar medium from the one which will remain locked in the remnant. The effective value of the mass cut is still very difficult to predict on the basis of theoretical explosive models. However, by reminding that the light curve of a type II supernova is largely powered by the amount of ^{56}Ni ejected, it is possible to infer the mass cut by determing at what mass layer the global amount of ^{56}Ni fits the "observed" abundance: it goes without saying that this mass layer will be largely dependent on the assumed time delay.

We have at this point all the instruments to understand how the pre-supernova structure influences the final yields. First of all it is clear that the final mass-radius relation plays a key role because it fixes the total amount of mass which will be located in each of the volumes defined above and hence the total amount of mass which will by processed by each of the four different burning conditions. Figure 1 shows, as an example, the final mass-radius relation (thick solid lines) for a 13 $M_\odot$ and a 25 $M_\odot$ of solar metallicity; the four vertical dotted lines mark the four "key" radii. It is immediatly evident that the amount of mass which will be exposed to any of the four burnings is much larger in the 25 $M_\odot$ than in the 13 $M_\odot$: for example the amount of mass which will be processed in the QSE condition is $\simeq 0.2$ $M_\odot$ for the 25 $M_\odot$ and $\simeq 0.06$ $M_\odot$ for the 13 $M_\odot$. The total amount of mass which will undergo any of the four burning stage is $\simeq 3$ $M_\odot$ for the 25 $M_\odot$ while it is only $\simeq 0.25$ $M_\odot$ for the 13 $M_\odot$. These mass-radius relations are the result of the superimposition of many successive (central and shell) hydrostatic burnings which obviously regulate the progressive contraction and heating of the core. Hence any uncertainty present in the computation of the various burning phase may reflect on the final mass-radius relation. Among the other factors let us mention the treatment of the convection and the cross section of the $^{12}C(\alpha,\gamma)^{16}O$ process as two of the major factors which may alter the quoted relation. Convection largely plays the major role in driving the evolution of a stellar model because it determines both the size of the convective regions and the efficiency of the mixing in these zones. The $^{12}C(\alpha,\gamma)^{16}O$ cross section contributes, together to the detailed treatment of the convection, to determine the $^{12}C/^{16}O$ ratio which is left by the central He burning: this ratio is of crucial importance since it determines the amount of fuel which will be available for both the C and Ne burnings and hence both the possible formation of a convective core in these phases and the final location of their shells. Also a possible

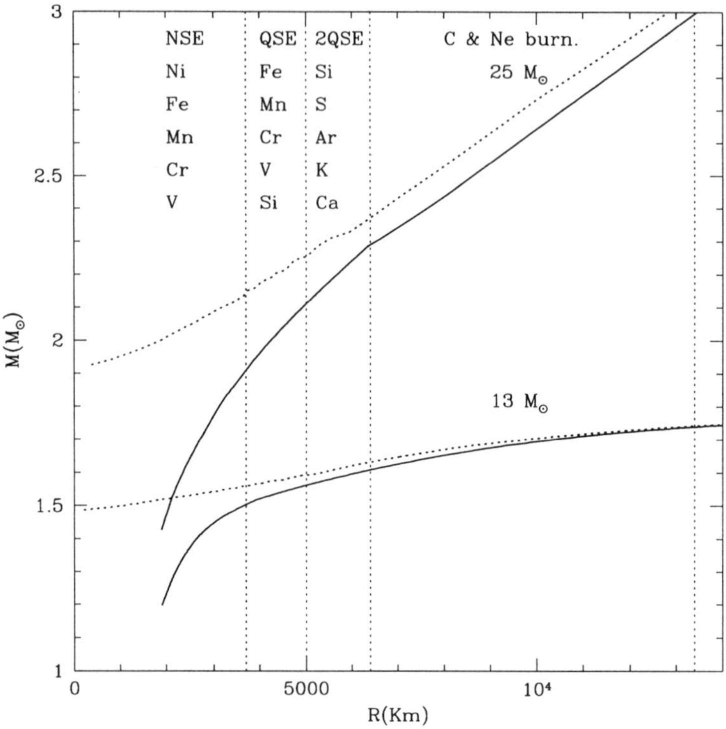

Figure 1. Mass-radius relation for two solar metallicity stellar models. Each solid line represents this relation at the moment of the collapse of the inner core while each dotted line is obtained having assumed 0.5 sec of time delay between the collapse and the rejuvenation of the shock. The four vertical lines mark the regions in which complete Si burning (NSE), incomplete Si burning (QSE), explosive O burning (2 QSE) and explosive C and Ne burnings occur. The main elements produced in each zone are reported.

time delay between the collapse and the rejuvenation of the shock would alter the mass-radius relation. Figure 1 shows as thick dotted lines how the mass-radius relation for the two quoted masses changes by imposing a delay of .5 sec. It is clear from the figure that the effect of the delay is that of reducing the amount of matter which will fall within each of the various volumes. This effect is due to the occurrence that the speed of the various layers scales inversely with the distance from the center.

If the mass-radius relation essentially determines the total amount of matter subject to a given explosive burning, the Y_e profile determines its specific chemical composition. As an example, for a temperature of 5 10^9 K, a density $\rho = 10^8$ and Y_e=.5, the five most abundant isotopes (by mass fraction) are ^{56}Ni=0.63, ^{55}Co=0.11, ^{52}Fe=0.07, ^{57}Ni=.06 and ^{54}Fe=.05, while the most abundant ones for Y_e=.49 are ^{54}Fe=0.28, ^{56}Ni=0.24, ^{55}Co=0.16, ^{58}Ni=.11 and ^{57}Ni=.08. Also the final Y_e profile will depend significanlty on both the previous hydrostatic evolution of the star and on the possible time delay. In particular, once again, the final Y_e profile will be largely influenced by the efficiency of convection. To clarify this point let us remind that during the evolution of a star there are essentially four key points in which the neutronization of the matter changes significantly (i.e Y_e reduces significantly). The first one is the H-burning phase in which two out of four protons become neutrons; the second one occurs when (during the He burning) the ^{14}N left by the CNO cycle is converted in ^{18}O through the well known chain $^{14}N(\alpha,\gamma)^{18}F(\beta^+)^{18}O$; the third one occurs during the O-burning and it is due to a series of efficient weak processes while the fourth one extends from the Si ignition onward and it is due to a very large number of weak processes which become efficient (see Chieffi, Limongi and Straniero 1998 and Limongi, Straniero and Chieffi 2000). The first two episodes of neutronization leave a Y_e profile flat within all the CO core because they do not depend on the physical conditions in which the burning occurs but only on the complete conversion of the H into He first and of the ^{14}N in ^{18}O later. By the way, the further conversion of ^{18}O in ^{22}Ne does not change Y_e. The two further episodes of neutronization, on the contrary, depend significantly on the mass (and unfortunately on the stellar model). The neutronization which occurs during the Oxygen burning obviously reduces Y_e only in the region where the burning has been efficient, i.e. up to the external border of the last O-convective shell, and hence it will depend on the initial mass of the star (i.e. on its previous evolutionary history) and, in turn, on the specific stellar model adopted. Also the last episode, which begins with the Si-ignition, modifies the neutronization up to the external border of the last Si convective shell. The interplay between these two burning shells will determine the final Y_e profile. Since the outer border of the Si shell marks the region "highly" neutronized which is usually assumed to remain locked in the remnant left by the explosion, two different possibilities exist: either the outer border of the last Si convective shell can extend beyond the outer border of the last O convective shell or not. In the first case the Y_e profile which will be exposed to the explosion will be essentially flat and hence it will not be altered by a possible infall. In the second case, on the contrary the Y_e profile will critically depend on the absolute location of the two borders; moreover in this case a possible infall may drastically change the Y_e which will be present in the volumes affected by the various kind of burnings discussed above. Obviously the Y_e profile will loose importance in the more external regions where no equilibrium condition exists. In these outer regions the final yields largely depend on the chemical stratification of the presupernova model.

By combining the properties of the presupernova models to those of the explosion it is possible to to identify which are the key parameters (and connected uncertainties) which influence the chemical composition of each of the zones described above. In particular:

a) the total mass and the chemical composition of the matter confined between the mass cut and the radius corresponding to 5 10^9 K (i.e. the region in which full NSE is reached) is mainly influenced by 1) the mass cut location itself, 2) the kind of freeze out, 3) the time delay, 4) the mass-radius relation and 5) the Y_e profile;

b) the matter coming from the region interested by the QSE condition will depend on essentially the same uncertainties which operate in the previous region but for the mass cut, whose detailed location should not alter the matter ejected by this region.

c) the matter exposed to the third regime (i.e., to the explosive Oxygen burning) depends on: 1) slightly on the time delay, 2) the mass-radius relation and 3) on $Y_{\rm e}$ (actually it affects mainly the odd elements, see Limongi, Chieffi and Straniero this volume);

d) the mass coming from the region suject to the C and Ne explosive burnings will depend on both the mass-radius relation and the detailed chemical composition preexisting the explosion.

e) the mass located more than 13400 Km from the center will totally depend on the presupernova evolution and not on the explosion.

A further parameter which must be mentioned and which alters the total amount of mass present in all of the four characteristic zones is the total energy of the shock since the radii corresponding to the four key temperatures scale as $E^{0.5}$.

The region which is less affected by both the presupernova model and the details of the explosion certainly is the explosive O-burning one. Either the total mass and the yields of the elements which are preferentially produced in this region, i.e. Si, S, Ar and Ca, may be considered a more robust theoretical prediction than that of the other elements. This is because they are affected neither by the mass cut location nor by the preexplosive chemical composition while they only marginally depend on Y_e, the mass-radius relation and the energy of the explosion. Hence, in our opinion, any of these elements is a better temporal tracer of the chemical evolution of the galaxy than the Iron or Oxygen. The theoretical prediction of both of them is, in fact, *highly* uncertain: the iron comes from the most internal layers, i.e. the ones mostly affected by the mass cut, the time delay and the Y_e value, while the Oxygen largely depends on the combined effects of two very uncertain quantities: the size of the convective core and the efficiency of the $^{12}C(\alpha,\gamma)^{16}O$ cross section.

We therefore want to conclude this short review on the yields coming from massive stars by strongly suggesting the use of one or more (or even a combination) of the "golden group" elements (Si, S, Ar and Ca) to trace the temporal changing of the various nuclear species instead of the (largely used up to now) Iron or Oxygen.

As a final note, let us mention that a detailed analysis of the presupernova evolution of stellar models of solar and lower metallicity may be found, respectively, in Limongi, Straniero and Chieffi (2000) and Straniero, Chieffi and Limongi (2000).

Acknowledgements

One of us (A.C.) thanks the Astronomical Observatory of Rome and its Director, Prof. Roberto Buonanno, for the generous ospitality at Monteporzio Catone. This paper has been partially supported by the MURST (COFIN98).

References

1. Arnett, W.D. (1973), *Ann. Rev. of Astron. and Astroph.*, **vol.11**, p.73
2. Arnett, W.D., Truran J.W. and Woosley, S.E. (1971), *Ap.J.*, **165**, 87
3. Aufderheide, M., Baron, E. and Thielemann, F.K. (1991), *Ap.J.*, **370**, 630
4. Bethe, H.A. (1990), *Rev. Mod. Phys.*, **62**, 801
5. Bodansky, D., Clayton D.D. and Fowler, W.A. (1968a), *Phys. Rev. Lett*, **20**, 161
6. Bodansky, D., Clayton D.D. and Fowler, W.A. (1968b), *Ap.J.Suppl.*, **148**, 16
7. Chieffi, A., Limongi, M. and Straniero, O. (1998), *Ap.J.*, **502**, 737
8. Clayton D.D. and Woosley, S.E. (1969), *Ap.J.*, **157**, 1381
9. Clifford, F.E., and Tayler, R.J. (1965), *Mem. R.A.S.*, **69**, 22
10. Hix, W.R. and Thielemann, F.K. (1996), *Ap.J.*, **460**, 869
11. Hoyle, F. (1946), *M.N.R.A.S.*, **106**, 343
12. Limongi, M., Straniero, O. and Chieffi, A. (2000), *Ap.J. Submitted*
13. Mezzacappa, A., Calder, A.C., Bruenn, S.W., Blondin, J.M., Guidry, M.W., Strayer, M.R. and Umar, A.S. (1998), *Ap.J.*, **495**, 911
14. Michaud, G. and Fowler, W.A. (1972), *Ap.J.*, **173**, 154
15. Straniero, O., Chieffi, A. and Limongi, M. (2000), *in preparation*
16. Thielemann, F.K., Hashimoto, M. and Nomoto, K., (1990), *Ap.J.*, **349**, 222
17. Thielemann, F.K., Nomoto, K. and Hashimoto, M. (1996), *Ap.J.*, **460**, 408
18. Thielemann, F.K., Rauscher, T., Freiburghaus, C., Nomoto, K., Hashimoto, M., Pfeiffer, B. and Kratz, K.L. (1999), *in Nuclei and Particle Physics, eds. J. Hirsh and D. Page (Cambridge University Press)*,
19. Truran, J.W. (1968a), *Ap.Space Sci.*, **2**, 391
20. Truran, J.W. (1968b), *Ap.Space Sci.*, **2**, 384
21. Truran, J.W., Arnett, W.D. and Cameron, A.G.W. (1967), *Can.J.Phys.*, **45**, 2315
22. Truran, J.W., Cameron, A.G.W. and Gilbert, A. (1966), *Can.J.Phys.*, **44**, 563
23. Weaver, T.A. and Woosley, S.E. (1980), *Ann. NY Acad. Sci.*, **336**, 335
24. Woosley, S.E., Arnett, W.D. and Clayton, D.D. (1973), *Ap.J.Suppl.*, **26**, 231
25. Woosley, S.E. and Weaver, T.A. (1982), *in Essays on Nuclear Physics, ed. C.A. Barnes, D.D. Clayton, and D.N. Schramm (Cambridge:Cambridge University Press)*, p.377
26. Woosley, S.E. and Weaver, T.A. (1995), *Ap.J.Suppl.*, **101**, 181

ASYMPTOTIC GIANT BRANCH STARS

Evolution and Nucleosynthesis

O. STRANIERO
Osservatorio Astronomico di Collurania, 64100 Teramo, Italy
M. LIMONGI
Osservatorio di Roma, Via Osservatorio 2, 00040 Monte Porzio (RM), Italy
A. CHIEFFI
Istituto di Astrofisica Spaziale (CNR), Roma, Italy
R. GALLINO
Dipartimento di Fisica Generale, Universita' di Torino, 10125 Torino, Italy
AND
M. BUSSO
Osservatorio Astronomico di Torino, 10025 Torino, Italy

1. Introduction

Hydrostatic nucleosynthesis modifies the internal composition of low and intermediate mass stars ($M \leq 8\ M_{\odot}$), but only in few cases these changes may affect the overall composition of the Galaxy. Their C-O core, where most of the nucleosynthesis occurred, is left as a dead white dwarf or is completely modified by explosive nucleosynthesis in close binary system during SNIa explosion. However, these stars can take part in the chemical evolution of the Galaxy when the external convection penetrates into an inner region previously enriched with the products of the nuclear burning (dredge up). This chemically modified envelope is then dispersed into the interstellar medium by stellar winds. A partial H-burning may also occur directly in the deepest zones of the convective envelope, if its temperature exceeds $20 - 30{\cdot}10^6$K.

In low and intermediate mass stars, dredge up episodes have been found in three different phases, namely the red giant branch (RGB), the early asymptotic giant branch (E-AGB) and the thermally pulsing asymptotic

F. Matteucci and F. Giovannelli (eds.), The Evolution of the Milky Way, 429–443.

giant branch (TP-AGB). CNO-burning at the base of the convective envelope has been found in intermediate mass TP-AGB models. A growing amount of theoretical and observational hints prove that AGB stars have played an important role in the chemical evolution of the Galaxy. In this paper we focus our attention in reviewing the nucleosynthesis occurring during the TP-AGB phase.

2. Thermally pulsing AGB models

All stars with $M < M_{up}^{'}$ (i.e. the minimum stellar mass for a non degenerate carbon ignition) will experience an AGB phase. Concerning low mass stars, they reach the AGB having already experienced a first dredge up episode during the RGB phase. Then, their envelope has been already enriched with the typical ashes of the H-burning: with respect to the original composition, ^{4}He, ^{3}He, ^{13}C, ^{17}O, and ^{14}N are enhanced, while ^{12}C, ^{15}N, ^{18}O, (possibly ^{16}O) are depleted. Also Li and Be are almost fully destroyed. In intermediate mass stars the first dredge up occurs only if the metallicity is large enough. For example at $Z = 0.02$ all stars with mass lower than $M_{up}^{'}$ experience a first dredge up, but this does not occurs at $Z = 0.001$ for $M > 3\ M_{\odot}$ (Dominguez et al. 1999). Theoretical stellar models show that during the first part of the AGB, usually referred as early AGB (Iben & Renzini, 1983), H-burning may cease and the external convection may extend inward (in mass) in the region modified by the CNO burning. This is the second dredge up, which is usually found in stellar models with $M \geq$ 4-5 $M_{\odot}$ (this value depends on the metallicity, see e.g. Dominguez et al. 1999). In this phase the surface abundance of helium and nitrogen increase. Later on, when the He shell get closer to the H/He discontinuity, He-burning dies down and H-burning restarts. From this moment the star enters in the TP-AGB phase. Thermal pulses occur when the H-burning shell accumulates enough He below it, so that He reignites. Despite matter is largely non degenerate, a runaway occurs, because the thermodynamical timescale needed to locally expand the stellar gas is much longer than the nuclear burning timescale of the 3α reaction. The theory of these non degenerate He flashes was firstly illustrated by Schwarzschild & Härm (1965). Thus the local temperature initially increases and, in turn, the He-burning luminosity rapidly increases (up to $10^8\ L_{\odot}$, depending on stellar mass). During this phase, a convective shell appears that extends from the He-burning shell up to the H/He discontinuity (see Figure 1).

Then the products of the 3α reaction and those of the neutron capture nucleosynthesis are mixed over the whole intershell region. When the expansion has progressed far enough, the temperature of the He shell decreases and quiescent He-burning settles on. The variations of the H and

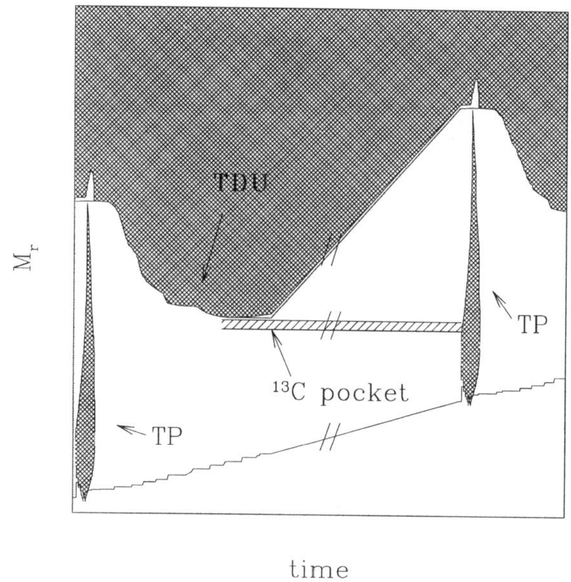

Figure 1. Evolution of the internal structure of a 5 $M_\odot$ $Z = 0.02$ model through the 22th and the 23th thermal pulses (TP). Shaded areas represent convective zones. The two solid lines indicate the locations of H- and He-burning shells. The location of the possible thin ^{13}C pocket (see text) is also reported. The interpulse is here only schematically drawn. Note that its duration is much longer than that of the thermal pulses. In this particular case the lifetime of the first thermal pulse is 22 yr, while the duration of the interpulse is 6794 yr.

He-burning luminosities for a 5 $M_\odot$ model of $Z = 0.02$ are reported in Figure 2 and 3, respectively.

The expansion powered by the He flash pushes outward the layers of material located above the He-burning shell. Temperature and density at the base of the H-rich envelope decrease, until the H-burning shell dies down. This is the right moment for the third dredge up (TDU). In fact at the H/He discontinuity the energy flux coming from the quiescent He shell is still growing so that the local radiative gradient increases. On the other hand, due to the low density, the ratio of gas pressure to radiation pressure decreases and the adiabatic gradient approaches its minimum value (i.e. that of a pure photon gas, namely $\nabla_{ad} = 0.25$). Then the Schwarzschild criterion for the convective instability is here fulfilled and the envelope penetrates (in mass) trough the H/He discontinuity. If the dredge up is deep enough to overlap the region previously mixed by the convective shell, fresh helium, carbon and heavy *s*-elements are brought to the surface. Finally, when the He-burning luminosity drops below the surface luminosity, the material that were pushed outward falls back again and warms up until

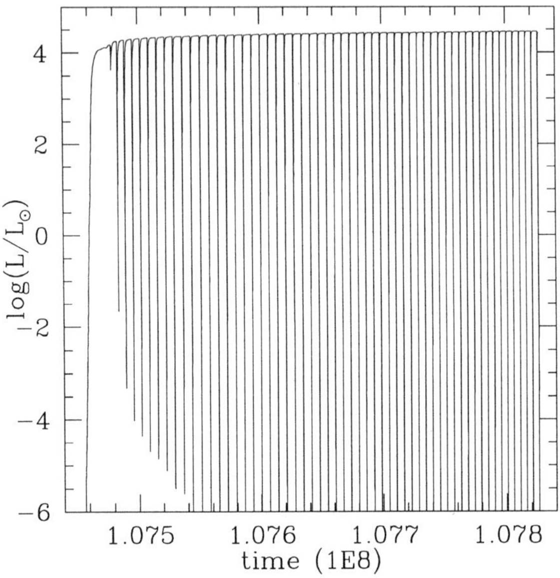

Figure 2. Variation of the H-burning luminosity for the 5 $M_{\odot}$ $Z = 0.02$ model.

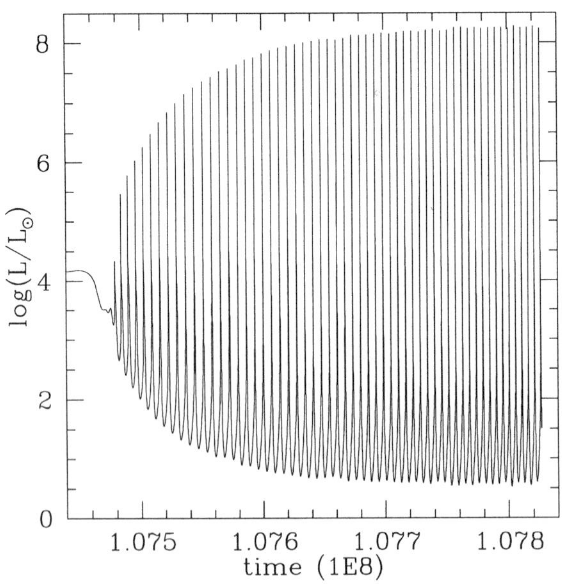

Figure 3. Variation of the He-burning luminosity for the 5 $M_{\odot}$ $Z = 0.02$ model.

hydrogen reignites.

In between two successive pulses (interpulse phase) the H-burning provides the energy required to support the weight of the envelope. In this phase the inner edge of the convective envelope always remains very close to the region where H-burning takes place. For the more massive AGB stars (i.e $M \geq 5\ M_\odot$) the temperature at the base of the convective envelope may be high enough that CNO burning can take place. In Figure 4 we show the variation of this temperature for a 7 $M_\odot$ $Z = 0.02$ model during the TP-AGB phase (from Straniero et al. 2000). After few TPs it reaches $80 \cdot 10^6$ K. This is the well known hot bottom burning (HBB, Sugimoto 1971, Iben 1973, Renzini & Voli 1981).

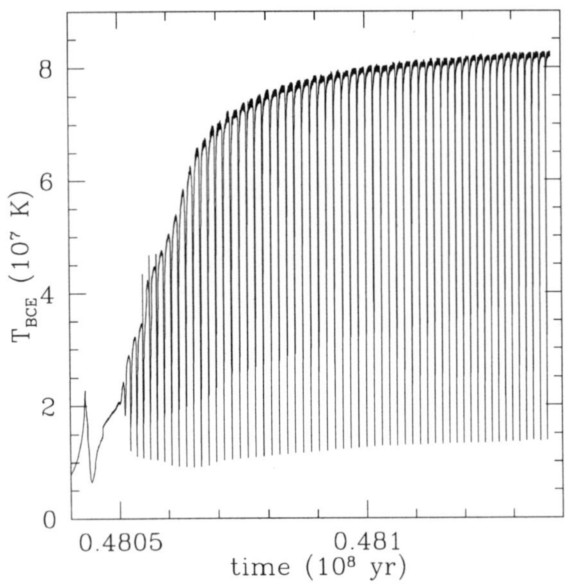

Figure 4. Temperature at the base of the convective envelope for the 7 $M_\odot$ $Z = 0.02$ model.

The duration of the AGB depends on the mass loss rate. The available data indicate that this rate ranges between 10^{-7} and 10^{-4} $M_\odot$/yr. It is not well clarified if during the AGB there is a continuous increase of the mass loss rate rather than an abrupt transition from a quiet wind and a violent (dynamical) envelope ejection. We have already recalled that immediately after the TP, namely during the quiescent He-burning, the ratio of the gas pressure to the radiation pressure (β) decreases in a thin layer located around the H/He discontinuity. Recently Sweigart (1998) found that when the envelope mass is reduced enough (how much depends on the core mass and metallicity), β becomes 0. This implies that the local stellar luminosity exceeds the Eddington luminosity inducing a dynamical instability of the envelope. We have found a similar situation in a model of 5 $M_\odot$ $Z = 0.02$ (Straniero et al. 2000). We cannot say if this instability can indefinitely grow up to the complete envelope removal. However we believe that this possibility deserves a further investigation.

This is a brief description of the major occurrences of the TP-AGB phase (for further details see Iben & Renzini 1983). In the following we will focus our attention on those features which make thermally pulsing stars an important site for the chemical evolution of the Galaxy.

3. *s*-process nucleosynthesis: the source of neutrons

The key ingredient to activate the *s*-process nucleosynthesis is the neutron source. Two neutron sources were firstly identified by Cameron (1954, 1960): the ^{13}C source, as due to the ^{13}C(α,n)^{16}O reaction and the ^{22}Ne source, as due to the ^{22}Ne(α,n)^{25}Mg reaction.

It was early understood that nucleosynthesis induced by slow neutron captures (*s*-processes) could occur in TP-AGB stars (Schwarzschild & Härm 1967; Sanders 1967; Ulrich & Scalo 1972; Sackmann, Smith & Despain 1974). Various mechanisms were proposed to activate the neutron source. The basic idea was that at the peak of the thermal pulse, the upper part of the convective shell could cross the H/He discontinuity bringing some protons inside the He intershell, where the mass fraction of carbon is about 0.2. Thus the reactions ^{12}C(p,γ)^{13}N(β^+,ν)^{13}C(α,n)^{16}O would release neutrons. However, detailed stellar models show that an insufficient amount of protons are captured in this way. In fact, since the H-burning shell remains active during the thermal runaway, the penetration of the convective shell into the H-rich envelope is strongly inhibited by an entropy barrier (Iben 1976). Note that, since the material in the intershell has been previously processed by the H-burning shell, some ^{13}C is already present in this region. However, the amount of ^{13}C left by the CNO burning is about one order of magnitude lower than that needed to reproduce the cosmic distribution of the *s* elements.

A completely different scenario was proposed by Iben (1975a,b). He noted that, as a consequence of the H-burning, the amount of nitrogen into the intershell is equal to the sum of the abundances (by number) of the original CNO elements. During the pulse, this ^{14}N is converted to ^{22}Ne trough the following chain: ^{14}N(α,γ)^{18}F(β^+,ν)^{18}O(α,γ)^{22}Ne. If the temperature at the base of the convective envelope is large enough (i.e. $\sim 3.2 \cdot 10^8$ K or larger) the ^{22}Ne(α,n)^{25}Mg reaction provides a significant neutron flux. Iben (1975a) firstly demonstrated that this condition is fulfilled in intermediate mass stars, in particular in a model of 7 $M_\odot$.

However, in low mass AGB stars the temperature at the base of the convective shell is too low for an efficient activation of the ^{22}Ne neutron source (Iben & Truran 1978, Straniero et al. 1997). In addition, the amount of neutrons produced in this way depends on the original CNO. For this reason, if at solar (or larger) metallicity the ^{22}Ne source is very efficient, this is no more true at lower Z. On the contrary, the ^{13}C pocket is a primary neutron source. It depends on the amount of carbon, which is produced in situ by the 3α reaction, and on the amount of protons diffused from the envelope. This does not mean that the nucleosynthesis activated by the ^{13}C neutron source does not depend on Z. In fact the lower the metallicity the lower is the amount of seeds nuclei (and poisons). It simply implies that the expected variation with the metallicity of the s-elements production induced by the ^{13}C neutron source would be different than that induced by the ^{22}Ne neutron source. This occurrence may be used to distinguish between these two cases. The combination of photometric and spectroscopic observations of AGB stars have clearly demonstrated that low mass stars, rather than intermediate mass stars, are among the main contributors to the s-process synthesis in the Galaxy, and that the $^{13}C(\alpha,n)^{16}O$ is the most reliable neutron source (Smith & Lambert 1985, 1986, 1990; Lambert et al. 1995, Busso et al. 2000). For this reason the search for the activation of the ^{13}C source in low mass AGB stars was intensified. The first attempt was still made by Iben and coworkers (Iben 1982; Iben & Renzini 1982a,b; Hollowell & Iben 1989). They proposed that the mixing of protons into the He/C rich intershell does not occurs during the thermal pulse, but during the post-flash dip, namely the period which elapses between the TDU and the reignition of the H-burning shell. In fact, at this time the entropy barrier disappears and the H/He discontinuity is directly in contact with the He/C rich zone. Iben (1982) investigated two possible mechanisms. The first one is based on atomic diffusion induced by the steep chemical gradient left by the TDU. By evaluating the timescale of this phenomenon, he showed that atomic diffusion is a promising possibility to move enough protons from the envelope into the intershell. The second process concerns the effect of carbon recombination, which occurs when the temperature drops below 10^6 K. The region immediately adjacent to the convective envelope meets this condition at the time of the post-flash dip. As a consequence of the carbon recombination, opacity increases and a semiconvective layer, capable to mix down protons and to mix up carbon, could form. In both cases a ^{13}C pocket, located below the deeper position attained by the base of the convective envelope at the time of the TDU, would result by the activation of the $^{12}C(p,\gamma)^{13}C$ reaction (see figure 1).

Once the ^{13}C pocket has been built up, the neutron production can start if the temperature rises up to about 10^8 K. In the original scheme proposed by Iben & Renzini (1982a) and reviewed by Hollowell & Iben (1989), it was assumed that the ^{13}C pocket is ingested by the successive convective pulse. In such a case the $^{13}C(\alpha,n)^{16}O$ reaction and the s-process nucleosynthesis would take place during the thermal runaway. However, Bazan & Lattanzio (1993) showed that the energy produced by the neutron source, and by the following neutron captures (about 8 MeV for each neutron), may be so large that the main characteristics of the thermal pulse could be substantially modified. As a matter of fact, Straniero et al. (1995) found that during the interpulse the temperature in the ^{13}C pocket rapidly increases up to $90 - 100 \cdot 10^6$ K, so that the ^{13}C is completely burned before the onset of the subsequent pulse, with no consequences on the overall stellar structure. Thus the s-process nucleosynthesis does not occur during the convective pulse, but takes place in radiative condition during the interpulse. The new scenario is substantially different with respect to the old one. First of all the duration of the whole nucleosynthesis episode is much larger: namely a few 10^4 yr to be compared with about 10 yr of the convective nucleosynthesis. Moreover, the temperature is lower and, in turn, the resulting neutron density is lower: typically 10^7 cm^{-3}, to be compared with several 10^9 cm^{-3} of the convective nucleosynthesis.

Despite this progress, the old question of the identification of the actual mechanism of the formation of the ^{13}C pocket still remains alive. Recently a new effort has been addressed in the search of this mechanism. Herwig et al. (1997), by using some results of the hydrodynamical calculation of Freytag et al. (1996), have developed a diffusion algorithm to describe the mechanical overshoot of material from the bottom of the convective envelope into the adjacent stable region. They applied this scheme to the computation of a 3 $M_{\odot}$ of solar composition and found that, when the TDU penetrates into the region previously mixed by the convective shell, a ^{13}C pocket forms. Similarly Langer et al. (1999) calculate AGB models taking into account for rotational induced mixing. Also in this case the authors adopt a diffusion algorithm and again a ^{13}C pocket is found.

Thus it seems that the most recent AGB stellar models can account for the ^{13}C neutron source. Note that the condition which makes more probable the formation of the ^{13}C pocket is the occurrence of the TDU. All these updated AGB models meet this condition. Straniero et al. (1997) showed that adopting the latest improvement of the input physics (opacity, equation of state and the like) and thanks to a better numerical precision, as allowed by the modern generation of computers, the TDU is obtained when the core mass is larger than 0.61 $M_{\odot}$, provided that the envelope mass does not drop below $0.4 - 0.5$ $M_{\odot}$. When TDU occurs, the H-rich envelope and the He/C rich intershell are put directly in contact. This condition is maintained until H-burning restarts. The time elapsed from the maximum penetration of the convective envelope and the H reignition is about 5000 – 6000 yr for low mass stars (i.e. 1.5 – 3 $M_{\odot}$) and about 500 – 1000 yr for intermediate mass stars. Thus a lot of physical mechanisms may be invoked to explain the diffusion of few protons (namely $10^{-5} - 10^{-6}$ $M_{\odot}$ of hydrogen) from the envelope into the intershell. Atomic diffusion, semiconvection, mechanical overshoot or rotational induced mixing are examples of those phenomena whose lifetime could be short enough. Note that, since the larger the mass the lower the duration of the post-flash dip, a larger ^{13}C pocket could form in low mass stars than in intermediate mass stars.

4. Nucleosynthesis in low mass AGB stars

In this section we summarize the results of our computation of neutron capture nucleosynthesis in low mass AGB stars, as recently revised by Gallino et al. (1998). This nucleosynthesis is based on stellar models with mass $1.5 \leq M/M_{\odot} \leq 3$ and solar composition (Straniero et al. 1997). Some unpublished models for lower metallicity (Z = 0.001, 0.006 and 0.01) are also used. In these models the first TDU episode was found to occur after about 9 – 10 TPs at $Z = 0.02$ and after 4 – 5 TPs at $Z = 0.001$. The amount of He/C rich material dredged up initially increases until an asymptotic value is attained. Then, when the envelope mass is substantially eroded by the mass loss and by the growth of the He core, the strength of the TDU decreases. The maximum mass dredged up is $1.5 \cdot 10^{-3}$ and $3 \cdot 10^{-3}$ $M_{\odot}$ for the models of 1.5 and 3 $M_{\odot}$, respectively. In all cases, when the residual envelope mass drops down to 0.4 – 0.5 $M_{\odot}$, the TDU disappears.

As discussed in the previous section, we believe that the ^{13}C neutron source could be at work in low mass AGB stars experiencing TDU. Although the mechanisms that allow protons from the envelope to be diffused into the He/C rich are not completely understood, the scheme derived from the most recent stellar models seems in good agreement with many observational constraints (e.g., Lambert et al. 1995).

At the beginning of the interpulse, when hydrogen reignites, the protons diffused in the intershell are rapidly captured by the abundant ^{12}C, thus forming a ^{13}C pocket. Then, the temperature increases until the $^{13}C(\alpha,n)^{16}O$ reactions start. The ^{13}C in the pocket is slowly (but completely) burned before the onset of the next thermal pulse (Straniero et al. 1995; Straniero et al 1997; Herwig et al. 1997). Thus, neutron capture synthesis takes place during the interpulse. When the successive TP starts, the highly *s*-enriched material in the thin ^{13}C pocket is ingested by the convective shell and diluted over the whole intershell region. Then, the next TDU brings fresh ^{12}C and *s* elements at the surface and fresh seeds in the innermost part of the convective envelope. Since a given convective shell partially overlaps the previous one, different fractions of the matter in the intershell have been exposed to different integrated neutron fluxes.

During the TP, the maximum temperature at the base of the convective shell does not exceeds $3 \cdot 10^8$ K, thus the ^{22}Ne(α,n)^{25}Mg is only marginally activated. This second burst of neutrons has minor effects on the bulk of the *s*-elements production, but is important for those branchings along the *s*-process path that are sensitive to the neutron density (Käppeler et al. 1990; Arlandini et al. 1999).

The nucleosynthesis results obtained by Gallino et al. (1998) for different metallicities are illustrated in Figure 5. Note how low mass AGB stars can

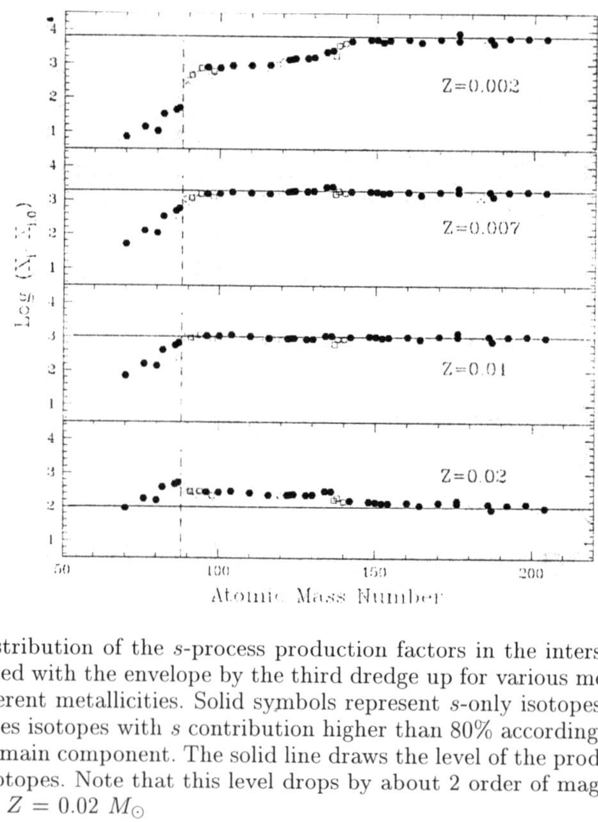

Figure 5. Distribution of the *s*-process production factors in the intershell matter cumulatively mixed with the envelope by the third dredge up for various models with $M = 2\ M_\odot$ and different metallicities. Solid symbols represent *s*-only isotopes, whereas open squares indicates isotopes with *s* contribution higher than 80% according to the classical analysis of the main component. The solid line draws the level of the production factor of the heaviest isotopes. Note that this level drops by about 2 order of magnitude between $Z = 0.002$ and $Z = 0.02\ M_\odot$

account for the so called main component of the solar distribution of the *s* elements (i.e. nuclei in the mass range $85 \leq A \leq 205$). Although the ^{13}C is a primary neutron source, there is an evident correlation of the *s*-element production with the metallicity. In particular, at low Z the heavy tail of the main component is overproduced. This occurrence may be understood by noting that at low Z more neutrons are available for each seed. As a consequence, a significant contribution to the so called strong component (i.e. $A > 205$; not shown in Figure 5) derives from low-metallicity and low-mass AGB stars (Gallino et al. 1998). In particular, ^{208}Pb and ^{209}Bi are strongly overproduced for $Z \leq 2{\cdot}10^{-3}$.

These models show that in low mass AGB stars the maximum temperature at the base of the convective envelope, during the interpulse, is too low for HBB to occur (it never exceeds 10^7 K). Thus, their contribution to the chemical evolution of the Galaxy is mainly due to the modification of the envelope composition induced by the third dredge up. In summary they are capable to supply the main products of H- and He-burnings, hence primary helium, carbon, as well as *s* elements.

Recently, it has been found that some intrinsic C stars (i.e. AGB stars having C/O ≥ 1), for which a low ^{12}C/^{13}C is measured, are also lithium rich (Abia & Isern 1996, 1997). Since the typical mass of these stars would be around 2 $M_{\odot}$, HBB cannot be responsible of these features. A possible explanation is that in these stars some extramixing, perhaps induced by differential rotation, could activate the Cameron-Fowler mechanism as well as could bring to the surface the ashes of the internal CN burning. (Wasserburg, Boothroyd & Sackmann 1995; Ventura, D'Antona & Mazzitelli 1999; see also Ventura in these Proceedings). Li is produced by the following chain ^{3}He(α,γ)^{7}Be(e^-,ν)^{7}Li that requires a temperature of at least $30 - 40 \cdot 10^6$ K. However, at this temperature Li is quickly destroyed by the reaction ^{7}Li(p,α)^{4}He. Thus the Li production might occur only if the mixing is fast enough to move the ^{7}Be in a cooler external region before electron capture on ^{7}Li takes place.

5. Nucleosynthesis in intermediate mass AGB stars

The exponential decay of the mass function reduces the possible contribution of intermediate mass AGB stars to the chemical evolution of the Galaxy. Nevertheless their lifetime is so short (in between 40 and 100 Myr for masses ranging between 7 and 5 $M_{\odot}$) that they could have important consequences for the early epoch of the Galaxy.

The third dredge up occurs rather early in thermally pulsing intermediate mass models, but the penetration of the convective envelope into the intershell decreases when the mass increases. We found that the amount of material dredged up is $1.3 \cdot 10^{-3}$ and $2.2 \cdot 10^{-4}$ $M_{\odot}$ for models of 5 and 7 $M_{\odot}$, respectively (Straniero et al. 2000). Note that also the mass extension of the intershell is lower for the more massive models. For these reason and because the duration of the post-flash dip is much shorter than that found in low mass stars, we usually assume that a smaller ^{13}C pocket forms (Vaglio et al. 1998). With this exception, the ^{13}C burning and the following neutron captures synthesis proceed with the same characteristics already found in low mass stars. In particular, in spite of the fact that the interpulse duration is much lower than in low mass stars, all the ^{13}C in the pocket is fully burned before the next TP. However, at variance with low mass stars, the temperature at the base of the convective shell is large enough that a significant ^{22}Ne burning occurs. This second burst of neutrons is characterized by a very high peak average neutron density (10^{11} cm^{-3}). As an example in Figure 6 we report the computed distribution of the s elements in a 5 $M_\odot$ of solar composition after about 50 TPs (from Vaglio et al. 1998). Note taht the distribution of s-processed isotopes shown in the figure is the result of the concurrent action of the two neutron sources.

The main conclusion of our studies of the s-process nucleosynthesis in intermediate mass stars is that they cannot be the source of the main component in the solar system. However, they provide a significant contribution to the formation of the Sr-Y-Zr peak in the Galaxy. Note that, as a consequence of the TDU, the amount of ^{14}N left by the H-burning shell in the intershell increases after each pulse. Since this ^{14}N is converted to ^{22}Ne in the convective shell, the ^{22}Ne(α,n)^{25}Mg becomes in part a primary neutron source.

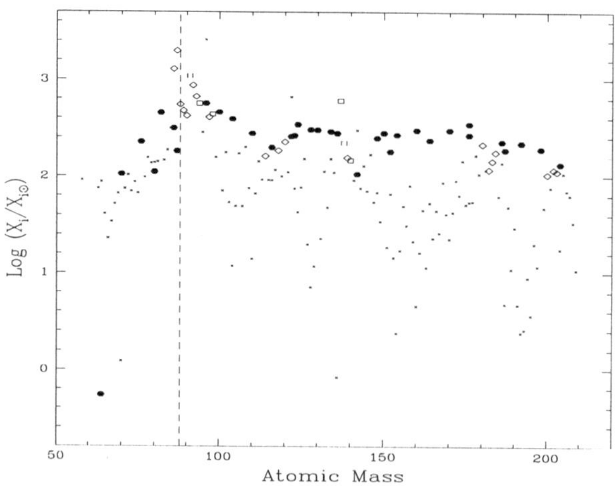

Figure 6. Distribution of the s-process production factors in the intershell matter cumulatively mixed with the envelope by the third dredge up for a 5 $M_\odot$ $Z = 0.02$ model. Solid symbols represent s-only isotopes, whereas open squares and diamonds refer to isotopes with s contribution higher than 80% and between 60 % and 80 %, respectively. Note the overproduction of ^{86}Kr, ^{87}Rb, ^{96}Zr, ^{122}Sn and ^{137}Ba, which indicates that important branchings along the s path are open in intermediate mass stars.

The occurrence of the HBB is another important event for the nucleosynthesis taking place in these stars. As a consequence of the third dredge up, the carbon abundance in the envelope increases. In low mass AGB stars of solar chemical composition, a C star is obtained after about 20 –

25 TPs (Straniero et al. 1997). However, if the temperature at the base of the convective envelope is larger than $6 \cdot 10^7$ K, carbon is converted to nitrogen. Thus, HBB prevents the formation of bright C stars. Note that, since fresh carbon is added to the envelope at each TDU, the final nitrogen yield of an intermediate mass star is not limited by the total amount of the original CNO. At each TDU about 10^{-4} $M_\odot$ of carbon is dredged up. The total number of TDU episodes is limited by the mass loss rate so that a conservative value might be in between 50 – 100. Then, if carbon is fully converted into nitrogen, the final yield would be of the order of 10^{-2} $M_\odot$ (for a more accurate estimation of this yield see Forestini & Charbonnel 1997).

HBB produces other important consequences on the envelope composition (e.g., Sackmann & Boothroyd 1992; Lattanzio et al. 1996; Forestini & Charbonnel 1997). First of all, it induces a lithium production, which is still due to the Cameron-Fowler mechanism. A depletion of the ^{3}He is evidently connected to this Li enrichment. The isotopic distribution of carbon, nitrogen and oxygen are significantly modified: ^{13}C, ^{14}N and ^{17}O increase, whereas ^{15}N and ^{18}O are strongly depleted. As a consequence of the activation of the Mg-Al cycle, both TDU and HBB cause an increment of ^{26}Mg, ^{26}Al and ^{27}Al. On the contrary ^{19}F and ^{25}Mg are depleted by the HBB, but they increase when TDU occurs.

6. Conclusion

Many details of the theory of AGB stars remain to be worked out. As an example, hydrodynamical phenomena and rotation, which are usually neglected in stellar model computations, may affect the predicted amount of dredged material. However, many spectroscopic and photometric observations of AGB stars in the Milky Way and the Magellanic Clouds confirm the present nucleosynthesis scenario. Measurements of s elements at the critical branches of the neutron capture chain provide precise informations on the physical environment of the s-process nucleosynthesis. As an example, Lambert et al. (1995) show that the observed low Rb/Sr ratio would imply a low neutron density, in agreement with the radiative ^{13}C burning. The absence of bright C stars (Wood, Bessell & Fox 1983) and the relative abundance of Li-rich bright AGB stars (Smith & Lambert 1989) confirm the occurrence of HBB in intermediate mass AGB stars. For a more detailed comparison between model predictions ad observations of AGB stars we suggest the paper of Busso et al. (2000).

Let us finally recall the interesting constraints on the isotopic compositions of C, N, O, Mg, Al, Si and their connection with s-element production coming from the analysis of circumstellar grains found in meteorites, which are believed to be formed in the envelope of AGB stars (Clayton 1988; Boothroyd, Sackmann & Wasserburg 1994; Hoppe et al. 1997; Choi, Wasserburg & Huss 1999; Gallino, Busso, & Lugaro 1997 and references therein).

This work was partially supported by the MURST Italian grant Cofin98.

References

1. Abia, C., Isern, J.: 1996, *ApJ* **460**, 443
2. Abia, C., Isern, J.: 1997, *MNRAS* **289**, L11
3. Arlandini, C., Käppeler, F., Wisshak, K., Gallino, R., Lugaro, M., Busso, M., Straniero, O.: 1999, *ApJ* **525**, 886.
4. Bazan, G., Lattanzio, J.C.: 1993, *ApJ* **409**, 762
5. Boothroyd, A.J., Sackmann, I.-J., Wasserburg, G.J.: 1994 *ApJ* **430**, L77
6. Busso, M., Gallino, R., Lugaro, M., Straniero, O., Smith, V.V., Lambert, D.L.: 2000, in *The Changes in Abundances in Asymptotic Giant Branch Stars*, eds. F. D'Antona and R. Gallino, *Mem. SAIt*, in press.
7. Cameron, A.G.W.: 1954, *Phys. Rev.* **93**, 932.
8. Cameron, A.G.W.: 1960, *A. J.* **65**, 485.
9. Choi, B.-G., Wasserburg, G.J., Huss, G.R.: 1999, *ApJ* **522**, L133.
10. Clayton, D.D.: 1988, *ApJ* **334**, 191.
11. Clayton, D.D., Fowler, W.A., Hull, T.C., Zimmerman, B.A.: 1961. *Ann. Phys.* **12**, 121.
12. Dominguez, I. Chieffi, A., Limongi, M., Straniero, O.: 1999, *ApJ* **524**, 226.
13. Gallino, R., Arlandini, C., Busso, M., Lugaro, M., Travaglio, C, Straniero, O., Chieffi, A., Limongi, M.: 1998, *ApJ* **497**, 388.
14. Gallino, R., Busso, M., & Lugaro, M.: 1997, in *Astrophysical Implications of the Laboratory Study of Presolar Material*, eds. T. J. Bernatowicz and E. Zinner, *AIP Conf. Proc.* **402**, Woodbury: American Institute of Physics, p. 115.
15. Herwig, F., Blöcker, T., Schönberner, D., El Eid, M.: 1997, *A&A* **324**, L81.
16. Hollowell, D., Iben, I.Jr.: 1989, *ApJ* **340**, 966.
17. Hoppe, P., Annen, P., Strebel, R., Eberhardt, P., Gallino, R., Lugaro, M., Amari, S., Lewis, R. S.: 1997 *ApJ* **487**, L101.
18. Iben, I.Jr.: 1973, *ApJ* **185**, 209.
19. Iben, I.Jr.: 1975a, *ApJ* **196**, 525.
20. Iben, I.Jr.: 1975b, *ApJ* **196**, 549.
21. Iben, I.Jr.: 1976, *ApJ* **208**, 165.
22. Iben, I.Jr.: 1982, *ApJ* **260**, 821.
23. Iben, I.Jr., Renzini, A.: 1982a, *ApJ* **259**, L79.
24. Iben, I.Jr., Renzini, A.: 1982b, *ApJ* **263**, L188.
25. Iben, I.Jr., Renzini, A.: 1983, *ARAA* **21**, 271.
26. Iben, I.Jr., Truran, J.W.: 1978, *ApJ* **220**, 980.
27. Forestini, M., Charbonnel, C.: 1997, *A&AS* **123**, 241.
28. Freytag, B., Ludwig, H.-G, Steffen, M.: 1996, *A&A* **313**, 497.
29. Käppeler, F., Gallino, R., Busso, M., Picchio, G., Raiteri, C. M.: 1990 *ApJ* **354**, 630.
30. Lambert, D.L., Smith, V.V., Busso, M., Gallino, R., Straniero, O.: 1995, *ApJ* **450**, 302.
31. Langer, N., Heger, A., Wellstein, S., Herwig, F: 1999, *A&A* **346**, L37.
32. Lattanzio, J., Frost, C., Cannon, R., Wood, P.: 1996, *Mem SAIt* **67**, 729.
33. Renzini, A., Voli, M.: 1981, *A&A* **94**, 175.
34. Sackmann, I.-J, Boothroyd, A.J.: 1992, *ApJ* **393**, L21.
35. Sackmann,I.-J, Smith, R.L., Despain, K.H.: 1974, *ApJ* **187**, 555.

36. Sanders, R.H.: 1967, *ApJ* **150**, 971.
37. Schwarzschild, M., Härm, R.: 1965, *ApJ* **142**, 855.
38. Schwarzschild, M., Härm, R.: 1967, *ApJ* **150**, 961.
39. Smith, V.V., Lambert, D.L.: 1985, *ApJ* **294**, 362.
40. Smith, V.V., Lambert, D.L.: 1986, *ApJ* **311**, 843.
41. Smith, V.V., Lambert, D.L.: 1989, *ApJ* **345**, 375.
42. Smith, V.V., Lambert, D.L.: 1990, *ApJS* **72**, 387.
43. Straniero, O., Gallino, R., Busso, M., Chieffi, A., Limongi, M., Salaris. M.: 1995, *ApJ* **440**, L85.
44. Straniero, O., Chieffi, A., Limongi, M., Gallino, R., Busso, M., Arlandini, C.: 1997, *ApJ* **478**, 332.
45. Straniero, O., Chieffi, A., Limongi, M., Gallino, R., Busso, M., Dominguez, I.: 2000, in *The changes in abundances in Asymptotic Giant Branch stars*, eds. F. D'Antona and R. Gallino, *Mem. SAIt*, in press.
46. Sugimoto, D.: 1971, *Prog. Theor. Phys.* **45**, 761.
47. Sweigart, A.V.: 1998, *IAUS* **191**, 533.
48. Ulrich, R.K., Scalo, J.M.: 1972, *ApJ* **176**, L37.
49. Vaglio, P., Gallino, R., Busso, M., Travaglio, C., Straniero, O., Chieffi, A., Limongi, M., Lugaro, M., Arlandini, C.: 1998, in *Nuclei in the Cosmos V*, eds. N. Prantzos and S. Harissopulos, Paris: Editions Frontières, p. 223.
50. Ventura, P., D'Antona, F., Mazzitelli, I: 1999 *ApJ* **524**, L111.
51. Wasserburg, G.J., Boothroyd, A.J., Sackmann, I.-J: 1995, *ApJ* **447**, L37.
52. Wood, P.R., Bessell, M.S., Fox, M.W.: 1983, *ApJ* **272**, 99.

YIELDS FROM TYPE IA SUPERNOVAE

F.-K. THIELEMANN, F. BRACHWITZ,
Department of Physics and Astronomy, University of Basel, CH-4056 Basel, Switzerland

K. NOMOTO, H. UMEDA,
Department of Astronomy and Research Center for the Early Universe, University of Tokyo, Tokyo 113-0033, Japan

K. IWAMOTO,
Department of Physics, Nihon University, Tokyo 101-8308, Japan

K. LANGANKE, G. MARTINEZ-PINEDO,
Institute of Physics & Astronomy, University of Aarhus, DK-8000 Aarhus C, Denmark

AND

D. J. DEAN, W. R. HIX AND M. S. STRAYER
Physics Division, Oak Ridge National Laboratory, Oak Ridge, TN 37831-4576, USA

Abstract. Supernovae of both types represent the most important nucleosynthesis contributors to galactic evolution and SNe Ia represent the major source of Fe-group nuclei. Our theoretical understanding is still burdened with many uncertainties, but the link between SNe Ia and exploding white dwarfs in binary stellar systems seems certain. The favored systems are accreting white dwarfs, approaching the (maximum stable) Chandrasekhar mass before contraction and central ignition. Major uncertainties are related (i) to the accretion rate in the binary system which determines the carbon ignition density, (ii) the flame speed after central ignition, (iii) a possible transition from a subsonic deflagration to a supersonic detonation, and (iv) a possible dependence of (i)-(iii) on the metallicity of the object or the accreted matter. The present overview tries to clarify the link between model uncertainties and their nucleosynthesis imprint.

F. Matteucci and F. Giovannelli (eds.), The Evolution of the Milky Way, 445–459.

1. Introduction

There are strong observational and theoretical indications that SNe Ia are thermonuclear explosions of accreting white dwarfs (Wheeler et al. 1995, Höflich & Khokhlov 1996, Nomoto et al. 1997, Livio 1999, Nomoto et al. 2000). Theoretical considerations lead to two options, (I) the so-called single degenerate or (II) double degenerate scenarios, involving either one or two C+O white dwarf(s) in a binary stellar system. The double degenerate scenario is related to the merging of two C+O white dwarfs with a combined mass exceeding the Chandrasekhar limit (Iben & Tutukov 1984). The single degenerate scenario relates to accretion of H or He via mass transfer from the binary companion (e.g. Nomoto 1982a). (a) In the case of He accretion at low rates, He detonates at the base of the accreted layer before the system reaches the Chandrasekhar mass (Nomoto 1982b, Woosley & Weaver 1994, Livne & Arnett 1995, Arnett 1996). (b) High accretion rates of H cause quasi-stable H-burning as well as subsequent He-burning in the accreted matter. This leads to an increase of the C+O white dwarf mass close to the Chandrasekhar mass limit before contraction and central ignition of C sets in. Both, the Chandrasekhar mass white dwarf models (b) and sub-Chandrasekhar mass models (b) have been considered so far (Nomoto et al. 1984, Canal 1997, Höflich et al. 1998) and the issue has recently experienced some progress. Observational features of SNe Ia in early phases clearly indicate that Chandrasekhar models give a more consistent picture than the sub-Chandrasekhar models of helium detonations (Höflich & Khokhlov 1996, Nugent et al. 1997, Livio 1999).

But even if the model type is agreed upon, there exist many unresolved aspects, leading to two possible approaches: (i) Self-consistent, first-principle calculations of supernovae which suffer, however, from either the analysis of instabilities and the multidimensional treatment of conductive and convective burning fronts which are related to the speed of deflagrations, or the possible transition from deflagrations to detonations. There has been major progress in recent years on these topics, but a fully resolved multidimensional treatment of the burning front propagation has not been performed, yet. (ii) Thus, when being interested in the outcome and nucleosynthesis yields of such events typically a more pragmatic approach is taken where ignition densities, the burning front speed, and the deflagration/detonation transition density are treated as free parameters. Variations of SN Ia properties like maximum brightness and lightcurve decay time (Phillips et al. 1990, Hamuy et al. 1995, Riess et al. 1999, Perlmutter et al. 1999) have to be understood within this framework (Höflich & Khokhlov 1996, Dominguez & Höflich 1999). [Metallicity dependences via the IMF, the C/O ratio in the resulting white dwarfs or the accreted matter can possibly affect galactic evolution as well cosmological use (Höflich et al. 1998, 1999, Umeda et al. 1999ab, Yoshi et al. 1996, Kobayashi et al. 1998, 2000). Very early time spectra could possibly help to identify the metallicity (Hatano et al. 1999, Lentz et al. 1999).]

By making use of the option (ii), i.e. parametrized explosion models, one can predict nucleosynthesis yields, but they are functions of the parameters employed. Details of the Fe-group composition in Chandrasekhar mass models depend on variations in the parameters mentioned above: central ignition densities, subsonic deflagration front speeds, and possible transitions from deflagrations to supersonic detonations (Thielemann et al. 1997, Iwamoto et al. 1999), which can provide some improvements over the traditional model W7 (Nomoto et al. 1984, Thielemann et al. 1986). This relates to the overall agreement with solar abundances based on a mix of SNe II and averaged SNe Ia contributions. However, direct abundance determinations in early and late time spectra as well as remnants of individual SNe Ia can also give constraints on the variation of these parameters. One interesting feature is e.g. the observed Ni/Fe ratio (Ruiz-Lapuente 1997, Liu et al. 1997, Mazzali et al. 1998, Dupke & White 1999), connected mostly to the deflagration speed. Ignition densities are related to accretion rates. Recent findings of supersoft X-ray sources make them potential progenitors of SN Ia events with high accretion rates and low ignition densities (Nomoto & Kondo 1991, Hachisu et al. 1996, 1999ab, Li & van den Heuvel 1997).

All such calculations should of course be performed with the best available input physics. A general overview of involved nuclear properties can be found in Käppeler et al. (1998) and Hoffman et al. (1999). Recent advances in electron capture rates for Fe-group (pf-shell) nuclei have been utilized in order to relate ignition densities and the Fe-group composition in a correct manner (Brachwitz et al. 1999).

2. Ignition and Burning Front Propagation

For Chandrasekhar mass models carbon ignition in the central region leads to a thermonuclear runway. High accretion rates cause a higher central temperature and pressure, favoring lower ignition densities. A flame front then propagates at a subsonic speed as a *deflagration wave* due to heat transport across the front (Nomoto et al. 1984). Here the most uncertain quantity is the flame speed which depends on flame front instabilities of various scales (Dominguez & Höflich 1999). Multi-dimensional hydro simulations of the flame propagation have been attempted by several groups, though the results are still preliminary (Khokhlov 1995, Niemeyer & Hillebrandt 1995, Niemeyer & Woosley 1997, Reinecke et al. 1999, Niemeyer 1999, Niemeyer et al. 1999, Livne 1999). These simulations have suggested that a carbon deflagration wave might propagate at a speed $v_{\rm def}$ as slow as a few percent of the sound speed $v_{\rm s}$ in the central region of the white dwarf. The nucleosynthesis consequences witness the actual burning front velocities and can thus serve as a constraint. After an initial deflagration in the central layers, the deflagration is assumed to turn into a detonation at lower densities (Khokhlov 1991, Woosley & Weaver 1994, Livne 1999). We will later discuss independent nucleosynthesis constraints for this deflagration/detonation transition.

In the deflagration wave, electron captures enhance the neutron excess. The amount of electron capture depends on both $v_{\rm def}$ (influencing the time duration of matter at high temperatures, and with it the availability of free protons) and the central density of the white dwarf ρ_{ign} (increasing the electron chemical potential or Fermi energy). The resultant nucleosynthesis in slow deflagrations (Khokhlov 1991) has some distinct features compared with faster deflagrations like W7 (Nomoto et al. 1984), thus providing important constraints on these two parameters. Initially slower deflagrations cause an earlier expansion of the outer layers with respect to the arrival of the burning front (information of the central ignition propagates with sound speed) and lead to lower burning densities for the outer deflagration and detonation layers. Our main aim is here to find the "average" SN I-a conditions responsible for their nucleosynthesis contribution to galactic evolution, especially for the Fe-group composition (Iwamoto et al. 1999).

This situation is complicated by the fact that electron capture on intermediate mass and Fe-group nuclei plays an important role in explosive burning. Electron capture affects the central electron fraction Y_e, which determines the composition of the ejecta from such explosions. Up to present, astrophysical tabulations based on shell model matrix elements were only available for light nuclei in the sd-shell (Fuller et al. 1980, 1982, 1985, Oda et al. 1994, Aufderheide et al. 1994). Recently new Shell Model Monte Carlo (SMMC) and large-scale shell model diagonalization calculations have also been performed for pf-shell nuclei (Dean et al. 1998, Martinez-Pinedo et al. 1999). These lead in general to a reduction of electron capture rates in comparison with previous, more phenomenological, approaches. Making use of the new shell model based rates, first results were reported by Brachwitz et al. (1999) for the composition of Fe-group nuclei produced in the central regions of SNe Ia in combination with possible changes in the constraints on model parameters like ignition densities ρ_{ign} and burning front speeds v_{def}.

The transition from a deflagration to a detonation (delayed detonation model) leads to a change in the ratio of incomplete Si-burning and,complete Si-burning with either normal freeze-out or alpha-rich freeze-out. The latter is less alpha-rich than in SNe II due to the smaller entropies or higher densities which leaves an imprint on the Fe-group composition. We know from low metallicity stellar observations that the average Fe-group yields of SNe II differ from those of SNe Ia (e.g. in the elemental Cr/Fe, Mn/Fe or also O-Ca/Fe ratio). We will see that this fact gives the strongest constraints on the transition density ρ_{tr} from deflagrations to detonations.

3. The Fe-Group Composition in Slow Deflagrations

In order to show the effect of ignition densities we adopt two models with central densities of ρ=1.37 (C) and 2.12×10^9 g cm^{-3} (W) at the onset of thermonuclear runaway and assume that a slow (S) deflagration propagates with speeds $v_{\rm def}/v_{\rm sound}$ = 0.015 (WS15, CS15), 0.03 (WS30,CS30) or 0.05 (CS50) and consider also two extreme cases of fully and initially laminar flame fronts (WLAM,WSL).

Fig.1 from Iwamoto et al. (1999b) summarizes the major options to change Y_e values in the central part of SNe Ia explosions: (i) the burning front speed v_{def}, determining the Y_e gradient and (ii) the central ignition density ρ_{ign}, being inversely proportional to Y_e. These features hold true as long as Y_e values in explosive burning do not drop below 0.44, when competing β^--decays have also to be taken into account (case WSL and WLAM). Y_e values of 0.47-0.485 lead to dominant abundances of ^{54}Fe and ^{58}Ni, values between 0.46 and 0.47 produce dominantly ^{56}Fe, values in the range of 0.45 and below are responsible for ^{58}Fe, ^{54}Cr, ^{50}Ti, ^{64}Ni, and values below 0.43-0.42 are responsible for ^{48}Ca. The intermediate Y_e-values 0.47-0.485 exist for all models of the set shown in Fig. 7, but the related mass zones which experience these conditions depend on the Y_e-gradient and thus v_{def}. Whether the lower vales with Y_e<0.45 are attained, depends on the central ignition density ρ_{ign}. Therefore, ^{54}Fe and ^{58}Ni are indicators of v_{def} while ^{58}Fe, ^{54}Cr, ^{50}Ti, ^{64}Ni, and ^{48}Ca are a measure of ρ_{ign}.

If SNe Ia are responsible for at least 50 % of the Fe-group nuclei in galactic evolution (see e.g. Matteucci 1996 and the discussion given in I-wamoto et al. 1999), the averaged SN Ia contribution should in principle not exhibit isotopic overabundances with respect to ^{56}Fe by more than a factor of 2, independent of the theoretical models which try to explain SNe Ia. Observational uncertainties might lead to larger deviations. However, recent findings reported by Dupke & White (1999) of Ni/Fe ratios in SN Ia remnants close to three times solar have to be analyzed with possible selection effects in mind, perhaps not showing a metallicity average. If we follow the above given arguments on averaged type Ia yields, and permit a limiting factor of 3 due to model and nuclear uncertainties, we can obtain limits for the model parameters v_{def} and ρ_{ign}.

^{54}Fe and ^{58}Ni can be produced within the factor of 3 uncertainty limits

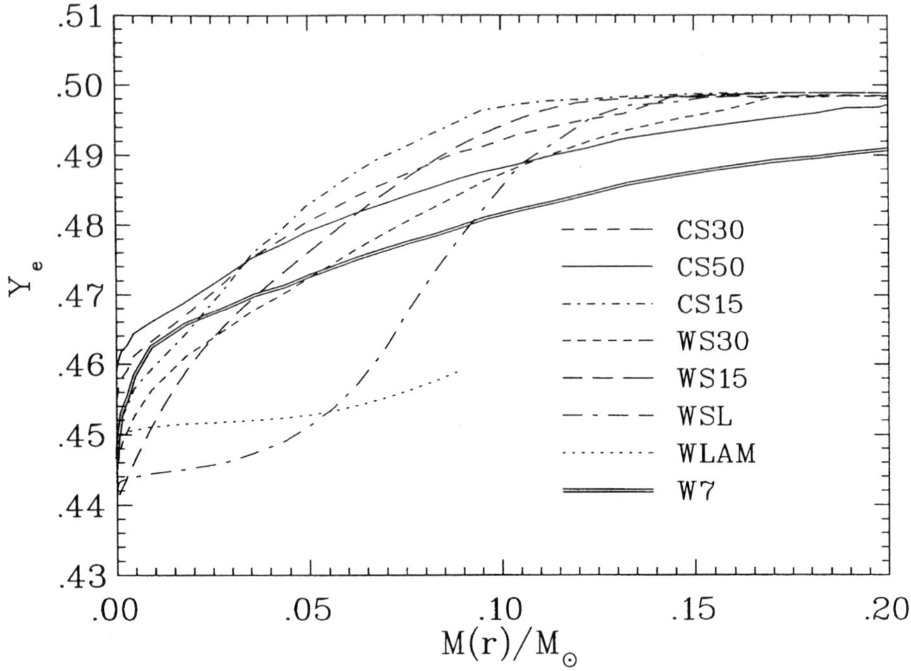

Figure 1. Y_e after freeze-out of nuclear reactions measures the electron captures on free protons and nuclei. Small burning front velocities lead to steep Y_e-gradients which flatten with increasing velocities (see the series of models CS15, CS30, and CS50 or WS15, WS30, and W7). Lower central ignition densities shift the curves up (C vs. W), but the gradient is the same for the same propagation speed. Only when the Y_e from electron captures is smaller than for stable Fe-group nuclei, subsequent β^--decays will reverse this effect (WSL and WLAM).

for the cases WS15, WS30, CS15, CS30, i.e. favoring v_{def} of 1.5-3% of the sound speed. ^{54}Cr is produced within a factor of two in CS15 and CS30, but apparently overproduced in WS15 and WS30, due to the too low central Y_e's. ^{50}Ti is produced close to solar values for CS15, slightly underproduced by CS30, well produced for WS30, and clearly overproduced in WS15 (for full details see Iwamoto et al. 1999 and the top of Fig.??). This exercise argues for CS15 and CS30 to be better models than WS15, WS30, W7 or CS50, in terms of avoiding overproduction of neutron-rich elements. The conclusions to be drawn from these results are that (i) v_{def} in the range 1.5-3% of the sound speed is preferred (cases 15 and 30 over 50), and (ii) ignition

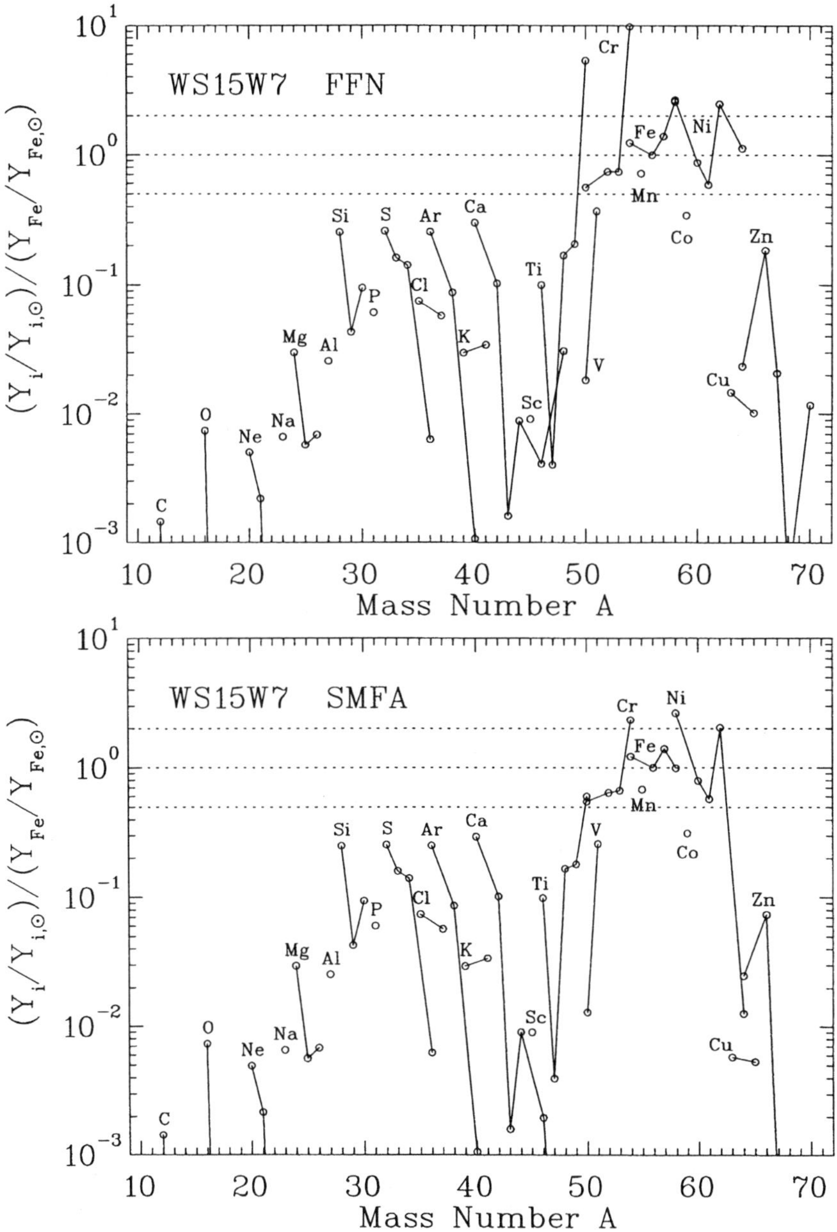

Figure 2. Ratio of abundances to solar predicted in model WS15 for different electron capture rate sets. Isotopes of one element are connected by lines. The ordinate is normalized to ^{56}Fe. Intermediate mass elements exist, but are underproduced by a factor of 2-3 for SNe Ia models in comparison to Fe-group elements. For FFN rates the Fe-group does not show a composition close to solar. Especially ^{54}Cr and ^{50}Ti are strongly overproduced by more than a factor of 3. The change from FFN rates (top) to SMFA (bottom) reduces the overproduction over solar strongly.

densities $\rho_{ign}<2\times10^9$ g cm^{-3} provide apparently a better agreement with the solar abundances of very neutron-rich species (case C rather than W). The latter interpretation seems to change, however, with improved electron capture rates on pf-shell nuclei (Dean et al. 1998, Martinez-Pinedo et al. 1999) and which permit ignition densities slightly in excess of $\rho_{ign}=2\times10^9$ g cm^{-3} (see bottom of Fig. 2, i.e. SMFA). In either case, ignition densities in the range $1.5-2\times10^9$ g cm^{-3} are in agreement with identifying the recently discovered supersoft X-ray sources as SN Ia progenitor systems with hydrogen accretion rates close to $\dot{M}=10^{-7}$M$_\odot$ (Li & van den Heuvel 1997).

4. The Effect of Improvements in Fe-group Electron Capture Rates

As the electron gas in white dwarfs is degenerate, characterized by high Fermi energies for the high density regions in the center, electron capture on intermediate mass and Fe-group nuclei plays an important role in explosive burning. Electron capture affects the central electron fraction Y_e, which determines the composition of the ejecta from such explosions. Recently new Shell Model Monte Carlo (SMMC) and large-scale shell model diagonalization calculations have also been performed for pf-shell nuclei (Dean et al. 1998, Martinez-Pinedo et al. 1999). These lead in general to a reduction of electron capture rates in comparison with previous, more phenomenological, approaches. Making use of these new shell model based rates, we present the first results from Brachwitz et al. (1999) for the composition of Fe-group nuclei produced in the central regions of SNe Ia and possible changes in the constraints on model parameters like ignition densities ρ_{ign} and burning front speeds v_{def}.

Four rate sets were employed: (i) The original FFN rates by Fuller et al. (1980, 1982) as a benchmark for further comparisons; (ii) inclusion of the electron captures rates calculated within the SMMC method by Dean et al. (1998), replacing the corresponding FFN rates. SMMC rates were used for the parent nuclei ^{45}Sc,^{48,50}Ti, ^{51}V,^{50,52}Cr,^{55}Mn,$^{54-56,58}$Fe, and ^{55,57,59}Co, ^{56,58,60}Ni, otherwise rates were taken from FFN; (iii) to simulate potential modification of the rates not provided by the SMMC method, we also multiplied the FFN electron capture and beta-decay rates within the Fe-group nuclei by factors, derived from comparison between FFN and rates from large-scale shell model diagonalization calculations by Martinez-Pinedo et al. (1999, labeled with SMFA). (iv) A further option is to treat even-even (ee), odd-A (oa), and odd-odd (oo) nuclei in different ways, in order to test the sensitivity of the models and the importance of the rates of particular nuclei. Such calculations are denoted by SMFA with the corresponding extension ee, oa, oo or by combinations, e.g. ee+oa. With these modifications of the electron capture rates we recalculated the nucleosynthesis for the SN Ia models WS15 and CS15.

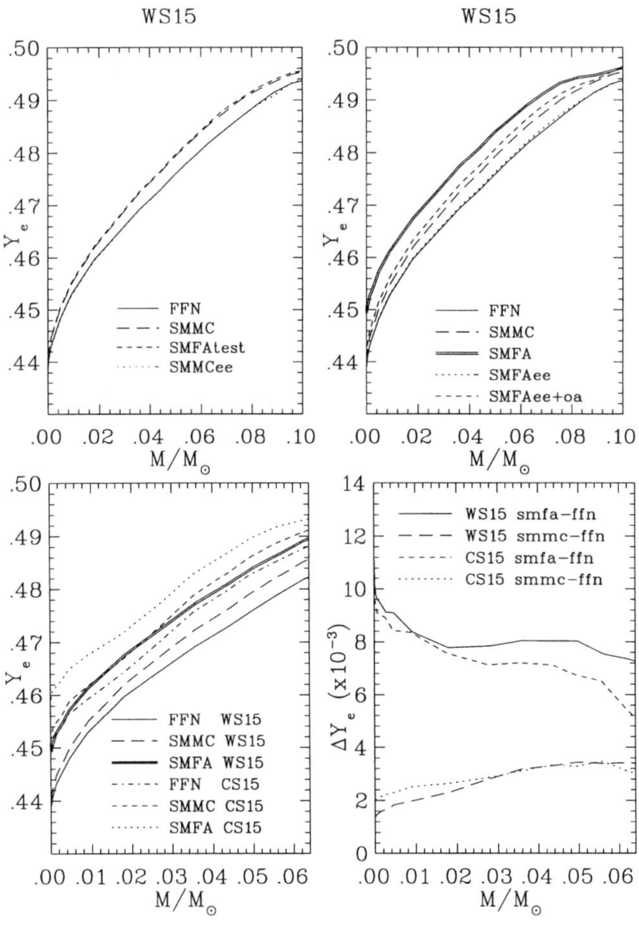

Figure 3. Y_e, the total proton to nucleon ratio and thus a measure of electron captures on free protons and nuclei, after freeze-out of nuclear reactions, as a function of radial mass for different models and electron capture rates. Also the Y_e-difference ΔY_e between various cases is shown at the bottom right (d). A detailed discussion of the changes with different electron capture rate sets [FFN=Fuller et al. 1980, 1982), SMMC=Dean et al. 1998), and SMFA=Martinez-Pinedo et al. (1999)] is given in the text.

The resulting Y_e-curves (Fig. ??b) displays a small Y_e shift between SMMC and SMFAee+oa, and a larger Y_e-shift between SMFAee+oa and SMFA. Therefore, the inclusion of odd-odd nuclei has the largest influence on the Y_e difference between SMFA and SMMC. Thus, the rate change for odd-A nuclei is mostly responsible for the Y_e-shift between FFN and SMMC, and the inclusion of odd-odd nuclei causes the largest part of the Y_e-shift between SMMC and SMFA. This makes clear that the changes in the electron capture rates for odd-A and odd-odd nuclei are responsible for the Y_e difference between SMFA and FFN, while the contribution of even-even nuclei is negligible, an assertion which was directly tested by case SMFAee (Fig. ??b).

Notice, however, that the changes for a given model (here WS15 and CS15) lead to almost parallel Y_e-curves in the intermediate Y_e-range responsible for the major abundances of ^{54}Fe and ^{58}Ni. This can also be seen in the close to constant ΔY_e-curves in (Fig. ??d). Thus, a change in electron capture rates does (to first order) not affect the Y_e-gradient of a model. Iwamoto et al. (1999) showed that the Y_e gradient is determined by v_{def} . Therefore, we can conclude that the consequences for the permitted range of burning front speeds remain the same. In Iwamoto et al. (1999) we determined this range v_{def}/v_s to be of the order 0.015-0.03. The central neutronization is dependent on ρ_{ign}, as shown in the previous subsection, and on the set of electron capture rates employed (see Fig. ??c). If the trend as experienced between models C (CS15 $\rho_{ign} = 1.37 \times 10^9$ g cm^{-3}) and W (WS15, $\rho_{ign} = 2.12 \times 10^9$ g cm^{-3}) continues in a similar way, we would expect a central Y_e-value comparable to that of WS15 with FFN rates for $\rho_{ign} = 2.6 \times 10^9$ g cm^{-3} when utilizing SMFA. This corresponds to a permitted ignition density increase by about a factor of 1.24 when shifting from FFN to SMFA rates.

5. Deflagration to Detonation Transition

Finally, if a deflagration turns into a detonation, the transition density ρ_{tr} affects the total amount of ^{56}Ni, the intermediate mass elements Si-Ca, and the ratio of different explosive Si-burning regimes like alpha-rich freeze-out to incomplete Si-burning (see Fig. 4 taken from Iwamoto et al. 1999). The ^{58}Ni and ^{54}Fe plateaus indicate the regions of alpha-rich freeze-out vs. incomplete Si-burning. The most obvious consequence of choosing different transition densities is the amount of ^{56}Ni produced in a SN Ia event (DD2 and DD1 stand for deflagration/detonation transitions occurring at densities of 2.2 and 1.7 $\times 10^7$ g cm^{-3}). Höflich & Khokhlov (1996) found from light curve modeling and spectra that the typical ^{56}Ni mass should be in the range 0.5-0.7 $M_\odot$. This agrees with the original fast deflagration model W7 (Nomoto et al. 1984, Thielemann et al. 1986). Among the DD models it would ask for a value somewhere between DD1 and DD2 (closer to DD2). The amount of Si-Ca in comparison to Fe is too large in DD1 models in order to compensate during galactic evolution for the well known overproduction of Si-Ca in SNe II (Matteucci 1996). Si/Fe ratios in SN Ia models put constraints on permitted Ia/(II+Ib) frequencies in order to obtain a solar mix combined with SN II contributions. DD2 seems to be closest to the present observational limits for this ratio by Cappelaro et al. (1997, see also van den Bergh & Tammann 1991).

Small transition densities favor larger amounts of matter which experience incomplete Si-burning. Low metallicity constraints (McWilliam 1997) require some overproduction of Mn (and Cr) in SNe Ia. These elements are mostly made as ^{55}Co and ^{52}Fe (decaying to Mn and Cr), which are favorably produced in incomplete Si-burning and would also require a deflagration/detonation transition between DD1 and DD2. (One should, however, realize that a fast deflagration like W7 can simulate this as well and on that these numbers would have to be rescaled or reinterpreted in multi-D calculations, Livne 1999.) Thus, combining all requirements on transition densities from total ^{56}Ni-yields, Si/Fe and Ia/(II+Ib) ratios, as well as specific elements favored in incomplete Si-burning, we would argue for a transition density close to 2×10^7 g cm^{-3}, i.e. between models DD1 and DD2. One should, however, be careful with these constraints based on spherically symmetric approximations of the burning front. Full 3D calculations could possibly produce the required ratio of matter from incomplete Si-burning and complete Si-burning with alpha-rich freeze-out in a different way (Livne 1999).

6. Conclusions and Outlook

SNe Ia represent the major source of Fe-group nuclei, i.e. they have to explain the "solar" Fe-group composition in combination with SN II yields, in particular abundances of nuclei like e.g. ^{50}Ti, ^{52,54}Cr, ^{55}Mn, ^{54,56}Fe, and ^{58}Ni. Major uncertainties involved in Chandrasekhar mass models are: (i) the companion star of the accreting white dwarf (which determines the accretion rate and the carbon ignition density ρ_{ign}), (ii) the flame speed of the deflagration front after central C-ignition v_{def}, and (iii) the density ρ_{tr} at which a possible transition from deflagrations to detonations occurs. Nucleosynthesis calculations for relatively slow deflagrations with a variety of deflagration speeds and ignition denisities, combined with delayed detonations, can provide several constraints for these parameters as they apply to "average" progenitor systems. The advantage is that different isotopes give independent clues on the parameters (i)-(iii). The Si-Ca/Fe ratio is another measure of the deflagration/detonation transition density and is directly related to the SN Ia/SN II ratio.

We have tested the effect of improved electron capture rates for Fe-group (pf-shell) nuclei on the conclusions drawn from the considerations

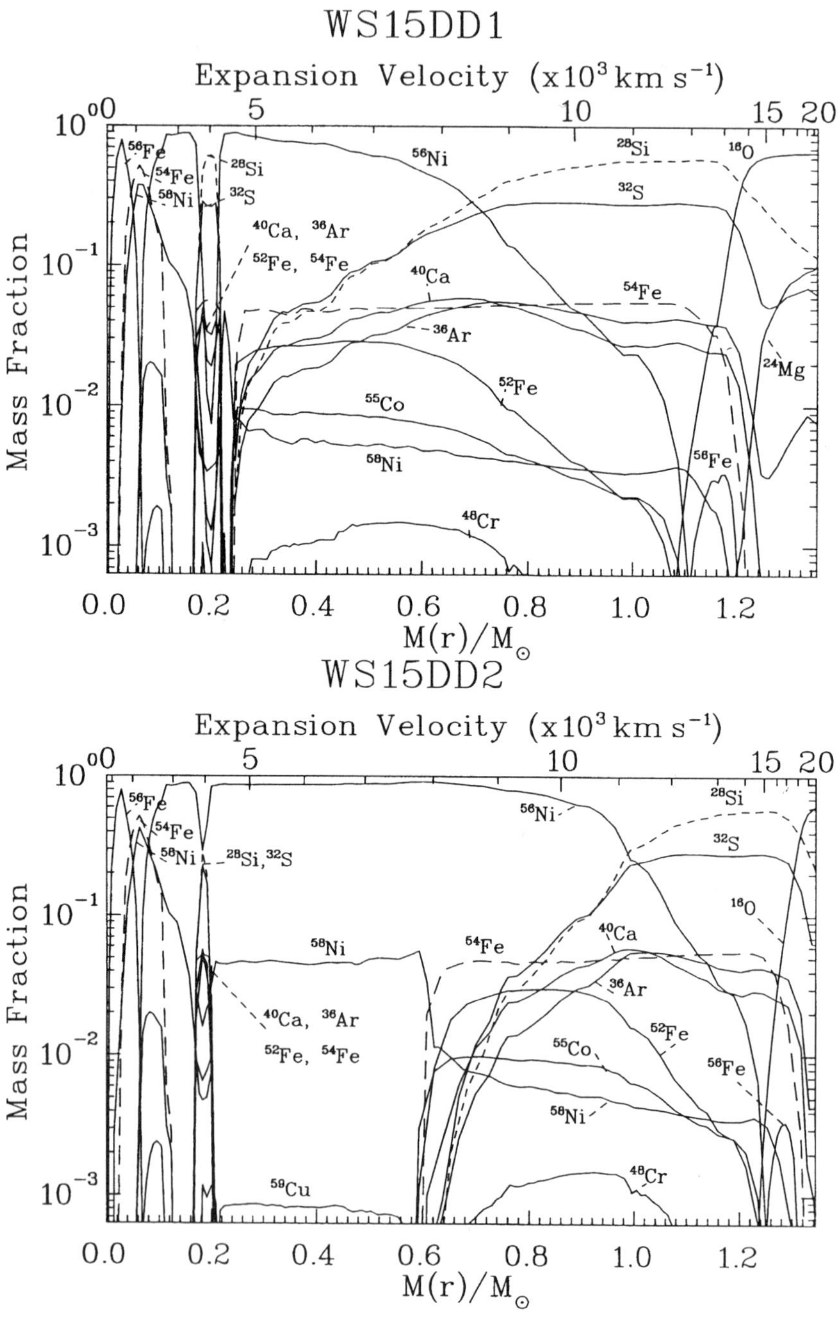

Figure 4. Composition of delayed detonation models WS15DD1 and DD2 as a function of expansion velocity or radial mass $M(r)$. In the series DD1-DD2 we see a decrease in the total amount of intermediate mass elements (Si-Ca), an increase in ^{56}Ni, and a change of the ratio between matter experiencing an alpha-rich freeze-out (indicated by the ^{58}Ni-plateau) and incomplete Si-burning (^{54}Fe-plateau).

listed above. A systematic analysis of metallicity effects on SN Ia events and their outcome is still pending.

This work has been supported in part by the Swiss Nationalfonds (2000-53798.98), the US Department of Energy (DOE contracts DE-AC05-96OR22464 and DE-FG02-96ER40983), the Danish Research Council, the grant-in-Aid for COE research (07CE2002) of the Ministry of Education, Science, and Culture in Japan, and a fellowship of the Japan Society for the Promotion of Science for Japanese Junior Scientists (6728). Some of us (KN and FKT) thank the Aspen Center for Physics for hospitality and inspiration during the 1999 Type Ia supernova program.

References

1. Arnett, W.D. 1996, *Nucleosynthesis and Supernovae*, Princeton Univ. Press
2. Aufderheide, M., Fushiki, I., Woosley, S.E., Hartmann, D. 1994, Ap. J. Suppl. 91, 389
3. Brachwitz, F., Dean, D.J., Hix, W.R., Iwamoto, I., Kishimoto, N., Langanke, K., Martínez-Pinedo, G., Nomoto, N., R. Strayer, M.S., Thielemann, F.-K. 1999, ApJ, in press (astro-ph/0001464)
4. Canal, R., 1997, in *Thermonuclear Supernovae*, eds. P. Ruiz-Lapuente, R. Canal, J. Isern, Kluwer Academic Publishers, p. 257
5. Cappellaro, E., Turatto, M., Tsvvetkov, D.Yu., Bartunov, O.S., Pollas, C., Evans, R., & Hamuy, M. 1997, A&A, 322, 431
6. Dean, D.J. et al. 1998, Phys. Rev. C58, 536
7. Dominguez, I., Höflich, P. 1999, ApJ, in press, astro-ph/9908204
8. Dupke, R.A., White, R.E. 1999, ApJ, in press
9. Fuller, G.M., Fowler, W.A., Newman, M. 1980, Ap. J. Suppl. 42, 447
10. Fuller, G.M., Fowler, W.A., Newman, M. 1982, Ap. J. Suppl. 48, 279
11. Fuller, G.M., Fowler, W.A., Newman, M. 1985, Ap. J. 293, 1
12. Hachisu, I., Kato, M., & Nomoto, K. 1996, ApJ 470, L97
13. Hachisu, I., Kato, M., & Nomoto, K. 1999a, ApJ 519, in press, 1999b, submitted
14. Hamuy, M., Phillips, M.M., Maza, J., Suntzeff, R.A., Aviles, R. 1995, AJ, 109, 1
15. Hatano, K., Branch, D., Fisher, A., Baron, E., Filippenko, A.V. 1999, ApJ, in press
16. Höflich, P., Khokhlov, A. 1996, Ap. J., 457, 500
17. Höflich, P., Wheeler, J.C., Thielemann, F.-K. 1998, Ap.J. 495, 617
18. Höflich, P., Nomoto, K., Umeda, H., Wheeler, J.C. 1999, ApJ, in press, astro-ph/9908226
19. Hoffman, R.D., Woosley, S.E., Weaver, T.A., Rauscher, T., Thielemann, F.-K. 1999, ApJ 521, 735
20. Iben, I.Jr., Tutukov, A.V. 1984, Ap. J. Suppl. 54, 335
21. Iwamoto, K., Brachwitz, F., Nomoto, K., Kishimoto, N., Umeda, H., Hix, W.R., Thielemann, F.-K. 1999, ApJS 125, 439
22. Käppeler, F., Thielemann, F.-K., Wiescher, M. 1998, Ann. Rev. Nucl. Part. Sci. 48, 175
23. Khokhlov, A.M. 1991, A&A 245, 114; 245, L25
24. Khokhlov, A.M. 1995, Ap. J. 449, 695
25. Kobayashi, C., Tsujimoto, T., Nomoto, K., Hachisu, I., & Kato, M. 1998, ApJ, 503, L155
26. Kobayashi, C., Tsujimoto, T., & Nomoto, K. 2000, ApJ, in press (astro-ph/9908005)
27. Lentz, E.J., Baron, E., Branch, D., Hauschildt, P.H., Nugent, P.E., 1999, submitted to ApJ

28. Li, X.D., van den Heuvel, E.P.J. 1997, Astron. Astrophys. 322, L9
29. Livio, M. 1999 in *Type Ia Supernovae: Theory and Cosmology*, Cambridge Univ. Press, in press
30. Livne, E. 1999, ApJ, in press, astro-ph/9910471
31. Livne, E., Arnett, W.D. 1995, Ap. J. 452, 62
32. Liu, W.H., Jeffery, D.J., Schultz, D.R. 1997, Ap. J. 483, L107
33. Martinez-Pinedo, G., Langanke, K., & Dean, D.J. 1999, ApJ, in press
34. Matteucci, F. 1996, Fund. Cosm. Phys. 17, 283
35. Mazzali, P., Cappelaro, E., Danziger, I.J., Turatto, M., Benetti, S. 1998, ApJ, 499, L49
36. McWilliam, A. 1997, Ann. Rev. Astron. Astrophys. 35, 503
37. Niemeyer J.C., Hillebrandt W. 1995, Ap. J., 452, 769
38. Niemeyer, J.C., Woosley, S.E. 1997, Ap. J. 475, 740
39. Niemeyer, J.C., Bushe, W. K., Ruetsch, G.R. 1999, ApJ 524, 290
40. Niemeyer, J.C. 1999, ApJ 523, L57
41. Nomoto, K. 1982a, ApJ 253, 798
42. Nomoto, K. 1982b, ApJ 257, 780
43. Nomoto, K., Thielemann, F.-K., Yokoi, K. 1984, Ap. J. 286, 644
44. Nomoto, K., Kondo, Y. 1991, Ap. J. 367, L19
45. Nomoto, K., Iwamoto, K., Kishimoto, N. 1997, Science 276, 1378
46. Nomoto, K., Umeda, H., Hachisu, I., Kato, M., Kobayashi, C., Tsujimoto, T. 2000 in *Type Ia Supernovae: Theory and Cosmology*, Cambridge Univ. Press, eds. J. Niemeyer & J.W. Truran, in press (astro-ph/9907386)
47. Nugent, P., Baron, E., Branch, D., Fisher, A., Hauschildt, P.H. 1997, Ap. J. 485 , 812
48. Oda, T., Hino, M., Muto, K., Takahara, M., Sato, K. 1994, At. Data Nucl. Data Tables 56, 231
49. Perlmutter, S. et al. 1999, ApJ, 517, 565
50. Phillips, M.M., Hamuy, M., Maza, J., Ruiz, M.T., Carney, B.W., Graham, J.A. 1990, PASP, 102, 299
51. Reinecke, M., Hillebrandt, W., Niemeyer, J.C. 1999, Astron. Astrophys., submitted
52. Riess, A.G. et al. 1999, AJ, 117, 707
53. Ruiz-Lapuente, P. 1997, in Thermonuclear Supernovae, eds. P. Ruiz-Lapuente, R. Canal, J. Isern, (Kluwer Academic Publishers), 681
54. Thielemann, F.-K., Nomoto, K., Yokoi, K. 1986, A&A 158, 17
55. Thielemann, F.-K., Nomoto, K., Hashimoto, M. 1996, Ap. J. 460, 408
56. Thielemann, F.-K. Nomoto, K., Iwamoto, K., Brachwitz, F. 1997, in *Thermonuclear Supernovae*, eds. P. Ruiz-Lapuente, R. Canal, J. Isern, Kluwer Academic Publishers, p. 485
57. Umeda, H., Nomoto, K., Yamaoka, H., Wanajo, S. 1999a, ApJ, 513, 861
58. Umeda, H., Nomoto, K., Kobayashi, C., Hachisu, I., Kato, M. 1999b, ApJ 522, L43
59. van den Bergh, S., & Tammann, G. 1991, ARA&A 29, 363
60. Wheeler, J.C., Harkness, R.P., Khokhlov, A.M., Höflich, P. 1995, Phys. Rep., 53, 221
61. Woosley, S.E., Weaver, T.A. 1994, Ap. J., 423, 371
62. Yoshi, Y., Tsujimoto, T., & Nomoto, K. 1996, ApJ, 462, 266

Speaker: F.K. THIELEMANN

R. GALLINO: What are the effects of the initial metallicity on SNIa nucleosynthesis? In particular the effects on ^{58}Fe and ^{64}Ni?

F.-K. THIELEMANN: There is a possible implicit metal dependence on the progenitor systems, including possible variations of the C/O ratio of the WD (see e.g. talks by I Dominguez and C. Kobayashi). I cannot comment on that dependence. However, assuming the same WD dwarf progenitor with unchanged C/O ratio, the metallicity of CNO nuclei turns after H and He-burning to ^{22}Ne which is a N=Z+2 nucleus. This has an effect on Y_e throughout the star (essentially for M(r)> 0.2 $M_\odot$ where Y_e is not changed due to electron captures). This affects directly ^{54}Fe and ^{58}Ni in α-rich freeze-out and incomplete Si-burning zones. ^{58}Fe and ^{64}Ni result from zones where a much smaller Y_e is attained due to electron capture in M(r)< 0.2 $M_\odot$.

I do not see any metallicity dependence (explicit) for these inner regions.

THE EVOLUTION OF ZERO METAL INTERMEDIATE MASS STARS

I. DOMÍNGUEZ[1]
Dpto. de Física Teórica y del Cosmos
Universidad de Granada, 18071 Granada, Spain
O. STRANIERO[2]
Osservatorio Astronomico di Collurania
I-64100 Teramo, Italy
M. LIMONGI[3]
Osservatorio Astronomico di Roma
Via Frascati 33, I-00040 Monteporzio Catone (Roma), Italy
AND
A. CHIEFFI[4]
Istituto di Astrofisica Spaziale (CNR)
Via fosso del Cavaliere, I-00133 Roma, Italy

1. Abstract

It is thought that the first stars formed from pregalactic clouds deprived of metals. Subsequent stellar nucleosynthesis has provided for all the observed elements from Carbon to Uranium. The first stars may have expanded a wide range of masses. In the present study we focus on intermediate mass stars with masses greater than 5 $M_{\odot}$ and Z=0. It is found, contrary to previous claims, that these stars experience the TP-AGB phase. During the AGB the downward *incursions* of the envelope increase the envelope metal content and nearly double the initial Helium.

2. Introduction

In the eighties a great effort was devoted to study the role of Pop. III stars and all the different aspects of the problem were considered. At that time cosmology experimented substantial theoretical advances but due to the lack of observations many results and predictions were speculative.

F. Matteucci and F. Giovannelli (eds.), The Evolution of the Milky Way, 461–466.

Nowadays the new telescopes give information up to redshift z=7 and with the New Generation of Space Telescopes and various satellites it would be possible to test and constraint the scenarios of galaxy formation, supernovae rates at high redshifts and so on.

Surveys looking for very metal poor stars are going on since 1981 and by now around 70 of these objects have been identified (Bond 1981, Bessel and Norris 1984, Beers, Preston and Shectman 1992; Ryan, Norris and Bessel 1991; Ryan, Norris and Beers 1996; Sneden et al. 1994, Primas et al. 1994, Ryan et al. 1991, Carney and Peterson 1991, Molaro and Castelli 1990, Molaro and Bonifacio 1990).

The first stars produced the first heavy elements in the Universe, maybe also produced the heavy elements observed in the Lyman-alpha forest clouds at redshifts z=2 to 4. These metals may have also played a role in the formation of galaxies, acting as coolants. They could as well cause enough ionizing radiation to reheat the universe at redshifts $z \geq 5$, as it is required by the observations.

In order to understand the role played by population III we have to know its *elusive* initial mass function (IMF). Recently Nakamura and Umemura (1999) have obtained that the typical mass of Pop. III is around $3M_{\odot}$, that may grow by accretion up to $16M_{\odot}$. Previously Yoshii and Saio (1986) found that the peak of the Pop. III IMF is around $4\text{-}10M_{\odot}$. If these predictions are right intermediate mass stars were abundant in the early universe.

In this paper we firstly present the evolution of intermediate mass zero metal stars with masses over $5M_{\odot}$ including the TP-AGB phase. The only previous work in which the evolution of a zero metal intermediate mass star is followed further than the early AGB (E-AGB) is that of Chieffi and Tornambè (1984). In particular they studied the evolution of a Z=0, $5M_{\odot}$ star. Instead of the usual thermal pulses, they found that, after some small instabilities, both shells advance contemporaneously and experience a steady burning. These results were confirmed by Fujimoto et al. (1984). Contrary to these works, we obtain that these stars experience the TP-AGB phase, behaving as more metal rich stars.

3. The models

All the evolutionary models presented in this work have been computed by means of the latest version of the FRANEC (Frascati RAphson Newton Evolutionary Code; release 4.7). We recall that the nuclear burning and the physical evolution are coupled and that a time dependent mixing scheme is adopted. The nuclear network includes 41 isotopes (269 reactions) for the H burning and 26 isotopes (147 reactions) for the He burning. In addition a reduced set of nuclear species and related reactions have been added for the Carbon burning, namely 9 isotopes (8 reactions), just to identify the value of M_{up}. For a detailed description of the code see Chieffi, Limongi & Straniero (1998)

We present here the evolution of a 5, 6 and 7 $M_\odot$ (Z=0 and Y=0.23) from the pre-main sequence up to the AGB. We have also followed the evolution of an 8 $M_\odot$ up to an off center Carbon ignition. In this first investigation no mass loss has been assumed.

4. Central H and He burning phases

The qualitative behaviors of our models before the AGB phase does not substantially differ from the ones already known (Ezer 1961, 1972, 1981; Ezer and Cameron 1971; Wagner 1974; D'Antona and Mazzitelli 1982; Castellani, Chieffi and Tornambè 1983; Tornambè and Chieffi 1986; Cassisi and Castellani 1993; Cassisi, Castellani and Tornambè 1996). Let us just recall the main characteristic of the various phases.

During the pre-main sequence these stars contract till the conditions for H burning via the p-p chain are achieved. The resulting pre-main sequence lifetime is longer for the zero metal models, as compared with the same masses having larger metallicities (i.e. in which the CNO cycle can occur). The ZAMS is located at higher L and T_{eff}. A small convective core develops (half the size of the one corresponding to a *normal* metallicity star), but the H burning extends far outside this core (it covers almost the 80% of the total mass of the star). The H ignition does not stop the contraction and the temperature continues to rise, until the 3α reactions start and some Carbon is produced. Note that only a small quantity of ^{12}C ($X_C \sim 10^{-11}$) is sufficient to switch the H burning to the more energetic CNO cycle. As a consequence of this new (and more efficient) burning regime, the local luminosity increases and the convective core grows in mass. Near the end of the central H burning the convective core disappears. During the major part of the He burning the star remains in the blue side of the HR diagram, so that no dredge-up occurs. For this reason, these stars enter the AGB with the original surface composition.

5. The early-AGB phase

In table 1 we have summarized some properties of our models at the end of the early AGB (E-AGB), namely (from column 1 to 6): the total mass, the mass of the He core, the surface abundances (mass fraction) of ^{4}He, ^{12}C, ^{14}N and ^{16}O.

Note the very large amount of He brought to the surface (the initial one was 0.23). This He was mainly produced during the extended central H burning (see previous section). At variance with *normal* stars for which the

TABLE 1.

M($M_\odot$)	$M_H(M_\odot)$	^{4}He	^{12}C	^{14}N	^{16}O
At the end of the 2nd dredge-up					
5.0	0.886	0.365	$9.47 \cdot 10^{-10}$	$4.29 \cdot 10^{-10}$	$4.99 \cdot 10^{-11}$
6.0	0.933	0.367	$8.44 \cdot 10^{-8}$	$8.29 \cdot 10^{-10}$	$1.17 \cdot 10^{-10}$
7.0	0.988	0.367	$1.88 \cdot 10^{-6}$	$1.41 \cdot 10^{-9}$	$2.71 \cdot 10^{-9}$
After 25 TPs					
5.0	0.906	0.365	$1.05 \cdot 10^{-5}$	$1.73 \cdot 10^{-6}$	$9.57 \cdot 10^{-7}$

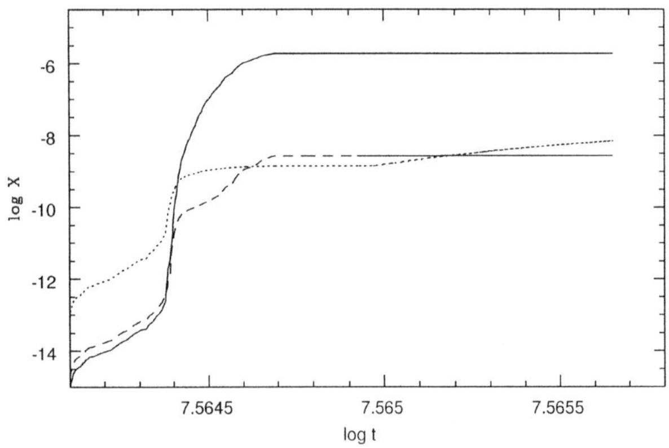

Figure 1. Evolution of the surface mass fraction of ^{12}C (solid line), ^{14}N (dashed line) and ^{16}O (dotted line), in the 7 $M_\odot$.

total amount of CNO is not modified by the 2nd dredge-up, here there is an important rise of all these elements. In fact, the surface Carbon abundance comes from the 3α, which were active during the H-burning. Such a Carbon was subsequently partially burned by the CNO, allowing a certain production of Oxygen and Nitrogen. In figure 1 and 2 we report the variation of the surface abundance of the CNO for the 7 and 5 $M_\odot$, respectively.

6. The TP-AGB phase

It is commonly believed that a zero metal star of intermediate mass does not experience the usual thermal instabilities (thermal pulse or TP), which characterize the AGB evolution, unless its core mass is lower than a critical value (Fujimoto et al. 1984). This is certainly true if these stars can maintain the original (no metals) composition in the envelope, so that the shell

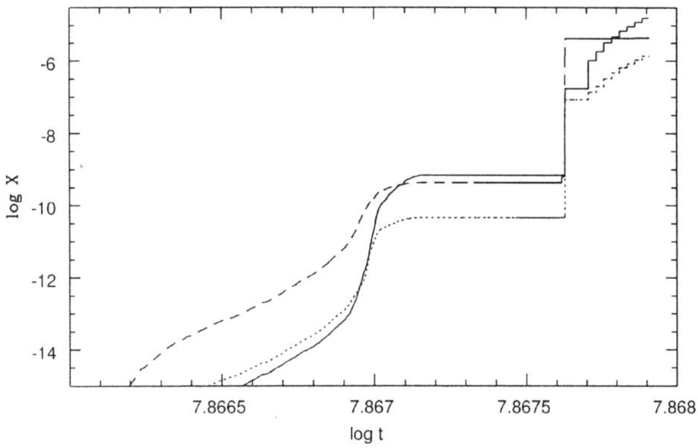

Figure 2. Evolution of the surface mass fraction of ^{12}C (solid line), ^{14}N (dashed line) and ^{16}O (dotted line), in the 5 $M_\odot$. The first episode of dredge-up occurs during the E-AGB. The second one (around log t=7.8676) marks the onset of the TP-AGB phase (see text).

CNO burning cannot take place. In our models of $M \geq 6$ $M_\odot$, the surface abundance of CNO, after the 2nd dredge up (actually the first), is large enough to allow a *normal* CNO burning and, in turn, to enter the TP-AGB phase. The evolutionary history of these thermally pulsing AGB stars is similar to the one found in more metal rich AGB stars. On the contrary, in the 5$M_\odot$, the onset of the TP phase presents some peculiarities. In such a case the amount of CNO nuclei left by the 2nd dredge-up is not sufficient to ensure an efficient H burning. A first period is characterized by weak pulses without the formation of convective shells. However the strength of these weak pulses increases and, after about 9 of them, a convective shell appears above the He burning region. Immediately after, another convective episode, initially confined at the base of the H rich envelope, penetrates the H/He discontinuity and overlaps the previous one. Then a lot of Carbon is dredged up (see figure 2, log t = 7.8676) and the H-burning experience a rather strong flash. Later on the evolution proceeds as in the case of the more massive sequences. In table 1 (last row) we have shown the surface composition of the 5$M_\odot$ star model after 25 thermal pulses.

7. Summary

We obtain, contrary to previous results, that zero metal intermediate mass stars ($M \geq 5$ $M_\odot$) experience the thermal instabilities that characterize the TP-AGB phase. The reason is that the Carbon abundance in the envelope at the end of the E-AGB phase is not the original one (i.e. Z=0). During the E-AGB the convective envelope penetrates and dredges up the ^{12}C, previously produced by the 3α reaction which operates within the H burning shell. This ^{12}C provides the necessary catalyst for an efficient CNO burning and allows the occurrence of the thermal pulses. The $5M_{\odot}$ is a limiting case. The ^{12}C enhancement of the envelope is not enough and the 3α reactions must provide part of the total energy during most of the AGB lifetime. A further episode of convective shells and dredge-up increases the ^{12}C and the star enters the *normal* TP-AGB evolution.

The metal content of the envelope increases during 2nd dredge-up (actually the first) and during the TP phase (up to $Z \sim 10^{-5}$). The surface Helium increases too, namely up to Y≈0.37, almost the double of the initial value. We are now analyzing in more details the nucleosynthesis occurring during the TP-phase to derive a better estimation of the yields of Pop. III stars. The result of this investigation will be presented in a forthcoming paper.

Acknowledgements

This work was partially supported by the MURST italian grant Cofin98, by the MEC spanish grant PB96-1428, by the Andalusian grant FQM-108 and it is part of the ITALY-SPAIN integrated action (MURST-MEC agreement) HI98-0095 .

References

1. Beers, T.C., Preston, G.W., Shectman, A., 1992, AJ 103 1987
2. Bessell, M.S., Norris, J., 1984, ApJ 285 622
3. Bond, H.E., 1981, ApJ 248 606
4. Carney, B.W., Peterson, R.C., 1981, ApJ 245 238
5. Cassisi, S., Castellani, V., Tornambè, A., 1996, ApJ 459 298
6. Cassisi, S., Castellani, V., 1993, ApJSS 88 509
7. Castellani, V., Chieffi, A., Tornambè, A., 1983, ApJ 272 249
8. Chieffi, A., Tornambè, A., 1984, ApJ 287 745
9. Chieffi, A., Limongi, M., Straniero, O., 1998, ApJ 737 762
10. D'Antona F., Mazzitelli, I., 1982, AA 115 L1
11. Ezer, D., 1961, ApJ 133 159
12. Ezer, D., Cameron, A.G.W., 1971, APSS 14 399
13. Ezer, D., 1972, APSS 18 226
14. Fujimoto, M.Y., Iben, I.J., Chieffi, A., Tornambè, A., 1984, ApJ 287 749
15. Molaro, P., Castelli, F., 1990, AA 228 426
16. Molaro, P., Bonifacio, P., 1990, AA 236 L5
17. Nakamura, F., Umemura, M., 1999, ApJ 515 239
18. Primas, F., Molaro, P., Castelli, F., 1994, AA 290 885
19. Ryan, S.G., Norris, J.E., Bessels, M.C., 1991, AJ 102 303
20. Ryan, S.G., Norris, J.E., Beers, T.C., 1996, ApJ 471 254
21. Ryan, S.G., Norris, J.E., Beers, T.C., 1999, ApJ 523 654
22. Sneden, C., Preston, W.G., McWilliam, A. Searle, L., 1994, ApJ 431 L27
23. Sparks, W.M., Endal, A.S., 1980, ApJ 237 130 ApJ 474 1
24. Tornambè, A., Chieffi, A. 1986, MN 220 529
25. Wagner, L.R., 1974, ApJ 191 173
26. Yoshii, Y., Saio, H., 1986, ApJ 301 587

THE EFFECT OF EXTERNAL POLLUTION ON THE EVOLUTION OF LOW-MASS METAL-FREE STARS.

S. CASSISI[1]
[1]*Osservatorio Astronomico di Collurania, Teramo, Italy*
A. WEISS[2]
[2]*Max-Planck-Institut für Astrophysik, Garching, Germany*
M. SALARIS[3]
[3]*Liverpool John Moores University, Birkenhead, United Kingdom*
AND
H. SCHLATTL[2]
[2]*Max-Planck-Institut für Astrophysik, Garching, Germany*

Abstract. The evolution of low-mass metal-free Population III stars accounting for the most updated physical inputs and for atomic diffusion has been investigated. In particular, we test the efficiency of pollution by the encounter with gas clouds, combined with particle diffusion, in order to verify if such mechanism could be a relevant external source of CNO-elements. The possibilities to discriminate between a Pop. II and a polluted Pop. III field star are also briefly discussed.

1. Introduction

It is widely accepted that the primordial gas from which the galaxies formed contained only hydrogen and helium. It is then natural to ask whether any of the first stars that formed from this material are still observable as "Population III" stars of ultralow or zero metallicity. The logical place to seek for such objects is, of course, in the halo of our Galaxy. However, even if the search for Population III stars has a rather long history, only a few stars of extremely low metallicity, $[Fe/H] \leq -4.0$ have been found (Norris et al. 1993 and reference therein). This apparent dearth of extremely metal deficient stars and the observed abundances pattern in field halo stars have

F. Matteucci and F. Giovannelli (eds.), The Evolution of the Milky Way, 467–472.

been, sometimes, shown to be consistent with the existence of a much larger group of Pop. III stars which have been disguised by external pollution.

One can in fact expect that in a real world a very metal poor structure, traveling through the Milky Way, can accrete matter from the gas clouds it encounters. If this occurs, the accreted matter is mixed with the pristine material of the star in the surface convective zone, and then the heavy-elements abundances in this zone increase. Moreover, since it is well known that the peculiar evolutionary properties of Pop. III stars are strongly related to the lack of CNO elements in the original stellar matter (Cassisi & Castellani 1993 - hereinafter CC93 - and reference therein), the external pollution mechanism could have significant consequences also on the evolutionary history of the star if the CNO elements in the accreted matter should have the chance to reach the H-burning layers. It is also important to analyze the temporal evolution of the surface heavy-element abundances in polluted stars, in order to check the possibility to yet identify a real Pop. III star by means of its observed spectrum. Until now, all theoretical investigations devoted to study pollution in Pop.III stars have provided some hints of the evolution of a polluted zero metal object by simply extrapolating the well-known behavior of "normal" Pop. II stars (Fujimoto et al. 1995). For these reasons we have decided to explore the theoretical scenario for Pop. III stars accounting in a homogeneous and self-consistent framework for both atomic diffusion and external pollution.

2. The evolutionary computations.

The evolutionary properties of extremely metal-poor stellar structures have been the subject of accurate investigations since the early 1970s (see CC93 and references therein). We refer to the quoted papers for a detailed description of the main evolutionary properties of Pop. III objects.

In the present work, we decided to consider a "true" metal free ($Z = 0$) chemical composition and a helium abundance equal to $Y = 0.23$. The evolutionary computations have been carried out by adopting a nuclear network accounting for both H- and He-burning, but neglecting any possible unconventional nuclear-reaction branches (Schlattl et al. 1999). We have considered stellar masses ranging from $0.8M_\odot$ to $1.2M_\odot$, and all the evolutionary sequences have been followed from the initial Zero-Age Main Sequence until the ignition of He-burning at the tip of the Red Giant Branch (see Weiss et al. (2000) for more details).

For each mass we have considered two different cases without external pollution: one *without atomic diffusion*, which we denote as 'canonical' and one *with atomic diffusion*. The second case we call 'standard' following the contemporary definition of the physics for the standard solar model. In addition, we have computed some models taking into account both atomic diffusion *and* external He and heavy-element pollution.

2.1. THE 'STANDARD' SCENARIO.

The comparison between our canonical models with the ones by CC93 shows that a fine agreement exists between the two computations, until the early evolution along the sub-giant branch (SGB).

In fact, present models show the presence of an evolutionary event, peculiar of true Pop. III stars, which occurs during the transition to the RGB: a flash in the H-burning region, which leads to a blue loop in the H-R Diagram. The same result has been obtained also by Fujimoto et al. (1990) by computing a $1M_{\odot}$ metal-free model. The physical reasons for the occurrence of the flash is the following: close to the central H-exhaustion, the core strongly contracts, increasing temperature and density, thereby counterbalancing the effect of the continuous hydrogen depletion on the energy-production rate. Due to this increase in both central density and temperature, the efficiency of the 3α reaction is enhanced and thus, the number of carbon nuclei grows. As a consequence, the energy delivered by the CNO cycle strongly increases during this phase. A thermal runaway is produced and a flash occurs which produces the loop of the track in the H-R diagram. In this phase the star develops a convective core. The ensuing expansion of the inner stellar regions produces a significant decrease in both density and temperature, which reduces the 3α nuclear rate and, as a consequence, also the energy produced by the CN-conversion. When the central abundance of hydrogen drops to zero the convective core disappears.

By analizing the evolutionary tracks, one can easily notice that the CNO flash along the SGB is missing in the CC93 computations, because they have assumed a non-zero primordial abundance of heavy elements: although very low ($Z = 10^{-10}$), this is sufficient to allow the CNO-cycle to operate, such that the central temperature remains low enough to avoid the thermal runaway. In this sense, *a metal abundance of the order of* $Z = 10^{-10}$ *seems to be a "critical" metallicity for getting the thermal runaway along the SGB.*

The comparison between the canonical and standard model accounting for atomic diffusion shows the expected changes. However, it is worth noticing that the efficiency of atomic diffusion during the MS evolution of low-mass stars increases when decreasing the metallicity, since the diffusion in the stellar envelopes is larger due to the thinner convective envelopes. Therefore, for fixed diffusion coefficients the effect of atomic diffusion is largest in metal-free stars. This is a quite important point to bear in mind when we will discuss the effect of external pollution on metal-free stars.

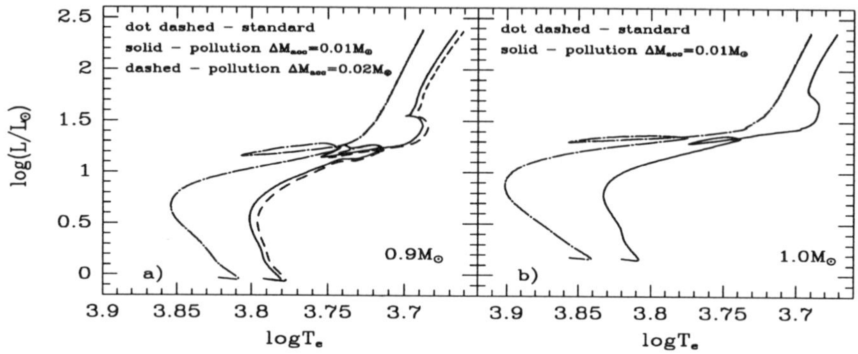

Figure 1. (a) The evolutionary paths in the H-R diagram for a $0.9M_\odot$ stellar structure, computed under different assumptions about the amount of accreted metal rich matter. (b) as (a) but for the $1M_\odot$ model

2.2. THE 'NON-STANDARD' SCENARIO.

One might explain the lack of objects with a Pop. III metallicity by assuming that the surface has been metal-enriched by accretion of metal-rich material through encounters with interstellar gas clouds. This scenario we have investigated quantitatively. The amount of material accreted by a star depends on several parameters as, for instance, the relative velocity between the star and the gas cloud, the cloud parameters and so on. We explored, if and when atomic diffusion is able to bring CNO elements below the edge of the convective envelope and into the H-burning regions. We have used a very simple accretion model by assuming that instantaneous accretion of all the metal-rich matter has happened immediately before the star reached the ZAMS. By adopting this approach, we maximize the "efficiency" of the combination of both processes (accretion + diffusion). Following the estimates by Yoshii (1981), we have modified the chemical composition of the outermost $0.01M_\odot$ of a model, by adopting as the chemical composition of the "accreted" matter $Z = 0.01$ and $Y = 0.25$.

In Fig. 1, we have plotted the evolutionary tracks in the H-R diagram for both the 'polluted' models and the standard one. The main result is that the accreted carbon never reaches the nuclear burning regions. The occurrence of the CNO-flash along the SGB – an event originated in the deep interior – is not affected at all by the accretion of metals. *This is an indirect proof that atomic diffusion is not able to bring CNO-elements into the nuclear burning region. Otherwise one would see an increase in the CNO-cycle burning rate during the previous MS-evolution, eventually developing into a thermal runaway.*

More details on the evolutionary properties can be found in Fig. 2: during the SGB evolution, a thin convective region of $\approx 0.02\,M_\odot$ appears in the envelope and basically all metals diffused during the MS evolution are dredged back to the surface (top panel). The increase in envelope metallicity is accompanied by a decrease in effective temperature (middle). Then the convective envelope goes even deeper, mixing therefore only matter with original zero metallicity; this results in a slight decrease of the surface abundances of the individual heavy elements. During the thermal runaway, $T_{\rm eff}$ increases and the convective envelope disappears, leaving the surface abundances unchanged for a short while. When the runaway stops and the star starts to climb up the RGB, the outer convection sets in again, this time reaching much deeper layers. At this point the surface metal abundances decrease even more (more primeval matter is mixed into the convective envelope), while the stellar track moves toward lower effective temperatures because of the progressively growing convective region. When the star is settling on its Hayashi track, the metal abundance in the convective envelope is still decreasing; this produces the "kink" which appears at the base of the RGB in Fig. 1. Up to this point our calculations indicate that diffusion is not able to transport the accreted metals into the nuclear burning regions. The last possibility for this occurrence takes place during the RGB evolution, where the hydrogen-burning shell might be able to pass through regions previously reached by the envelope convection at its maximum extension ($1^{\rm st}$ dredge-up). However, in all models we have computed, the distance in mass between the inner point reached by the convective envelope during its maximum penetration and the location of the hydrogen-burning shell at the He-flash is never less than $\approx 0.1M_\odot$, so this possibility is completely ruled out. This evolutionary scenario is not significantly changed when accounting for a larger amount of metal-rich accreted matter (see Fig. 2).

3. Conclusions.

We investigated the possibility that a low-mass Pop. III star – due to accretion of metal-rich matter through encounters with molecular clouds – could increase its original amount of CNO elements in the core to a level which allows the ignition of the CN cycle, thus modifying its evolutionary behavior from the one characteristic of metal-free object to the one typical of extremely metal-poor stars.

Our calculations show that neither atomic diffusion during the MS phase nor outer convection during the RGB evolution, are able to bring CNO-elements into the nuclear burning region. So, in spite of the remarkable changes of the heavy-element abundances in the outer layers, the evolutionary behavior of a polluted Pop. III star is always regulated by the

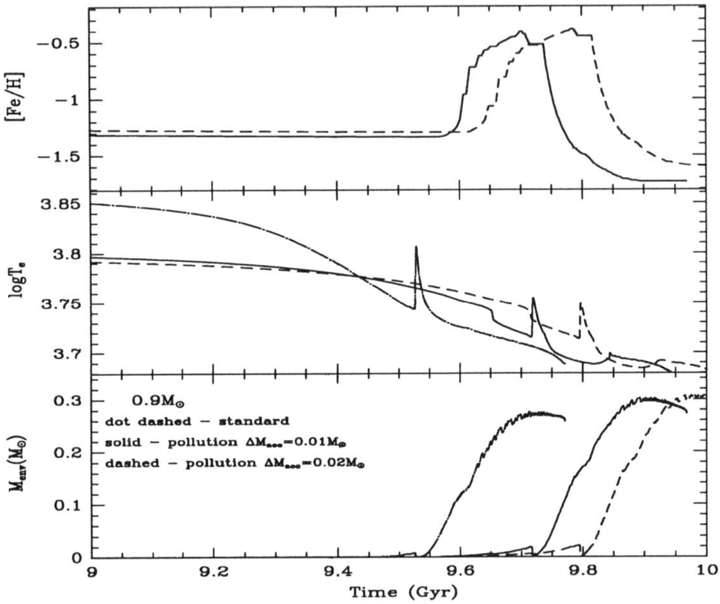

Figure 2. Evolution of the surface abundance of iron, the effective temperature, and the size of the convective envelope for a $0.9M_{\odot}$ model computed taking into account different amounts of accreted metal-rich matter

original chemical composition. Therefore, from the point of view of an 'external' observer it is extremely difficult to discriminate between a polluted Pop. III field star and a very metal poor Pop. II one.

The analysis of the surface abundance ratios does not appear to be of very much help. This means that, if the chemical pollution of the stellar surface is effective, the still surviving Pop. III stars could be all disguised as extremely Pop. II stars with no chance to discriminate between "true" extremely metal-poor stars and polluted Pop. III objects.

References

1. Cassisi, S. & Castellani, V. 1993, ApJS, **88**, 509
2. Fujimoto, M.Y., Iben, I.Jr. & Hollowell, D. 1990, ApJ, **349**,580
3. Fujimoto, M.Y., Sugiyama, K., Iben, I.Jr. & Hollowell, D. 1995, ApJ, **444**,175
4. Hollowell, D., Iben, I.Jr. & Fujimoto, M.Y. 1990, ApJ, **351**,245
5. Norris, J.E., Peterson, R.C. & Beers, T. 1993, ApJ, **415**, 797
6. Schlattl, H., Weiss, A., Cassisi, S. & Salaris, M. 1999, Proceeding of the MPA/ESO workshop *The First Stars*, A. Weiss, T. Abel & V. Hill eds., *in press*
7. Weiss, A., Cassisi, S., Schlattl, H. & Salaris, M. 2000, ApJ, *in press*
8. Yoshii, Y. 1981, A&A, **97**, 280

EXPLOSIVE NUCLEOSYNTHESIS OF MASSIVE STARS THE EFFECT OF THE METALLICITY

M. LIMONGI[1]
Osservatorio Astronomico di Roma
Via Frascati 33, I-00040, Monteporzio Catone, Roma, Italy
e-mail: marco@nemo.mporzio.astro.it

A. CHIEFFI[2]
Istituto di Astrofisica Spaziale (CNR)
Via Fosso del Cavaliere I-00133, Roma, Italy
e-mail: achieffi@ias.rm.cnr.it

AND

O. STRANIERO[3]
Osservatorio Astronomico di Collurania
I-64100, Teramo, Italy
e-mail: straniero@astrte.te.astro.it

1. Introduction

In this talk we discuss the influence of the initial chemical composition of a massive star on its final explosive nucleosynthesis.

At present our computations include the presupernova evolution of a grid of massive stars in the range $13-25$ M$_\odot$ for three initial compositions, i.e., $Z = 2 \cdot 10^{-2}$ ($Y = 0.28$), $Z = 10^{-3}$ ($Y = 0.23$) and $Z = 0$ ($Y = 0.23$). The code we are using is the FRANEC (version 4.7); it has been extensively presented in Chieffi and Straniero (1989) and Chieffi, Limongi and Straniero (1998), The adopted nuclear network includes 40 isotopes (and 269 reactions), from neutron to ^{28}Si, for the H burning, 87 isotopes (and 627 reactions), from neutron to ^{68}Zn, for the He burning and 179 isotopes (and 1672 reactions), from neutron to ^{68}Zn, for the more advanced nuclear burnings.

The changings in the presupernova chemical composition due to the explosive nucleosynthesis are taken into account by evolving each mass zone of the presupernova mantle by assuming a temporal changing of both

F. Matteucci and F. Giovannelli (eds.), The Evolution of the Milky Way, 473–479.

the temperature and the density obtained by assuming 1) the shock to be radiation dominated; 2) the expansion to be adiabatic; 3) the density decline to follow an exponential low and finally 4) the final kinetic energy to be 1.2 foes (1 foe = 10^{51} erg). The nuclear network adopted into the explosive nucleosynthesis calculations is the same used for the presupernova evolution (i.e. the ones including 179 nuclear species). More details on these computations will be addressed in a forthcoming paper (Straniero, Chieffi and Limongi 2000).

For sake of semplicity in this paper we will address the differences among models of metallicity $2 \cdot 10^{-2}$ and 10^{-3}.

2. The Presupernova Models

The final nucleosynthetic explosive yields of a massive star depend on the presupernova model essentially through the final $M - R$ and $M - Y_e$ relations, where Y_e is the electron fraction. In fact, since the shock temperature, that is the maximum temperature the matter reaches when the shock travels the envelope, is a function of both the radius and the energy of the explosion (see Woosley and Weaver 1980 for more details), for a fixed energy of the explosion the radial profile of the presupernova model defines the mass involved in each explosive burning (see Chieffi, Limongi and Straniero this volume). Figure 1 shows that the radial profile between the outer edge of the iron core and the outer edge of the CO core (that is the zone which is significantly affected by the explosive nucleosynthesis) of models having the same mass (and a metallicity $Z = 0.02$ and $Z = 0.001$ respectively) are rather similar among them, but for the two 20 $M_\odot$ models which show slightly different radial profiles. By the way, this occurrence is due to the fact that the evolution of the solar metallicity 20 $M_\odot$ is slightly different from its corresponding 10^{-3} metallicity model (see Straniero, Chieffi and Limongi 2000). As a consequence of the similar radial profile the mass involved in each explosive burning are rather similar too (See Figure 2). The other important quantity of the presupernova model that influences significantly the final explosive yields is the electron fraction profile. In fact, since, in general, the Nuclear and/or Quasi Statistical Equilibirum (NSE and/or QSE) distribution is a function of the temperature, density and electron fraction, the degree of neutronization is crucial when the shock temperature is high enough for the matter to reach such a condition, i.e., in explosive Si burning. Figure 3 shows that in all the models the various explosive burnings occur essentially within the zones processed up to the carbon burning during the hydrostatic evolution (i.e., within the last carbon convective shell) and that the electron fraction within this zone, due to the transformation of ^{14}N into ^{22}N, is obviously larger in the 10^{-3} metal-

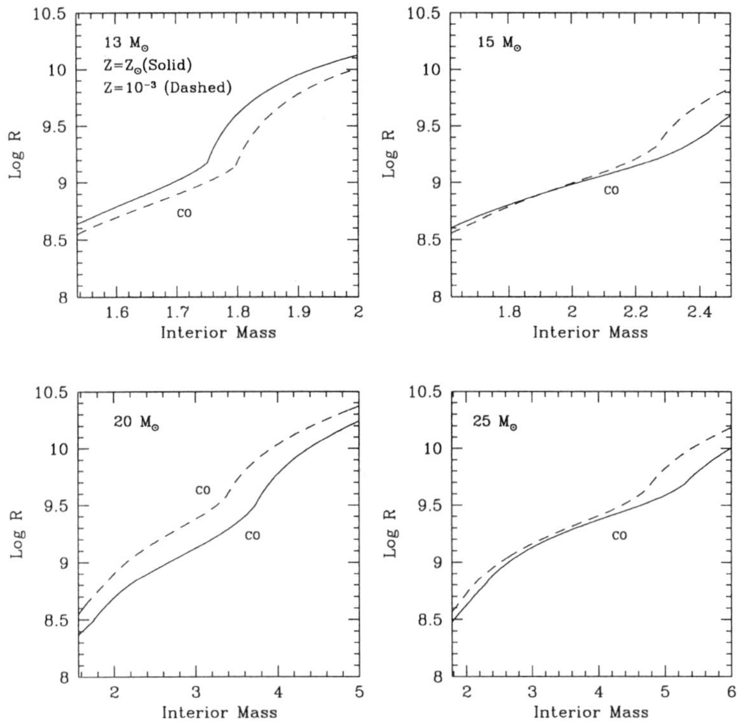

Figure 1. Presupernova Mass-Radius relation of the zone affected by the explosive nucleosynthesis, i.e., between the outer edge of the iron core and the outer edge of the CO core (labeled with 'CO' in the figure). The solid lines refer to solar metallicity models while the dashed lines refer to the 10^{-3} metallicity models.

licity models than in the solar metallicity ones. Finally, the presupernova chemical composition is important only in the zones which do not reach the NSE (or QSE) because of the low shock temperature or do not experience at all any explosive burning.

3. Explosive Yields

In this section we discuss the effect of the differences and similarities among the presupernova models of different metallicity on the final explosive yields. For sake of semplicity we consider, for each element, the most abundant isotope and discuss separately the even and the odd elements. In order to outline the effect of the metallicity the final yield do not include the initial abundance. All the unstable isotopes have been decayed into their parent nuclei.

Figure 4 shows that the effect of the metallicity on the even elements is rather small, except for ^{20}Ne and ^{24}Mg in the case of the two 20 $M_{\odot}$ models.

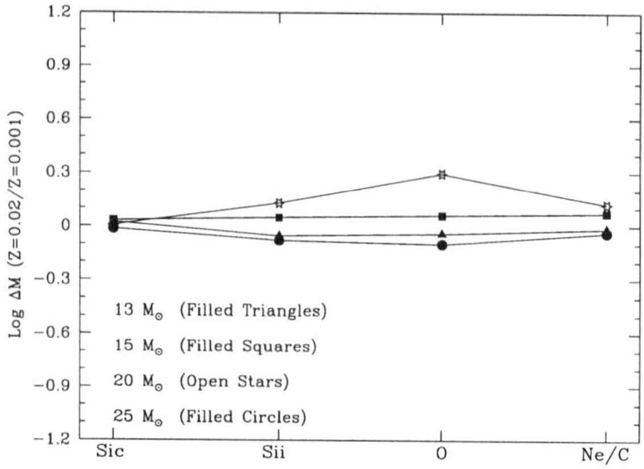

Figure 2. Comparison of the mass involved in each explosive burning between the solar and the 10^{-3} metallicity models.

Such an occurrence can be understood once the site in which every isotope is procuded is known.

^{16}O is produced by He burning and then partially destroyed by explosive O burning; since the He core masses sligthly depend on the metallicity (because of the counterbalancing effect of increasing both the metallicity and the helium content), the final yield of ^{16}O will show a mild dependence on the metallicity too.

The intermediate mass elements (e.g., Si, S, Ar and Ca) are produced by explosive Oxygen burning. Since during explosive O burning the matter reaches a QSE distribution the final yields of these elements will depend on both the neutronization of the presupernova model (within this zone) and the amount of mass processed by explosive O burning. However, since the QSE abundance of all the alpha elements mildly depend on the neutronization, the final yields of the intermediate mass elements would scale directly with the mass involved in the explosive O burning (see Figure 2), which, as we have already discussed above, mildly depends on the metallicity.

The heavier elements are produced by explosive Si burning, where the matter reaches the nuclear statistical equilibrium and hence they behave as the intermediate mass elements.

The case of ^{20}Ne and ^{24}Mg is slightly more complex and requires a more extensive discussion which will be addressed in Straniero, Chieffi and Limongi (2000).

Figure 5 shows that the case of the odd elements is much more complex than the case of the even ones. In fact, at variance with the even elements, they show a significant dependence on both the metallicity and the mass. In particular, increasing the metallicity they are on average overproduced

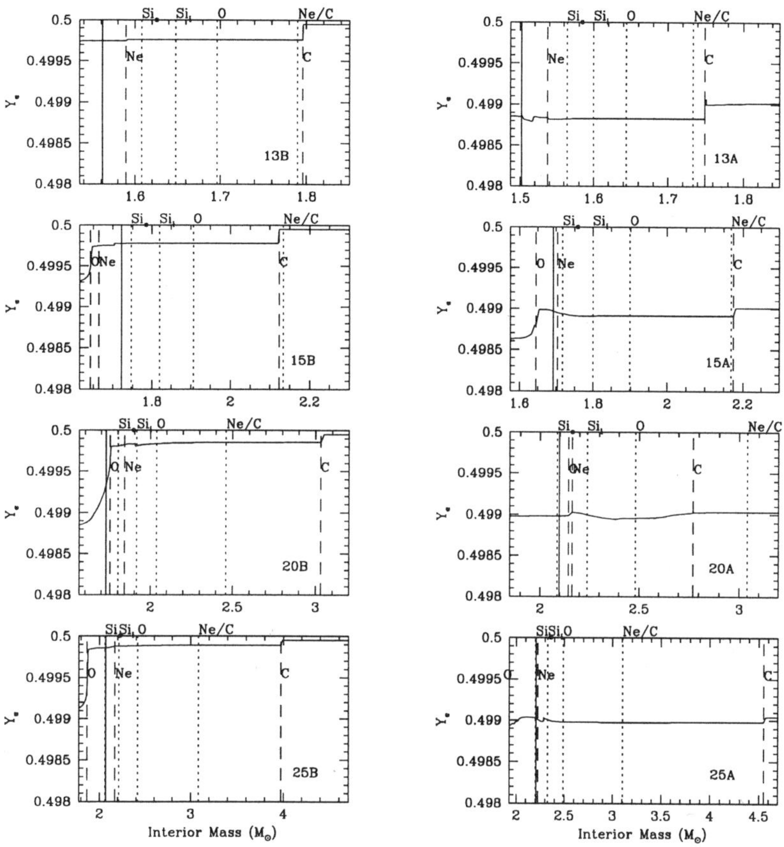

Figure 3. Presupernova electron fraction profile of the two sets of models. The dashed vertical lines mark the outer border of the zones which have been processed up to a certan (labeled) hydrostatic burning while the dotted vertical lines mark the outer edge of the zones processed by the various (labeled) explosive burnings.

by a factor of ~ 4 with a scatter in mass ranging from a factor of ~ 2 (in the case of P, V, and Mn) up to a factor of ~ 4 (except for Sc and Co which show a much large scatter). This occurrence is due to the fact that both the hydrostatic and explosive nucleosynthesis of all these elements is a strong function of the initial metallicity, the explosive nucleosynthesis depending on the initial mettallicity through the final electron fraction profile. As a consequence both the elements produced by the hydrostatic burnings, and more or less destroyed by the explosive nucleosynthesis (like, e.g., Na, Al, Sc and Co), and the ones produced only by the explosive burnings (like, e.g., P, Cl, V, and Mn) show large differences although the presupernova evolution and the behaviour of the explosion are rather similar among models of the same mass.

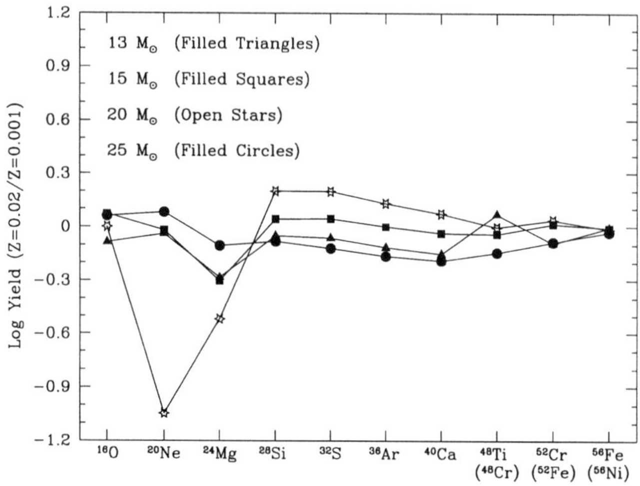

Figure 4. Comparison of the final explosive yields of the even elements between the solar and the 10^{-3} metallicity models. The isotopes within parenthesis are the unstable nuclei which contribute significantly to the final yield of their corresponding stable isotopes.

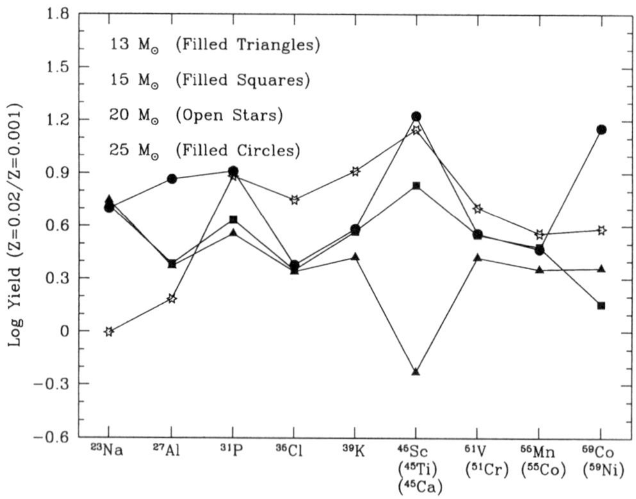

Figure 5. Comparison of the final explosive yields of the odd elements between the solar and the 10^{-3} metallicity models. The isotopes within parenthesis are the unstable nuclei which contribute significantly to the final yield of their corresponding stable isotopes.

References

1. Chieffi, A., Limongi, M. and Straniero, O. (1998), *ApJ*, **502**, 737
2. Chieffi, A. and Straniero, O. (1989), *ApJS*, **71**, 48
3. Straniero, O., Chieffi, A. and Limongi, M. (2000), in preparation
4. Weaver. T.A. and Woosley, S.E. (1980), *Ann. NY Acad. Sci.*, **336**, 335

Speaker: M. LIMONGI

A. MAEDER: I am a bit surprised that you do not account for the effect of mass loss. I would expect mass loss and its dependence on metallicity to be more important than any other metallicity dependent effect.

M. LIMONGI: I perfectly agree with you, but we decided to begin by computing a first set of models without any mass loss. In the next future we will compute also evolutions with mass loss.

P. VENTURA: You said that the He-burning cores of models computed with different metallicities for a fixed mass are similar. Within which limits do you get it?

M. LIMONGI: The maximum variation we obtain is less than 10%, and occurs for the two 25 $M_\odot$ stars.

CHEMICAL YIELDS
FROM LOW- AND INTERMEDIATE-MASS STARS

P. MARIGO

Department of Astronomy, University of Padova
Vicolo dell'Osservatorio 5, 35122 Padova, Italy

Abstract

We present new sets of chemical yields from low- and intermediate-mass stars with $0.8M_{\odot} \leq M \leq M_{\rm up} \sim 5M_{\odot}$, and three choices of the metallicity, $Z = 0.02$, $Z = 0.008$, and $Z = 0.004$ (Marigo 2000, in preparation). These are then compared with the yields calculated by other authors on the basis of different model prescriptions, and basic observational constraints which should be reproduced.

1. Surface chemical abundances

In this work we predict the changes in the surface abundances of several elements (H, ^{3}He, ^{4}He, ^{12}C, ^{13}C, ^{14}N, ^{15}N, ^{16}O, ^{17}O, and ^{18}O), for a dense grid of stellar models with initial masses from $\sim 0.8M_{\odot}$ to $M_{\rm up} = 5M_{\odot}$, and three choices of the original composition, i.e. $[Y = 0.273, Z = 0.019]$, $[Y = 0.250, Z = 0.008]$, $[Y = 0.240, Z = 0.004]$.

Various processes may concur to alter the surface chemical composition of a star, namely: i) the first dredge-up occurring at the base of the RGB; ii) the second dredge-up taking place during the E-AGB only for stars with $M > (3.5-4.0)M_{\odot}$; iii) the third dredge-up experienced by stars with $M > (1.2-1.5)M_{\odot}$ during the TP-AGB phase; and iv) hot-bottom burning in the most massive AGB stars with $M > (3.5-4.0)M_{\odot}$.

Predictions for first and second dredge-up are taken from Padova stellar models with overshooting (Girardi *et al.* 2000), whereas for the TP-AGB phase the results of synthetic calculations are adopted (Marigo *et al.* 1996, 1998, 1999). The reader should refer to these works for all the details.

F. Matteucci and F. Giovannelli (eds.), The Evolution of the Milky Way, 481–486.

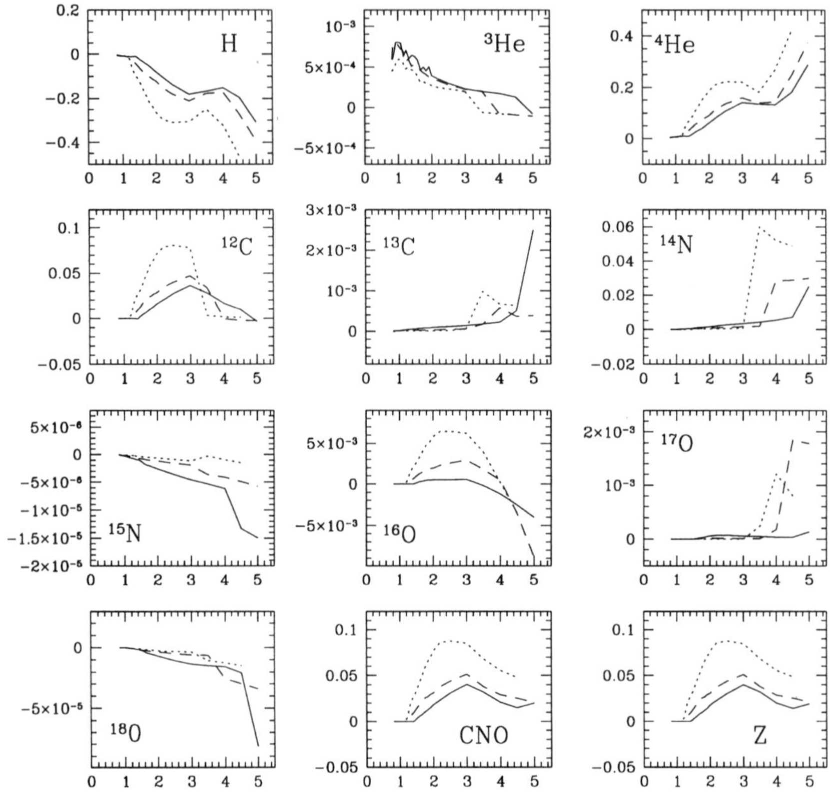

Figure 1. Net chemical yields (in $M_\odot$) as a function of the initial mass (in $M_\odot$) of the star. The solid, dashed, and dotted lines correspond to metallicities $Z = 0.019$, $Z = 0.008$, and $Z = 0.004$, respectively. The adopted mixing length parameter is $\alpha = 1.68$.

2. Stellar yields

Yields from low- and intermediate-mass stars are determined by the wind contributions during the RGB and AGB phases. In these calculations mass loss is described by the Reimers' prescription ($\eta = 0.45$) for the RGB phase, and by the Vassiliadis & Wood (1993) formalism for the AGB phase. Yields for the elements under consideration are shown in Fig. 1 as a function of the initial stellar mass, for three choices of the metallicity. Major positive contributions correspond to ^{4}He, ^{12}C, and ^{14}N. Complete sets of stellar yields, distinguishing both the secondary and primary components of the CNO contributions, will be available in Marigo (2000, in preparation).

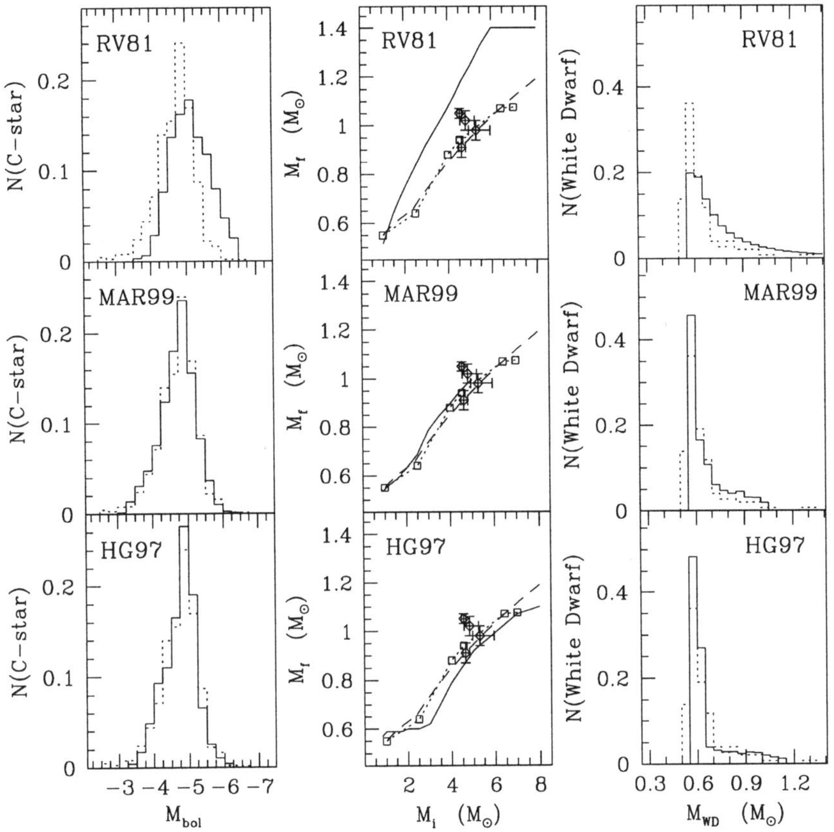

Figure 2. Constraints on AGB models. Observations are shown with dotted and dashed lines. Theoretical predictions of different authors are presented with solid lines. Left panels: The carbon star luminosity function in the LMC. The observed data are from Costa & Frogel (1996). Middle panels: The initial-final mass relation in the solar neighbourhood. The observed data are taken from Herwig 1996 (empty squares connected by dotted line), and Jeffries 1997 (squares with error bars). Right panels: The white dwarf mass distribution in the solar neighbourhood. Observations are from Bergeron *et al.* (1992), and Bragaglia *et al.* (1995).

3. Comparison with previous synthetic AGB models

Chemical yields crucially depend on the adopted mass-loss and nucleosynthesis prescriptions. In Fig. 2 a comparison is made between different synthetic AGB models - namely: Renzini & Voli 1981 (RV81), van de Hoek & Groenewegen 1997 (HG97), and this work (MAR99) - on the basis of some key-observables. All these constraints are satisfactorily reproduced by both calibrated HG97 and MAR99 models, whereas RV81 results are quite dis-

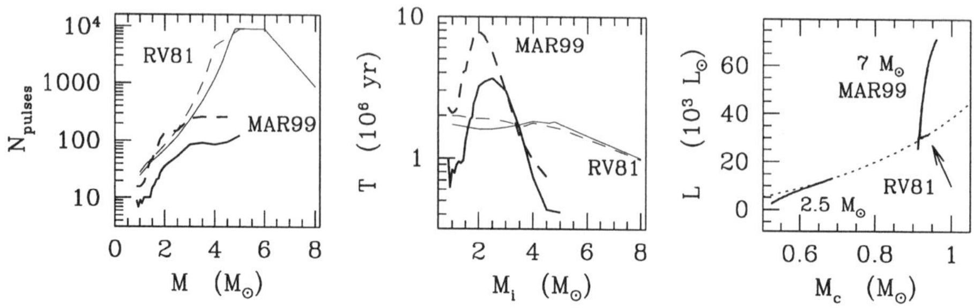

Figure 3. Comparison of model predictions. Left panel: Number of thermal pulses as a function of the initial stellar mass, for metallicities $Z = 0.02$ (solid line), and $Z = 0.004$ (dashed line). Middle panel: TP-AGB lifetimes as a function of the stellar initial mass. Right panel: Luminosity evolution of a $7M_\odot$ TP-AGB model with hot-bottom burning as a function of the core mass (solid lines), as predicted by MAR99 and RV81 (according to their Eq. (15) for the same M_c). The standard $M_c - L$ relation, followed by the 2.5 $M_\odot$ model, is shown by the dotted line.

crepant. Specifically, the uncalibrated RV model predicts too few faint and too many bright carbon stars, together with a sizeable excess of massive white dwarfs (with $M > 0.7M_\odot$). This can be explained considering the lower efficiency of both the third dredge-up and mass-loss (Reimers' law with $\eta = 0.333 - 0.666$) adopted by RV81.

The effect of different mass-loss prescriptions is also clear from Fig. 3. The most massive AGB stars are expected to suffer a huge number of thermal pulses (hence dredge-up episodes) in the RV81 model ($\sim 10^4$), about two order of magnitude more than in the MAR99 model ($\sim 10^2$, left panel). As a consequence, the duration of the TP-AGB phase for these models is affected in the same direction, being much longer in RV81 than in MAR99 (middle panel of Fig. 3)

A final remark concerns the process of hot-bottom burning suffered by the most massive AGB stars ($M > (3.5-4)\,M_\odot$). Its treatment determines the temperature at the base of the convective envelope and the related nucleosynthesis, as well as the luminosity evolution of these stars. To this respect, it is worth noticing that the break-down of the $M_c - L$ relation (first pointed out by Blöcker & Schönberner 1991) is included by MAR99 (see Marigo 1998), whereas in RV81 and HG97 this overluminosity effect is not taken into account (right panel of Fig. 3).

4. Yields from single stellar populations

In order to compare stellar yields with different values of $M_{\rm up}$ (i.e. $8M_\odot$ for RV81 and HG97 classical models, and $5M_\odot$ for MAR99 model with

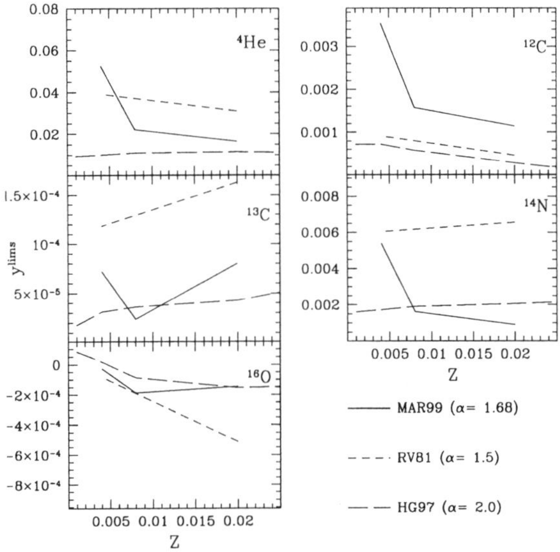

Figure 4. Integrated yield contributions from low- and intermediate-mass stars as a function of the metallicity, as defined by Eq.(4). The mixing-length parameters (α) adopted by the authors are indicated.

overshooting), we calculate the quantities

$$y_k^{\rm lims} = \frac{\int_{0.8}^{M_{\rm up}} m p_k(m)\phi(m)dm}{\int_{0.8}^{100} m\phi(m)dm} \tag{1}$$

where $p_k(m)$ is the fractional yield of the element k produced by a star of initial mass m. These quantities express the relative chemical contribution from low- and intermediate stars belonging to a given simple stellar population. They are shown in Fig. 4 as a function of the metallicity for the three sets here considered.

Differences show up both in metallicity trends and absolute values of $y_k^{\rm lims}$. Compared to previous calculations, MAR99 yields show a pronounced dependence on the metallicity, i.e. yields increase with decreasing Z. Conversely, the RV81 and HG97 sets present quite weak trends with Z.

The metallicity dependence can be explained as follows. On one side, AGB lifetimes of low-mass stars increase at decreasing metallicities, as mass-loss rates are expected to be lower. This fact leads to a larger number of dredge-up episodes. Moreover, both the onset and the efficiency of the third dredge-up are favoured at lower metallicities. These factors concur to produce a greater enrichment in carbon. On the other side, hot-bottom burning in more massive AGB stars becomes more efficient at lower metallicities, leading to a greater enrichment in nitrogen. The combination of all factors favours higher positive yields of helium at lower Z.

As far as the single elemental species are concerned, we can notice:

- MAR99 yields of ^{4}He are larger than those by HG97, due to the earlier activation of the third dredge-up and, likely, to a greater efficiency/duration of hot-bottom burning in our models. Predictions by RV81 show no significant trend with Z, and higher positive yields (due to the quite low mass-loss rates adopted).
- MAR99 yields of ^{12}C are systematically higher than those of RV81 and HG97 because of the earlier onset (and average greater efficiency than in RV81) of the third dredge-up.
- The dominant contribution to the yields of ^{14}N comes from hot-bottom burning in the most massive AGB stars. Differences in the results reflect different efficiencies of nuclear reactions and AGB lifetimes. In particular, according to MAR99 the production of ^{14}N, mainly of primary synthesis, is favoured at lower Z.

References

1. Bergeron, P., Saffer, R.A., Liebert, J. (1992) A spectroscopic determination of the mass distribution of DA white dwarfs, *Astrophysical Journal*, **Vol. no. 394**, pp. 228–247
2. Bragaglia, A., Renzini A., Bergeron P. (1995) Temperatures, gravities, and masses for a sample of bright DA white dwarfs and the initial-to-final mass relation, *Astrophysical Journal*, **Vol. no. 443**, pp. 735–752
3. Costa, E., Frogel, J.A. (1996) Carbon Stars in the Large Magellanic Cloud: Luminosities, Colors, and Implications for the History of Star Formation, *Astronomical Journal*, **Vol. no. 112**, pp. 2607–2633
4. Girardi, L., Bressan, A., Bertelli, G., Chiosi, C. (2000) Evolutionary tracks and isochrones for low- and intermediate-mass stars: from 0.15 to 7 $M_\odot$, and from $Z = 0.0004$ to 0.03, *Astronomy & Astrophysics Supplement series*, in press
5. Herwig F. (1996) in Stellar Evolution: What Should Be Done, *32nd Liège Int. Astrophys. Coll.*, eds. A. Noels et al., p. 441
6. van den Hoek, L.B., Groenewegen, M.A.T. (1997) New theoretical yields of intermediate mass stars, *Astronomy & Astrophysics Supplement series*, **Vol. no. 123**, pp. 305–328 (HG97)
7. Jeffries R.D. (1997) On the initial-final mass relation and maximum mass of white dwarf progenitors, *Monthly Notices of the Royal Astronomical Society*, **Vol. no. 288**, pp. 585–588
8. Marigo, P., Bressan, A., Chiosi, C. (1996) The TP-AGB phase: a new model, *Astronomy & Astrophysics*, **Vol. no. 313**, pp. 545–564
9. Marigo, P. (1998) Envelope burning over-luminosity: a challenge to synthetic TP-AGB models, *Astronomy & Astrophysics*, **Vol. no. 340**, pp. 463–475
10. Marigo, P., Bressan, A., Chiosi, C. (1998) TP-AGB stars with envelope burning, *Astronomy & Astrophysics*, **Vol. no. 331**, pp. 564–580
11. Marigo, P., Girardi, L., Bressan, A. (1999) The third dredge-up and the carbon star luminosity functions in the Magellanic Clouds, *Astronomy & Astrophysics*, **Vol. no. 344**, pp. 123–142
12. Renzini, A., Voli, M. (1981) Advanced evolutionary stages of intermediate-mass stars. I, *Astronomy & Astrophysics*, **Vol. no. 94**, pp. 175–193 (RV81)
13. Vassiliadis, E., Wood, P.R. (1993) Evolution of low- and intermediate-mass stars to the end of the asymptotic giant branch with mass loss, *Astrophysical Journal*, **Vol. no. 413**, pp. 641–657

AGB STARS NEW MODELS: LITHIUM PRODUCTION IN THE GALAXY AND MAGELLANIC CLOUDS

P. VENTURA
Osservatorio Astronomico di Roma
Via Frascati - 00040 - MontePorzio Catone (RM)

1. Introduction

The discovery in the Galaxy and in the Magellanic Clouds of several Asymptotic Giant Branch (AGB) sources which display a surface lithium abundance ($\log(\epsilon(^7Li))$ up to 4, where $\log(\epsilon(^7Li)) = \log(^7Li/H) + 12$) which is much larger than those commonly observed in galactic giants ($\log(\epsilon(^7Li)) < 0.5$) (Abia et al. 1993; Smith & Lambert 1989, 1990; Smith et al. 1995) is a clear hint on the fact that AGB stars manufacture lithium, despite the fact that this latter is heavily destroyed within the interiors of stars via the reaction $^7Li + p \longrightarrow 2\alpha$.

A mechanism which might lead to lithium production within AGB stars during the phase of thermal pulses has been suggested by Cameron & Fowler (1971): the temperature at the base of the external envelope of these sources should become sufficiently hot to ignite "hot bottom burning", with the consequent production of Beryllium, which might decay into lithium in the outermost layers of the star, where lithium produced might survive to proton fusion due to the low temperatures of those layers. Since the results obtained are strongly dependent on the ratios among the time scales of the various processes involved (mixing, nuclear reactions), this problem must necessarily be faced by adopting a diffusive scheme for chemical mixing, where mixing of chemicals and nuclear burning are self-consistently coupled (Sackmann & Boothroyd 1992).

To achieve a good agreement between the results coming from theoretical models and the observational picture available it is however necessary to know with a large precision the total luminosity of the sources observed, so that in this preliminary phase we turned our attention to the stars ob-

F. Matteucci and F. Giovannelli (eds.), The Evolution of the Milky Way, 487–493.

served in the Large Magellanic Cloud (LMC) in order to tune some of the input physics parameters adopted to compute the evolution of our models.

In this paper we show that a minimum mass loss rate is required to achieve a good agreement with the observations, while the oveshooting parameter and the convective model adopted, though influencing the borders of the interval of masses involved in this process, do not modify the general lithium vs. luminosity trend, which is the only reliable information coming from the observations.

We finally show that those AGB sources with masses in proximity of the lower limit to ignite carbon in a no degeneracy regime give the major contribute to the global enrichment of the interstellar medium, and present the relation between the initial mass of the star and the lithium yield.

2. Chemical and physical inputs

The evolutions of the models were computed starting from the pre-main sequence phase by using the ATON2.0 code, a complete description of which can be found in Ventura et al. (1998). We adopted a metallicity typical of the LMC, namely Z=0.01.

Convection is addressed within the Full Spectrum of Turbulence (FST) model (Canuto & Mazzitelli 1991,1992), which is to be preferred to the old traditional mixing length theory for its larger physical consistency; a further advantage is that the results obtained in terms of lithium production have been shown to be completely independent of any free parameter (Mazzitelli et al. 1999).

Overshooting from the borders of convective regions is modeled by assuming an exponential decay of convective velocities with an e-folding distance which is assumed to be proportional to the pressure scale height H_p, with a constant of proportionality for which it was derived a conservative estimate by fitting the main sequence of some young open clusters: $\zeta = 0.02$.

Mass loss has been modeled according to Blöcker (1995). The "superwind" phase is reproduced by multiplying the canonical Reimers' formula by a power function of the luminosity:

$$\dot{M} = 4.83 \cdot 10^{-9} M^{-2.1} L^{2.7} \dot{M}_R$$

where $\dot{M}_R$ is the usual Reimers' mass loss rate given by:

$$\dot{M}_R = 4 \cdot 10^{-13} \eta_R \frac{LT}{M}$$

One of the goals of the present work is to tune the free parameter η_R.

3. Lithium sinthesis in AGB models

Lithium production within the interiors of AGB stars is well explained via the Cameron-Fowler mechanism. The minimum temperature to ignite the latter is $T_{bce} = 4 \cdot 10^7 K$. The larger T_{bce} the larger is the rate of Beryllium production, so that larger abundances of lithium are expected. In any case this process is stopped when T_{bce} exceeds $\sim 7 \cdot 10^7$K, in which case Beryllium burns mainly according to the reaction $^7Be + p \longrightarrow 2\alpha$.

The lower limit of the range of masses involved is determined by the condition that at a certain stage of the evolution T_{bce} might exceed $4 \cdot 10^7$K, and in the present computations we found that the minimum mass to achieve this condition is $M = 3.5 M_\odot$. As for the upper limit, it is clear that stars which reach central temperatures large enough to ignite Carbon in conditions of no degeneracy do not experience thermal pulses, thus have no chances of achieving the afore mentioned mechanism for lithium production. We found that the upper limit for this set of computations is $M = 6 M_\odot$.

We did also some computations of models with solar metallicity ($Z = 0.02$), and found that in that case the range of initial masses involved is $3.7 M_\odot \leq M \leq 6.5 M_\odot$. These are the expexted masses of the progenitors of the galactic lithium rich sources observed in the solar neighbourhood.

3.1. THE INFLUENCE OF OVERSHOOTING

Before entering the discussion we tried to understand how much the results obtained can depend on the overshooting parameter ζ assumed in our computations. Since the choice $\zeta = 0.02$ is to be considered as a lower limit, we did some tests with ζ increased by 50%, i.e. $\zeta = 0.03$. We have verified that in this case the upper AGB limit lowers to $5.8 M_\odot$, and that a $3 M_\odot$ failed to ignite the Cameron-Fowler mechanism. On the basis of these results we can claim that the interval $3.5 M_\odot \leq M \leq 6 M_\odot$ is well extablished within $0.5 M_\odot$, even considering all the uncertainties connected with the determination of the overshooting parameter.

3.2. TUNING THE MASS LOSS RATE

Mass loss strongly influences the evolution of AGB stars expecially at the end of their life, triggering a general cooling of the outer layers and a decline of the total luminosity. Fig.1 shows the comparison between the evolutions of two $5 M_\odot$ models computed with two values of the η_R parameter entering the Reimers' formula, namely $\eta_R = 0.01$ and $\eta_R = 0.05$. First, we note that with the lowest η_R we reach larger luminosities but, more important for our analysis, we see that the general lithium vs. luminosity trend is largely modified.

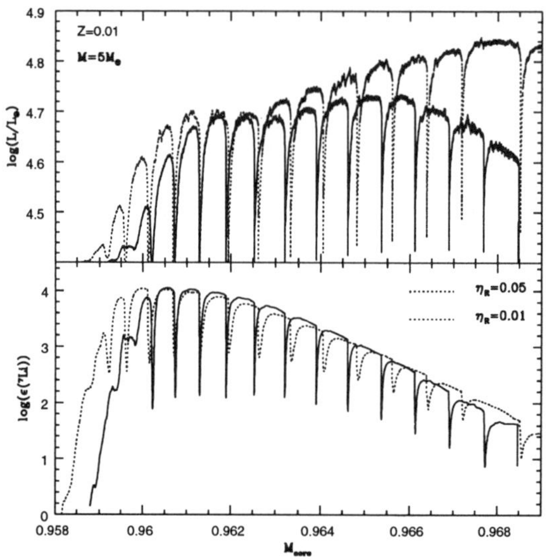

Figure 1. Variation with core mass of luminosity (**top**) and surface lithium (**bottom**) of two $5M_{\odot}$ models computed with different mass loss rates.

In the model with the larger mass loss rate, soon after $\log(\epsilon(^7Li))$ starts to decrease, the luminosity declines to lower values due to the strong mass loss rate. On the other hand, in the $\eta_R = 0.01$ case, the luminosity remains still large when lithium is no longer present in the envelope, with the consequence that for a long phase the star should be observed at large luminosities with negligible amounts of surface lithium. Our numerical simulations show that by adopting the $\eta_R = 0.01$ value we would expect at least half of the most luminous AGB sources with $-7 \leq M_{bol} \leq -6.5$ to have negligible lithium within their envelope, which is just the opposite of what comes from the observations, which show that all the most luminous AGB stars in the LMC are lithium rich (Smith et al. 1995).

On the basis of the above results we conclude that at least a value $\eta_R = 0.05$ is required to have a good agreement between our theoretical models and the observations of lithium rich AGB stars in the LMC: this leads to mass loss rates of $\sim 2 \cdot 10^{-4} M_{\odot} yr^{-1}$ at $M_{bol} \sim -6.8$. Recently, van Loon et al. (1999) found an empirical maximum mass loss rate for LMC AGBs, which shows that at the same luminosity given above some sources are observed with $\dot{M} \sim 4 \cdot 10^{-4} M_{\odot} yr^{-1}$. We conclude that values $0.05 \leq \eta_R \leq 0.1$ should lead to a global agreement between theoretical models and observations. We stress here that with different relationships between the initial mass of the star and the core mass at the begininning

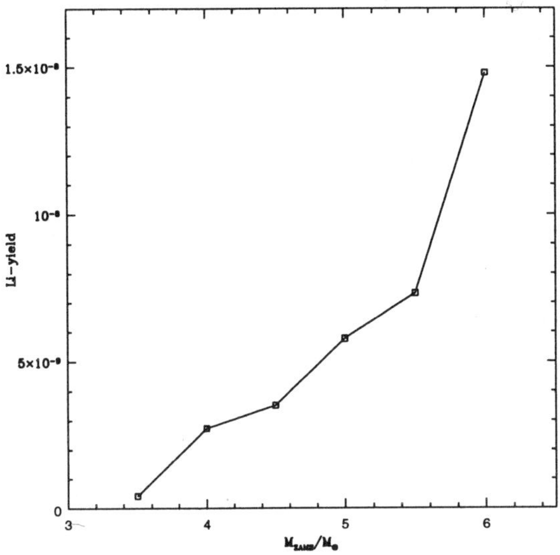

Figure 2. Variation with initial mass of the total amount of lithium ejected by our models.

of the AGB phase, or between the core mass and the luminosity (which in these phases critically depend on the convective model adopted) the same value of η_R would lead to different mass loss rates.

4. Lithium yield

The variation with the initial mass of the lithium yield of the stars can be seen in fig.2. We find an almost linear overall trend, with the only exception of the $6M_\odot$ models, which gives a lithium yield more than double that of the $5.5M_\odot$. This is due to the fact that the $6M_\odot$ achieves the ignition of the Cameron - Fowler mechanism well before the beginning of thermal pulses, so that the largest values of $\log(\epsilon(^7Li))$ are attained when the luminosity of the star (hence the mass loss rate) is approximately at its maximum values.

We therefore reach the important conclusion that among this class of objects the stars close to the AGB limit seem to be the major contributors to the lithium enrichment of the interstellar medium.

The values shown in fig.2 are to be considered as lower limits, since they are obtained by adopting $\eta_R = 0.05$. In case of $\eta_R = 0.1$ (which seems to be the largest possible value) the yields of fig.2 would be approximately doubled. Complete results will be presented in a forthcoming paper (Ventura et al. 2000).

The lithium yield is, contrary to the lithium vs. luminosity trend, dependent both on the overshhoting parameter ζ and on the convective model adopted. In case of a less efficient convection (which we would obtain by assuming an MLT treatment) the minimum mass necessary to achieve hot bottom burning would be larger, so that the overall yield would be reduced. An analogous behaviour is expected with lower ζ's.

5. Conclusions

In this paper we have presented AGB models of intermediate mass in order to find out some constraints on some macro-physics input parameters by fitting the lithium vs. luminosity trend wich is observed in the LMC. We find that the range of initial masses involved is approximately $3.5M_\odot \leq M \leq 6M_\odot$, and that this interval, within $0.5M_\odot$, is independent of the overshooting distance assumed. For typical galactic AGB stars in the solar neighbourhood we expect the above range to be $3.7M_\odot \leq M \leq 6.5M_\odot$.

The results confirm that it is not possible to have clear informations on the overshooting distance or on the validity of the convective model adopted, since these quantities do not influence the global relationship between lithium and luminosity of the models. The only quantity upon which it has been possible to put some constraints is the mass loss rate, for which a lower limit is required to avoid the presence of several large luminosity sources with negligible amounts of surface lithium, which are not observed. By adopting the Blöcker's recipe for $\dot{M}$ we find that a minimum value of the free parameter $\eta_R = 0.05$ entering Reimers' formula is required to allow our models to fit the observations.

We find that lithium yield, as expected, is a growing function of the initial mass of the star, but that masses around $6M_\odot$, which is the upper AGB limit, confirm to be the largest contributors to the lithium enrichment of the interstellar medium.

References

1. Abia C., Isern J. & Canal R. (1993), *A&A*, **275**, 96
2. Blöcker T. (1995), *A&A*, **297**, 727
3. Cameron A.G. & Fowler W.A. (1971), *ApJ*, **164**,111
4. Canuto V.M.C. & Mazzitelli I. (1991), *ApJ*, **370**, 295
5. Canuto V.M.C. & Mazzitelli I. (1992), *ApJ*, **389**, 724
6. Mazzitelli I., D'Antona F. & Ventura P. (1999), *A&A*, **348**, 846
7. Sackmann I.J. & Boothroyd A.I. (1992), *ApJ*, **392**, L71
8. Smith V.V. & Lambert D.L. (1989), *ApJ*, **345**, L75
9. Smith V.V. & Lambert D.L. (1990), *ApJ*, **361**, L69
10. Smith V.V., Plez B. & Lambert D.L. (1995), *ApJ*, **441**, 735
11. Van loon J.Th., Groenewegen M.A.T., de Koter A., et al., (1999), *A&A*. **351**, 559
12. Ventura P., Zeppieri A., Mazzitelli I. & D'Antona F. (1998), *A&A*, **334**, 953

Speaker: P. VENTURA

R. GALLINO: What is the reason for Li to decrease so sharply after any thermal pulse and then to decrease when hot bottom burning is no more at work?

P. VENTURA: It is simply that the temperature at the base of the envelope is not large enough to ignite the Cameron-Fowler mechanism, but still enough to trigger lithium burning due to proton fusion.

S. RYAN's Comment: There are a number of metal-poor halo dwarfs and giants which show C and s- element excesses resembling pollution from an AGB companion. Some of these objects also show Li. It will be interesting to have AGB models for very metal-poor stars to see whether these can explain the observations.

R. GALLINO's Comment: If the primary AGB star is of low mass ($M \leq 2M_{\odot}$) then Li may be produced by cool bottom processes (or "extramixing") according to Boothroyd and Sockmann (ApJ about 1998).

OBSERVATIONAL CONSTRAINTS TO THE EVOLUTION OF MASSIVE STARS

N. PANAGIA
ESA/Space Telescope Science Institute
3700 San Martin Drive, Baltimore, MD 21218, USA
E-mail panagia@stsci.edu

Abstract. We consider some aspects of the evolution of massive stars which can only be elucidated by means of "indirect" observations, *i.e.* measurements of the effects of massive stars on their environments. We discuss in detail the early evolution of massive stars formed in high metallicity regions as inferred from studies of HII regions in external galaxies.

1. Introduction

Massive stars play a crucial role in the evolution of galaxies and the whole Universe, because they are the primary sources of radiative ionization and heating of the diffuse medium, they provide most of the nucleosynthesis products to boost the metal content of galaxies and the intergalactic medium, and they constitute a major supply of kinetic energy for galaxies, both through stellar winds during their quiescent phases and, eventually, in the form of fast ejecta from supernova explosions.

Therefore, it is fundamental to reach a proper understanding of the formation processes, the detailed properties and the evolution of massive stars. Despite the fact that their high luminosities make them "easy" targets for detailed observational studies, many aspects and properties of massive star evolution are far from being fully understood. This is because in any stellar generation, massive stars constitute a small fraction of the newly formed stars (say, less than 1% by number), they are "elusive" in that their lifetimes are very short (say, less than 10–20 million years), and often they are heavily obscured by the parent molecular clouds where they were formed, making even their identification rather cumbersome. As a consequence it is not easy to cover all evolutionary phases with *direct* observations of a statistically significant sample of objects.

F. Matteucci and F. Giovannelli (eds.), The Evolution of the Milky Way, 495–501.

One can overcome these difficulties and gain additional insights by considering phenomena that *indirectly* can provide hints and clues to the problem. In other words, besides studying individual massive stars, one can look at the effects that these stars have on their environments (*e.g.*, HII regions, circumstellar nebulae, SNRs), and infer from there what the stars were doing in special phases of their evolution (*e.g.*, formation, LBV and pre-SN phases, etc.) that would not be accessible in other ways.

Thus, one can use radio observations of supernovae, which probe the circumstellar material ejected by the progenitor stars several thousand years before explosion, to study the very last phases of their evolution. These phases represent a tiny fraction of a massive star lifetime, $\sim$0.1%, and, therefore, they are extremely difficult to reveal and study with direct observations. Although this is an interesting aspect, we are not going to review it here, but rather we refer the reader to recent papers (Montes *et al.* 1998, Panagia *et al.* 1999, Weiler *et al.* 1999).

Here, we consider and discuss one particular aspect of the evolution of massive stars, namely their formation and early evolution in high metallicity environments. We will show that observations of HII regions in external galaxies show that the ionization of He is much lower than that of H when the O/H ratio in the gas is appreciably higher than solar. This implies that at high metallicities either very massive stars ($M > 25\ M_{\odot}$) do not form, or they never reach their expected ZAMS location.

2. Ionized Helium in the Milky Way

There is clear observational evidence in the H II regions of our Galaxy that the fractional abundance of ionized helium $n(He^{+})/n(H^{+})$ is not a monotonic function of the galactocentric radius. Moving outwards from the Galactic Center, the ionized He abundance is found to increase in the inner Galaxy, then it attains a maximum near the solar circle, and finally drops in the outer Galaxy (*e.g.*, Mezger & Wink 1983 and references therein). The negative gradient in the outer galaxy reflects a genuine decrease in the He abundance in the outward direction (*e.g.*, Panagia 1980; Güsten & Mezger 1982). The positive gradient in the inner Galaxy instead is an effect of the radial metallicity gradient which produces a systematic variation of the spectrum of the ionizing radiation (Panagia 1980).

The fractional ionization of helium is extremely sensitive to the most energetic part of the radiation field powering an H II complex. Hence, it can provide valuable information on the presence and the abundance of the most massive ($m \geq 20\ M_{\odot}$) stars, which are responsible for most of the radiation with energy in excess of 24.6 eV. Therefore, it is a powerful tool to study how the details of the star formation process vary in different physical environments. There are several mechanisms through which a higher metallicity lowers the He ionization in an H II region:

• The relative number of He-ionizing photons in the stellar spectrum is reduced because of both a stronger line blanketing in the 200–500 Å wavelength range, and a higher continuum opacity.

• The stellar radius becomes larger and the effective temperature decreases for a star of given mass, because of the increased continuum opacity in the sub-atmospheric layers of the star.

• The upper cut-off of the Initial Mass Function (IMF, m_U, may be shifted to lower masses (*e.g.*, Kahn 1974; Shields & Tinsley 1976).

• A higher metallicity may induce a steeper IMF (i.e. a larger value of the slope α of the IMF $N(m) \propto m^{-\alpha}$) at least for $m > 10$ M$_\odot$, where the bulk of the ionizing radiation is produced (*e.g.*, Terlevich & Melnick 1983).

In the first two cases, metallicity acts "directly" on the radiation field of the ionizing star cluster, by modifying the stellar spectra without affecting the star formation processes. In the third and fourth case instead, metallicity acts "indirectly" and the changes in the radiation field result from changes in the properties of the IMF.

Panagia (1980) demonstrated that the combined effects of at least the first three processes are needed to explain the He ionization in the Milky Way. Moreover, these processes appear to account for the observed gradient of the effective temperature of the ionizing radiation inferred from the fitting of theoretical models to observations of low-metallicity objects (Talent 1980; Campbell 1988).

3. Ionized Helium in External Galaxies

Considering external galaxies, several authors (*e.g.*, Pagel 1986, Viallefond 1988, Robledo-Rella & Firmani 1991) have suggested that a systematic change of the IMF with metallicity is required by observations. Others (*e.g.*, Fierro, Torres - Peimbert & Peimbert 1986, McGaugh 1991) have come to the opposite conclusion, and the controversy is still open. A thorough assessment of this subject is now possible and necessary.

We have considered a large sample of extragalactic H II regions which provides an extensive coverage of a very wide metallicity range (almost a factor of 100), and includes galaxies with a variety of morphological types and luminosities. Such a sample is in many respects much more homogeneous than any sample of galactic H II regions. All the H II regions observed are large (diameter D > 50 pc), tenuous ($n_e < 500$ cm^{-3} as derived from the [S II] I(6717)/I(6731) ratio) and must be ionized by large OB associations.

Here, we limit our analysis to data published as of February 1992. (A more complete investigation, including the discussion of data published as of December 1999, is in progress and will be completed soon; Lenzuni and Panagia 2000, in preparation). Thus, our sample currently includes 287 H II regions in 46 spiral and irregular galaxies with positive detections of the [O II] lines at 3726 and 3729 Å (usually unresolved), of the [O III] lines at 4959 and 5007 Å , and of at least one of the He I lines. Additional observations were also collected for 87 "Blue Compact Galaxies" (BCG's). None of these objects is resolved into individual H II regions, the observations being relative to the entire galaxy or, possibly, to its central, brightest parts. These galaxies appear to be undergoing a stage characterized by a collective mode of star-formation. Their spectra are heavily dominated by H II region-like emission, hence they can be treated for our purposes as giant, isolated, extragalactic H II regions.

4. Analysis and Discussion

Ionized helium abundances are shown in Figure 1 as a function of oxygen abundances, for all of the H II regions in spiral and irregular galaxies and the Blue Compact Galaxies of our sample. The long baseline in metallicity offers a unique opportunity to constrain *both* the abundance of primordial helium Y_P *and* to the $\Delta Y/\Delta Z$ gradient, thus fully determining the "helium enrichment curve" (HEC) which relates the total abundance of helium to the abundance of oxygen. Considering that evolutionary effects always decrease the He^+/H^+ ratio because the aging of a stellar cluster results in a softening of the ionizing radiation field, and that young clusters are observationally favoured because they are intrinsically brighter than older clusters, the HEC can be derived by determining the upper envelope of the distribution shown in Figure 1. Among the class of curves $Y = Y_0 + \Delta Y/\Delta Z \times Z$ the best fit to the upper envelope of the observations is obtained for $Y_0 = 0.243$ and $(\Delta Y/\Delta Z)_\odot = 3.2$ (see dashed curve in Fig. 1). This relation is consistent with the observational results of Pagel *et al.* (1992) as well as with Maeder's (1992) theoretical models.

An inspection to Figure 1 reveals that the observed He^+/H^+ ratio appears to be almost constant up to solar O abundance ($\log(O/H)_\odot+12 \simeq 8.8$) and then it declines rather quickly for higher metallicities. This is a clear sign that He is progessively less ionized as the O abundance increases, and implies that the mean radiation temperature of the ionizing stars becomes lower than about 38,000 K around $\log(O/H) \simeq 8.5$.

We find that mechanisms through which metallicity acts "directly" on the radiation field of the ionizing star cluster are not enough to explain the observed gradient of the He ionization fraction with O abundance. Most of the effect appears instead to be due to "indirect" mechanisms, *i.e.* a marked deficiency of hot stars with increasing metal abundances. There are at least

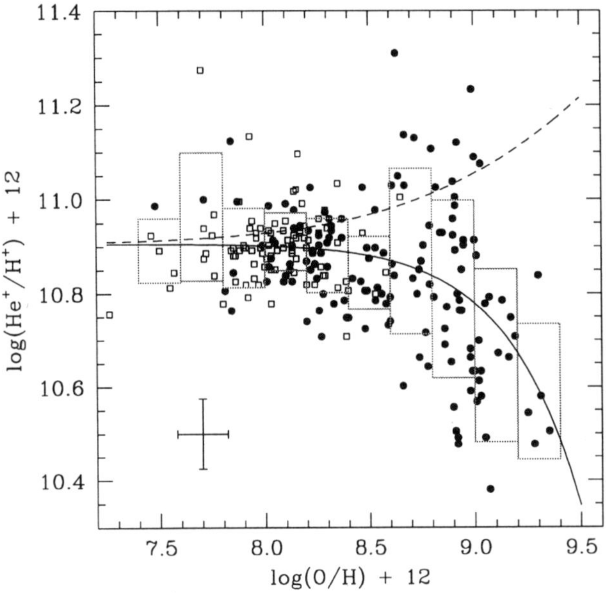

Figure 1. He^+/H^+ ratios (by number) as a function of the O abundances for the 287 H II regions in 46 Spiral and Irregular galaxies (filled symbols) and for the 87 Blue Compact Galaxies (open squares) of our sample. The cross in the lower left corner represents the typical uncertainty of the data. The boxes correspond the $\pm 1\sigma$ deviations from the averages calculated within 0.2 dex intervals in log(O/H). The dashed line represents the helium enrichment curve (HEC) and the solid line is a model calculation in which the He ionization decreases exponentially with increasing O abundance.

three possible scenarios to explain this fact:
1. The IMF slope becomes steeper for higher metallicities.
2. The IMF upper cutoff moves to lower masses for higher metallicities.
3. The most massive stars become progressively unable to provide ionizing radiation, either because at high metallicities the remnant of their pre-MS cocoons remains optically thick over most of a star's lifetime, or because pulsational instabilities prevent the most massive stars from reaching their expected ZAMS surface conditions.

Our model calculations show that varying only the slope of the IMF, *i.e.* point (1), does not give a satisfactory fit to the data because the resulting ionization decline would be too shallow. On the other hand, point (2), *i.e.* a systematic variation of the IMF upper mass cut-off with metal abundance ($m_U \propto Z^{-\beta}$, $\beta > 0$) can reproduce the observed trend of the He ionization, with $m_{U\odot} = 48$ $M_\odot$ and $\beta = 0.60$. Point (3) could also account for the observations provided that the invoked effects are indeed capable to produce the sharp decline of He ionization as observed.

From the observational point of view there are no direct studies to conclusively discriminate between points (2) and (3). Observations of massive stars near the Galactic Center, such as the Pistol star, the Sickle and the Quintuplet clusters (*e.g.,* Figer *et al.* 1999 and references therein) seem to favor the third possibility, because they are so bright ($log(L/L_\odot) > 6$) that they must be quite massive. On the other hand, one may argue that those clusters are so close to the Galactic Center that tidal forces may drastically affect the dynamical processes that lead to the formation of stars and, therefore, they may not be representative of normal situations. The ideal investigation to clarify this issue should include nebular and stellar spectroscopy of a large sample of HII regions in galaxies which display marked effects of incomplete He ionization, such as M51 or M83. Another discriminant between hypotheses (2) and (3) is that if a lowering of the IMF upper cutoff is *the* explanation (*i.e.* point (2)), then the frequency of Wolf-Rayet stars relative to early type stars is expected to be abnormally low in the high-Z regions because in this case the reduced ionization is entirely due to the lack of truly massive stars that are expected to end up as WR stars in their final stages of evolution.

References

1. Campbell A., 1988, *Ap. J.* **335**, 644
2. Figer, D., McLean, I.S., Morris, M., 1999, *Ap. J.* **514**, 202
3. Fierro J., Torres-Peimbert S., Peimbert M., 1986, *P.A.S.P.* **98**, 1032
4. Güsten R., Mezger P., 1982, *Vistas in Astronomy* **26**, 159
5. Kahn F.D., 1974, *Astr. Ap.* **37**, 149
6. Maeder A., 1992, *Astr. Ap.* **264**, 105
7. McGaugh S.S., 1991, *Ap. J.* **380**, 140
8. Mezger P.G., Wink J.E., 1983, in *Primordial Helium*, eds. P.A. Shaver, D. Kunth and K. Kjär (ESO, Garching, Germany)
9. Montes, M.J., Van Dyk, S.D., Weiler, K.W., Sramek, R.A., Panagia, N., 1998, *Ap. J.* **506**, 874
10. Pagel B.E.J., 1986, in *Highlights of Astronomy* Vol. 7, ed. J.P. Swings (Reidel, Dordrecht, The Netherlands)
11. Pagel B.E.J., Simonson E.A., Terlevich R.J., Edmunds M.G., 1992, *M.N.R.A.S.* **255**, 325
12. Panagia N., 1980, in *Radio Recombination Lines*, ed. P. A. Shaver (Reidel, Dordrecht, The Netherlands)
13. Panagia, N., *et al.*, 1999, *Mem S.A.It.*, in press
14. Robledo-Rella V., Firmani C., 1990, *Rev. Mexicana Astron. Astrof.* **21**, 236
15. Shields G.A., Tinsley B., 1976, *Ap. J.* **203**, 66
16. Terlevich R.J., Melnick J., 1983, *M.N.R.A.S.* **195**, 839
17. Viallefond F., 1988, in *Galactic and Extra-Galactic Star Formation*, eds. R. Pudritz and M. Fich (Reidel, Dordrecht, The Netherlands)
18. Weiler, K.W., *et al.*, 1999, in *"The Largest Explosions since the Big Bang: Supernovae and Gamma Ray Bursts"*, eds. M. Livio, K. Sahu & N. Panagia, (CUP, Cambridge, England), in press

Speaker: N. PANAGIA

A. MAEDER: The lack of stars close to the ZAMS could well be due to the fact that the very massive stars only become visible when they have somehow moved away from the ZAMS. This would be quite consistent with recent pre-MS models with accretion.

N. PANAGIA: This is an attractive possibility that probably can account for most of the observed effects.

VI-FORMATION AND CHEMICAL EVOLUTION OF THE MILKY WAY

Danny Vanbeveren, Simonetta Fabrizio, Letizia Stanghellini and Giuseppe Bono

THE CHEMICAL EVOLUTION OF THE MILKY WAY

MONICA TOSI
Osservatorio Astronomico di Bologna
Via Ranzani 1, I-40127 Bologna, Italy

1. Introduction

The field of chemical evolution modeling of the Galaxy is experiencing in the last years a phase of high activity and important achievements. There are, however, several open questions which still need to be answered. In this review I will try to summarize what have been the most important achievements and what are some of the most urgent questions to be answered.

The reason for the recent increase of activity and success of chemical evolution models is probably two-folding. First of all, on the observational side, the last decade has witnessed a tremendous improvement in the quality and in the amount of data on the major Galactic features, like the chemical abundances and abundance ratios in stellar and gaseous objects of various types, the density distributions of gas and stars in different Galactic regions, etc.: Fundamental data which provide stringent constraints on evolution models. In addition, also on the theoretical side there has been a recent blooming of new studies, with several new groups working on stellar nucleosynthesis to derive reasonable yields for stars of all mass and of several initial metallicities, and taking into account as much as possible the large uncertainties affecting the latest evolutionary phases. If we consider that for almost two decades the only usable set of yields for low and intermediate mass stars was that provided by Renzini & Voli (1981), while now we can choose among those by Forestini & Charbonnel (1997), van den Hoek & Groenewegen (1997), Boothroyd & Sackman (1998) and Marigo (1998 and this volume), all published in the last two years, it is apparent that we have entered an era of great interest in stellar nucleosynthesis studies.

These circumstances have favoured the appearance in the literature of an increasing number of good chemical evolution models computed by an increasing number of people. Nowadays there are several models able to

F. Matteucci and F. Giovannelli (eds.), The Evolution of the Milky Way, 505–515.

satisfactorily reproduce all the major observational constraints, not only in the solar neighbourhood but also in the whole Galaxy. Only in the last few months one could count at least four different groups who have presented models in fairly good agreement with the data: Boissier & Prantzos (1999, hereinafter BP), Chang et al. (1999), Chiappini et al. (1999, CMP) and Portinari & Chiosi (1999, PC).

2. Major Results

Before analysing the various results, it is important to recall that standard chemical evolution models follow the large-scale, long-term phenomena and can therefore reproduce only the average trends, not the cloud-to-cloud, star-to-star fluctuations. To put it in Steve Shore's words: *They are a way to study the climate, not the weather, in galaxies.* This can be considered a limitation of the models, but is the obvious price to pay to avoid introducing too many free parameters that would make it much more difficult to infer the overall evolutionary scenario with sufficient reliability. As well known, we have not yet been able to find a unique scenario for the most probable evolution of the Milky Way (see e.g. Tosi 1988a), but we are converging toward a fairly limited range of possibilities for the involved parameters (initial mass function, IMF, star formation rate, SFR, gas flows in and out of the Galaxy).

Thanks to the improvements both on the observational and on the theoretical sides, good chemical evolution models of the Milky Way nowadays can reproduce the following list of observed features:

- Current distribution with Galactocentric distance of the SFR (e.g. as compiled by Lacey & Fall 1985);
- current distribution with Galactocentric distance of the gas density (see e.g. Tosi, 1996, BP and references therein);
- current distribution with Galactocentric distance of the star density (see e.g. Tosi, 1996, BP and references therein);
- current distribution with Galactocentric distance of element abundances as derived from HII regions and from B-stars (e.g. Shaver et al. 1983, Smartt & Rollerston 1997);
- distribution with Galactocentric distance of element abundances at slightly older epochs, as derived from PNe II (e.g. Pasquali & Perinotto 1993, Maciel & Chiappini 1994, Maciel & Köppen 1994);
- age-metallicity relation not only in the solar neighbourhood but also at other distances from the center (e.g. Edvardsson et al. 1993);
- metallicity distribution of G-dwarfs in the solar neighbourhood (e.g. Rocha-Pinto & Maciel 1996);

- local Present-Day-Mass-Function (PDMF, e.g. Scalo 1986, Kroupa et al. 1993);
- relative abundance ratios (e.g. [O/Fe] vs [Fe/H]) in disk and halo stars (e.g. Barbuy 1988, Edvardsson et al. 1993, Israelian et al. this volume).

As mentioned above, the most recent examples of how good models can fit the above list of observed Galactic features are given by BP, Chang et al. (1999), CMP and PC (see also in this book the contributions by Chiappini, by Portinari and by Prantzos).

If one bears in mind that the free parameters involved in the computation of standard chemical evolution models are essentially the IMF, the law for the SFR, and those for gas flows in and out of the Galaxy, it is clear that the number of observational constraints is finally sufficient to put significant limits on the parameters. In fact, if we compare the results of all the models in better agreement with the largest set of empirical data, we see that they roughly agree on the selection of the values for the major parameters. The conclusions that can be drawn from such comparison are:
• **IMF**: after several sophisticated attempts (e.g. CMP) to test if a variable IMF could better fit the data, it is found, instead, that a roughly constant IMF is most likely, even if the exact slopes and mass ends are still subject of debate.
• **SFR**: it cannot be simply and linearly dependent only on the gas density; a dependence on the Galactocentric distance is necessary, either implicit (e.g. through the total mass density as in Tosi 1988a or in Matteucci & François 1989) or explicit (e.g. as in BP). We don't know however what is its actual behaviour (see e.g. Portinari, this volume) or even if it should be considered as fairly continuous or significantly intermittent as recently suggested by Rocha-Pinto et al. (1999).
• **gas flows**: all the models in better agreement with the data invoke no or negligible galactic winds and a substantial amount of infall of metal poor gas (not necessarily primordial, e.g. Tosi 1988b, Matteucci & François 1989) and there are increasing observational evidences on this phenomenon (see also Burton, this volume). We have no empirical information, however, on the spatial and temporal distribution of the accretion process: uniform or not ? continuous or occurring in one, two or several episodes ? (e.g. Beers & Sommer-Larsen 1995, Chiappini et al. 1997, Chang et al. 1999).

3. Open Questions

It is apparent from the summary presented above that, in spite of the wealth of good data and models described in the previous sections, the scenario of the Milky Way evolution is not completely clear. There are still several issues we don't understand, including some of conspicuous im-

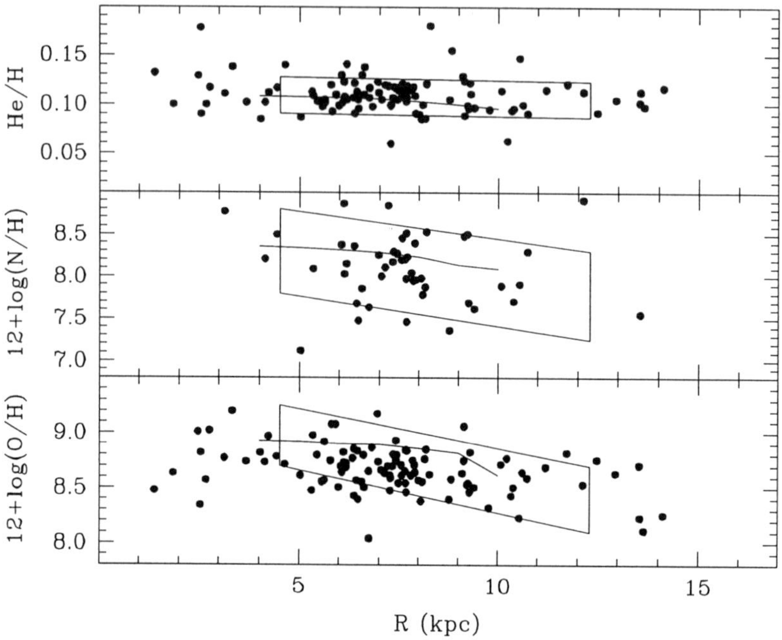

Figure 1. Radial distribution of the He, N and O abundances as derived from observations of PNe of type II (see text for references) and from the predictions of Tosi's (1988a) model 1 for the Galaxy medium 2 Gyr ago.

portance. Among these, I consider of special interest the evolution of the abundance gradients and that of CNO isotopes.

3.1. ABUNDANCE GRADIENTS

Thanks to the recent results by Smartt & Rollerstone (1997) we finally know that young objects (HII regions and B-stars) all show the same metallicity distribution with Galactocentric distance and a fairly steep negative gradient. All the models in better agreement with the Galaxy constraints are able to reproduce this distribution (see Tosi 1996, Chiappini, Portinari and Prantzos in this volume).

Slightly older objects, such as PNe of type II whose progenitors on average are 2 Gyr old, show similar abundances and possibly flatter gradients (e.g. Maciel & Köppen 1994). Good models of Galaxy evolution reproduce well not only the present abundance distribution, but also the distributions derived from PNeII observations. For instance, Fig.1 shows the predictions of the best of models of type 1 in Tosi's (1988a) set for the He, N and O abundance distributions with Galactic radius 2 Gyr ago. The adopted

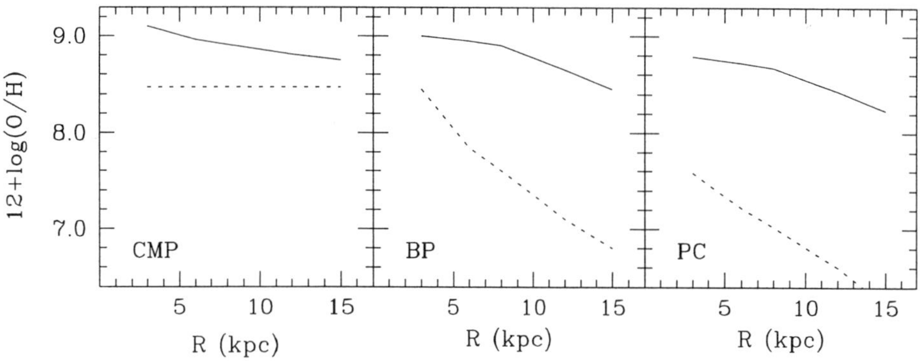

Figure 2. Oxygen gradients predicted by the models from Chiappini et al. (1999, left), Boissier & Prantzos (1999, center), and Portinari & Chiosi (1999, right). The solid curves refer to the present epoch and the dotted ones to 1 Gyr after the disk formation.

stellar yields are Marigo's (1998 and this volume) for low and intermediate mass stars and Limongi et al. (this volume) for massive stars. The data points correspond to the PNeII measures by Pasquali & Perinotto (1993) and the open boxes sketch the distribution of the values derived by Maciel & Chiappini (1994) and Maciel & Köppen (1994). The data sets are in perfect agreement with each-other and the model predictions fit well their average distributions.

When we consider earlier epochs, the predictions from different models diverge, despite the common assumption that the Galaxy is initially formed of primordial gas. For instance, the three models which are presented in this volume by Chiappini, Portinari and Prantzos, and that are in fairly good agreement with all the observational constraints, predict the gradient evolutions schematically described in Fig.2 (see BP, CMP and PC for more details). The initial distribution of oxygen with galactic radius in the left panel is totally flat, becomes initially slightly positive, then turns to negative and steepens with time, reaching at the present epoch the observed slope of -0.08 dex/kpc; vice versa, the gradient at 1 Gyr in the central panel is negative and quite steep and then slowly flattens with time, particularly in the inner galactic regions, reaching finally the observed slope at the present time; the same trend occurs in the right panel, but with different absolute abundances. If one compares (e.g. Tosi 1996) all the models able to reproduce the observed Galactic features, it is easy to understand that they present all the possible varieties of gradient evolution: from slopes initially positive becoming first flat and then increasingly negative, to slopes initially flat and then becoming increasingly negative, to slopes initially negative and then becoming increasingly flat.

The reason for such a variety of gradient evolutions is the strong dependence of the radial slope on the radial variations of the ratio between ISM enrichment from stars (i.e. SFR) and ISM dilution from metal poor gas (i.e. initial conditions and/or infall of metal poor gas). Regions with higher SFR have larger enrichment, but can remain relatively metal poor if they contain or accrete large amounts of metal poor gas. It is then sufficient to have different initial conditions or different assumptions on the temporal behaviours of the SFR and of the infall rate to obtain quite different abundance gradients at the various epochs.

The following few examples of possible scenarios give an idea of the sensitivity of the gradient evolution to the boundary conditions:

- If the efficiency in the chemical enrichment of the inner Galactic regions at early epochs is low (for instance because the SFR is low and/or there is a high amount of primordial gas), then the early radial distribution of the heavy elements is flat. And to reach the observed present slope it has to become negative and steepen with time.
- If, instead, the enrichment efficiency in the inner regions at early epochs is high (for high SFR or low gas mass), then the early gradient is negative and steep. And to reach the present slope it has to flatten with time.
- If at late epochs the acretion (infall) of metal poor gas is stronger in the outer than in the inner regions, then the gradient tends to steepen with time because of the increasing dilution for increasing galactocentric distance.
- If at late epochs the inner regions exhaust their gas, then the metallicity saturates there and the inner gradient becomes increasingly flat with time.

All these scenarios are plausible: how can we understand which are the right ones ? If we knew the right history of the abundance gradients we would also know what is the most likely evolution of the Galactic disk. Unfortunately, despite their accuracy, the observational data already available on open clusters and on field stars are not yet sufficient to clearly distinguish whether the abundance gradients were steeper or flatter at early epochs. Open clusters are probably the best candidates to provide such information, thanks to their visibility at large distances and to the relative ease to derive their age and metallicity, but as described by Bragaglia (this volume, and references therein) the number of clusters treated homogeneously is still too small.

3.2. EVOLUTION OF CNO ISOTOPES

The CNO isotopes are important because they are stable, diffused and largely studied, since they provide the seeds for the production of heavier elements. In particular, the stellar nucleosynthesis of the carbon and oxygen isotopes is examined in detail in most of the most recent studies. Nonetheless, it is not completely clear yet how they should behave during the Galaxy evolution. The problem was already pointed out twenty years ago by Penzias (1980), who noticed that the observed decrease of the local $^{18}O/^{17}O$ from the solar to the local ISM value and the corresponding increase of $^{16}O/^{18}O$ were difficult to interpret. In fact, chemical evolution models predicted (Tosi 1982) $^{18}O/^{17}O$ to remain roughly constant in the last 4.5 Gyr and $^{16}O/^{18}O$ to steadily decrease. Those predictions were based on simple arguments on the relative enrichment of primary and secondary elements produced by stars of different masses, and have been confirmed by subsequent studies based on nucleosynthesis studies of solar metallicity stars (e.g. Prantzos et al. 1996).

These results for the carbon and oxygen isotopic ratios are represented by the solid line in Fig.3. The left hand panels show the time behaviour of the isotopic ratio in the solar neighbourhood as predicted by models and as observed in the sun and in the local ISM, which are assumed to be representative of the average local ratios 4.5 Gyr ago and now, respectively. The right hand panels show the present distribution with Galactocentric distance as predicted by the same models and as derived from radio observations of molecular clouds. The solid line corresponds to the same model presented in Fig.1 (Tosi-1), assuming the yields for solar initial metallicity computed by Boothroyd & Sackman (1998), by Forestini & Charbonnel (1997) and by Woosley & Weaver (1995) for low, intermediate and high mass stars, respectively. Qualitatively similar results were obtained by Prantzos et al. (1996) adopting the solar yields by Marigo et al. (1996), Renzini & Voli (1981) and Woosley & Weaver (1995). It is apparent that while the predictions for $^{12}C/^{13}C$ and $^{16}O/^{17}O$ are in fair agreement with the data, the time behaviour of the oxygen isotopic ratios involving ^{18}O is inconsistent with them. There have been several speculations on how this impasse could be overcome, with suggestions that either the theory or the data or both might be wrong or misinterpreted (see e.g. Prantzos et al. 1996, Tosi 1996, Wielen & Wilson 1998), but no solution has been found yet.

One possibility is that it is not correct to adopt solar yields also for the earlier epochs, when stars were certainly metal poorer. Now that stellar yields are available also for lower metallicities, we expect to find an improvement in the comparison between model predictions and observed

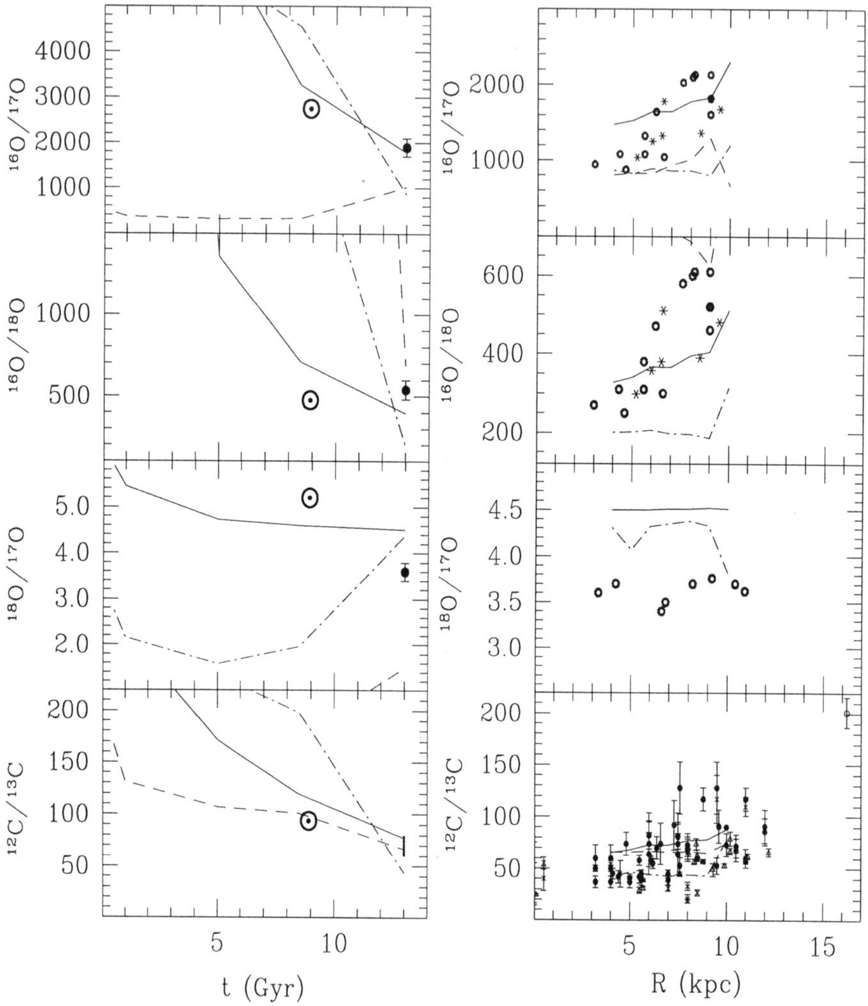

Figure 3. Carbon (bottom panels) and oxygen (three top panels) isotopic ratios. Left panels: evolution in the solar neighbourhood. Right panels: current distributions with Galactocentric distance. The solar symbol represents the solar ratio derived from Anders & Grevesse 1989; all the other data are from radio observations of molecular clouds (see Sandrelli et al. 1998, for references). All the curves refer to Tosi (1988a) model 1 but assuming different stellar yields as described in the text.

ratios. Unfortunately, this is definitely not the case, as clearly shown by the dashed and dash-dotted lines in Fig.3. The dash-dotted curve represents the same model as the solid curve, with the same sources for the yields, but adopting the low metallicity yields at earlier epochs and the solar ones only when the ISM reaches Z=0.02. It is apparent that, rather than improving the agreement with the data, this curve worsens the fit, both for the local evolution and for the current distribution with Galactocentric distance. This result is strongly dependent on the adopted yields and we may hope that different nucleosynthesis studies would provide more consistent predictions, but so far no set of stellar yields is able to reproduce all the shown observed distributions. Some of the available yields do improve the results on one isotopic ratio, but worsen the results on other ratios, as exemplified by the dashed lines, showing the predictions of the same model when Marigo's (this volume) metallicity dependent yields are adopted for low and intermediate mass stars and Limongi's et al. (this volume) for massive stars: the data on the carbon isotopic ratio are now well reproduced, but the predicted oxygen ratios are definitely inconsistent with the data.

I will then conclude this short description of the state of the art in Galactic chemical evolution models by emphasizing that, despite the great work that has been done by observers and theoreticians to improve the number and the quality of the observational and theoretical constraints, further efforts on both sides are needed to shed light on several unclear issues. In particular, it would be important to derive accurate chemical abundances in stars and clusters of different ages and Galactic locations and to study in better detail the stellar nucleosynthesis in stars of all masses and initial metallicities.

I warmly thank S.Chieffi, M.Limongi, P.Marigo and O.Straniero for providing their yields in advance of publication and S.Sandrelli for help. Conversations with them and with C.Chiappini, F.Matteucci and N.Prantzos were very fruitful.

References

1. Anders, E. & Grevesse, N. (1989) Abundances of the elements - Meteoritic and solar, *Geochimica and Cosmochimica Acta*, **53**, 197
2. Barbuy, B. (1988) Oxygen in 20 halo giants, *Astron.Astrophys.*, **191**, 121
3. Beers, T.C. & Sommer-Larsen, J. (1995) Kinematics of metal-poor stars in the Galaxy, *Astrophys.J.Suppl.*, **96**, 175
4. Boothroyd, A.I. & Sackman, I.-J. (1999) The CNO Isotopes: Deep Circulation in Red Giants and First and Second Dredge-up, *Astrophys.J.*, **510**, 232
5. Boissier, S. and Prantzos, N. (1999) Chemo-spectrophotometric evolution of spiral galaxies: I. The model and the Milky Way, *Mon.Not.R.Astr.Soc.*, **307**, 857, BP
6. Chang, R.X., Hou, J.L., Shu, C.G., Fu, C.Q. (1999) Two-component model for the chemical evolution of the Galactic disk, *Astron.Astrophys*, in press, astro-ph/9903317
7. Chiappini, C., Matteucci, F., Gratton, R. (1997) The Chemical Evolution of the Galaxy: The Two-Infall Model, *Astrophys.J.* **477**, 765
8. Chiappini, C., Matteucci, F., Padoan, P. (1999) The Chemical Evolution of the Galaxy with variable IMF, astro-ph/9908207, CMP
9. Forestini and Charbonnel (1997) Nucleosynthesis of light elements inside thermally

pulsing AGB stars: I. The case of intermediate-mass stars, *Astron.Astrophys.Supp*, **123**, 241

10. Edvardsson, B., Andersen, J., Gustafsson, B., Lambert, D.L., Nissen, P.E., Tomkin, J. (1993) The Chemical Evolution of the Galactic disk. I. Analysis and results, *Astron.Astrophys.*, **275**, 101
11. Kroupa, P., Tout, C.A., Gilmore, G. (1993) The distribution of low-mass stars in the Galactic disc, *Mon.Not.R.Astr.Soc.*, **262**, 545
12. Lacey, C.G. and Fall, S.M. (1985) Chemical evolution of the Galactic disk with radial gas flows, *Astrophys.J.*, **290**, 154
13. Maciel, W.J. and Chiappini, C., (1994) Abundances and radial gradients from disk Planetary Nebulae: He, N, C, and Cl, *Astrophys.& Sp.Sc.*, **219**, 231
14. Maciel, W.J. and Köppen, J. (1993) Abundance gradients from disk Planetary Nebulae: O, Ne, S, and Ar, *Astron.Astrophys.* **282**, 436
15. Marigo, P. (1998) PhD Thesis, University of Padova
16. Matteucci, F. and François, P. (1989) Galactic Chemical Evolution: Abundance Gradients of Individual Elements, *Mon.Not.R.Astr.Soc.*, **239**, 885
17. Pasquali, A. and Perinotto, M. (1993) Chemical behaviour of planetary nebulae and galactic abundance gradients, *Astron.Astrophys.*, **280**, 581
18. Penzias, A.A. (1980) Nuclear Processing and Isotopes in the Galaxy, *Science*, **208**, 663
19. Portinari, L. and Chiosi, C. (1999) On star formation and chemical evolution in the Galactic disc, *Astron.Astrophys* in press, astro-ph/9908326, PC
20. Prantzos, N., Aubert, O., Audouze, J. (1996) Evolution of carbon and oxygen isotopes in the Galaxy, *Astron.Astrophys*, **309**, 760
21. Renzini, A. and Voli, M. (1981) Advanced evolutionary stages of intermediate-mass stars. I - Evolution of surface compositions, *Astron.Astrophys.*, **94**, 175
22. Rocha-Pinto, H.J. and Maciel, W.J. (1996) The metallicity distribution of G dwarfs in the solar neighbourhood, *Mon.Not.R.Astr.Soc.*, **279**, 447
23. Rocha-Pinto, H.J., Scalo, J.M., Maciel, W.J., Flynn, C. (1999) An intermittent star formation history in a normal disk galaxy: the Milky Way, *Astrophys.J. Lett.* in press, astro-ph/9908328
24. Sandrelli, S., Visco, A., Tosi, M. (1998) The stellar yields - Galactic chemical evolution connection: a re-analysis, in *Tour symposium on nuclear astrophysics III*, M.Arnould et al. eds (AIP Conf.Proceedings 425, Woodbury, NY) p.581
25. Scalo, J.M. (1986) The stellar initial mass function, *Fund.Cosmic Phys.* **11**, 1
26. Shaver, P. A., McGee, R. X., Newton, L. M., Danks, A. C., Pottasch, S. R. (1983) The galactic abundance gradient, *Mon.Not.R.Astr.Soc.*, **204**, 53
27. Smartt, S. and Rollerstone, W. (1997) The Galactic Oxygen Abundance Gradient, *Astrophys.J.Lett.*, **481**, L47
28. Tosi, M. (1982) CNO isotopes and Galactic chemical evolution, *Astrophys.J.*, **254**, 699
29. Tosi, M. (1988a) Models of Galactic chemical evolution: the problem of uniqueness, *Astron.Astrophys.*, **197**, 33
30. Tosi, M. (1988b) The effect of metal-rich infall on galactic chemical evolution, *Astron.Astrophys.*, **197**, 47
31. Tosi, M. (1996) Comparison of Chemical Evolution Models for the Galactic Disk, in From Stars to Galaxies: The Impact of Stellar Physics on Galaxy Evolution, C. Leitherer, U. Fritze-von-Alvensleben, and J. Huchra eds, *ASP Conf.Ser.* **98**, 229
32. van den Hoek, L. B. and Groenewegen, M. A. T. (1997) New theoretical yields of intermediate mass stars, *Astron.Astrophys.Suppl.*, **123**, 305
33. Wielen, R. and Wilson, T.L. (1987) The evolution of the C, N, and O isotope ratios from an improved comparison of the interstellar medium with the Sun, *Astron.Astrophys*, **326**, 139
34. Woosley, S.E. and Weaver, T.A. (1995) The Evolution and Explosion of Massive Stars. II. Explosive Hydrodynamics and Nucleosynthesis, *Astrophys.J.* **101**, 181

Speaker: M. TOSI

W. MACIEL: From one of your plots your model apparently predicts some He/H gradient. Is that true? In our earlier work on planetary nebulae we derived a small gradient, but we have recently taken into account the contamination of the progenitor star nucleosynthesis, and the gradient completely disappears.

M. TOSI: The He gradient predicted by the models for 2 Gyr ago is about -0.02 - -0.03 dex/Kpc. In my models it is flatter at earlier epochs.

W. MACIEL: In your review of chemical evolution models you have not mentioned chemodynamical models (Hensler et al.). Is there a reason for it?

M. TOSI: I have reviewed only "standard" chemical evolution models because my feeling is that chemo-dynamical models are not yet accurate enough on the chemistry side and in general are not able yet to reproduce the list of observational constraints. I believe however that they represent an important step forward in galaxy evolution models.

F. MATTEUCCI: You did not show the predicted gas distributions for those models predicting the abundance gradients. A way of discriminating among models is just to see if they fit at the same time the abundance gradients, the gas distribution and the star formation rate distribution. Can you comment on that?

M. TOSI: Unfortunately the very different evolution of the gradient slope refers also to models which do reproduce all the list of observational constraints, including the gas and SFR radial distribution.

N. PRANTZOS: My question is the same as Francesca's. Do all these models reproduce also the present profiles of gas, stars, SFR? Since their assumptions are different (some form the disk inside-out, some the opposite way) I would expect different results...

M. TOSI: Just as an example: your models, Matteucci and Chiappini, and mine all reproduce fairly well the observed SFR, gas and star density distributions (as well as many other constraints), but they predict opposite evolutions of the gradient slopes.

RECONSTRUCTING THE STAR FORMATION HISTORY OF THE GALAXY

X. HERNANDEZ
Osservatorio Astrofisico di Arcetri,
Largo E. Fermi 5, 50125 Firenze, Italy

D. VALLS–GABAUD
Laboratoire d'Astrophysique, UMR CNRS 5572,
Observatoire Midi-Pyrénées,
14 Av. E. Belin, 31400 Toulouse, France

AND

G. GILMORE
Institute of Astronomy,
Madingley Road, Cambridge CB3 0HA, UK

Abstract.

The evolution of the star formation rate in the Galaxy is one of the key ingredients quantifying the formation and determining the chemical and luminosity evolution of galaxies. Many complementary methods exist to infer the star formation history of the components of the Galaxy, from indirect methods for analysis of low-precision data, to new exact analytic methods for analysis of sufficiently high quality data. We summarise available general constraints on star formation histories, showing that derived star formation rates are in general comparable to those seen today. We then show how colour-magnitude diagrams of volume- and absolute magnitude-limited samples of the solar neighbourhood observed by Hipparcos may be analysed, using variational calculus techniques, to reconstruct the local star formation history. The remarkable accuracy of the data coupled to our maximum-likelihood variational method allows objective quantification of the local star formation history with a time resolution of $\approx$ 50Myr. Over the past 3Gyr, the solar neighbourhood star formation rate has varied by a factor of $\sim$4, with characteristic timescale about 0.5Gyr, possibly triggered by interactions with spiral arms.

F. Matteucci and F. Giovannelli (eds.), The Evolution of the Milky Way, 517–530.

1. Star Formation Rates: Qualitative Deductions

Star formation rates and histories can be estimated in special cases from a combination of chemical evolution models and the total stellar mass formed into stars. Basically, this exploits the stellar evolutionary, and Type I supernova timescales for element production. One requires that stars formed at a rate consistent with their chemical distributions, whether a δ-function, a range in the products of both Type I and Type II supernaova, products of single supernovae with time for efficient mixing of the ISM, or whatever. Combining these albeit crude estimates of the *duration* of star formation with a calculation of the stellar mass formed, provides a star formation rate. Perhaps surprisingly, given the crude calculation, such derived rates are both similar to those determined more accurately today, and are all quite low.

TABLE 1. A summary of star formation rates, and durations of star formation, in some Galactic stellar populations. These values are derived from combination of chemical element scatter and masses.

Population	Duration	Formation Rate
	(years)	$\mathcal{M}_{\odot} yr^{-1}$
Globular cluster	$\leq 10^8$	≥ 0.01
ωCen	$\geq 10^8$	≤ 0.1
Halo, [Fe/H]≤ -2.0	$\leq 10^8$	~ 1
Halo, [Fe/H]~ -1.5	$\leq 10^9$	~ 1
Bulge; high [α/Fe]	few.10^8	10-100
Bulge; low [α/Fe]	few.10^9	10-100
Thick Disk	few.10^9	1-10
Current Disk	10^{10}	$\sim 1 - 10$
Inner Disk	?	?
Satellite dSph	many.10^9	$\leq 10^{-3}$
Assembly	early	
Infall	continuing?	$\sim$ 4Gyr

In cases where no spread in element ratios is seen, and there is no range in [Fe/H], star formation was plausibly complete before new chemical elements could be produced; perhaps globular clsuters are an example of this case. For field halo stars with [Fe/H]≥ -2, where a wide range in [Fe/H] but a very small range in [α/H] is seen, star formation must have continued for long enough for efficient mixing of supernova ejecta into the ISM. Since no products of Type I supernovae are seen, this brackets allowed star formation durations. By applying such qualitative considerations, we deduce that most of the Milky way formed at a star formation rate which is comparable to that of today. Only the Galactic Bulge, and the inner disk, where star formation histories remain very poorly known, are available to retrieve the Milky Way's place as a 'typical object' on the Madau plot. These simple calculations are summarised in Table (1).

2. Quantitative Determination of Star Formation Rates

Most attempts at deducing the past history of the star formation rate in the Galaxy have relied on indirect age indicators, typically chemical evolution models, using an age-metallicity relation, or from empirical correlations of stellar properties, such as chromospheric activity, with age. Because of the many uncertainties and observational biases inherent in these methods, many corrections must be applied to the samples before a SFH can be inferred. A good recent example, illustrating the strengths and complexities of the techniques, is Rocha-Pinto et al. (2000). Given all this, it is not too surprising to find conflicting results from independent analyses of the limited available data.

The precision in luminosities of field stars provided by the Hipparcos satellite gives us for the first time the data quality to allow application of a more objective method to infer the star formation history which produced the solar neighbourhood. Indeed, recent analytical developments show that it is now possible to reconstruct the star formation history which gave rise to the observed distribution of stars in the HR diagram for any (relatively) simple stellar population. The constraints in present application of the methodology are that one requires (i) a small scatter in the metallicity distribution, and known mean abundance, (ii) a colour-magnitude diagram extending below the turnoff age of interest, and (iii) appropriate stellar evolutionary tracks. Whilst previous reconstructions of the star formation history based on the inversion of CMDs were forced to rely on assumed parametric forms for the evolution of the star formation rate, we have developed and implemented a new and rigorous method which makes no a *priori* assumptions about the possible complexity of the star formation history.

This method is explained in detail in Hernandez et al. (1999, Paper I) and in Gilmore etal. (2000). These references also present results from the extensive simulations which were used to quantify the biases inherent in any reconstruction based on observational colour-magnitude data. The effects of unresolved binaries, of uncertainties in the stellar initial mass function, and of errors in the adopted system metallicity, were assessed. It was concluded that derivation of an absolute normalisation of the derived star formation rate is still not robust, given the uncertainties in the distribution of mass ratios in binaries, of the fraction of binaries in a given sample, of the IMF, and of metallicity distributions. However, a relative star formation history can be inferred, where the relative amplitudes of the variations in star formation rate are correct, provided the IMF, the properties of binaries and the distribution of metallicity do not evolve with time. With these provisos, the method was applied to a sample of HST CMDs of dSph galaxies of the Local Group (Hernandez, Gilmore, & Valls-Gabaud 2000, Paper II). A further paper extends the application to the Solar neighbourhood (Hernandez, Valls-Gabaud & Gilmore 2000, Paper III).

Full details of the method are presented in Paper I, and in Gilmore et al. (2000), and will not be repeated here. Very briefly, the method takes as inputs only the positions of n stars in a colour magnitude array, each having a colour c_i and luminosity l_i, with (uncorrelated) associated errors $\sigma(c_i)$ and $\sigma(l_i)$ respectively. Using the likelihood technique, we first construct the probability that the n observed stars resulted from some function $SFR(t)$. This is given by

$$\mathcal{L} = \prod_{i=1}^{n} \left(\int_{t_0}^{t_1} SFR(t)\, G_i(t)\, dt \right), \tag{1}$$

where

$$G_i(t) = \int_{m_0}^{m_1} \frac{\rho(m;t)}{2\pi\sigma(l_i)\sigma(c_i)} \exp\left(\frac{-D(l_i;t,m)^2}{2\sigma^2(l_i)} \right) \exp\left(\frac{-D(c_i;t,m)^2}{2\sigma^2(c_i)} \right) dm \tag{2}$$

In the above expression $\rho(m;t)$ is the density of points along the isochrone of age t, around the star of mass m. It is determined by convolving an assumed IMF with a stellar evolutionary track, defining the duration of the differential evolutionary phases for a star of mass m. The ages t_0 and t_1 are a minimum and a maximum age which need to be considered in the specific problem of interest, while m_0 and m_1 are a minimum and a maximum mass which need to be considered along each isochrone, typically $0.6\mathcal{M}_\odot$ and $20\mathcal{M}_\odot$. $D(l_i;t,m)$ and $D(c_i;t,m)$ are the differences in luminosity and colour, respectively, between the ith observed star and a general star of age and mass (m,t). We refer to $G_i(t)$ as the likelihood matrix, since each element represents the probability that a given star, i, was actually formed at time t with any mass.

Following the discussion of Paper I, we may write the condition that the likelihood has an extremal as the variation $\delta\mathcal{L}(SFR) = 0$, allowing a full variational calculus analysis to be applied. Developing first the product over i using the chain rule, and dividing the resulting sum by $\mathcal{L}$, one obtains:

$$\sum_{i=1}^{n} \left(\frac{\delta \int_{t_0}^{t_1} SFR(t)\, G_i(t)\, dt}{\int_{t_0}^{t_1} SFR(t)\, G_i(t)\, dt} \right) = 0 \tag{3}$$

Introducing the new variable $Y(t)$ defined as:

$$Y(t) = \int \sqrt{SFR(t)}\, dt \implies SFR(t) = \left(\frac{dY(t)}{dt} \right)^2$$

into Equation (2) we can develop the Euler equation to yield

$$\frac{d^2Y(t)}{dt^2} \sum_{i=1}^{n} \left(\frac{G_i(t)}{I(i)} \right) = -\frac{dY(t)}{dt} \sum_{i=1}^{n} \left(\frac{dG_i/dt}{I(i)} \right) \tag{4}$$

where

$$I(i) = \int_{t_0}^{t_1} SFR(t)\, G_i(t)\, dt$$

We have now transformed what was an optimisation problem, finding the function that maximises the product of integrals defined by equation (1), into an integro-differential equation with a boundary condition (at either t_o or t_1) which can be solved by iteration. Further details on the numerical aspects of the procedure are available in Paper I, and need not be repeated here.

This methodology has two important advantages over traditional maximum likelihood techniques:
(i) variational calculus allows a fully non-parametric approach, independent of one's astrophysical pre-conceptions; and
(ii) since the optimal star formation history $SFR(t)$ is solved for directly, the computational procedure is very fast, not requiring repeated CPU intensive comparisons between observed and synthetic diagrams.

3. Objective Determination of Recent Star Formation Histories

We now apply the methodolgy outlined above to the Solar Neighbourhood. A suitable data set can be derived from the Hipparcos catalogue (ESA 1997), providing absolute luminosities and colours which are, with careful selection, are almost bias-free. The sample is however restricted to bright stars, and so samples only a very small volume, and releatively young ages. We have derived from the Hipparcos catalogue a volume-limited survey of the solar neighbourhood. This selection is described in detail in Paper III, along with the kinematic and geometrical corrections that have to be made to correct the sample for its selection function.

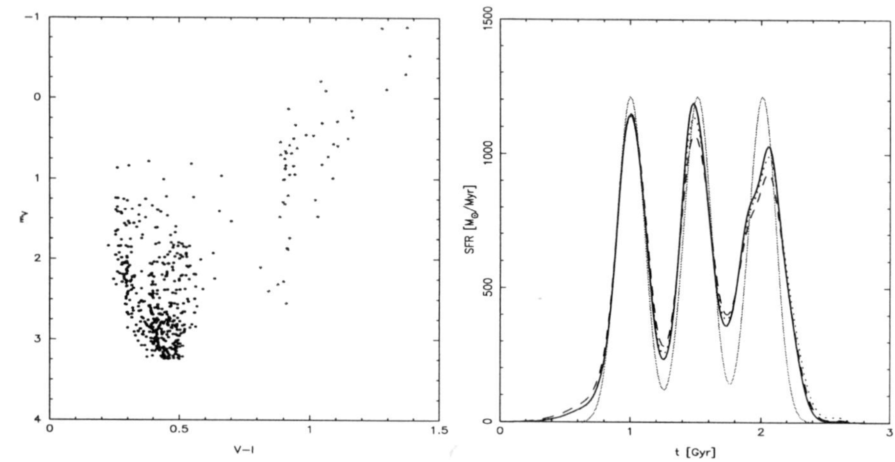

Figure 1. Proving the method with a series of 3 bursts of star formation. The synthetic CMD (left panel) results from the 3-burst star formation history shown by the dotted line in the right panel. The solid lines on the right panel are the last three iterations of the inversion, given no information except the CMD of the left panel, and an appropriate metallicity. An accurate reconstruction of the star formation history is apparent. This simulation utilises the same number of stars and the same observational uncertainties as in the Hipparcos sample.

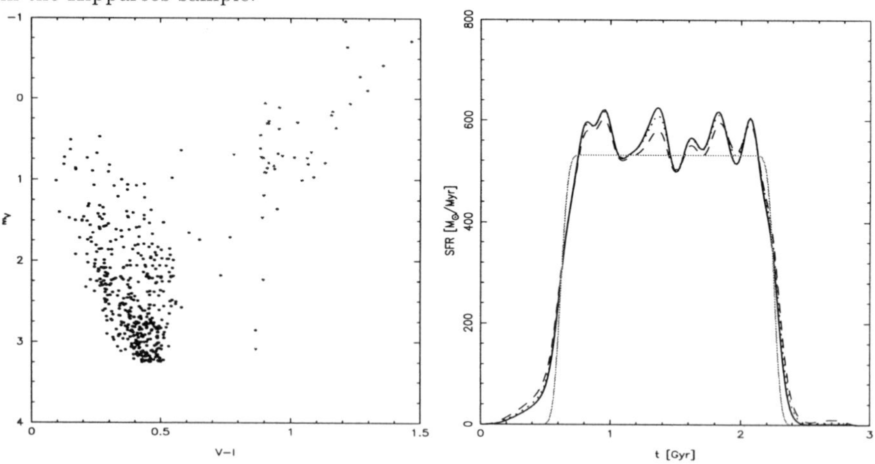

Figure 2. Same as Figure 1, except that the simulated CMD is derived from a constant SFR over several Gyr (dotted curve, right hand panel). Note that the CMD appears superficially very similar to the one presented in Fig. 1, even though the simulated star formation history is quite different. The method implemented here is able to distinguish and recover the correct (test) star formation histories.

In summary, we restrict consideration to non-variable stars with apparent magnitude brighter than $V = 7.9$ and error in parallax smaller than 20%. To avoid a wide distribution in the uncertainties in colour and luminosity the sample is further reduced to absolute magnitudes brighter than $M_V = 3.15$. This absolute magnitude-limited sample implies that only stars younger than about 3 Gyr are considered. Determination of the star formation history of the Galactic disk at earlier times awaits a deeper sample, such as will be naturally provided by the GAIA mission.

3.1. SIMULATIONS AND CALIBRATIONS

This analysis extends the temporal resolution of the method to much shorter times than used in the simulations described in Paper I. In order to check whether samples such as the one selected can in fact be inverted to infer their star formation history with extremely high temporal resolution, we performed a further extensive set of of tests. As before, these involved creation, convolution with an error and sampling function, and objective inversion, of colour magnitude data resulting from a variety of complex star formation histories. Figure 1 illustrates one such test, where a series of 3 bursts (right panel, dashed line) gives rise to the synthetic CMD shown on the left. The solid lines in the right panel give the last 3 iterations of the method, showing convergence to an accurate reconstruction of the input star formation history.

Similarly, Figure 2 shows a test with a constant star formation rate continuing over several Gyrs. The small fluctuations in the reconstructed star formation history arise from numerical instabilities created by the shot noise due to the small number of stars involved, about 450. This implies that a smoothing procedure has to be applied, resulting in a degradation in the effective time resolution to about 50 Myr. Note also that only stars bluewards of $V-I$=0.7 are used, to avoid unnecessary complications created by a small number of potentially older stars with an extremely poorly quantified selection function contaminating the reconstruction.

4. Star Formation History of the Solar Neighbourhood

Figure (3) shows the colour magnitude diagram for those stars in the volume-limited sample complete to $M_V < 3.15$ for stars in the Hipparcos catalogue having errors in parallax of less than 20% and with apparent magnitude $m_V < 7.25$ (left panel). The right panel of this figure shows the result of applying our inversion method to this data set. The dotted envelope spanning the best-fit star formation history quantifies the range of many reconstructions arising from different M_V selection cuts in the Hipparcos catalogue. This quantifies the plausible range of systematic errors likely to be present, due predominately to the small sample of stars, and the inevitably *a posteriori* selection function which must be applied to Hipparcos data. Reconstruction based on the $(M_V, B-V)$ diagram rather than $(M_V, V-I)$ gives essentially the same results, showing that the available isochrones are a good match to the photometric systems.

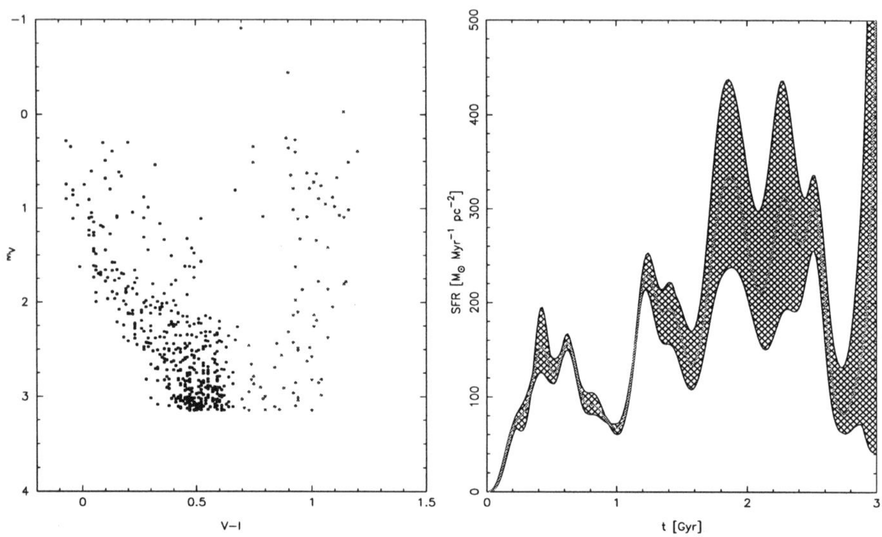

Figure 3. The left panel shows the colour magnitude data for the complete sample of nearby massive stars with good parallaxes from Hipparcos. The right panel presents a best-fit reconstruction of the star formation history (solid line), along with the envelope resulting from reconstructions with different M_V cuts, to assess the robustness of the inferred star formation history. The decrease to zero at recent times is an artefact of the very small sampling volume.

The derived local star formation history of the Solar Neighbourhood shown in Figure (3) shows what may be described as an underlying constant level of star formation activity, onto which is superimposed a strong, quasi-periodic variation with a period close to 0.5 Gyr. The high time resolution of our star formation history reconstruction makes it difficult to compare with the results derived from chromospheric activity studies (eg Rocha-Pinto et al. 1999), although qualitatively we do find the same activity at both 0.5 and above 2 Gyr, but not the decrease found by them between 1 and 2 Gyr.

Assuming the variable component really is a cyclic pattern superimposed on a base level, and that our brief time interval is sufficient to identify a true characteristic 'period', we may consider its meaning. One possible interpretation of a cyclic component, or at least of some temporal regularity in the star formation history of the solar neighbourhood, can be found in the density wave hypothesis (Lin and Shu, 1964) for the presence of spiral arms in late type galaxies. As the pattern speed and the circular velocity are in general different, the solar neighbourhood periodically crosses an arm region, where the increased local gravitational potential might possibly trigger an episode of star formation. In this case, the time interval Δt between encounters with an arm at the solar neighbourhood is

$$\Delta t = \frac{0.22\,\mathrm{Gyr}}{m} \left(\frac{\Omega}{29\,\mathrm{km\,s^{-1}kpc^{-1}}}\right)^{-1} \left|\frac{\Omega_p}{\Omega} - 1\right|^{-1}$$

where m is the number of arms in the spiral pattern. The classical value of the pattern speed, $\Omega_p = 0.5\,\Omega \approx 14.5$km s^{-1} kpc^{-1} would imply the rather surprising conclusion that interaction with a single arm ($m = 1$) would be enough to account for the observed regularity in the recent SFR history.

However, more recent determinations tend to point to much larger values for the pattern speed (e.g. Mishkurov et al. 1979, Avedisova 1989, Amaral and Lépine 1997) close to $\Omega_p \sim 23 - 24$ km s^{-1} kpc^{-1}, which would then imply that the regularity present in the reconstructed $SFR(t)$ would be consistent with a scenario where the interaction of the solar neighbourhood with a two-armed spiral pattern would have induced the star formation episodes we detect. This is reminiscent of the explanations put forward to account for the inhomogeneities observed in the HIPPARCOS velocity distribution function, where well-defined branches associated with moving groups of different ages (Chereul et al. 1999, Skuljan et al. 1999, Asiain et al. 1999) could perhaps be also associated with an interaction with spiral arm(s), although in this case the time scales are much smaller. Of course, other explanations are possible; for example the cloud formation, collision and stellar feedback models of Vazquez & Scalo (1989) predict a phase of oscillatory star formation rate behaviour as a result of a self-regulated star formation régime. Close encounters with the Magellanic Clouds have also been suggested to explain the intermittent nature of the star formation rate, though on longer time scales (Rocha-Pinto et al. 2000).

4.1. ARE THE RESULTS RELIABLE?

The first question that arises upon application of any new technique, especially one involving complex numerical optimisations, is how reliable is the reconstruction of the star formation history we have deduced? The most common procedure to compare a specific star formation history with an observed CMD has been to use the star formation history to generate a synthetic CMD, and then to compare this to the observations, using some statistical test to determine the degree of similarity between the two.

This method has a significant disadvantage in the present case, in that one is not comparing the 'true' star formation history with the data, but rather a particular realisation of the star formation history is being compared with the data. The distinction becomes arbitrary only when large numbers of stars are found in all regions of the colour magnitude diagram. This is not true in general, and especially not so in this application. Instead we follow a Bayesian approach, and adopt for use here the W statistic presented by Saha (1998). This is essentially

$$W = \prod_{i=1}^{B} \frac{(m_i + s_i)!}{m_i! s_i!}$$

where B is the number of cells into which the CMD is split, and m_i and s_i are occupation numbers, the numbers of points which the two distributions being compared have in each cell. This asks for the probability that two distinct data sets are random realisations of the same underling distribution.

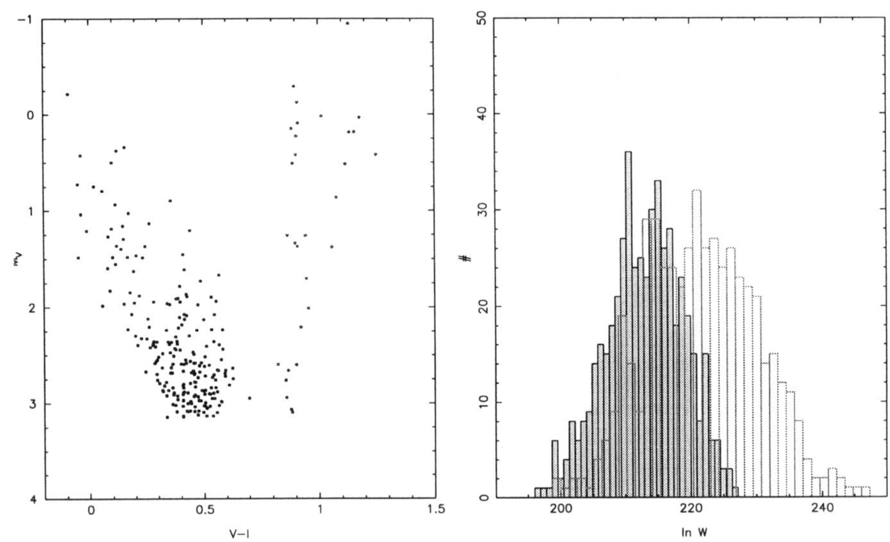

Figure 4. The left panel shows one synthetic colour magnitude distribution resulting from the reconstructed star formation history of Fig. 3 (left panel), generated with the same selection function as for the observed Hipparcos sample. The right panel shows the distribution of the W statistic for 500 model-model comparisons (solid lines), illustrating the range of the W statistic expected from sampling statistics, given a fixed underlying star formation function. The dashed line shows the W statistics resulting from 500 comparisons of simulated models and the actual Hipparcos data, from Figure (3). The overlap between the two sets of results shows that the derived star formation history of Figure (3) is indeed a statistically valid representation of the data.

In implementing this test we first produce a large number of random realisations of the colour magnitude diagram generated by our inferred star formation history, and then compute the W statistic between pairs in this sample of CMDs. This gives a distribution which is used to determine the range of values of W which is expected to arise in random realisations of the star formation history being tested. Next, we produce a new set of a large number of random realisations of the colour magnitude diagram generated by our inferred star formation history, and compute the W statistics between the observed data set and these new random realisations of the star formation history. This gives a new distribution of W, which can then be objectively compared to the one arising from the model-model comparison. This allows us to assess whether both data and modeled colour magnitude diagrams are compatible with a unique underling distribution.

Figure (4) shows one such synthetic colour magnitude diagram produced from our inferred star formation history for the solar neighbourhood, down to $M_V = 3.15$. This can be compared to the Hipparcos CMD complete to the same M_V limit shown in Figure (3). A visual inspection reveals approximately equal numbers of stars in each of the distinct regions of the diagram. Such a comparison is of little value, however, as noted in dicussion of the simulations summarised in Figures (1) & (2). A rigorous statistical comparison is also possible. The right panel of Figure (4) shows a histogram of the values of the W statistic for 500 random realisations of our inferred star formation history in a model-model comparison. This gives the probability density function for the W, given our star formation history. The dashed histogram presents the result of the comparison of 500 synthetic colour magnitude diagrams with the observed Hipparcos data set: both sets of W are compatible.

5. Conclusion

Star formation rates and histories can be estimated in special cases from a combination of chemical evolution models and the total stellar mass formed into stars. Basically, this exploits the stellar evolutionary, and Type I supernova timescales for element production. In cases where no spread in element ratios is seen, and there is no range in [Fe/H], star formation was plausibly complete before new chemical elements could be produced; perhaps globular clsuters are an example of this case. For field halo stars with [Fe/H]≥ -2, where a wide range in [Fe/H] but a very small range in $[\alpha/H]$ is seen, star formation must have continued for long enough for efficient mixing of supernova ejects into the ISM. Since no products of Type I supernovae are seen, this brackets allowed star formation durations. By applying such qualitative considerations, we deduce that most of the Milky way formed at a star formation rate which is comparable to that of today. Only the Galactic Bulge, and the inner disk, where star formation histories remain very poorly known, are available to retrieve the Milky Way's place as a 'typical object' on the Madau plot.

In the immediate Solar neighbourhood, and in dSph satellite galaxies, the high quality data and the simple stellar populations respectively allow us to be more quantitative. We have applied the objective variational calculus method for the reconstruction of star formation histories from observed colour magnitude data, developed in our Paper (I), to the data in the Hipparcos catalogue, yielding the star formation history of the solar neighbourhood over the last 3 Gyr. Surprisingly, a structured star formation history is obtained, showing a cyclic pattern with a period of about 0.5 Gyr, superimposed on some underlying star formation activity which increases slightly with age. No random bursting behaviour was found at the time resolution of 0.05 Gyr of our method. A first order density wave model for the repeated encounter of galactic arms could explain the observed regularity.

References

1. Asiain R., Figueras F., Torra J., 1999, A&A, 350, 434
2. Amaral L.H., Lépine J.R.D., 1997, MNRAS, 286, 885
3. Avedisova V.S., 1989, Astrophys. 30, 83
4. Chereul E., Crézé M., Bienaymé O., 1999, A&A Suppl. 135, 5
5. Gilmore G., Hernandez X., Valls-Gabaud D., 2000, in 'Astrophysical Dynamics', eds D. Berry, D. Breitschwerdt, A. da Costa & J. Dyson, Ap Sp Sci Special Issue, in press
6. Hernandez X., Valls-Gabaud D., Gilmore G., 1999, MNRAS, 304, 705 (paper I)
7. Hernandez X., Gilmore G., Valls-Gabaud D., 2000, MNRAS, in press (astro-ph/0001337; Paper II)
8. Hernandez X., Valls-Gabaud D., Gilmore G., 2000, MNRAS, in press (paper III)
9. Lin C.C., Shu F.H., 1964, ApJ, 140, 646
10. Mishkurov Y.N., Pavloskaya E.D., Suchkov A.A., 1979, AZh, 56, 268
11. Rocha-Pinto H.J., Scalo J., Maciel W.J., Flynn C., 1999, ApJ Lett submitted (astro-ph/9908328)
12. Rocha-Pinto H.J., Scalo J., Maciel W.J., Flynn C., 2000, A&A, submitted (astro-ph/0001383)
13. Saha P., 1998, AJ, 115, 1206
14. Vazquez E.C., Scalo J.M., 1989, ApJ, 343, 644

Speaker: G. GILMORE

P. WHITELOCK: As you probably know, Mike Irwin has been looking for and finding AGB carbon-stars in the halo. Obviously, every AGB star will represent a very large number of main-sequence-stars. So, are you saying this is an insignificant fraction of the halo?

G. GILMORE: Not insignificant, but all the surveys are consistent with about a 10 percent contribution from intermediate ages as metal rich stars in the outer halo field. The predictions suggest 100 percent, which is wrong. I note that Roddy Ibata and I discovered the Sgr dwarf by noticing in real time at the AAT that every very red (AGB) star had the same high velocity, and not even a single one was a real field star.

A. MAEDER: When you reconstruct the history of the SFR, is there no risk to confuse the blue stragglers, which are often very numerous, with some younger populations? How do you account for that?

G. GILMORE: The method can only use available isochrones. Thus BSS are interpreted as young stars. In this case astrophysical intuition is necessary in the interpretation. Of course, an analytical algorithm cannot provide that, so we must.

N. PRANTZOS: In deriving the SF history from CDM, how do you deal with e.g. stellar metallicity or binarity? Also: you derive a present-day SFR too low with respect to the observed one.

G. GILMORE: The stellar metallicity must be known independently, and in the present implementation must have a narrow spread. Effects of binarism can be modelled by simulations. The local SFR at very young times is affected by the very limited Hipparcos sample - young stars have such a small velocity dispersion that no small volume can provide a meaningful "average" result. For that we must wait for GAIA.

M. FEAST: In the LMC the Mira P-2 is very narrow. In the Baade windows (NGC 4522 and SGR I) time relation is broad showing a very significant depth in the line of sight. Can this explain the difficulty you have in fitting the C-M of bulge fields with the globular cluster sequences?

G. GILMORE: Line of sight depth effects are certain to be at least a part of the answer. Probably, the mix of this and metallicity spread is most of the explanation Available HST data are insufficient to answer this reliably in spite of any efforts. We are currently modelling our ISO imaging data for the inner bulge to investigate this further.

INHOMOGENEOUS CHEMICAL EVOLUTION OF THE GALACTIC HALO

T. TSUJIMOTO
National Astronomical Observatory, Japan

T. SHIGEYAMA
Reserch Center for the Early Universe, University of Tokyo, Japan

AND

Y. YOSHII
Institute of Astronomy, University of Tokyo, Japan

Abstract. We find that abundance patterns seen in extremely metal-poor stars with [Fe/H] $\lesssim -2.5$ follow those seen in the individual supernova remnants. This suggests that stars are made from individual supernova events. By incorporation of a sequence of supernova-induced star formation into the enrichment process of heavy elements, we describe the evolution of stellar abundances in the Galactic halo characterized by the inhomogeneous mixing of chemical compositions in dense shells and the ambient medium. Here we focus on the chemical evolution of r-process elements in the Galactic halo.

1. Fossil Imprints of Supernova Ejecta in Extremely Metal-Poor Stars

Recent observations and the analyses imply that the abundance pattern of an extremely metal-poor star with [Fe/H] $\left(\equiv \log(\mathrm{Fe/H}) - \log(\mathrm{Fe/H})_{\odot}\right)$ $\lesssim -2.5$ may retain information of a preceding single supernova (SN) event or at most a few SNe (McWilliam et al. 1995; Ryan, Norris, & Beers 1996). Assuming that the formation of extremely metal-poor stars is triggered by a single supernova remnant (SNR) and that the formed stars retain the abundance pattern of this SN, one could predict abundance patterns (including hydrogen) of these stars from theoretical SN models available

F. Matteucci and F. Giovannelli (eds.), The Evolution of the Milky Way, 531–539.

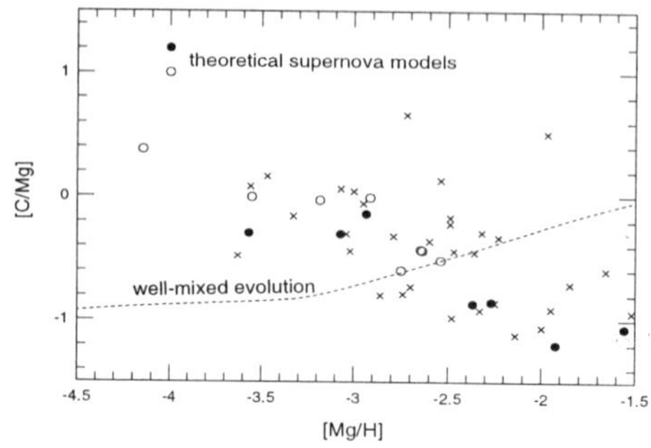

Figure 1. Comparison of the observed correlation of [C/Mg] – [Mg/H] (crosses) with the theoretical nucleosynthesis [C/Mg] ratio from SNe II as a function of the predicted [Mg/H] (filled circles: Tsujimoto et al. 1995; open circles: Woosley & Weaver 1995). The evolutionary change in [C/Mg] predicted by the one-zone chemical evolution model including the contribution from AGB stars is also shown by the dashed curve.

at present (e.g., Woosley & Weaver 1995; Tsujimoto et al. 1995). In this section, we will compare the abundance pattern thus obtained with observations to test whether the chemical enrichment by individual SNRs could explain the observed abundances on the surfaces of extremely metal-poor stars.

The yields of heavy elements not influenced by the mass cut between the forming neutron star (or black hole) and the ejected envelope are expected to be more reliably calculated in Type II supernova (SN II) models than those influenced by the mass cut. Therefore the observed abundance ratios of [C/Mg] with respect to [Mg/H] can tell whether extremely metal-poor stars were formed from individual SNe or from the well-mixed interstellar gas. The observed correlation of [C/Mg] with [Mg/H] (McWilliam et al. 1995) is shown in Figure 1. The dashed curve in the same figure is the correlation of these ratios predicted from the one-zone Galactic chemical evolution model in which the gas is distributed uniformly and the heavy elements are well-mixed within the zone. This obvious disagreement with the observation indicates that the mixing of elements did not take place at that time. The open and filled circles are the correlation of the abundance ratios in SNRs at the end of expansions calculated from two theoretical SN models (Shigeyama & Tsujimoto 1998). The coincidence of the two correlations must be a result from that most of the stars with $[Mg/H] \lesssim -2$ were formed from individual SNRs.

2. Iron Yields Deduced from Extremely Metal-Poor Stars

Figure 2a shows that the [Mg/Fe] ratios from both of the models do not follow the observations. This confirms that none of the two models correctly describes the dynamics and/or nucleosynthesis near the surface of the Fe core in progenitor stars during explosion. We will reverse the argument. Suppose that the trends in the observed abundance ratios, say [Mg/Fe], tell yields of SNe as a function of [Mg/H]. The χ^2 fit of the observational

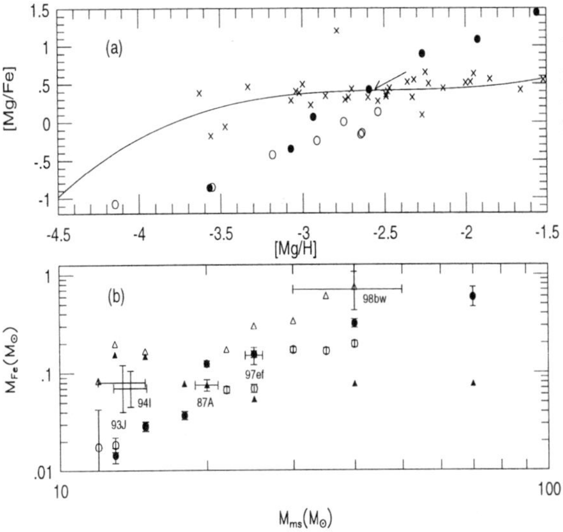

Figure 2. (*a*) The crosses are the observed [Mg/Fe] for stars plotted against the [Mg/H]. The solid curve shows the χ-square fit to the observed points with the errors. The open and filled circles show the same quantities in SNRs calculated from theoretical SN models. (*b*) The masses of Fe in SN models are shown by triangles. Circles show the masses of Fe in SN models with 1 σ error bars of the χ^2 fit. Points with error bars are the masses of ^{56}Ni, that eventually decays to ^{56}Fe derived from SN light curve analyses.

data to a cubic polynomial formula gives the solid curve in Figure 2a. This curve and [Mg/H] of each model predict the mass of Fe ($M_{\rm Fe}$) from a SN as a function of $M_{\rm ms}$. Filled circles in Figure 2b show the predicted Fe masses using Tsujimoto et al. (1995). These Fe masses are completely different from those given in the original models (triangles). Open circles are those from Woosley & Weaver (1995) and give a similar amount of Fe for each progenitor mass. Points with error bars also shown in the same figure are the Fe masses obtained from SN light curve analyses. Good fits of the Fe masses from the light curve analyses to those inferred from the abundance patterns demonstrate that this procedure to obtain the Fe yields works.

Following the same procedure for obtaining the Fe yields, we can derive the yields for other elements such as Ti, Cr, Mn, Co, Ni and r-process elements (Tsujimoto & Shigeyama 1998). The accuracy of our determination is guaranteed by the quite good agreement of the IMF-weighted abundance ratios with the observed plateau seen in the halo stars for $-2 <$ [Fe/H] < -1. The yields thus obtained are indispensable because theoretical SN II models to date can reliably predict the amounts of only a few elements.

3. Supernova-Induced Star Formation Model

The hypothesis that stars are born from each SNR sheds light on the entire process of early-epoch star formation. Galactic chemical evolution is envisioned as a successive sequence of SN explosions, shell formation, and resulting star formation. In this picture, the abundance pattern of each star is set not only by the heavy elements ejected from the SN explosion which directly preceeds local star formation, but also incorporates those elements already present in the interstellar gas which are swept up by the SNR. The predicted stellar abundance patterns are thus different from those of the gas at the time when stars form. This difference can be quite large in the early stage of Galactic evolution, when the metallicity in the gas is very low. The abundance ratios of low-metallicity stars are predicted to exhibit a large star-to-star scatter, depending in detail on the abundance patterns of SN ejecta with different progenitor masses.

We assume that the star-forming process is confined to separate clouds, which make up the entire halo. Each cloud is initially composed of metal-free Population III stars (Pop III) and gas which has yet to form stars. The mass fraction of such stars in our model is a parameter hereafter denoted by $x_{\rm III}$, the massive ones of which eventually explode as Pop III SNe to initiate chemical evolution. All stars of subsequent generations are assumed to form in SNR shells. The mass fraction of each shell which turns into stars is assumed to be constant, and is denoted by ϵ. Heavy elements ejected from an SN are assumed to be trapped and well-mixed within the SNR shell. Some of these elements go into stars of the next generation, and the rest is left in the gas that will be mixed with the ambient medium. The above process will repeat with increasing metallicity until SNRs can no longer sweep up enough gas to form shells. No stars will form from SNRs after this happens, and the process terminates. A formulation to describe the chemical evolution in a cloud is shown in Tsujimoto, Shigeyama, & Yoshii (1999a). An overview of the chemical evolution of the Galactic halo is as follows – (1) The metal-free Pop III stars form by some (as yet unspecified) mechanism in primordial gas clouds of the Galactic halo. (2) The most massive stars among them explode as SNe II, which trigger a series of star formation events. (3) Star formation terminates when dense shells cannot be formed from SNRs. (4) About 90% of the cloud mass remains unused in star formation and may fall onto the still-forming Galactic disk.

4. The Production Site for r-Process Elements

Europium (Eu), for which about 97% of its abundance in the solar system is of pure r-process origin, has been considered as one of the most useful elements for locating the astrophysical sites of r-process nucleosynthesis. However for most stars of extremely low metallicity ([Fe/H] $\lesssim -2.0$), obtaining an abundance measurement of Eu is a difficult observational task. For example, McWilliam et al. (1995) only obtained Eu abundances for 11 of the 33 metal-poor stars in their sample. This incompleteness implies that one might miss valuable insights to r-process production mechanisms by

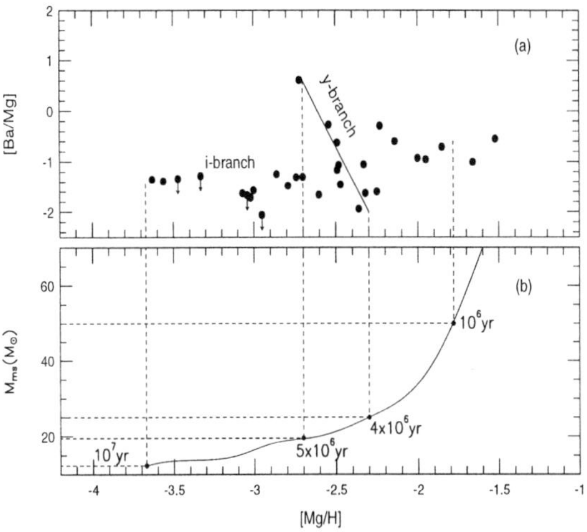

Figure 3. (*a*) The correlation of [Ba/Mg] with [Mg/H] for metal-poor stars (McWilliam 1998). The thick line is the left-side boundary to the y-branch representing the mass range of the r-process site. (*b*) SN II progenitor mass $M_{\rm ms}$ plotted as a function of [Mg/H] inside the shell swept up by the SNR. The stellar lifetimes are also shown for four reference masses of $M_{\rm ms}$ = 12, 20, 25, and $50M_\odot$. Dashed lines in the lower and upper panels illustrate how we assign $M_{\rm ms}$ to each star along the boundary.

concentrating exclusively on observations of Eu. McWilliam (1998) showed that the [Ba/Eu] ratios for stars with $-3 \lesssim$ [Fe/H] $\lesssim -2$ are consistent with pure r-process ratios, and concluded that Ba in extremely metal-poor stars is also of r-process origin.

Figure 3a illustrates the hypothesis we seek to defend, that the metal-poor stars of the Galaxy populate two separate branches in the [Ba/Mg] – [Mg/H] plane. The first branch extends rightward of the vertical boundary which begins at ([Ba/Mg], [Mg/H]) = (−2.0, −2.3) and ends at ([Ba/Mg], [Mg/H]) = (+0.6, −2.7). The second branch is horizontal from [Mg/H]$\sim$ −2.5 to −3.7 at a constant value of [Ba/Mg]$\sim$ −1.6. The data of Ryan et al. (1996) for seven of their extremely metal-poor stars show a similar trend in the [Ba/Mg]−[Mg/H] plane. We associate this feature, a bifurcation of the observed elemental ratios into two branches in the Mg abundance interval $-2.7 \leq$ [Mg/H] ≤ -2.3, with two distinct processes. The first branch, which we call the "y"-branch, is associated with the production of Ba and Mg from individual massive supernovae. The derived mass of Ba synthesized in SNe II is $8.5 \times 10^{-6} M_\odot$ for $M_{\rm ms} = 20M_\odot$, and $4.5 \times 10^{-8} M_\odot$ for $M_{\rm ms} = 25M_\odot$. We conclude that SNe II with $M_{\rm ms} \approx 20M_\odot$ are the dominant source of r-process nucleosynthesis in the early Galaxy. An SN-induced chemical evolution model with this $M_{\rm ms}$-dependent Ba yield creates the y-branch, reflecting the different nucleosynthesis yields of [Ba/Mg] for each SN II with $M_{\rm ms} \gtrsim 20M_\odot$. The second branch, which we call the "i"-branch, is associated with the elemental abundance ratios of stars which were formed in the dense shells of the interstellar medium swept up by SNe II with $M_{\rm ms} < 20M_\odot$ that do *not* synthesize r-process elements, and applies to stars with observed Mg abundances in the range [Mg/H]< -2.7.

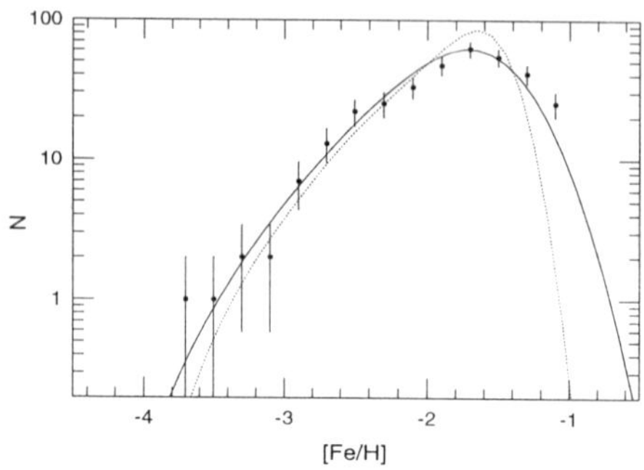

Figure 4. The frequency distribution of the long-lived halo field stars for [Fe/H] < -1 against the iron abundance, compared with the observations of Ryan & Norris (1991). The observational data for [Fe/H]> -1 are not shown because these data do not correspond to the number pure halo stars due to an enormous contamination by disk stars (see Ryan & Norris 1991). The model curves have been convolved with Gaussians having σ=0.3 dex (solid line) and σ=0.15 dex (dotted line), respectively.

The Ba abundances in these stars reflect those of the interstellar gas at the (later) time of their formation (see Tsujimoto, Shigeyama, & Yoshii 1999b).

The existence of a [Ba/Mg] i-branch strongly suggests that SNe II which are associated with stars of progenitor mass $M_{\rm ms} \leq 20M_{\odot}$ are infertile sources for the production of r-process elements.

5. Chemical Evolution of r-Process Elements in the Galactic Halo

The free parameters in our model are the mass fraction $x_{\rm III}$ of metal-free Pop III stars initially formed in each cloud, and the mass fraction ϵ of stars formed in a dense shell swept up by each SNR. A value of $x_{\rm III}$ sensitively determines the level of [Ba/Mg] for the i-branch, so we take $x_{\rm III} = 2.5\times10^{-4}$ to be consistent with the observed value [Ba/Mg]=-1.4 at [Mg/H]~ -3.7. We take $\epsilon = 4.3 \times 10^{-3}$ to reproduce the observed [Fe/H] distribution function of halo field stars for [Fe/H]< -1 (Fig. 4).

Adopting the result that SNe II with $M_{\rm ms} = 20 - 25M_{\odot}$ are the dominant site for r-process nucleosynthesis, and using the derived Ba yield dependent on $M_{\rm ms}$, we calculate the Ba evolution at early epochs of the Galactic halo. For later evolution, which pertains to the abundance range $-2 \lesssim$ [Mg/H] $\lesssim -1$, we add the contribution from s-process nucleosynthesis, assuming that the production site for early-epoch s-process elements is $2 - 3\ M_{\odot}$ AGB stars. The stellar IMF is assumed to be of the Salpeter form. The upper and lower mass limits for stars that explode as SNe II are taken to be $50M_{\odot}$ and $12M_{\odot}$, respectively.

Figure 5a is a predicted frequency distribution of stars in the [Ba/Mg]–[Mg/H] plane, normalized to unity when integrated over the entire area. In order to enable a direct comparison with the data, the frequency distribution has been convolved with a Gaussian with $\sigma = 0.2$ dex for [Ba/Mg] and $\sigma = 0.1$ dex for [Mg/H]. The [Ba/Mg] ratio "plateau" in the range of $-2 \lesssim$ [Mg/H] $\lesssim -1$ is obviously higher than that of i-branch ([Mg/H] $\lesssim -2.5$), which implies a significant contribution from s-process nucle-

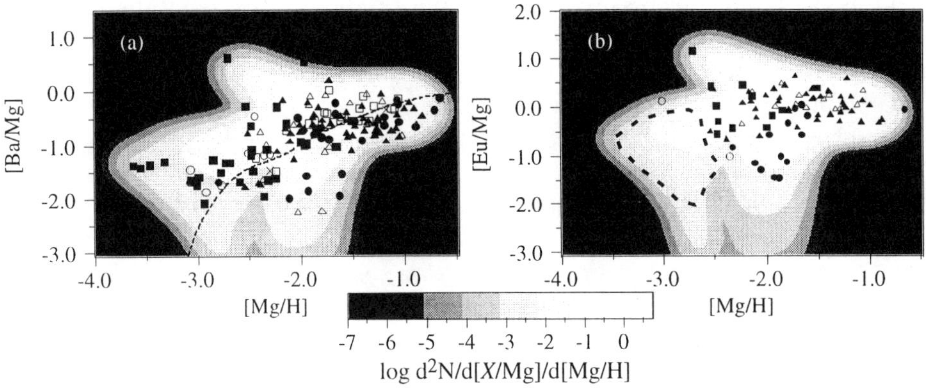

Figure 5. (*a*) The frequency distribution of the long-lived stars in the [Ba/Mg]-[Mg/H] plane. The symbols represent the data taken from several references. For comparison, the result of the conventional one-zone chemical evolution model is shown by dashed line assuming the r-process site of $8-10\ M_\odot$ supernovae. (*b*) The same as (*a*) but for Eu. The dashed curve encloses the predicted region in which i-branch stars must exist.

osynthesis. For comparison, we show by the dashed line in this figure the conventional one-zone chemical evolution model assuming that the production sites for r-process and s-process are $8-10\ M_\odot$ supernovae and $2-3\ M_\odot$ AGB stars, respectively. It is clear that the conventional model predicts a much smaller [Ba/Mg] than is consistent with the observed i-branch at $[\mathrm{Mg/H}] \lesssim -2.7$.

To summarize our hypothesis: The earliest stars, stars that were formed from the shells swept up by SNe II with $M_{\mathrm{ms}} = 20-25 M_\odot$, first create the vertical boundary seen in Figure 5a. Thereafter, the y-branch develops at higher [Mg/H] and converges to the plateau level of [Ba/Mg] at higher metallicities, as described in detail in Tsujimoto et al. (1999a). At the same time, the subsequently formed stars, arising from the SNRs with $M_{\mathrm{ms}} = 12-20 M_\odot$, do not produce Ba, and therefore have the lowest value of [Ba/Mg] at $[\mathrm{Mg/H}] \lesssim -2.5$. These stars populate the lower envelope of the i-branch, then fill in this branch at lower [Mg/H], but gradually *increasing* [Ba/Mg]. As the Mg abundance in the ISM becomes dominant over that ejected from SNe II, the i-branch develops at higher abundances, increasing both [Mg/H] and [Ba/Mg].

Because the upper mass limit in our present model for the Ba production site, $25 M_\odot$, is set below the canonical upper limit $\sim 50 M_\odot$ for stable stellar masses, we predict the existence of another i-branch in the range of $-2.3 \lesssim [\mathrm{Mg/H}] \lesssim -1.5$, which is made up of stars arising from the SNRs with $M_{\mathrm{ms}} = 25-50 M_\odot$.

As a final exercise, we also calculate the Eu evolution in the early Galaxy using the same models and assumptions we have employed for the prediction of Ba evolution. Figure 5b shows the predicted frequency distribution of stars in the [Eu/Mg]–[Mg/H] plane, after convolving with $\sigma = 0.2$ dex for [Eu/Mg] and $\sigma = 0.1$ dex for [Mg/H]. The Eu production site must be the same as for Ba, or at least share a part of SNe II sites where Ba is produced, because the rapid n-capture process cannot synthesize Eu without producing Ba. Therefore, we assume that Eu is also produced by SNe II with $M_{\rm ms} = 20 - 25M_\odot$. We scale the Ba yield by the pure r-process value [Ba/Eu]=−0.72 and obtain a predicted Eu yield of $M_{\rm Eu} = 1.3\times10^{-6}M_\odot$ for $M_{\rm ms} = 20M_\odot$, and $M_{\rm Eu} = 7.0\times10^{-9}M_\odot$ for $M_{\rm ms} = 25M_\odot$. Our calculation predicts the existence of a Eu i-branch at [Mg/H] $\lesssim -2.5$.

The i-branch should not exist for α- and Fe-peak heavy elements formed in the early Galaxy, because the entire range of SNe II progenitor masses will contribute to substantial amounts of such elements. However, the i-branch is expected to exist, though is not confirmed as of yet, in the case of the r-process elements such as La, Ce, Nd, and Eu.

References

1. McWilliam, A. (1998) Barium Abundances in Extremely Metal-poor Stars, *Astronomical Journal*, **Vol. 115**, pp. 1640–1647
2. McWilliam, A., Preston, G.W., Sneden, C. and Shectman, S. (1995) A Spectroscopic Analysis of 33 of the Most Metal-Poor Stars. I., *Astronomical Journal*, **Vol. 109**, pp. 2736–2799
3. Ryan, Norris, J.E. and Beers, T.C. (1996) Extremely Metal-poor Stars. II. Elemental Abundances and the Early Chemical Enrichment of the Galaxy, *Astrophysical Journal*, **Vol. 471**, pp. 254–278
4. Ryan, S.G. and Norris, J.E. (1991) Subdwarf Studies. III. The Halo Metallicity Distribution, *Astronomical Journal*, **Vol. 101**, pp. 1865–1878
5. Shigeyama, T. and Tsujimoto, T. (1998) Fossil Imprints of the First-Generation Sueprnova Ejecta in Extremely Metal-Poor Stars, *Astrophysical Jounal*, **Vol. 507**, pp. L135–L139
6. Tsujimoto, T., Shigeyama, T. and Yoshii, Y. (1999a) Chemical Evolution of the Galactic Halo through Supernova-Induced Star Formation and Its Implication for Population III Stars, *Astrophysical Journal*, **Vol. 519**, pp. L63–L67
7. Tsujimoto, T., Shigeyama, T. and Yoshii, Y. (1999b) Probing the Site for r-Process Nucleosynthesis with Abundances of Barium and Magnesium in Extremely Metal-Poor Stars, *Astrophysical Journal*, in press
8. Tsujimoto, T. and Shigeyama, T. (1998) New Insights into the Early Stage of the Galactic Chemical Evolution, **Vol. 508**, pp. L151–L154
9. Tsujimoto, T. et al. (1995) Relative Frequencies of Type Ia and Type II Supernovae in the Chemical Evolution of the Galaxy, LMC and SMC, *Monthly Not. Royal Astron. Soc.*, **Vol. 277**, pp. 945–958
10. Woosley, S.E. and Weaver, T.A. (1995) The Evolution and Explosion of Massive Stars. II. Explosive Hydrodynamics and Nucleosynthesis, *Astrophysical Journal Supplement Ser.*, **Vol. 101**, pp. 181–235

Speaker: T.TSUJIMOTO

A. MAEDER: Can you produce empirical yields for the other r-elements as you have made for M(Fe) vs. the initial stellar mass?

T. TSUJIMOTO: Yes. We can construct yields of any observed heavy element from each Type II Supernova, including α-, iron-group, and n-capture elements, as a function of the initial stellar mass of their progenitor star.

N. PRANTZOS: When you say that: "one massive star is formed for each SN remnant" is it in a *causal* or in a *statistical* sense? Because, after all, in OB associations, we see many stars formed simultaneously.

T. TSUJIMOTO: In a statistical sense. However, our results can be ascribed to a physical cause that star formation induced by supernova explosions is suppressed and self-regulated by the photoionization from an OB star. Here I should stress that we modelled the chemical evolution of the galactic "halo". In the "disk", it is likely that there is another site for star formation relating to the spiral arm, which might create the OB associations.

F. MATTEUCCI: How do you explain, in the context of your model, the fact that s-process elements show a much larger scatter than α-elements?

T. TSUJIMOTO: A scatter in the abundance ratios among low-metallicity stars reflects a different nucleosynthesis yield for each type II supernova with a different progenitor mass. As both the mass of ejected α-element and iron increase with increasing progenitor mass, the [α/Fe] ratios among halo field stars are roughly constant. In contrast, the mass dependence for n-capture elements is the reverse of those for α-elements and iron. Such a difference in the mass dependence of the ejected mass leads to large scatter of the observed [n-capture/Fe] ratios. Furthermore, the correlation of [n-capture/Fe] with [Fe/H] has two branches, i.e. y- and i-branches. The latter i-branch should not exist in the [α/Fe]-[Fe/H] plane. The existence of i-branch in [n-capture/Fe]-[Fe /H] plane is a second cause for a large scatter.

STELLAR YIELDS AND CHEMICAL EVOLUTION

The solar neighborhood as a calibrator

DANIEL THOMAS
Universitäts-Sternwarte München

LAURA GREGGIO
Osservatorio Astronomico di Bologna

AND

RALF BENDER
Universitäts-Sternwarte München

Abstract. Uncertainties in stellar nucleosynthesis and their impact on models of chemical evolution are discussed. Comparing the Type II supernova nucleosynthesis prescriptions from Woosley & Weaver (1995) and Thielemann, Nomoto, & Hashimoto (1996), it turns out that the latter predict higher Mg/Fe ratios that are more favorable in reproducing the observed abundance features of the Milky Way. Provided that chemical evolution models are calibrated on the solar neighborhood, they offer a powerful tool to constrain structure formation. In particular, galaxy formation models that yield star formation histories significantly longer than 1 Gyr fail to reproduce the super-solar Mg/Fe ratios observed in elliptical galaxies.

1. Introduction

Population synthesis models based on *solar* abundance ratios predict – for a given Fe index – Mg indices that are weaker than measured in the integrated spectra of elliptical galaxies (e.g. Worthey, Faber & González 1992 [33]; Davies, Sadler & Peletier 1993 [3]; Mehlert et al. 2000 [18]). Although the link from line indices to element abundance ratios is not straightforward (Greggio 1997 [9]; Tantalo, Chiosi, & Bressan 1998 [23]), the strong Mg absorption features are interpreted as an enhancement of α-elements. More quantitative studies find an average [Mg/Fe] overabundance of $0.3-0.4$ dex (Weiss, Peletier, & Matteucci 1995 [31]; Greggio 1997 [9]). This conclusion

F. Matteucci and F. Giovannelli (eds.), The Evolution of the Milky Way, 541–546.

gets further support from the detection of enhancement of Mg and other α-elements in the stars of the Milky Way bulge (McWilliam & Rich 1994 [17]). These findings imply short formation timescales and/or an initial mass function (IMF) that is flat at the high-mass end (Matteucci 1994 [15]). Scenarios that form elliptical galaxies in mergers of evolved spirals fail to reproduce such α-enhanced stellar populations unless a significant flattening of the IMF is assumed (Thomas, Greggio, & Bender 1999 [27]). Galaxy formation models that are based on hierarchical clustering lead to extended star formation histories in elliptical galaxies which produces Mg/Fe element ratios that are too low compared to the values quoted above (Bender 1996 [1]; Thomas 1999 [25]; Thomas & Kauffmann 2000 [28]).

In this paper we address the question of how robust the predictions from chemical evolution models are. For this purpose we analyze two recent Type II supernova (SNII) nucleosynthesis prescriptions (Woosley & Weaver 1995 [32], hereafter WW95; Thielemann, Nomoto, & Hashimoto 1996 [24] and Nomoto et al. 1997 [19], hereafter TNH96), that represent a crucial input for the determination of Mg/Fe element ratios as a function of timescales (see also Gibson 1997 [6]; Portinari, Chiosi, & Bressan 1998 [21]).

2. Stellar yields

The ratio of Mg to Fe is a measure for the contribution from high-mass stars to the chemical enrichment, because Mg is only produced in SNII (short-lived, massive stars), while a significant part of Fe comes from Type Ia supernovae (SNIa, low-mass, long-lived binary systems). The crucial input ingredients to calculate the enrichment of these elements are both the supernova rates and the stellar yields (nucleosynthesis and ejection per supernova, see Chieffi, this volume; Limongi, this volume). In the following we briefly comment on the SNII yields of Mg and Fe that are calculated by WW95 and TNH96 (see also Thomas, Greggio, & Bender 1998 [26]).

Magnesium Magnesium is produced in both hydrostatic (before the explosion) and explosive carbon burning. The synthesis thus depends upon the (still uncertain) parameters of stellar evolution (convection criteria, $^{12}C(\alpha,\gamma)^{16}O$-rate, etc.) and the modeling of the explosion. We find that the yields of Mg in the two sets of models are in good agreement for low initial stellar masses in the range $10-18\ M_\odot$. In a $20\ M_\odot$ star, instead, TNH96 produce significantly more Mg than WW95, likely because they adopt the Schwarzschild criterion for convection (WW95; Thomas et al. 1998 [26]). At the high-mass end ($30-70\ M_\odot$), the total mass and the composition of the ejecta mainly depend on the explosion energy and on the mass-cut, which leads to slightly higher Mg-yields in the TNH96 models.

Iron Since the iron *ejected* by SNeII is entirely produced during the explosion, the yield is very sensitive to details of the explosion model (total energy, mass-cut, fall-back effect, energy transport, etc.). Hence, particularly at higher masses, the Fe-yields from SNeII are very uncertain. For $M < 35\ M_\odot$, WW95 (model B) give higher Fe yields than TNH96.

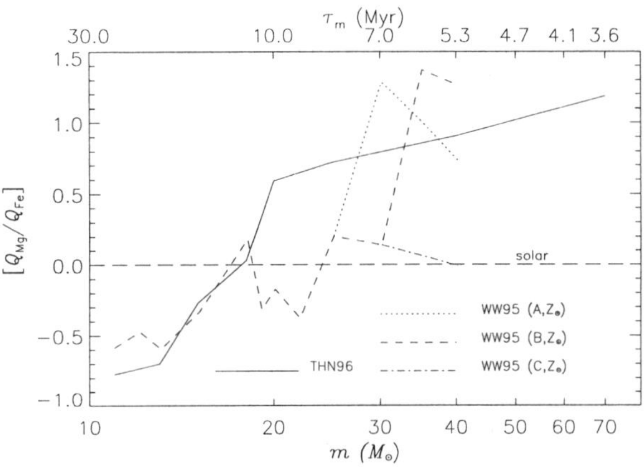

Figure 1. [Mg/Fe] ratio in the ejecta of SNII explosions as a function of initial stellar mass. WW95 with solar initial metallicity are considered. WW95 models A,B,C assume different explosion energies for the high-mass stars increasing from model A to C. Model B is equivalent to the prescription in TNH96.

Fig. 1 shows the resulting [Mg/Fe] ratio in the SNII ejecta for the two nucleosynthesis prescriptions WW95 and TNH96. For $M \sim 20\ M_\odot$, the [Mg/Fe] predicted by TNH96 is ~ 1 dex higher than WW95. At the high-mass end, WW95 (model B) produce larger Mg/Fe because of the very low Fe yield due to the fall-back effect.

3. The solar neighborhood

The significant discrepancies between the Mg/Fe yields from WW95 and TNH96 lead to very different conclusions on star formation timescales and IMF slopes (Thomas et al. 1998 [26], 1999 [27]). It is therefore crucial to calibrate the chemical evolution model on the abundance patterns of the solar neighborhood, for which the observational data provide the most detailed information. Here we show the impact of the above discrepancies in the stellar yields on the evolution of [Mg/Fe] with [Fe/H] in the solar neighborhood. The model description for the chemical evolution in the Milky Way is based on the standard infall model (Matteucci & Greggio 1986 [16]). The recipe for the rate of SNIa is taken from Greggio & Renzini (1983) [10], a Salpeter initial mass function ($x = 1.35$, $M_{\rm min} = 0.1\ M_\odot$, $M_{\rm max} = 70\ M_\odot$) is adopted. For more details see Thomas et al. (1998) [26].

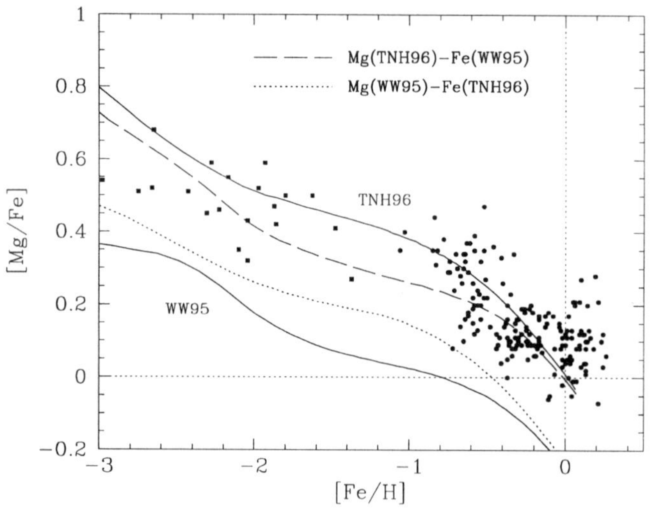

Figure 2. [Mg/Fe] ratio as a function of iron abundance [Fe/H] in the solar neighborhood. Data are from Magain 1989 (squares) and Edvardsson et al. 1993 (circles). The solid lines denote models using SNII-yields from WW95 and TNH96. The broken lines show the case for mixed stellar yields. Mg(TNH96)-Fe(WW95): dashed; Mg(WW95)-Fe(TNH96): dotted.

Fig. 2 shows observed stellar abundance ratios in the [Mg/Fe]-[Fe/H] plane. Models using the same input parameters but different sets of SNII nucleosynthesis prescriptions (WW95 and TNH96) are plotted as solid lines. With the TNH96-yields, the Mg-enhancement in the (metal-poor) halo stars ([Fe/H] < -1) is well reproduced. The sharp decrease of [Mg/Fe] at [Fe/H] ~ -1 implies the enrichment from SNIa to be delayed by ~ 1 Gyr (Pagel & Tautvaisiene 1995 [20], this volume) which is well matched with the SNIa model adopted here (see also Greggio 1996 [8]). Adopting WW95 yields without modifying the other input parameters, the resulting Mg/Fe ratios are well below the data for all metallicities (see also Timmes, Woosley, & Weaver 1995 [29]). Note that the strongest constraint on the SNII nucleosynthesis comes from the metal-poor halo stars, while the mismatch around solar metallicity could in principle be improved by adjusting the parameters of the star formation history and of the SNIa rate.

Models with mixed stellar yields, namely, Mg(TNH96)-Fe(WW95) and Mg(WW95)-Fe(TNH96) are shown by the dashed and dotted lines, respectively. This exercise demonstrates that the low Mg/Fe ratios deduced from the WW95 models is mainly due to the low Mg yields. Especially at the high metallicity regime, the large Fe SNII-yields from WW95 have only a small effect on the final abundance ratios as a considerable fraction (~ 60 per cent) of Fe comes from SNeIa.

We conclude that the impact of uncertainties in stellar yields on chemical evolution models is not negligible. Chemical evolution models that are applied to the Bulge or extragalactic systems should be carefully calibrated on the solar neighborhood.

4. Discussion

4.1. MODEL IMPROVEMENTS

An important requirement for the successful calibration of chemical evolution models is the accuracy and reliability of abundance and age determinations. Analyzing high quality spectra of galactic stars, Fuhrmann (1998) [5] demonstrates that the scatter in Mg/Fe found by Edvardsson et al. (1993) [4] can be reduced. His data also point towards a more pronounced separation of the various components of the Galaxy, that may require more sophisticated modeling (e.g., two-infall model, Chiappini, Matteucci, & Gratton 1997 [2]).

The determination of the exact timescale for the delayed Fe enrichment from SNIa depends on the star formation history that is constrained through the observationally defined age-metallicity relation. As the latter still suffers from a large scatter, it may be more promising in the future to directly adopt the star formation history that is determined with chromospheric ages (Rocha-Pinto et al. 2000 [22]).

Invoking a metallicity dependence of SNIa rates (Kobayashi et al. 1998 [11]) and the consideration of delayed and inhomogeneous mixing (Malinie et al. 1993 [13]; Thomas et al. 1998 [26]; Tsujimoto, Shigeyama, & Yoshii 1999 [30]) represent further improvements of chemical evolution models.

4.2. CONSTRAINT ON ELLIPTICAL GALAXIES

Unless one allows for a flattening of the IMF, the very high Mg/Fe ratios in elliptical galaxies prohibit the occurrence of late star formation, i.e. induced by a merger (Thomas et al. 1999 [27]). This constraint seems to be at odds with the finding that ellipticals exhibit a considerable scatter in Hβ line strengths (González 1993 [7]). However, composite stellar population models containing a small fraction of old metal-poor stars allow for an alternative interpretation: Such models succeed in reproducing the strong metallic *and* the strong Balmer lines observed in ellipticals without invoking young ages (Maraston, this volume; Maraston & Thomas 2000 [14]).

Acknowledgments

DT thanks F. Matteucci and F. Giovannelli for a very lively and fruitful workshop. This work was supported by the "Sonderforschungsbereich 375-95 für Astro-Teilchenphysik" of the Deutsche Forschungsgemeinschaft.

References

1. Bender, R.: 1996. In: R. Bender and R. L. Davies (eds.): *New Light on Galaxy Evolution*. Dordrecht, p. 181.
2. Chiappini, C., F. Matteucci, and R. Gratton: 1997. *ApJ* **477**, 765.
3. Davies, R. L., E. M. Sadler, and R. F. Peletier: 1993. *MNRAS* **262**, 650.
4. Edvardsson, B., J. Andersen, B. Gustafsson, D. L. Lambert, P. E. Nissen, and J. Tomkin: 1993. *A&A* **275**, 101.
5. Fuhrmann, K.: 1998. *A&A* **338**, 161.
6. Gibson, B. K.: 1997. In: *Cosmic Chemical Evolution*. Dordrecht.
7. González, J.: 1993. Phd thesis, University of California, Santa Cruz.
8. Greggio, L.: 1996. In: D. Kunth, B. Guiderdoni, M. Heydari-Malayeri, and T. X. Thuan (eds.): *The interplay between massive star formation, the ISM and galaxy evolution*. Gif-sur-Yvette Cedex-France, p. 89.
9. Greggio, L.: 1997. *MNRAS* **285**, 151.
10. Greggio, L. and A. Renzini: 1983. *A&A* **118**, 217.
11. Kobayashi, C., T. Tsujimoto, K. Nomoto, I. Hachisu, and M. Kato: 1998. *ApJ* **503**, L155.
12. Magain, P.: 1989. *A&A* **209**, 211.
13. Malinie, G., D. H. Hartmann, D. D. Clayton, and G. J. Mathews: 1993. *ApJ* **413**, 633.
14. Maraston, C. and D. Thomas: 2000. *ApJ*. submitted.
15. Matteucci, F.: 1994. *A&A* **288**, 57.
16. Matteucci, F. and L. Greggio: 1986. *A&A* **154**, 279.
17. McWilliam, A. and R. M. Rich: 1994. *ApJS* **91**, 749.
18. Mehlert, D., R. P. Saglia, R. Bender, and G. Wegner: 2000. *A&AS*. in press.
19. Nomoto, K., M. Hashimoto, F.-K. Tsujimoto, N. Kishimoto, Y. Kubo, and N. Nakasato: 1997. *Nucl. Phys. A* **616**, 79c.
20. Pagel, B. E. J. and G. Tautvaisiene: 1995. *MNRAS* **276**, 505.
21. Portinari, L., C. Chiosi, and A. Bressan: 1998. *A&A* **334**, 505.
22. Rocha-Pinto, H. J., J. Scalo, W. Maciel, and C. Flynn: 2000. *ApJ*. submitted, astro-ph/9908328.
23. Tantalo, R., C. Chiosi, and A. Bressan: 1998. *A&A* **333**, 419.
24. Thielemann, F.-K., K. Nomoto, and M. Hashimoto: 1996. *ApJ* **460**, 408.
25. Thomas, D.: 1999. *MNRAS* **306**, 655.
26. Thomas, D., L. Greggio, and R. Bender: 1998. *MNRAS* **296**, 119.
27. Thomas, D., L. Greggio, and R. Bender: 1999. *MNRAS* **302**, 537.
28. Thomas, D. and G. Kauffmann: 2000. In: I. Hubeny, S. Heap, and R. Cornett (eds.): *Spectrophotometric dating of stars and galaxies*. Dordrecht. PASP Conference Proceedings series, astro-ph/9906216.
29. Timmes, F. X., S. E. Woosley, and T. A. Weaver: 1995. *ApJS* **98**, 617.
30. Tsujimoto, T., T. Shigeyama, and Y. Yoshii: 1999. *ApJ* **519**, L63.
31. Weiss, A., R. F. Peletier, and F. Matteucci: 1995. *A&A* **296**, 73.
32. Woosley, S. E. and T. A. Weaver: 1995. *ApJS* **101**, 181.
33. Worthey, G., S. M. Faber, and J. J. González: 1992. *ApJ* **398**, 69.

THE INFLUENCE OF NOVA NUCLEOSYNTHESIS ON THE CHEMICAL EVOLUTION OF THE GALAXY

D. ROMANO
SISSA/ISAS
Trieste, Italy

AND

F. MATTEUCCI
Dipartimento di Astronomia, Università di Trieste,
Trieste, Italy

Abstract. We adopt up-to-date yields of ^{7}Li, ^{13}C, ^{15}N from classical novae and use a well tested model for the chemical evolution of the Milky Way in order to predict the temporal evolution of these elemental species in the solar neighborhood. In spite of major uncertainties due to our lack of knowledge of metallicity effects on the final products of explosive nucleosynthesis in nova outbursts, we find a satisfactory agreement between theoretical predictions and observations for ^{7}Li and ^{13}C. On the contrary, ^{15}N turns out to be overproduced by about an order of magnitude.

1. Introduction

Classical novae are sporadically injecting material processed by explosive hydrogen-burning nucleosynthesis into the interstellar medium (ISM). Both theoretical and observational evidence suggests that novae, although they have probably processed less than $\sim$ 0.3% of the Galactic matter, may be important producers of ^{7}Li, ^{13}C, ^{15}N, ^{17}O, ^{22}Na, and ^{26}Al (see Gehrz *et al.* 1998 for a review). The role of novae in the enrichment of the ISM has already been investigated by D'Antona & Matteucci (1991) (hereafter DM) and Matteucci & D'Antona (1991). In DM ^{7}Li production from novae is computed according to the results of Starrfield *et al.* (1978) concerning the explosive nucleosynthesis in nova outbursts and those of D'Antona & Mazzitelli (1982) concerning the nova secular evolution. The DM results

F. Matteucci and F. Giovannelli (eds.), The Evolution of the Milky Way, 547–555.

suggest that novae should contribute to more than 50% to the global ^{7}Li production, that ^{13}C and ^{15}N are produced proportionally to ^{7}Li in novae, and that no primary ^{13}C should originate from low-intermediate mass stars in order to fit the observational constraints in the DM scenario.

Here we reanalyse the influence of novae on the chemical evolution of the solar neighborhood, by using the results recently obtained by José & Hernanz (1998) (hereafter JH) from a grid of hydrodynamical nova models.

2. The chemical evolution model

The adopted chemical evolution model is that of Chiappini *et al.* (1997). In this model the Galaxy is formed by two distinct episodes of mass accretion: the first one builds the halo and thick-disk substructures; the second one, almost independent from the former and temporally delayed with respect to it, originates the thin-disk, mainly through accretion of matter of primordial chemical composition.

The explosive nucleosynthesis from nova outbursts is included in the model under quite straightforward assumptions on nova progenitors. We assume that at any given time t the rate of nova outbursts is a fraction α of the white dwarf (WD) formation rate at a previous time $t-\Delta t$ multiplyed by the number of outbursts per nova (n):

$$R_{outbursts}(t) = n\,\alpha \int_{0.8}^{8} \psi(t-\tau_m-\Delta t)\phi(m)\,dm.$$

We assume that all stars with initial mass between 0.8 and 8 $M_\odot$ end their life as WDs. τ_m is the lifetime of a star of mass m and $\Delta t = 1$ Gyr is the delay time required for the WD to cool and the first nova outburst to occur (Romano *et al.* 1999). The parameter α, constant in time, is fixed from the condition of reproducing the currently observed rate of nova outbursts in the Galaxy, namely $R_{outbursts}(t_{Gal}) \sim 30 \pm 30\%$ yr^{-1} (Della Valle 2000). The parameter $n = 10^4$ is the total number of outbursts suffered by the average nova (Bath & Shaviv 1978).

2.1. DETAILED NUCLEOSYNTHESIS PRESCRIPTIONS

2.1.1. ^{7}*Li synthesis*

The stellar sources of ^{7}Li considered are: carbon stars, massive AGB stars, Type II supernovae (SNe) and novae. For stars with initial masses in the range $2-5$ $M_\odot$ (carbon stars) and $5-8$ $M_\odot$ (massive AGB stars) we evaluate the masses ejected as newly produced ^{7}Li following DM and Matteucci *et al.* (1995). In particular, in Model A and Model B we adopt the same prescriptions on ^{7}Li as in the best model of Matteucci *et al.* (1995), whereas in Model C we set to zero the contribution from carbon stars and assume a lower ^{7}Li production from massive AGB stars.

For stars with initial mass $M > 10\ M_\odot$ we use either the whole yields of ^{7}Be + ^{7}Li tabulated by Woosley & Weaver (1995) (Model A and Model B) or the same yields but reduced by a factor of 2 (Model C).

Li production during thermonuclear runaways in novae is included in our model by averaging on 14 evolutionary sequences computed by JH, to which we refer for details about input physics, nucleosynthesis and related uncertainties. We assume that $\sim 30\%$ of nova systems accrete hydrogen rich matter onto the surface of ONeMg WDs, whereas the remaining accrete hydrogen rich matter onto the surface of CO WDs. One model (Model A) does not take into account novae as Li producers.

A further ^{7}Li source is identified in the Galactic cosmic ray (GCR) spallation on ISM nuclei. Therefore, we run a model (Model C + GCRs) in which we add the absolute ^{7}Li yields from GCRs as given by Lemoine *et al.* (1998) (their Table 1, case with lower-bound spectrum and $x = 1$).

2.1.2. *Carbon and nitrogen evolution*

TABLE 1. Models referring to CN isotope evolution (see text).

Model	novae	low-intermediate masses
Model 1	no	RV (without hot bottom burning)
Model 2	no	RV (with hot bottom burning)
Model 3	no	HG (constant mass loss)
Model 4	no	HG (variable mass loss)
Model 5	yes (average)	RV (without hot bottom burning)
Model 6	yes (average)	HG (constant mass loss)
Model 7	yes (average)	HG (variable mass loss)
Model 8	yes (min ^{13}C)	RV (without hot bottom burning)
Model 9	yes (min ^{13}C)	HG (constant mass loss)
Model 10	yes (min ^{13}C)	HG (variable mass loss)

In the mass range $0.8 - 8\ M_\odot$ we adopt either the old yields from Renzini & Voli (1981) (hereafter RV) (their case with or without hot bottom burning; *i.e.* their case with or without primary ^{13}C and ^{14}N production) or the most recent ones from van den Hoek & Groenewegen (1997) (hereafter HG) (their case with constant efficiency of mass loss along the AGB or their case with variable efficiency of mass loss along the AGB).

The nucleosynthesis prescriptions in the mass range $M > 10\ M_\odot$ are from Woosley and Weaver (1995) in all models.

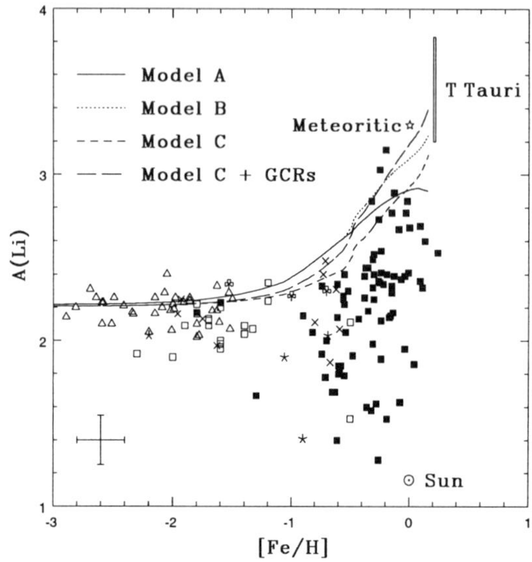

Figure 1. Theoretical predictions on A(Li) vs. [Fe/H] from four different models (prescriptions on Li synthesis as in Sect.2.1.1) compared to observations.

As far as CN production from novae is concerned, we run models without adding their contribution, models in which the average carbon and nitrogen production from 14 evolutionary sequences (JH) is included, and models in which only the evolutionary sequences corresponding to the minimum ^{13}C production are considered. The different models are listed in Table 1.

3. Results

3.1. A(LI) VS. [FE/H]

The ^7Li abundance is measured in stars of all metallicities and the observations indicate a plateau at low metallicities (PopII stars) followed by an increase in the ^7Li abundance in the more metal-rich stars. However, in metal-rich stars ^7Li can be strongly depleted (Fig.1) so the aim of chemical evolution models is to fit the upper envelope of the observed distribution, assuming that this represents the growth of ^7Li in time as a consequence of stellar and perhaps GCR production. In this framework, the abundance of ^7Li measured in PopII stars represents the primordial ^7Li abundance, although a small contribution from GCRs already at these low metallicities cannot be excluded. A recent and careful analysis of extant data in the A(Li) vs. [Fe/H] diagram (Romano *et al.* 1999) has revealed a possible extension of the plateau toward metallicities higher than previously thought,

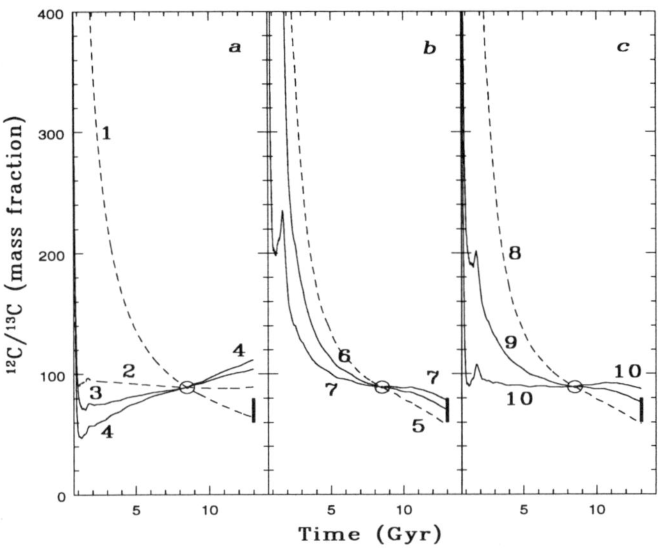

Figure 2. The temporal behaviour of the carbon isotope ratio as predicted by the model using different nucleosynthesis prescriptions (see Sect.2.1.2). In all cases the model $^{12}C/^{13}C$ ratio in the solar neighborhood 4.5 Gyr ago is normalised to its solar value.

[Fe/H] ~ -0.4 dex. Therefore, astrophysical sources able to synthesize ^{7}Li and restore it back into the ISM on long timescales should be preferred. Among these, novae are promising candidates: in fact, in our model the first WDs form after $\sim$ 600 million years from the beginning of Galaxy formation, then a billion year more has to elapse to allow them to cool at a level that ensures strong enough nova outbursts.

In Fig.1 we show the theoretical A(Li) vs. [Fe/H] obtained from the four different models discussed in Sect.2.1.1. Models taking into account novae as Li factories (Models B, C, and C + GCRs) predict that a substantial amount of Li should be returned to the ISM at late times, thus reproducing the steep rise suggested by the data. In particular, Model C + GCRs guarantees a very good fit to the observational data but it also predicts a too high contribution from GCRs to the solar ^{7}Li abundance ($\sim$ 45%). In conclusion, we suggest that the best model should be intermediate between Model B and Model C + GCRs.

3.2. THE CARBON ISOTOPE RATIO

The observations clearly show that the $^{12}C/^{13}C$ ratio decreases from the time of the Solar System formation up to now: $(^{12}C/^{13}C)_\odot = 89 \pm 2$ (Cameron 1982), $(^{12}C/^{13}C)_{\rm ISM} = 40-80$ (Crane & Hegyi 1988; Stahl *et*

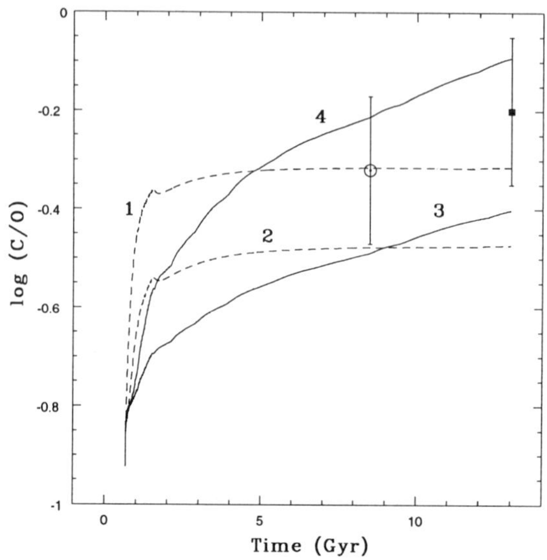

Figure 3. Theoretical log(C/O) vs. time from four different models (see text for details).

al. 1989). Values of $(^{12}C/^{13}C)_{ISM}$ between 60 and 80 seem to be the most likely ones (Centurión & Vladilo 1991; Langer & Penzias 1993).

Among models excluding novae as CN isotope producers, Model 1 predicts too low ^{13}C and ^{14}N at the time of Sun's formation, whereas all the others are unable to properly fit the decreasing trend of the carbon isotope ratio in the last 4.5 Gyr (Fig.2, panel *a*). However, Model 4 predicts a log(C/O) vs. time in good agreement with observations (see Fig.3). This is due to the larger ^{12}C production by low mass stars in the yields of HG compared to the old ones of RV.

Concerning ^{13}C the overall situation is not yet clear, but an extra source of this element seems to be required to improve the agreement with observations.

^{13}C and ^{15}N produced during a nova outburst are primary products since they form from C and O seeds which were synthesized during the previous evolution of the WD progenitor, starting from H and He. However, due to the long time taken for novae to expell their synthesized products, the abundances of ^{13}C and ^{15}N produced by novae behave like secondary products (Wilson & Matteucci 1992, and refs. therein). As a consequence of this, when including this late contribution into the chemical evolution model, the $^{12}C/^{13}C$ ratio in the solar neighborhood is expected to decrease from the time of the Solar System formation up to now. In Models 5, 6, and 7, where the nova contribution is added by averaging on the results from

TABLE 2. Elemental abundances (mass fraction) at the time of Sun's formation as predicted by the model under different input nucleosynthesis, compared to the observations (Grevesse & Noels 1993).

	$^{12}C_{\odot}$	$^{13}C_{\odot}$	$^{14}N_{\odot}$	$^{15}N_{\odot}$
Model 1	$2.56E-3$	$8.89E-6$	$5.02E-4$	–
Model 2	$1.75E-3$	$4.29E-5$	$1.30E-3$	–
Model 3	$1.66E-3$	$1.39E-5$	$9.79E-4$	–
Model 4	$3.14E-3$	$2.56E-5$	$1.46E-3$	–
Model 5	$2.60E-3$	$9.10E-5$	$6.49E-4$	–
Model 6	$1.69E-3$	$9.75E-5$	$1.11E-3$	$2.44E-5$
Model 7	$3.17E-3$	$1.08E-4$	$1.59E-3$	$2.44E-5$
Model 8	$2.58E-3$	$3.11E-5$	$5.69E-4$	–
Model 9	$1.67E-3$	$3.74E-5$	$1.05E-3$	$2.49E-5$
Model 10	$3.15E-3$	$4.87E-5$	$1.53E-3$	$2.50E-5$
Observed	$3.62E-3$	$4.03E-5$	$1.07E-3$	$3.92E-6$

14 evolutionary sequences (JH), ^{13}C turns out to be always overproduced (see Table 2). Therefore, we need some *ad hoc* assumptions: the average ^{13}C production from a single nova system during the overall evolutionary history of the solar neighborhood should be not larger than the minimum one suggested by JH (Models 8, 9, 10). A model intermediate between Model 9 and Model 10 should guarantee the best fit to all the observational constraints considered here (see Table 2 and Fig.2, panel *c*). Unfortunately, the solar ^{15}N abundance turns out to be overestimated by all the models.

Therefore, we explain the decreasing ^{12}C/^{13}C ratio in the local ISM as due to the primary nature of ^{12}C coupled to the primary + secondary nature of ^{13}C (see also Prantzos *et al.* 1996). A recent work by Palla *et al.* (1999) strongly supports the idea that the majority of the planetary nebula progenitors must have undergone an extra-mixing process, resulting in an enhanced ^{13}C abundance in the surface layers and leading to a ^{12}C/^{13}C ratio lower than predicted by standard stellar evolutionary models. This ^{13}C source is equivalent to that from novae since in both cases this extra source is acting at late times. In summary: we cannot conclude that novae are necessary to explain the ^{12}C/^{13}C ratio, but we can conclude that the inclusion of nova nucleosynthesis in the chemical evolution model for our galaxy improves the predictions for both ^{7}Li and ^{13}C evolution.

Acknowledgements

We thank the organizers for a very interesting meeting and C. Chiappini and N. Prantzos for many helpful comments. We acknowledge financial support from the Italian Ministry for University and for Scientific and Technological Research (MURST).

References

1. Bath, G.T. and Shaviv, G. (1978) *MNRAS*, **183**, 515
2. Cameron, A.G.W. (1982) in *Essays in Nuclear Astrophysics* C.A. Barnes, D.D. Clayton, N.D. Schramm (eds.), p. 23. Cambridge: Cambridge Univ. Press.
3. Centurión, M. and Vladilo, G. (1991) *A&A*, **251**, 245
4. Chiappini, C., Matteucci, F. and Gratton, R. (1997) *ApJ*, **477**, 765
5. Crane, P. and Hegyi, D.J. (1988) *ApJ*, **326**, L35
6. D'Antona, F. and Matteucci, F. (1991) *A&A*, **248**, 62 (DM)
7. D'Antona, F. and Mazzitelli, I. (1982) *ApJ*, **260**, 722
8. Della Valle, M. (2000), this proceedings
9. Gehrz, R.D., Truran, J.W., Williams, R.E. and Starrfield, S. (1998) *PASP*, **110**, 3
10. Grevesse, N. and Noels, A. (1993) in *Origin and Evolution of the Elements* N. Prantzos *et al.* (eds.), p. 13. Cambridge: Cambridge Univ. Press.
11. José, J. and Hernanz, M. (1998) *ApJ*, **494**, 680 (JH)
12. Langer, W.D. and Penzias, A.A. (1993) *ApJ*, **408**, 539
13. Lemoine, M., Vangioni-Flam, E. and Cassé, M. (1998) *ApJ*, **499**, 735
14. Matteucci, F. and D'Antona, F. (1991) *A&A*, **247**, L37
15. Matteucci, F., D'Antona, F. and Timmes, F.X. (1995) *A&A*, **303**, 460
16. Palla, F., Bachiller, R., Stanghellini, L., Tosi, M. and Galli, D. (1999) *A&A*, in press
17. Prantzos, N., Aubert, O. and Audouze, J. (1996) *A&A*, **309**, 760
18. Renzini, A. and Voli, M. (1981) *A&A*, **94**, 175 (RV)
19. Romano, D., Matteucci, F., Molaro, P. and Bonifacio, P. (1999) *A&A*, **352**, 117
20. Stahl, O., Wilson, T.L., Henkel, C. and Appenzeller, I. (1989) *A&A*, **221**, 321
21. Starrfield, S., Truran, J.W., Sparks, W.M. and Arnould, M. (1978) *ApJ*, **222**, 600
22. van den Hoek, L.B. and Groenewegen, M.A.T. (1997) *A&AS*, **123**, 305 (HG)
23. Wilson, T.L. and Matteucci, F. (1992) *A&ARv*, **4**, 1
24. Woosley, S.E. and Weaver, T.A. (1995) *ApJS*, **101**, 181

Speaker: D. ROMANO

C. ABIA: There is also evidence that low-mass AGB stars ($M \leq 2 - 3M_{\odot}$) can contribute to the galactic Li. May be you should consider this in your model.

D. ROMANO: We also computed a model taking into account Li production from stars in the mass range 2-5 $M_{\odot}$ and found their contribution almost negligible. So, our Best Model does not take this contribution into account.

THE EVOLUTION OF THE MILKY WAY DISK

N. PRANTZOS[1] AND S. BOISSIER[1]
Institut d'Astrophysique de Paris
98bis, Bd. Arago, 75014 Paris

1. Introduction

The Milky Way is a heterogeneous system, with at least three components (halo, bulge, disk) having very different chemical, photometric and kinematical properties. A reliable model for the evolution of the Milky Way accounting for those properties does not exist at present. In particular, it is not clear how the various components are related, e.g. whether the evolution of the halo has affected in an important way the one of the bulge or the disk (and vice versa); in fact, it seems that the halo evolution has been completely decoupled from the one of the disk, who evolved quite independently (e.g. Gilmore and Wyse 1998). Neither is it clear whether there has been significant interaction between the various parts of the disk, through large scale gas mouvements in the radial direction. Despite the development of various models, all these issues are still open.

In the following, we present a very simple model for the chemical evolution of the galactic disk, considering its as an ensemble of independently evolving, concentric, rings and formed by infall of primordial composition in an "inside-out" scheme. In our modelisation we are guided by phenomenology rather than theoretical principles and we try to construct a model that reproduces all the major observational constraints of the Milky Way with a minimum number of free parameters (see Boissier and Prantzos 1999, hereafter BP99, for details).

2. The Solar Neighborhood

In the solar neighborhood (defined as a cylinder of ~1 kpc radius at a distance R_S=8 kpc from the galactic center), the main observables relevant to chemical evolution are (for a detailed discussion of these observables see PP99).

F. Matteucci and F. Giovannelli (eds.), The Evolution of the Milky Way, 557–563.

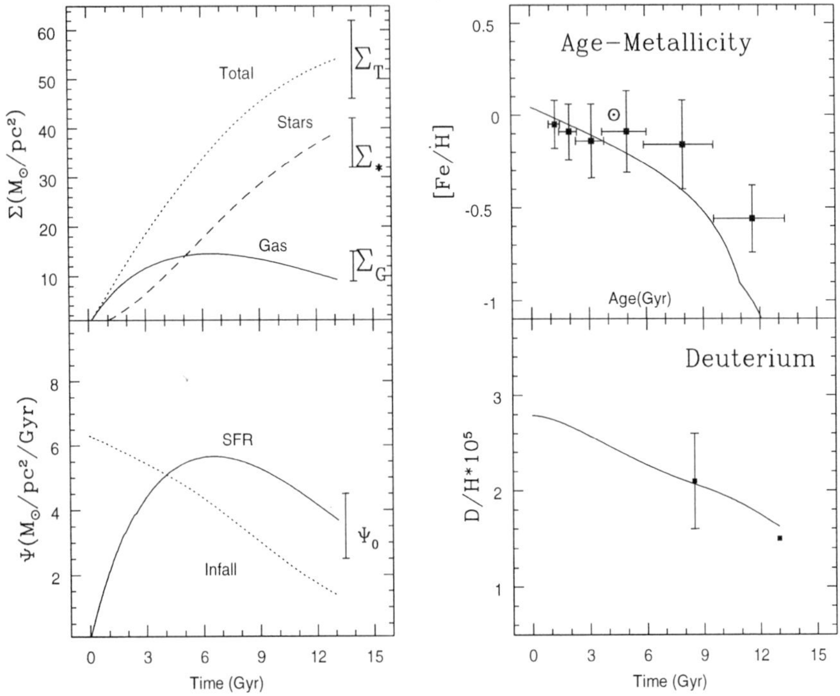

Figure 1. Evolution of various quantities in the solar neighborhood and comparison to observations, according to an infall model with time-scale of τ=7 Gyr and star formation rate $\Psi \propto \Sigma_{GAS}^{1.5}$ (from BP99).

i) the current surface densities of gas (Σ_G), stars (Σ_*) and total amount of matter (Σ_T), as well as the current star formation rate Ψ_0.

ii) the abundances of various elements and isotopes at solar birth ($X_{i,\odot}$) and today ($X_{i,O}$).

iii) the age-metallicity relationship, traced by the Fe abundance of long-lived, F-type stars.

iv) the oxygen vs. Fe (O-Fe) relationship, interpreted in terms of a delayed ($\sim$1 Gyr) appearence of SNIa, producing most of galactic Fe.

v) the metallicity distribution of long-lived G-type stars showing that very few of them were formed at [Fe/H]<-0.7 (1/5 solar) or [O/H]<-0.5 (1/3 solar).

vi) the present day mass function (PDMF), resulting from the stellar IMF and the SFR history; this observable is rarely considered in studies of the local chemical evolution. However, the resulting theoretical PDMF constitutes an important consistency check for the adopted SFR and IMF.

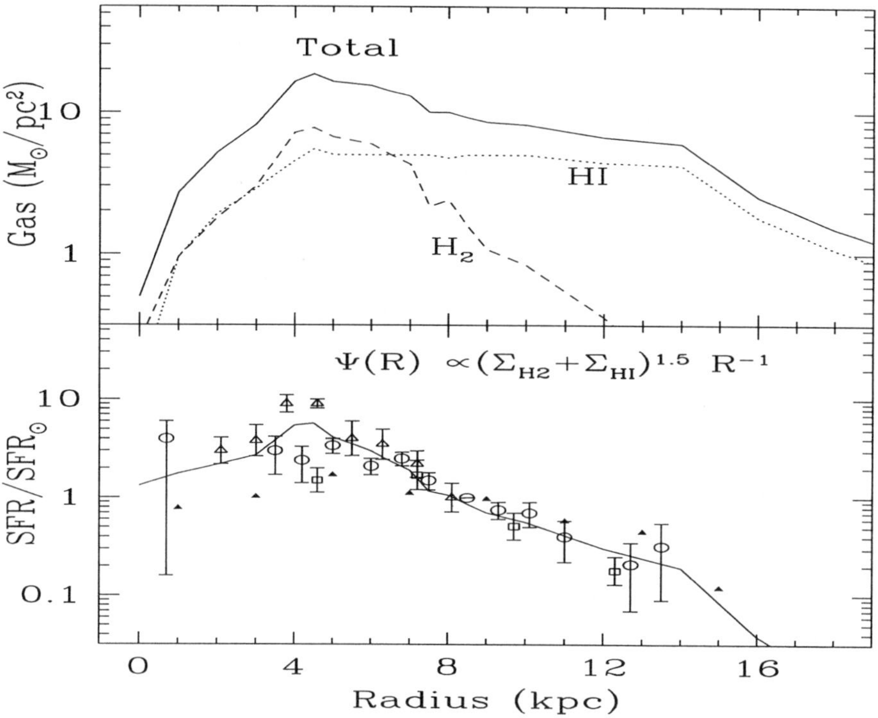

Figure 2. Upper panel: Gaseous profile $\Sigma_{GAS}(R)$ of the Milky Way disk. *Lower panel:* Star formation rate profile, assuming that $\Psi(R) \propto \Sigma_{GAS}(R)/R$ and comparison to various tracers of current SFR (from Boissier and Prantzos (1999).

It turns out that these observables can be reproduced with a simple model involving slow infall (on timescales of ~6-7 Gyr, e.g. Chiapini et al. 1997) and a star formation rate $\Psi \propto \Sigma^k_{GAS}$ with k in the 1.-2 range. The strongest constraint is clearly the differential metallicity distribution of G-dwarf stars, suggesting a long timescale for the formation of the local disk. This is corroboreted by the observed current local SFR, which is not very different from the past average one (see BP99).

3. The Milky Way disk

Contrary to the case of the Solar neighborhood, the available obsevations for the Milky Way disk offer information only about its current status, not its past history; ther is then much more freedom in modelling it. Still, it is meaningful to construct models, in so far as the number of free parameters is smaller than the number of obsevational constraints. Those constraints are:

i) The total mass of gas and stars in the disk ($M_G \sim$6-8 10^9 $M_\odot$ and $M_* \sim$4-5 10^{10} $M_\odot$, respectively); the total current SFR ($\sim$3-6 $M_\odot$ yr^{-1}) and the current supernova rates ($\sim$1-2 SNII/century and $\sim$0.2-0.4 SNI-a/century, respectively, as suggested by observations of external spirals; the uncertainty on the value of the Hubble constant and on the exact spectral type of the Milky Way - Sb or Sbc - prevents from giving a precise value). Notice that those quantities are often mentioned as constraints to (and predictions of) one-zone models of the Galaxy, assumed to reflect the evolution of the whole disk. This is obviously wrong, since the disk is a heterogeneous system, as the observed gradients suggest (see points iii-vi below). A multi-zone model with different SFR histories in its various zones should be obviously used. In fact, observations of external spirals give rather the SN frequency in SNU (i.e. in number of SN per century and per 10^{10} $L_{B\odot}$); knowledge of the current L_B for the Milky Way allows then to infer its current SN frequency. However, a succesful model of the Milky Way should reproduce in a self-consistent way *both* its SN frequency and L_B luminosity.

ii) The current gas profile, dominated by the molecular ring at galactocentric distance R$\sim$4-5 kpc and by HI of roughly constant surface density at distances 6-14 kpc.

iii) The stellar profile, exponentially decreasing outwards. The value of the characteristic scalelength is still under debate, but recent studies converge towards low scale-lengths, around 2.5-3 kpc. A recent analysis of the COBE data also points to a scalelength h=2.6 kpc (see BP99 and references therein). The combination of observables (ii) and (iii) leads to a gas fraction profile steeply decreasing in the inner disk, suggesting that the star formation efficiency has been larger in those regions than in the outer disk.

iv) The current SFR profile (traced by the surface density of pulsars and supernova remnants or the H_α emissivity profile), strongly decreasing outwards. Notice that the SFR profile does not follow the molecular or the total (molecular+atomic) one, i.e. the SFR is not simply proportional to some power of the gaseous profile.

v) The current metallicity profile, usually traced by oxygen observed in HII regions, young planetary nebulae, and B-stars, showing a gradient of d[O/H]$\sim$-0.08 dex/kpc.

Again, it turns out that the disk evolution can be reproduced with a simple model, involving both radially dependent infall (with timescales increasing outwards, forming the disk "inside-out") and radially dependent SFR efficiency (which can be theoretically justified from the theory of star formation induced by spiral waves in disks). The simple model has fewer parameters than constraints and predicts that:

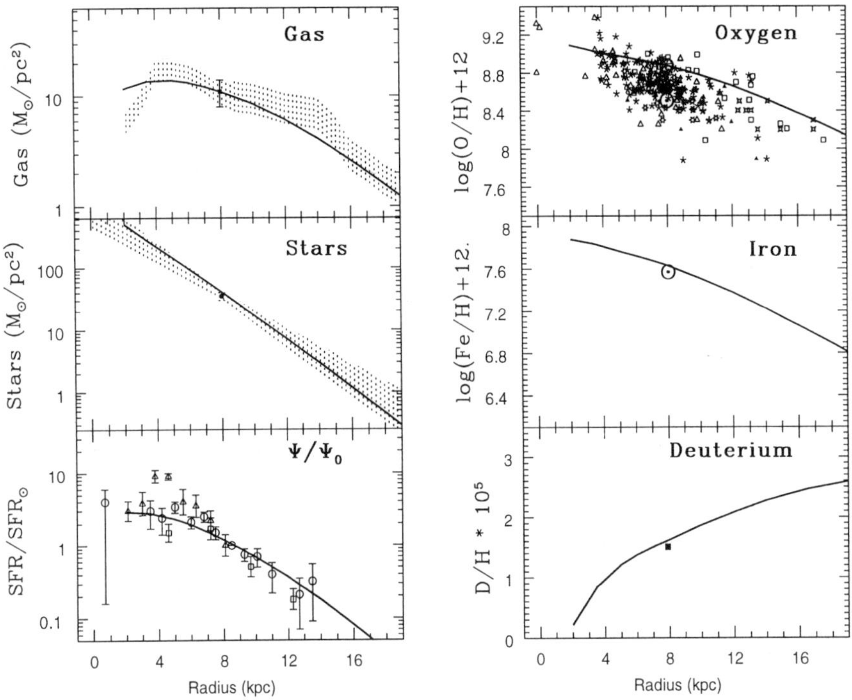

Figure 3. Radial profiles of various quantities in the Milky Way disk. Model results (*thick curves*) are plotted at an age of T=13 Gyr and are obtained for a model with SFR $\Psi(R) \propto \Sigma_{GAS}(R)/R$ and radially dependent infall timescale (i.e the disk is formed "inside-out"). From Boissier and Prantzos (1999).

i) there should exist a colour gradient in the disk; indeed, observations show that the disk scalelngth in the B-band is larger than in the K-band (see point iii above).

ii) the deuterium abundance profile (expected to be steeper in the inner disk than the corresponding oxygen one) is the most sensitive tracer of the past history of the disk (Prantzos 1996); observations with the FUSE-LYMAN satellite could help establish the D profile and reveal the disk evolution.

iii) abundance gradients should have been steeper in the past, because of the adopted "inside-out" formation scheme. In principle, this should be testable with observations of abundances in planetary nebulae of different ages, but in practice the uncertainties are still too large for such a test.

4. Conclusions

The theory of galaxy evolution is still in embryonic stage, because we do not yet have a clear understanding of the driver of that evolution, namely star formation in large scales. As a result, very few things can be stated with some degree of certainty on the evolution of external galaxies. In the case of the Milky Way, observations combined to some "toy models" make us feel that the following statements can be considered rather "safe":

- The solar neighborhood evolved slowly, on a time scale of several Gyr (because of the lack of metal poor stars in the G-dwarf metallicity distribution).

- There are no substantial amounts of dark matter locally (from the kinematics of stellar populations in the direction perpendicular to the disk).

- The star formation efficiency has been larger in the inner disk during the galactic history (from the observed gas fraction profile, steeply decreasing inwards).

- No important radial mixing, driven by bars, infall or viscous evolution, has taken place in the disk (because the observed abundance gradients are rather steep).

- The disk has been formed "inside-out" (because the observed scale-lengths, R_B ~4 kpc and R_K ~2.6 kpc indicate a more extended radial distribution for the young stellar populations).

Several important questions remain unanswered; here is a short list of some of them (to be combined to those of Tosi 2000).

- Is the Sun's composition typical of the local interstellar medium 4.5 Gyr ago? (solar CNO abundances are ~30% higher than those of gas and young stars in the nearby Orion nebula); was the Sun born in the metal-rich inner Galaxy and did it migrate from its birthplace to the current one, as some dynamical models suggest?

- What role (if any) played the galactic bar in shaping the gas and abundance profiles of the disk?

- How did the abundance gradients evolve? in "inside-out" schemes, they should obviously become flatter with time.

At present, it does not seem clear whether chemo-dynamical models could help with these (and many other!) questions. It seems that, once more, we shall have to rely on forthcoming observations to see some light...

References

1. Boissier S. and Prantzos N., (1999), *MNRAS*, **307**, 857
2. Chiapini C., Matteucci F. & Gratton R., (1997), *ApJ*, **477**, 765
3. Gilmore G. & Wyse R., (1998), *AJ*, **176**, 748
4. Prantzos N., (1996), *A&A*, **310**, 106
5. Tosi M., (2000), *this volume*

Speaker: N. PRANTZOS

W. MACIEL's Comment: Concerning the dispersion of the Edvardsson et al. age-metallicity relation, we have recently derived a new relation using a larger sample and chromospheric ages. This relation is similar to that of Edvardsson et al., but the dispersion is much lower.

THE EFFECTS OF A VARIABLE IMF ON THE CHEMICAL EVOLUTION OF THE GALAXY

C. CHIAPPINI[1]

Departamento de Astronomia - Observatório Nacional
R. Gal. José Cristino 77 - Rio de Janeiro - 20921-400 - Brazil

AND

F. MATTEUCCI[2]

Dipartimento di Astronomia - Universitá di Trieste
Via G. B. Tiepolo 11 - 34131 - Trieste - Italy

Abstract. In this work we explore the effects of adopting an initial mass function (IMF) variable in time on the chemical evolution of the Galaxy. In order to do that we adopt a chemical evolution model which assumes two main infall episodes for the formation of the Galaxy. We study the effects on such a model of different IMFs. First, we use a theoretical one based on the statistical description of the density field arising from random motions in the gas. This IMF is a function of time as it depends on physical conditions of the site of star formation. We also investigate the behaviour of the model predictions using other variable IMFs, parameterized as a function of metallicity.

Our results show that the theoretical IMF when applied to our model depends on time but such time variation is important only in the early phases of the Galactic evolution, when the IMF is biased towards massive stars. We also show that the use of an IMF which is a stronger function of time does not lead to a good agreement with the observational constraints suggesting that if the IMF varied this variation should have been small. Our main conclusion is that the G-dwarf metallicity distribution is best explained by infall with a large timescale and a constant IMF, since it is possible to find variable IMFs of the kind studied here, reproducing the G-dwarf metallicity but this worsens the agreement with other observational constraints.

F. Matteucci and F. Giovannelli (eds.), The Evolution of the Milky Way, 565–572.

1. Introduction

Observational constraints are of fundamental importance to build a realistic chemical evolution model. With respect to these constraints the last years have been of crucial importance (Pagel 1997) and, in the case of the Milky Way, the new observational data required a revision of the previous chemical evolution models (see Pagel and Tautvaisiene 1995 and Chiappini et al. 1997, hereinafter CMG, for a discussion on this point).

In particular, Gratton et al. (1996, 1999) showed that the distribution of the abundances of α-elements to Fe for a large homogeneous sample of stars in the solar neighbourhood seems to indicate a short timescale for the evolution of the halo and thick disk phases and a sudden decrease in the star formation in the epoch preceding the formation of the thin disk. An analogous result was found by Fuhrmann (1998) for the [Mg/Fe] ratio. Moreover, Beers and Sommer-Larsen (1995) have shown that the thick disk population extends to very low metallicities. Those are very important new information which stimulated us (CMG) to consider a different picture of Galaxy formation than those previously adopted.

Previous models, in fact, (e.g. Matteucci and François 1989) were based on a pure colapse picture where the disk formed from gas shed from the halo. These models, however, are difficult to conciliate with the new results discussed above indicating a discontinuity between halo and thin disk. We than suggested the so-called *Two-Infall Model*, a model that assumes that the Galactic thin disk was not only formed from gas shed from the halo but was formed mainly from extra-galactic gas. We assume two main infall episodes, the first one is responsible for the formation of the population made of that fraction of the halo and thick disk stars which originated from a fast dissipative collapse, such as suggested by Eggen et al. (1962). The second infall episode forms the thin disk component with a timescale much longer than that of the halo formation.

In this new picture the disk was formed slowly (with a timescale of 7-8 Gyrs at the solar vicinity) and from inside-out (Matteucci and François 1989). A direct consequence of that is that at high redshift we should expect to see smaller disks in size (Roche et al. 1998). This long timescale for the formation of the thin disk at the solar vicinity, required to produce a good fit of the obseved G-dwarf metallicity distribution (Rocha-Pinto and Maciel 1996) was also confirmed by recent chemical evolution models (eg. Portinari et al. 1998, Prantzos and Silk 1998, Chang et al. 1999) as well as by chemo-dynamical models (eg. Hensler 1998) and is also in agreement with the results showed in this conference by Carraro (2000).

The two-infall model adopted a constant IMF. On the empirical grounds there is at present no clear direct evidence that the IMF in the Galaxy has varied with time. A detailed discussion about possible observed variations in the IMF in different environments is given by Scalo (1998), but such variations are comparable with the uncertainties still involved in the IMF determinations. The present uncertainties in the observational results prevent any conclusion against the idea of an universal IMF.

However, a variable IMF, which formed relatively more massive stars during the earlier phases of the evolution of the Galaxy compared to the one observed today in the solar vicinity, has often been suggested as being one of the possible solutions for the G-dwarf problem (namely the deficiency of metal-poor stars in the solar neighborhood when compared with the number of such stars predicted by the simple model). Such an IMF would also be physically plausible from the theoretical point of view if the IMF depends on a mass scale such as the Jeans mass. Given the uncertainties in both theoretical and observational grounds, the proposed IMFs can in principle be tested by means of a detailed chemical evolution model.

An example of a theoretical approach to the IMF problem is the one proposed by Padoan et al. (1997 - hereinafter PNJ). Since random motions are probably ubiquitous in sites of star formation, PNJ suggested to describe the formation of protostars as the gravitational collapse of Jeans masses in a density distribution shaped by random supersonic motions (but see Scalo et al. 1998). This IMF was already tested in models of elliptical galaxies by Chiosi et al. (1998) and they concluded that a strongly varying IMF could be suitable for such galaxies.

Our goal was to address the question of what time-dependent IMF properties are allowed in order to still match the observational constraints when adopting the two-infall model (CMG). To do this we tested different IMFs in our chemical evolution code, going from those where the variability is contained in the slope of the power-law, assumed to be a function of the metallicity (parametrizations adopted by Scully et al. 1996, Matteucci and Tornambé 1987), to the one by PNJ which predicts also a change in the stellar mass which contributes most to the IMF as a function of time.

A detailed description of the *Two-Infall* model as well as of the PNJ IMF and the hypothesis needed to introduce it in our chemical evolution code are discussed in detail in Chiappini et al. 2000a (hereinafter CMP; see also CMG and PNJ).

2. Results

2.1. THE SOLAR VICINITY

The two-infall chemical evolution model, published by CMG (hereinafter Model A) has been modified to test the IMF proposed by PNJ (hereinafter Model B and C) and the other two IMFs (Models D - Matteucci and Tornambé 1987 and E - Scully et al. 1996). We address the reader to the CMP paper where a detailed description of each one of the IMFs adopted here can be found.

The predictions of models A and B turn out to be very similar concerning the observed properties of the solar vicinity (tables 2 and 3 of CMP). This is due to the fact that the PNJ IMF, when applied to our Galaxy, does not vary much over most of the Galactic lifetime which is a consequence of our assumptions of constant molecular cloud temperature during the Galactic evolution. In fact, if the cloud temperature and density have varied strongly over the history of the Galaxy, the predicted IMF will also vary more and this would certainly worsen the agreement with the observational constraints considered here.

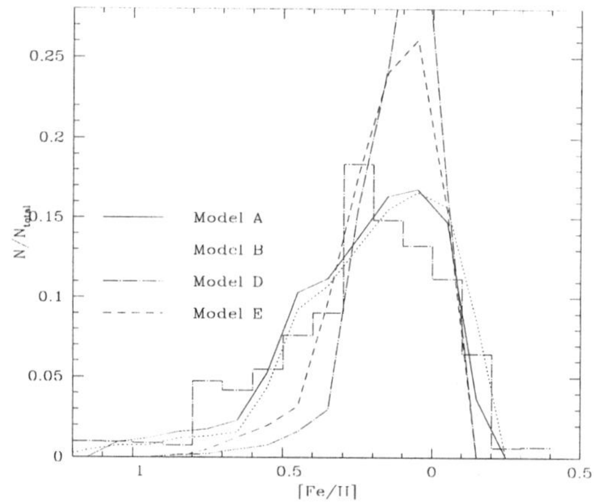

Figure 1. The G-dwarf metallicity distribution predicted by models A, B, D and E. The data are from Rocha-Pinto & Maciel (1996)

On the other hand, the other two IMFs (models D and E) vary substantially over the entire Galactic evolution. As a consequence of that, these IMFs do not give a good agreement with the solar vicinity observational constraints. Figure 1 shows the predicted G-dwarf metallicity distribution for models D and E compared with models A and B. Those models predict too few metal-poor stars and a higher solar metallicity peak than the observed one. The predicted solar abundances are also not in agreement with the observed ones (Table 3 of CMP).

It is worth noting that here we adopted a long timescale for the formation of the solar vicinity (8 Gyrs) and that a better agreement with the G-dwarf observed metallicity distribution could in principle be achieved by adopting these variable IMFs in a closed box model scenario. However, as recently showed by Martinelli and Matteucci (1999), models which can fit the G-dwarf metallicity distribution with a more variable IMF and which adopt a shorter infall timescale for the solar vicinity formation do not give good agreement with the other observed properties. Moreover, in a close box model the predicted gradients along the disk would be essentially flat.

From the comparison with the solar vicinity observational constraints we can confirm the result by Matteucci & Tornambé (1985) that only a constant IMF or an IMF that varied only at early times can be in agreement with the solar vicinity properties.

2.2. THE GALACTIC DISK

The IMF in model B combined with an inside-out cenario for the disk formation predicts a higher number of low mass-stars towards the galactic center, where the metallicity is higher (see Figure 8 of CMP). A direct consequence of this fact is the predicted flatter gradient with respect to the one of Model A. Figure 2a shows the oxygen abundance gradient as predicted by model A (CMG), model B (adopting the PNJ IMF) and model C (same as B but with star formation efficiency increasing with decreasing galactocentric distance). Model B predicts a flat gradient between 6 and 10 kpc, and a small negative one in the inner part of the Galaxy ($R < 6kpc$). Model C, which adopts an increasing star formation efficiency ν towards the galactic center, predicts a steeper gradient inside the solar circle. In this case a the bimodal abundance gradient, steeper in the inner part and flatter outside, similar to the one found by CMG, is recovered. However, as it can be seen in Figure 2b, a model with a higher star formation efficiency in the central parts (Model C) consumes the gas very fast thus reaching the threshold density value very soon. As a consequence, model C predicts a flat gas density radial profile at variance with observations, whereas Model B predicts a gas density distribution very similar to the one predicted by model A and in better agreement with observations.

Figure 2b also shows models D and E predictions for the radial gas profile. From this figure it can be seen that the only two models in good agreement with the observed radial profile are models A and B. However, as shown before, model B predicts even flatter abundance gradients than model A. Models D and E predict a flatter radial gas profile than the observed one but in better agreement with the observations than the one predicted by model C. Moreover, the oxygen abundance gradient predicted by models D and E (see figure 11 of CMP) has a slope which is similar to the one obtained with model A. In any case, those two models do not give predictions in agreeement with the solar vicinity constraints as already mentioned in the previous section.

This is a very important result showing that only model A, with a

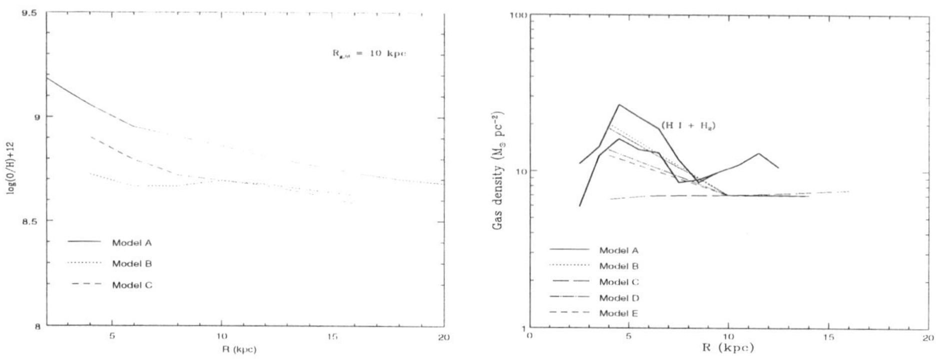

Figure 2. a) Oxygen abundance gradient predicted by models A, B and C; b) Total surface gas density distribution given by Rana (1991) compared with the one predicted by models A, B, C, D and E

constant IMF, is in agreement simultaneously with the solar vicinity and the disk observational constraints. The abundance gradients predicted by model A, although slightly flatter than the observed ones, are steeper in the inner parts of the Galactic disk and also steepen with time, in agreement with the recent results by Maciel & Quireza (1999) (but see Chiappini et al. 2000b for a detailed discussion on abundance gradients).

3. Conclusions

The main results are summarized below:
a) We tested the IMF proposed by PNJ in a model for the chemical evolution of the Milky Way (CMG) and we showed that this IMF gives good agreement with the observed properties of the solar vicinity. However, such an agreement is due to the fact that this IMF when applied to the two-infall model shows a time variation that is important only in the early phases of Galactic evolution. This in turn is due to the simplifying assumptions adopted here, like neglecting the dependence of the IMF on the gas temperature which would produce more sharply varying IMF. In these early phases the IMF is biased towards massive stars.
c) the PNJ IMF combined with the inside-out picture for the thin disk formation predicts a gradient flatter than the one predicted by a model which adopts a constant IMF. This situation cannot be reversed by changing the SFR because in this case the abundance gradient is recovered but the gas density profile is destroyed.
d) Models which adopt IMFs strongly dependent on metallicity, thus simulating a dependence also on the gas temperature, are not in agreement with the most important observational constraints of the solar vicinity and pre-

dict radial gas profiles at variance with observations, therefore they should be rejected.

e) We conclude that a constant IMF and the assumption of a continuous infall onto the Galactic disk is still the best way to explain the observational constraints in the Milky Way including the G-dwarf metallicity distribution. A probable source of the required infall could be the HVCs (see Burton and Braun 1999 for a discussion on this point).

References

1. Beers, T., Sommer-Larsen, J. 1995, ApJS, 96, 175
2. Burton, B., Braun, R.. 1999, Nature, 402., p. 359
3. Carraro, G. 2000 (this volume)
4. Chang, R.X., Hou, J.L., Shu, C.G., Fu C.Q. 1999, A&A, 350, 38
5. Chiappini, C., Matteucci, F., Padoan, P. 2000a, ApJ (in press - CMP)
6. Chiappini, C., Matteucci, F., Romano, D. 2000b (in preparation)
7. Chiappini, C., Matteucci, F. & Gratton, G. 1997, ApJ, 477, 765 (CMG)
8. Chiosi, C., Bressan, A., Portinari, L. & Tantalo, R. 1998, A&A 339, 355
9. Eggen, O. J., Lynden-Bell, D., Sandage, A. R. 1962 ApJ 136, 748
10. Fuhrmann, K. 1998, A&A, 338, 161
11. Gratton, R., Carretta, E., Matteucci, F. & Sneden, C. 1996, in Formation and Evolution of the Galactic Halo - Inside and Out, eds. Morrison H. and Sarajedini, A., ASP Conf. Series 92, 307
12. Gratton, R., Carretta, E., Matteucci, F. & Sneden, C. 1999 (in preparation)
13. Larson, R. B. 1998, MNRAS, 301, 569
14. Hensler, G. 1998, in *Galaxy evolution: Connecting the distant Universe with the local fossil record*, Meudon, ed. M. Spite., Editions Frontières (in press)
15. Maciel, W. J.& Quireza, C. 1999, A&A, 345, 629
16. Martinelli, A. & Matteucci, F. 1999, A&A (in press)
17. Matteucci, F. & François, P. 1989, MNRAS, 239, 885
18. Matteucci, F. & Tornambé, A. 1985, A&A, 142, 13
19. Padoan, P., Nordlund, A. A. & Jones, B. J. T. 1997, MNRAS, 288, 145 (PNJ)
20. Pagel, B. E. J. 1997, Nucleosynthesis and chemical evolution of galaxies, Cambridge University Press
21. Pagel, B. E. J. & Tautvaisiene, G. 1995, MNRAS, 276,505
22. Pardi, C., Ferrini, F., Matteucci, F. 1995, ApJ 444, 207
23. Portinari, L., Chiosi, C., Bressan, A. 1998, A&A, 334, 505
24. Prantzos, N., Silk, J. 1998, ApJ, 507, 229
25. Rocha-Pinto, H. J., Maciel, W. J. 1996, MNRAS 279, 447
26. Roche, N., Ratnatunga, K., Griffiths, R. E., Im, M. & Naim, A. 1998, MNRAS 293, 157
27. Scalo, J. M. 1998, 38th Hertmonceux Conference, eds. G. Gilmore and D. Howell, ASP Conf. Series, Vol. 142, p.201
28. Scalo, J. M., Vazquez-Semadeni, E., Chappell, D. & Passot, T. 1998, ApJ, 504, 835
29. Scully, S., Cassé, M., Olive, K. A. & Vangioni-Flam, E. 1996, ApJ, 462, 960

Speaker: C. CHIAPPINI

A. MAEDER: Observationally there is no evidence of differences of IMF with metallicity. However, from the study of populations of massive stars in galaxies, it is not excluded that there is a positive correlation between a high SFR (as in starbursts) and a top-heavy IMF.

C. CHIAPPINI: Yes, I agree. My conclusions on the IMF are only relative to the Milky Way, in other words we can exclude a variable IMF for our galaxy, but not variations in the IMF in different galaxies (e.g. elliptical or starburst galaxies).

S. SMARTT: Is there any observational evidence for an IMF significantly different to normal? It would appear to me that all the observational data indicate that the IMF (both low and high mass) is incredibly universal. Hence, arbitrarily varying the IMF maybe problematic.

C. CHIAPPINI: There is no conclusive observational evidence for IMF variations (e.g. Scalo, 1998), although a top heavy IMF is often suggested in order to explain some observations related to giant elliptical galaxies (e.g. Faber et al., 1992; Chiosi et al., 1998) and it is also expected. On theoretical grounds there are various suggestions of a variable IMF (e.g. Larson, 1998; Padoan et al, 1997). In this work we tested the Padoan et al. IMF which was based on numerical experiments of highly supersonic random flows. In this sense we did not arbitrarily varied the IMF, but we tested a theoretical one.

M. TOSI's Comment: Comments on previous questions by Chieffi and Smartt.

1) Infall was introduced in models with the aim of diluting the metals. Several years ago Francesca and I have independently shown that the dilution effect is mantained if the infall metallicity is $Z_{inf} = 0.2Z_{\odot}$ (which is the observed infall abundance).

2) It is not an observational direct evidence, but an indirect one: in IZW18, the metal poorest blue compact galaxy known so far, the Colour-Magnitude diagrams of its stellar populations can be reproduced only by assuming a flat IMF (i.e. with many more massive stars than in other galaxies).

METALLICITY EFFECT ON TYPE IA SUPERNOVAE AND GALACTIC AND COSMIC CHEMICAL EVOLUTION

C. KOBAYASHI[1], T. TSUJIMOTO[2], AND K. NOMOTO[1]

[1] *Department of Astronomy, School of Science, University of Tokyo, Bunkyo-ku, Tokyo 113-0033, Japan*

[2] *National Astronomical Observatory, Mitaka, Tokyo 181-8588, Japan*

1. Introduction

There exist two distinct types of supernova explosions: One is Type I-I supernovae (SNe II), which are the core collapse-induced explosions of short-lived massive stars ($\gtrsim 8M_\odot$) and produce more O and Mg relative to Fe (i.e., [O/Fe] > 0), and the other is Type Ia supernovae (SNe Ia), which are the thermonuclear explosions of accreting white dwarfs (WDs) in close binaries and produce mostly Fe and little O. The exact companion stars of the WDs have not been identified but must be relatively long-lived low-mass stars.

The role of these two types of supernovae in the chemical evolution of galaxies can be seen in the [O/Fe]-[Fe/H] relation: Metal-poor stars with [Fe/H] $\lesssim -1$ have [O/Fe] ~ 0.45 on the average, while disk stars with [Fe/H] $\gtrsim -1$ show a decrease in [O/Fe] with increasing metallicity. Such an evolutionary change in [O/Fe] against [Fe/H] has been explained with the early heavy-element production by SNe II and the delayed enrichment of Fe by SNe Ia. Conversely, chemical evolution models can constrain the nature of progenitor systems of SNe Ia, and the lifetime of SN Ia progenitors was estimated as long as $0.5 - 3$ Gyr ([?]).

Here, we construct more realistic SN Ia model by introducing the evolutionary scenario of SNe Ia progenitors (section 2), and test the model with the chemical evolution in the solar neighborhood (section 3). As a future test of the SN Ia model, we make a prediction for the cosmic supernova rate history (section 4).

F. Matteucci and F. Giovannelli (eds.), The Evolution of the Milky Way, 573–578.

2. Type Ia Supernova Progenitor System

The progenitors of the majority of SNe Ia are most likely the Ch mass WDs, although the sub-Ch mass models might correspond to some peculiar subluminous SNe Ia. For the evolution of accreting WDs toward the Ch mass, two scenarios have been proposed: One is a double-degenerate (DD) scenario, i.e., merging of double C+O WDs with a combined mass surpassing the Ch mass limit, and the other is a single-degenerate (SD) scenario, i.e., accretion of hydrogen-rich matter via mass transfer from a binary companion. The issue of DD versus SD is still debated, although theoretical modeling has indicated that the merging of WDs does not make typical SNe Ia (e.g., [?]).

Our SD scenario has two progenitor systems: One is a red-giant (RG) companion with the initial mass of $M_{\rm RG,0} \sim 1 M_\odot$ and an orbital period of tens to hundreds days ([?]; [?]; [?]). The other is a near main-sequence (MS) companion with an initial mass of $M_{\rm MS,0} \sim 2-3 M_\odot$ and a period of several tenths of a day to several days ([?]; [?]). In our SD scenario, a C+O white dwarf accretes H-rich meterials from the companion star and grows its mass to the Ch mass to explode as an SN Ia. Optically thick winds from the mass accreting WD play an essential role in stabilizing the mass transfer and escaping from forming a common envelope. The optically thick winds are driven by a strong peak of OPAL opacity at $\log T({\rm K}) \sim 5.2$. Since the peak is due to iron lines, the optically thick winds depend strongly on the iron abundance ([?]).

The metallicity effect on SNe Ia is clearly demonstrated by the size of regions to produce SNe Ia in the diagram of the initial orbital period versus initial mass of the companion star (see Figure 2 of [?]). The SN Ia regions are much smaller for lower metallicity because the wind becomes weaker. The wind velocity depends on not only iron abundance [Fe/H] but also WD mass $M_{\rm WD}$ in the sense that the wind velocity is higher for larger $M_{\rm WD}$ (see Figure 1 of [?]). Since only WD with the initial mass of $M_{\rm WD,0} \leq 1.2 M_\odot$ can produce an SN Ia and there are only few WDs with the mass of $M_{\rm WD,0} \gtrsim 1.1 M_\odot$, the mass of the most massive WD which can produce SNe Ia is considered to be $\sim 1 M_\odot$. In order that WDs with $1 M_\odot$ produce SNe Ia, [Fe/H] should be larger than ~ -1.1. Thus, we assume that no SN Ia occur at [Fe/H] ≤ -1.1 in our chemical evolution model. The initial mass ranges of the companion stars are $0.9 M_\odot \lesssim M_{\rm RG,0} \lesssim 1.5 M_\odot$ for the WD+RG system and $1.8 M_\odot \lesssim M_{\rm MS,0} \lesssim 2.6 M_\odot$ for the WD+MS system. The distribution function of companions is approximated by a power-law mass spectrum with a slope $x = 0.35$. The fraction of primary stars of $3-8 M_\odot$ which eventually produce SNe Ia is set to be $b = 0.05$ for the MS+WD system and 0.02 for the RG+WD system, which are adjusted to reproduce the chemical evolution in the solar neighborhood.

3. The Chemical Evolution in the Solar Neighborhood

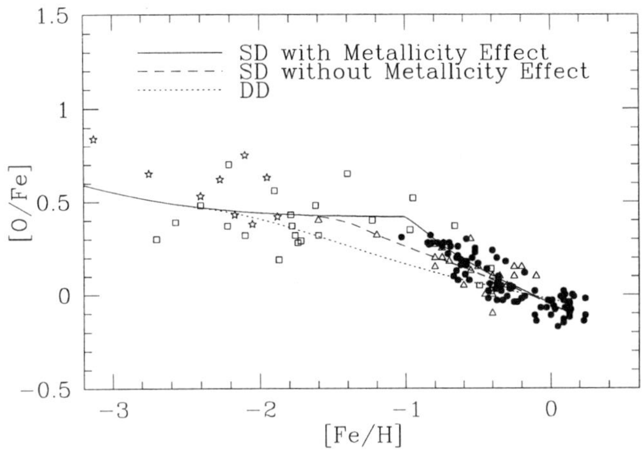

Figure 1. The evolutionary change in [O/Fe] against [Fe/H] for three SN Ia models. The dotted line is for the DD scenario, and the other lines are for our SD scenario with (solid line) and without (dashed line) the metallicity effect on SNe Ia. See [?] for observational data sources.

We use the chemical evolution model that allows the infall of material from outside the disk region. The infall rate, the star formation rate (SFR), and the initial mass function are given by [?], and the nucleosynthesis yields of SNe Ia and II are taken from [?]. For the SD scenario, the lifetime of SNe Ia is determined from the main-sequence lifetime of the companion star. For the DD scenario, we adopt the distribution function of the lifetime of SNe Ia by [?], majority of which is $\sim 0.1 - 0.3$ Gyr.

Figure 1 shows the evolutionary change in [O/Fe] for three SN Ia models. The dotted line is for the DD scenario. The other lines are for our SD scenario with (solid line) and without (dashed line) the metallicity effect on SNe Ia. i) In the DD scenario the lifetime of the majority of SNe Ia is shorter than 0.3 Gyr. Then the decrease in [O/Fe] starts at [Fe/H] ~ -2, which is too early compared with the observed decrease in [O/Fe] starting at [Fe/H] ~ -1. ii) For the SD scenario with no metallicity effect, the companion star with $M \sim 2.6 M_{\odot}$ evolves off the main-sequence to give rise to SNe Ia at the age of ~ 0.5 Gyr. The resultant decrease in [O/Fe] starts too early to be compatible with the observations. iii) For the metallicity dependent SD scenario, SNe Ia occur at [Fe/H] $\gtrsim -1$, which naturally reproduce the observed break in [O/Fe] at [Fe/H] ~ -1.

4. Cosmic Supernovae Rate

Galaxies that are responsible for the cosmic SFR have different timescales for the heavy-element enrichment, and the occurrence of supernovae depends on the metallicity therein. Therefore we should calculate the cosmic supernova rate by summing up the supernova rates in spirals (S0a-Sa, Sab-Sb, Sbc-Sc, and Scd-Sd) and ellipticals with the ratio of the relative mass contribution. The relative mass contribution is obtained from the observed relative luminosity proportion and the calculated mass to light ratio in B-band ([?]). The photometric evolution is calculated with the spectral synthesis population database taken from [?]. We adopt $H_0 = 50$ km s^{-1} Mpc^{-1}, $\Omega_0 = 0.2$, $\lambda_0 = 0$, and the galactic age of $t_{\rm age} = 15$ Gyr.

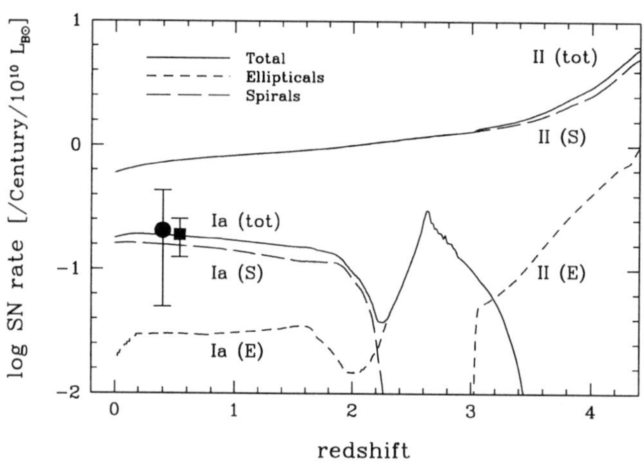

Figure 2. The cosmic supernova rates (solid line) as the composite of ellipticals (short-dashed line) and spirals (long-dashed line). The upper three lines show SN II rates, the lower three lines show SN Ia rates. Observational data sources: circle, [?]; square, [?].

First, we make a prediction of the cosmic supernova rates by using the galaxy models which are constructed to meet the observational constraints of cluster galaxies. We assume that elliptical galaxies are formed by a single star burst and stop the star formation at $t \sim 1$ Gyr due to the supernova-driven galactic wind, while spiral galaxies are formed by a relatively continuous star formation. The infall rate and the SFR are given by [?]. These models can well reproduce the present gas fraction, color, supernova rates, and the color evolution in cluster ellipticals.

Figure 2 shows the cosmic supernova rates in cluster galaxies. The SN Ia rate in spirals drops at $z \sim 2$ because of the low-metallicity inhibition of SNe Ia. We can precisely test the metallicity effect by finding this drop of the SN Ia in spirals, if high-redshift SNe Ia at $z \gtrsim 1.5$ are observed with the detection of their host galaxies by using the Next Generation Space Telescope. In ellipticals, the chemical enrichment takes place so early that the metallicity is large enough to produce SNe Ia at $z \gtrsim 2$. The two peaks of SN Ia rates at $z \sim 2.5$ and $z \sim 1.5$ come from the MS+WD and RG+WD binary systems, respectively. The SN Ia rate in ellipticals decreases at $z \sim 3$, which is determined from the shortest lifetime of SNe Ia of ~ 0.5 Gyr. Thus, the total SN Ia rate decrease at the same redshift as ellipticals, i.e., $z \sim 3$.

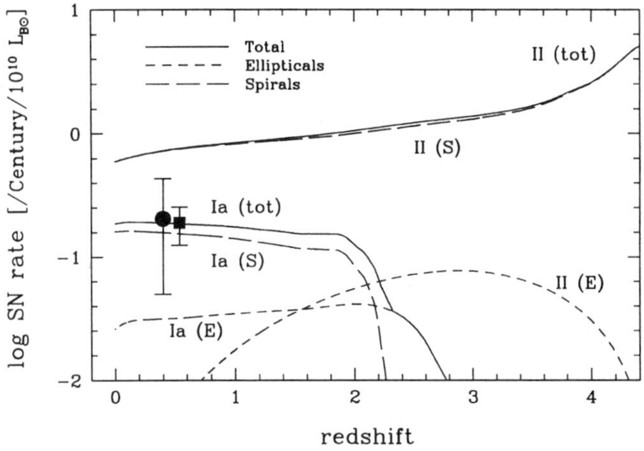

Figure 3. The same as Figure 2, but for the formation epochs of ellipticals span at $1 \lesssim z \lesssim 4$, which is corresponding to field ellipticals.

We also predict the cosmic supernova rates for the case that the formation of ellipticals in fields are protracted, that is, the formation epoch of ellipticals spans over the wide range of redshifts. The observed spectra of ellipticals in the Hubble Deep Field suggest that the formation of ellipticals is protracted in fields, that is, the formation epochs of ellipticals span in the wide range of redshifts ([?]). To predict the cosmic star formation history in field, we then combine the SFRs of ellipticals and spirals by assuming the distribution function of the formation epoch $z_{\rm f}$ as $\propto \exp(-((z-2)/2)^2)$ in the range of $0 \leq z \leq 5$.

Figure 3 shows the cosmic supernova rates in field galaxies. As well as Figure 2, the SN Ia rate in spirals drops at $z \sim 2$. Although the timescale of star formation each elliptical is as short as in Figure 2, the star formation takes place more gradually if SFR is averaged over the whole of the field ellipticals. The averaged SN Ia rate decreases at $z \sim 2.5$ acting in contact with the averaged SFR. Thus, the total SN Ia rate decreases gradually from $z \sim 2$ to ~ 3.

The rate of SNe II in ellipticals evolves following the SFR without time delay. Thus, SNe II may be observed in low-redshift ellipticals. The difference of SN II and Ia rates between cluster and field ellipticals reflects the difference in the star formation histories in different environments.

5. Conclusions and Discussion

We introduce a metallicity dependence of the SN Ia rate in the Galactic and cosmic chemical evolution models. In our scenario involving a strong wind from WDs, few SNe Ia occur at $[Fe/H] \lesssim -1$. Our model successfully reproduces the observed chemical evolution in the solar neighborhood. We make the following predictions that can test this metallicity effect. 1) SNe Ia are not found in the low iron abundance environments such as dwarf galaxies and the outskirts of spirals. 2) The SN Ia rate in spirals drops at $z \sim 2$ due to the low-iron abundance, while SNe Ia can be found at $z \gtrsim 2$ in cluster ellipticals, where the timescale of metal enrichment is sufficiently short. 3) If the formation of field ellipticals is protracted to lower redshifts, the SNe Ia rate decreases from $z \sim 2$ in the field ellipticals.

Acknowledgements

This work has been supported in part by the grant-in-Aid for Scientific Research (08640336) and COE research (07CE2002). C.K. thanks to the Japan Society for Promotion of Science for a financial support. We would like to thank I. Hachisu and M. Kato for providing us with their new results, and T. Kodama for providing us with the database of simple stellar population spectra.

References

1. Yoshii, Y., Tsujimoto, T., & Nomoto, K. 1996, *Astrophys. J.*, 462, 266
2. Nomoto, K., Yamaoka, H., Shigeyama, T., Kumagai, S., & Tsujimoto, T. 1994, in Supernovae, Les Houches Session LIV, ed. S.A. Bludman et al. (Amsterdam: North-Holland), 199
3. Hachisu, I., Kato, M., & Nomoto, K. 1996, *Astrophys. J.*, 470, L97
4. Hachisu, I., Kato, M., & Nomoto, K. 1999, *Astrophys. J.*, 522, 487
5. Hachisu, I., Kato, M., Nomoto, K., & Umeda, H. 1999, *Astrophys. J.*, 519, 314
6. Li, X. -D., & van den Heuvel, E. P. J. 1997, *Astron. Astrophys.*, 322, L9
7. Kobayashi, C., Tsujimoto, T., Nomoto, K., Hachisu, I, & Kato, M. 1998, *Astrophys. J.*, 503, L155
8. Tsujimoto, T. et al. 1995, *Mon. Not. R. Astr. Soc.*, 277, 945
9. Tutukov, A. V., & Yungelson, L. R. 1994, *Mon. Not. R. Astr. Soc.*, 268, 871
10. Kobayashi, C., Tsujimoto, T., & Nomoto, K. 1999, *Astrophys. J.*, submitted (astro-ph/9908005)
11. Kodama, T., & Arimoto, N., 1997, A&A 320, 41
12. Pain, R., et al. 1996, *Astrophys. J.*, 473, 356
13. Pain, R. 1999, talk at the Type Ia Supernova workshop (Aspen Center for Physics)
14. Franceschini, A., Silva, L., Fasano, G., Granato, G. L., Bressan, A., Arnouts, S. & Danese, L. 1998, *Astrophys. J.*, 506, 600

ON THE ABUNDANCE GRADIENTS IN THE GALACTIC DISC

L. PORTINARI
Dipartimento di Astronomia, Università di Padova
Vicolo dell'Osservatorio 5, I-35122 Padova (PD), Italy

1. Introduction

Negative radial gradients of metallicity in spiral discs have long been established. For the disc of the Milky Way, recent results on bright blue stars (Smartt & Rolleston 1997, Gummersbach et al. 1998) show an oxygen gradient around –0.07 dex/kpc, in good agreement with the one derived from nebular studies (Shaver et al. 1983, Fich & Silkey 1991, Vilchez & Esteban 1996, Simpson et al. 1995, Afflerbach et al. 1997, Rudolph et al. 1997, Maciel & Köppen 1994). The apparent discrepancy of the past years is thus overcome; see also Smartt (2000, this book).

In the light of these recent results, we studied the metallicity gradient of the Galactic disc with a chemical evolution model including different star formation laws and radial gas flows (Portinari & Chiosi 1999, 2000).

2. The star formation process and the inside–out scenario

For galactic discs, a variety of formulations for star formation (SF) is suggested in literature, following three main lines of thought.

1. The simplest SF law is that by Schmidt (1959), with a star formation rate (SFR) proportional to the amount of available gas:

$$SFR \propto \sigma_g^{\kappa}$$

   where σ_g is the surface gas density. In our models S15 we adopt this SF law, with an empirical exponent $\kappa = 1.5$ (Kennicutt 1998).
2. A SF process triggered by spiral density waves through shocks in spiral arms and/or by gravitational instabilities in the gaseous disc is described by a SF efficiency inversely proportional to the galactocen-

F. Matteucci and F. Giovannelli (eds.), The Evolution of the Milky Way, 579–585.

tric radius (Oort 1974, Wyse & Silk 1989, Wang & Silk 1994):

$$SFR \propto \left(\frac{R}{R_\odot}\right)^{-1} \sigma_g^\kappa$$

The favoured empirical exponent in this case is $\kappa = 1.0$ (Kennicutt 1998), which we adopt in our models O10.

3. A self–regulating SF induced by the gravitational settling of gas onto the disc plane, balanced by the energy feed–back from massive stars is described by a SF efficiency exponentially decreasing with radius (Talbot & Arnett 1975, Dopita & Ryder 1994):

$$SFR \propto \sigma_T(R)^\alpha \, \sigma_g(R)^\beta$$

where σ_T is the total surface density, which follows an exponential radial profile. In our models DR we adopt the prescription by Dopita & Ryder (1994), with $\alpha = 1/3$ and $\beta = 5/3$.

All of these theoretical SF laws have some observational support (Kennicutt 1998, Dopita & Ryder 1994).

With the aim of drawing fairly general conclusions, independent of the particular choice for the SF recipe, we apply in turn all the various options for the SF law described above, and we analyse their effects on the predicted radial profiles of gas and metallicity for the Galactic disc (see Portinari & Chiosi 1999).

Our models S15a, O10a and DRa are open models with the same, uniform infall time-scale, where the different SF laws are adopted, so as to compare the effects of the sole SF prescriptions. No SF law turns out to be able, by itself, to reproduce the observed metallicity gradient all over the disc. Models S15a and O10a predict a flat abundance gradient, while model DRa predicts the right slope in the outer parts, but it flattens out in the inner regions (Fig. 1).

Some additional "dynamical" assumption is needed to reproduce the overall gradient, a typical possibility being that the formation of the Galactic disc proceeded inside–out, i.e. that the accretion time-scale is shorter in the inner than in the outer regions (e.g. Matteucci & François 1989). In models S15b, O10b, DRb we impose this scheme in order to reproduce the overall gradient (Fig. 1). However, even with this assumption, models adopting a Schmidt or an Oort–type law hardly reconcile with the observed metallicity gradient: to obtain a good agreement with the data, in models S15b and O10b we must impose an extreme radial variation of the infall time-scale, leading to accretion time-scales in the outer regions far longer than a Hubble time, which is unacceptable. Only the Dopita & Ryder SF law can give a good fit to the data, combined with a plausible inside–out scheme (model DRb).

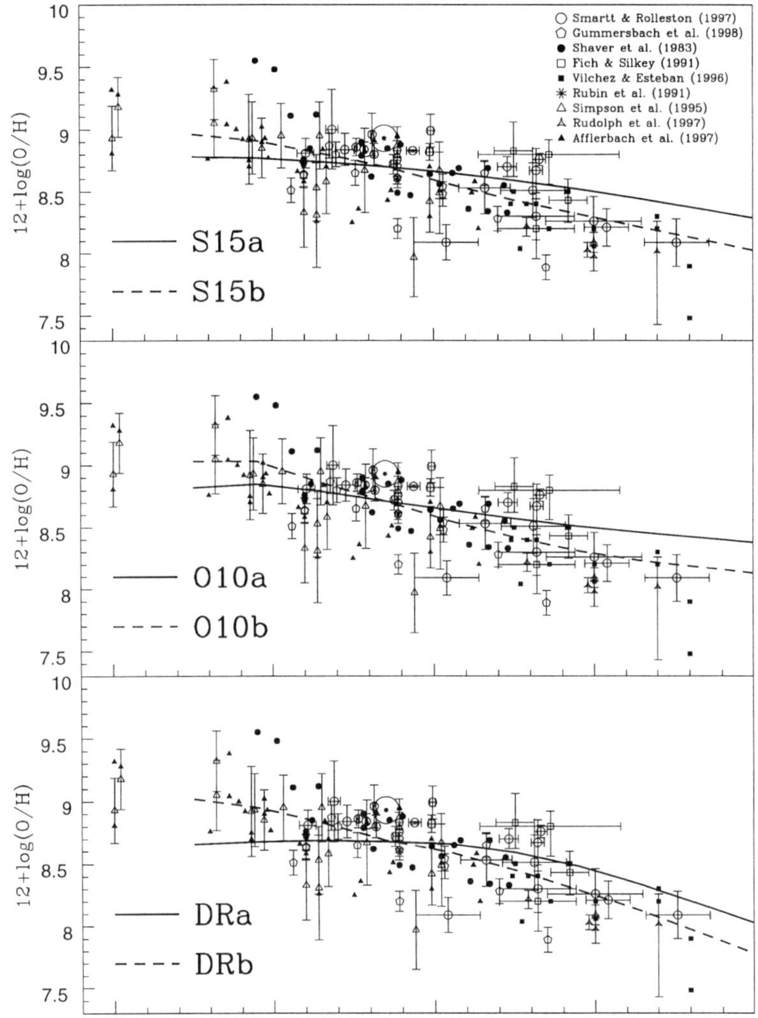

Figure 1. Predicted radial oxygen gradient from the various models (see text) compared to observational data.

Besides, no matter the adopted SF law, all of these models share a common problem: they cannot reproduce, at the same time, both the metallicity gradient and the radial gas profile, in particular the peak corresponding to the molecular ring around 4–6 kpc (Fig. 2). This latter feature is likely to be a consequence of the dynamical influence of the Galactic bar, which causes gas to drift and accumulate at suitable resonance points. To describe these effects, radial gas flows must be included in the model.

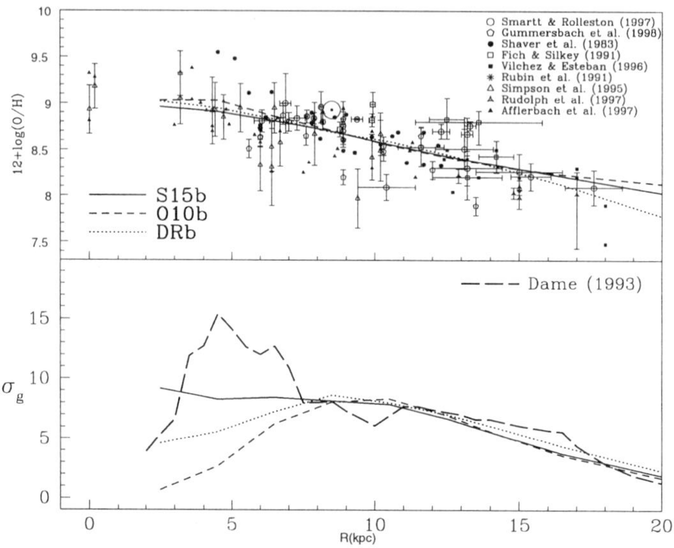

Figure 2. Chemical models reproducing the observed metallicity gradient do not reproduce the observed gas profile.

3. The role of radial gas flows and of the Galactic bar

We developed a chemical model with radial gas flows following an arbitrary velocity profile, which can describe both inflows and outflows along the disc (Portinari & Chiosi 2000). Radial inflows provide an alternative mechanism to the inside–out scenario to explain the metallicity gradient. In fact, models with radial inflows can reproduce the gradient with no need to vary the infall time-scale; at the same time they can provide a high gas fraction in the inner regions, at odds with static models (Fig. 3). In general, it turns out that even low radial flow velocities (≤ 1 km/sec) have sizeable effects upon model predictions concerning the metallicity gradient and moreover concerning the gas distribution (Portinari & Chiosi 2000).

Velocity profiles for radial drifts simulating the dynamical influence of the Galactic bar help to reproduce the peak in the gas distribution around 4 kpc. From the relevant literature, two representative scenarios for the structure of the bar can be outlined. **Case A**: slow radial inflows all over the disc pile up gas inward, down to $r = 3.5 - 4$ kpc; here, the bar CR radius is found and the gas is quickly swept inward toward an ILR, which causes the drop in the gas profile at 3.5 kpc (e.g. Fig. 3). **Case B**: the bar CR is around 2.5 kpc and its OLR around 4.5 kpc, so that all the gas from $r = 2.5$ kpc tends to outflow and pile up around the resonance at 4.5 kpc.

Our models are just qualitative: they do not describe the detailed dynamical formation and evolution of the bar and of the related gas flows

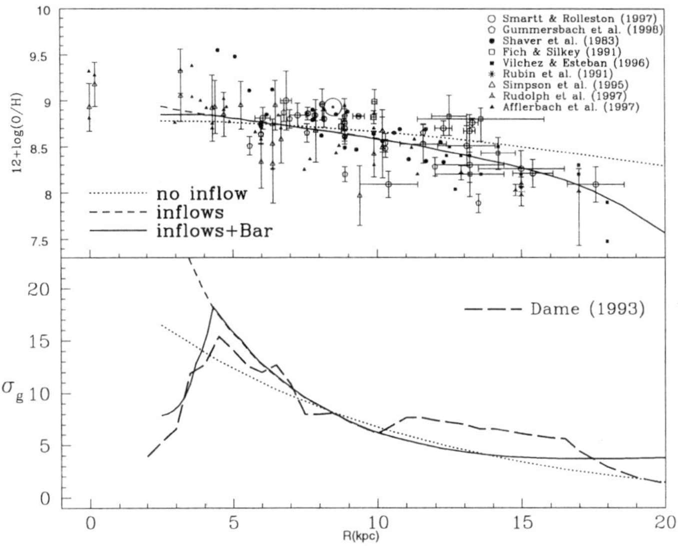

Figure 3. *Dotted line*: the static model S15a without a varying infall time-scale do not reproduce the metallicity gradient. *Dashed line*: models with radial inflows can reproduce the gradient while maintaining a high gas density in the inner regions. *Solid line*: models with inflow velocity profiles simulating the dynamical influence of the bar can reproduce the gas peak of the molecular ring For further details, see Portinari & Chiosi (2000).

to form the molecular ring; they simply assume some flow velocity profile mimicking the influence of the bar. Nevertheless, they provide interesting general indications. First of all, only introducing the effects of the bar it is finally possible to reproduce the radial gas profile properly. And in all models radial inflows all over the disc seem necessary to replenish the inner regions with enough gas, so that the observed molecular ring can form under the influence of the bar. Therefore, models with radial inflows seem to be favoured.

4. Conclusions

All the various SF laws suggested for spiral discs need to be combined with some further "dynamical assumption" to reproduce the observed metallicity gradient. Such assumption can be the inside–out formation scenario and/or radial gas inflows. In particular, a Schmidt law with $\kappa = 1.5$ or an Oort-type law with $\kappa = 1.0$ (Kennicutt 1998) are excluded in the framework of simple static models, even in the inside–out scenario; if these SF laws are assumed, radial flows are necessary to reproduce the metallicity gradient.

Moreover, radial flows are needed to reproduce the detailed gas distribution over the disc, since static models cannot account at the same time both for the metallicity gradient and for the molecular ring at 4–6 kpc. This latter feature is likely due to the dynamical influence of the Galactic bar, and in fact radial gas flows mimicking these effects allow to reproduce the molecular ring together with the metallicity gradient.

The relationship between the Galactic bar and the observed metallicity gradient is still an open riddle. In fact, barred galaxies are known to have systematically shallower gradients, probably due to radial mixing induced by bars (e.g. Martin & Roy 1994). Our Galaxy hosts a bar, and yet its gradient is rather typical for a normal spiral. It has been suggested that this apparent discrepancy can be solved if the Galactic bar is very young (<1 Gyr, Gummersbach et al. 1998), so that only little radial mixing has taken place so far; but this is in contrast with other age estimates (e.g. Sevenster 1997, 1999). An alternative explanation might be that the bar is just too "small" or "weak", so that, independently of its age, we do not expect it to influence the gradient out to the outskirts of the disc, but only in the inner regions (say, within its OLR at 4–6 kpc). To investigate this issue, detailed velocity profiles as predicted from full gas–dynamical models of the Galactic bar must be adopted in the chemical models; work is in progress in this direction.

Acknowledgments This study is financed by the Italian MURST, contract cofin. 9802192401.

References

1. Afflerbach A., Churchwell E., Werner M.W., 1997, ApJ 478, 190
2. Dame T.M., 1993, in *Back to the Galaxy*, S. Holt, F. Verter (eds.), p. 267
3. Dopita M.A., Ryder S.D., 1994, ApJ 430, 163
4. Fich M., Silkey M., 1991, ApJ 366, 107
5. Gummersbach C.A., Kaufer A., Schäfer D.R., et al., 1998, A&A 338, 881
6. Kennicutt R.C., 1998, ApJ 498, 541
7. Maciel W.J., Köppen J., 1994, A&A 282, 436
8. Martin P., Roy J.-R., 1994, ApJ 424, 599
9. Matteucci F., François P., 1989, MNRAS 239, 885
10. Oort J.H., 1974, IAU Symp. 58, 375
11. Portinari L., Chiosi C., 1999, A&A 350,827
12. Portinari L., Chiosi C., 2000, A&A submitted
13. Rudolph A.L., Simpson J.P., Haas M.R., et al., 1997, ApJ 489, 94
14. Schmidt M., 1959, ApJ 129, 243
15. Sevenster M., 1997, PhD thesis, Leiden University (the Netherlands)
16. Sevenster M., 1999, MNRAS in press (astro-ph 9907319)
17. Shaver P.A., McGee R.X., Newton M.P., et al., 1983, MNRAS 204, 53
18. Simpson J.P., Colgan S.W.J., Rubin R.H., et al., 1995, ApJ 444, 721
19. Smartt S.J., Rolleston W.R.J., 1997, ApJ 481, L47
20. Talbot R.J., Arnett D.W., 1975, ApJ 197, 551
21. Vilchez J.M., Esteban C., 1996, MNRAS 280, 720
22. Wang B., Silk J., 1994, ApJ 427, 759
23. Wyse R.F.G., Silk J., 1989, ApJ 339, 700

Speaker: L. PORTINARI

M. TOSI: Have you also tried to reproduce gradients and gas distributions with SFR $\propto {\sigma_g}^{1.5} R^{-1}$ or SFR $\propto {\sigma_g}^{1.5} {\sigma_{tot}}^n$ as done by Prantzos and Matteucci, respectively? Do you succeed to reproduce both distributions as these authors do?

L. PORTINARI: The Talbot and Arnett SF law (SFR $\propto {\sigma_T}^{k-1} {\sigma_g}^k$, $k = 1.5$) behaves very similar to the Dopita and Ryder law, as expected since they stem from the same basic physical process of self-regulating SF. A SF law of the Oort-type with exponent 1.5 (SFR $\propto R^{-1} {\sigma_g}^{1.5}$) has an intermediate behaviour between the $k = 1.0$ case and the Dopita and Ryder's law. It predicts a gradient of -0.04 dex/Kpc, but the observed slope of -0.07 can be recovered combined with an inside-out formation scenario. The required variation of $\tau(r)$ is not too extreme and well acceptable.

FORMATION OF DISK GALAXIES: ON THE ANGULAR MOMENTUM PROBLEM, THE TULLY-FISHER RELATION AND MAGNETOHYDRODYNAMICS

JESPER SOMMER-LARSEN
Theoretical Astrophysics Center
Juliane Maries Vej
DK-2100 Copenhagen Ø, Denmark

Abstract. Two ways of possibly solving the angular momentum problem plaguing cold dark matter (CDM) *ab initio* simulations of disk galaxy formation are discussed: 1) Stellar feedback processes and 2) Warm dark matter (WDM) rather than CDM.

In relation to the chemical evolution of disk galaxies our simulations indicate that in case 1) the first generation of *disk* stars formed in disk galaxies like the Milky Way should have an abundance about two dex below solar, in fairly good agreement with the lowest observed abundance of the metal-weak tail of the Galactic thick disk. For the second case no such statements can be made without further assumptions about the star-formation history of the galaxies.

We find that the I-band Tully-Fisher relation can be matched by WDM disk galaxy formation simulations provided $(M/L_I) \sim 0.8$ for disk galaxies, which Sommer-Larsen & Dolgov (1999) argue is a reasonable value.

Finally it is discussed how the magnetic field strengths observed in galactic disks can be obtained through disk galaxy *formation*, as an alternative to the conventional dynamo hypothesis.

1. Introduction

The formation of galactic disks is one of the most important unsolved problems in astrophysics today. In the currently favored hierarchical clustering framework, disks form in the potential wells of dark matter halos as the baryonic material cools and collapses dissipatively. Fall & Efstathiou (1980) have shown that disks formed in this way can be expected to possess the

F. Matteucci and F. Giovannelli (eds.), The Evolution of the Milky Way, 587–597.

observed amount of angular momentum (and therefore the observed spatial extent for a given mass and profile shape), but only under the condition that the infalling gas retain most of its original angular momentum.

However, numerical simulations of this collapse scenario in the cold dark matter (CDM) cosmological context (e.g., Navarro & Benz 1991, Navarro & White 1994, Navarro, Frenk, & White 1995) have so far consistently indicated that when only cooling processes are included the infalling gas loses too much angular momentum (by over an order of magnitude) and the resulting disks are accordingly much smaller than required by the observations. This discrepancy is known as the *angular momentum problem* of disk galaxy formation. It arises from the combination of the following two facts: a) In the CDM scenario the magnitude of linear density fluctuations $\sigma(M) = \langle(\delta M/M)^2\rangle^{1/2}$ increases steadily with decreasing mass scale M leading to the formation of non-linear, virialized structures at increasingly early epochs with decreasing mass i.e. the hierarchical "bottom-up" scenario. b) Gas cooling is very efficient at early times due to gas densities being generally higher at high redshift as well as the rate of inverse Compton cooling also increasing very rapidly with redshift. a) and b) together lead to rapid condensation of small, dense gas clouds, which subsequently lose energy and (orbital) angular momentum by dynamical friction against the surrounding dark matter halo before they eventually merge to form the central disk. A mechanism is therefore needed that prevents, or at least delays, the collapse of protogalactic gas clouds and allows the gas to preserve a larger fraction of its angular momentum as it settles into the disk. Two such possible solutions are discussed in section 2.

In section 3 we present some new results from our WDM disk galaxy formation simulations on the Tully-Fisher relation and in section 4 we discuss how the magnetic field strengths of a few μG observed in galactic disks can be obtained via disk galaxy *formation*, as an alternative to disk dynamo amplification.

2. Towards solving the angular momentum problem

Two ways of possibly solving the angular momentum problem have recently been discussed in the literature: a) by invoking the effects of stellar feedback processes from either single, more or less uniformly distributed stars or starbursts and b) by assuming that the dark matter is "warm" rather than cold. Both options lead to the suppression of the formation of early, small and dense gas clouds, for a) because the small gas clouds may be disrupted due to the energetic feedback of primarily type II super-nova explosions and for b) simply because fewer of the small and dense gas clouds form in the first place for WDM free-streaming masses $M_{f,\mathrm{WDM}} \sim 10^{10}$-$10^{11} M_\odot$.

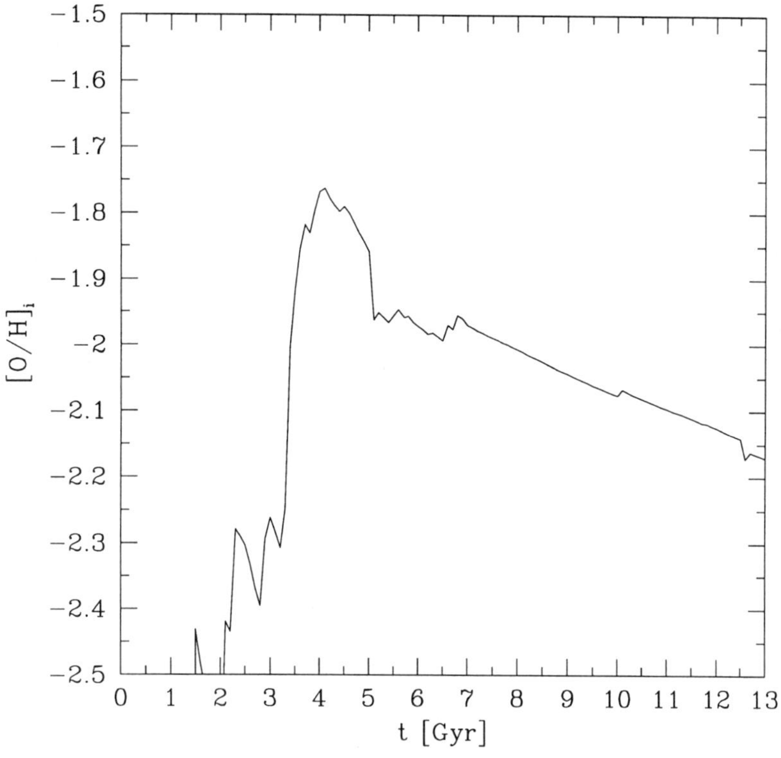

Figure 1. The initial disk oxygen abundance resulting from infall of a mixture of enriched and unenriched gas onto the disk of a forming, Milky Way sized model galaxy. This abundance should be representative of the oxygen abundance of the first generation of *disk* stars formed and is fairly consistent with the lowest found observationally for the metal-weak tail of the Galactic thick disk.

2.1. STELLAR FEEDBACK PROCESSES

Sommer-Larsen et al. (1999) showed that the feedback caused by a putative, early epoch of more or less uniformly distributed population III star formation was not sufficient to solve the angular momentum problem. Based on test simulations they showed, however, that effects of feedback from star-bursts in small and dense protogalactic clouds might do that. Preliminary results of more sophisticated simulations incorporating stellar feedback processes in detail indicate that this is at least partly the case. Considerable fine-tuning seems to be required, however: About 2-3% of the gas in the proto-galactic region of a forming disk galaxy should be turned into stars. If less stars are formed the feedback is not strong enough to cure the angular momentum problem and, vice versa, if more stars are formed during this fairly early phase of star-formation, the energetic feedback causes the formation of the main disks and thereby the bulk of the stars to be delayed too much compared to the observed star-formation history of the Universe.

This requirement of fine-tuning is advantageous, however, in relation to the early chemical evolution of disk galaxies, since the early star-formation histories of the galaxies are then well constrained. Furthermore, as it is possible to track the elements produced and ejected by (primarily) type II supernovae in the star-bursts one can determine the fraction of these elements, which ultimately settle on the forming disk and hence determine the rate and metallicity of the gas falling onto the disk. In Figure 1 we show the time evolution of the oxygen abundance in a forming disk as a result of infall of a mixture of enriched and unenriched gas (neglecting the contribution of ejecta from stars formed subsequently in the disk). We have assumed a Salpeter IMF with $M_{low} = 0.1M_{\odot}$ and $M_{up} = 60M_{\odot}$ and that a typical type II supernova ejects $\sim 2M_{\odot}$ of oxygen. This abundance can be regarded as the initial abundance of the disk, its value depending on when star-formation subsequently commenced in the disk (note that such two-epoch star-formation models have been advocated by, e.g., Chiappini, Matteucci & Gratton 1997). As can be seen from the figure this initial disk abundance is of the order $[O/H] \sim -2$. This is similar to the lowest abundance of the low-metallicity tail of the Galactic thick disk – see Beers & Sommer-Larsen (1995).

2.2. WARM DARK MATTER

Another, more radical way of solving the angular momentum problem is to abandon CDM altogether and assume instead that dark matter is "warm". Such a rather dramatic measure not only proves very helpful in this respect, as will be discussed below, but may also be additionally motivated: Recently, various possible shortcomings of the CDM cosmological scenario in relation to structure formation on galactic scales have been discussed in the literature: 1) CDM possibly leads to the formation of too many small galaxies relative to what is observed, i.e. the *missing satellites problem* (e.g., Klypin et al. 1999). 2) Even if galactic winds due to star-bursts can significantly reduce the number of visible dwarf galaxies formed, sufficiently many of the small and tightly bound dark matter systems left behind can still survive to the present day in the dark matter halos of larger galaxies like the Milky Way to possibly destroy the large, central disks via gravitation-

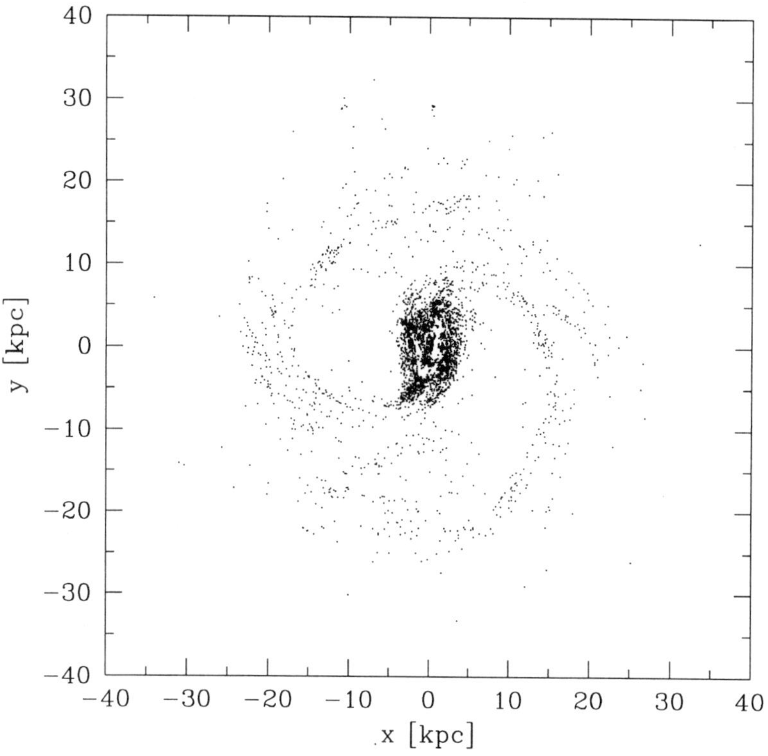

Figure 2. Face-on view of a disk galaxy with characteristic circular velocity $V_c \sim 300$ km/s formed in a warm dark matter simulation with no conversion of gas into stars (so the disk is purely gaseous). The mass of the disk is $M_{disk} \sim 2 \cdot 10^{11} M_\odot$ and its specific angular momentum is $j_{disk} \sim 2000$ kpc km/s.

al heating, as discussed by Moore et al. (1999a). 3) The dark matter halos produced in CDM cosmological simulations tend to have central cusps with $\rho_{DM}(r) \propto r^{-N}, N \sim 1-2$ (Dubinski & Carlberg 1991, Navarro et al. 1996, Fukushige & Makino 1997, Moore et al. 1998, Kravtsov et al. 1998, Gelato & Sommer-Larsen 1999). This is in disagreement with the flat, central dark matter density profiles (cores) inferred from observations of the kinematics of dwarf and low surface brightness galaxies (e.g., Burkert 1995, de Blok & McGaugh 1997, Kravtsov et al. 1998, Moore et al. 1999b, but see also van den Bosch et al. 1999).

The first two problems may possibly be overcome by invoking warm dark matter (WDM) instead of CDM: On mass scales less than the free-streaming mass, $M \lesssim M_{f,\mathrm{WDM}}$, the growth of the initial density fluctuations in the Universe is suppressed relative to CDM due to relativistic free-streaming of the warm dark matter particles. In conventional WDM theory these become non-relativistic at redshifts $z_{nr} \sim 10^6$-10^7 for $m_{\mathrm{WDM}} \sim$ 1 keV, which is the characteristic WDM particle mass required to give sub-galactic to galactic free-streaming masses. As a consequence of this suppression, fewer low mass galaxies (or "satellites") are formed cf., e.g., Moore et al. (1999a) and Sommer-Larsen & Dolgov (1999, SD99). The central cusps problem may be more generic (Huss et al. 1999 and Moore et al. 1999b), but WDM deserves further attention also on this point.

SD99 show that the angular momentum problem may be resolved by going from cold to warm dark matter, with characteristic free-streaming mass $M_{f,\mathrm{WDM}} \sim 10^{10}$-$10^{11} M_{\odot}$, and without having to invoke effects of stellar feedback processes at all. The reason why this kind of warm dark matter leads to a solution of the angular momentum problem is that because of the suppression of density fluctuations on sub-galactic scales relative to CDM the formation of a disk galaxy becomes a much more coherent and gentle process enabling the infalling, disk-forming gas to retain much more of its original angular momentum. In fact SD99 find it likely that the angular momentum problem can be *completely* resolved by going to the WDM structure formation scenario, which is more than can be said for the CDM+feedback approach so far. In Figure 2 we show a face-on view of a disk galaxy with characteristic circular velocity (where the rotation curve is approximately constant) $V_c \sim 300$ km/s formed in a WDM simulation (in this simulation gas was not converted into stars). Clearly it is no longer a problem to form extended, high angular momentum disks in fully cosmological simulations. In comparison the extent of typical disks formed in "passive" CDM simulations (i.e. simulations not incorporating the effects of stellar feedback processes) is less than 1 kpc – see, e.g., Sommer-Larsen et al. (1999).

Unlike the CDM+feedback solution, one does not get a constraint on the early star-formation histories of the proto-galaxies, so no statements about the abundance of the first generation of disk stars can be made without further assumptions.

SD99 discuss possible physical candidates for WDM particles and find that the most promising are neutrinos with weaker or stronger interactions than normal, majorons (light pseudogoldstone bosons), or mirror or shadow world neutrinos.

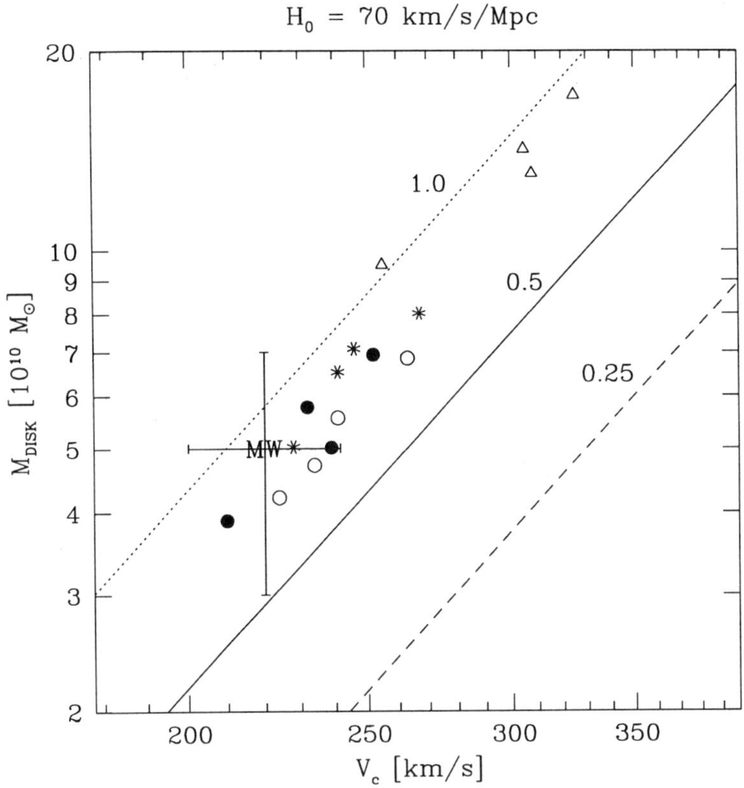

Figure 3. The mass vs. characteristic circular velocity "Tully-Fisher" relation for the final disks formed in 16 WDM simulations of Sommer-Larsen & Dolgov (for H_0=70 km/s/Mpc). Also shown is the observed I-band TF relation of Giovanelli et al. converted to mass assuming (M/L_I) = 0.25 (*dashed line*), 0.5 (*solid line*) and 1.0 (*dotted line*) and H_0=70 km/s/Mpc. Finally, the symbol "MW" with errorbars shows the likely range of the total, baryonic mass and characteristic circular velocity of the Milky Way.

3. The Tully-Fisher relation

In Figure 3 we show the cooled-out disk mass M_{disk} at redshift z=0 as a function of the characteristic circular velocity V_c of model galaxies formed in our WDM simulations (assuming a Hubble parameter H_0=70 km/s/Mpc). Also shown is the I-band Tully-Fisher (TF) relation of Giovanelli et al. (1997) converted to mass assuming I-band mass-to-light ratios (M/L_I)=0.25, 0.5 and 1.0 in solar units and H_0=70 km/s/Mpc. Finally, the baryonic mass of the Milky Way, estimated in a completely independent way, is shown

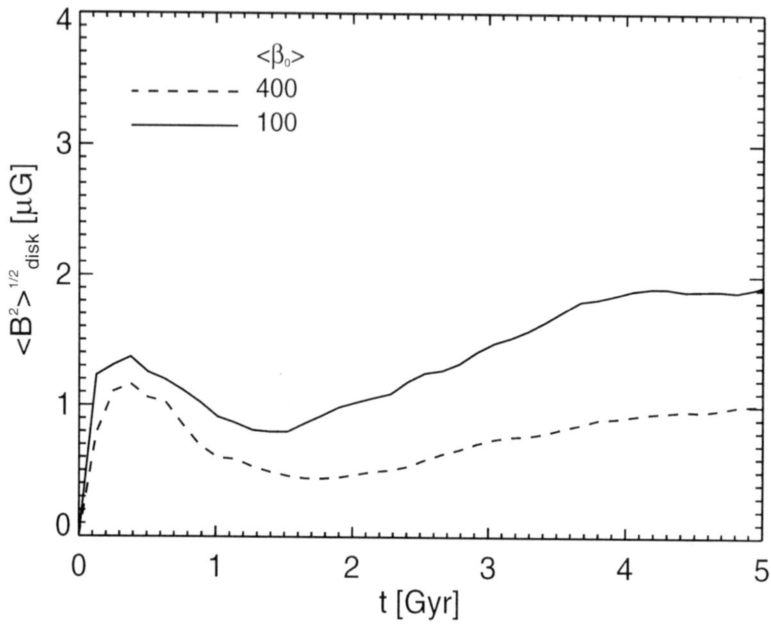

Figure 4. The temporal evolution of the average field strength in the disk for two values of $\langle\beta_0\rangle$.

(see SD99 for details). As can be seen from the figure we can match the slope of the TF relation very well assuming a constant (M/L_I). To get the normalization right a $(M/L_I) \sim 0.8$ is required. SD99 argue that this is quite a reasonable value in comparison with various dynamical and spectrophotometric estimates. Moreover, it is clearly gratifying that the Milky Way data point falls right on top of the theoretical as well as observational M_{disk}-V_c relations (for $(M/L_I) \sim 0.8$, H_0=70 km/s/kpc).

Steinmetz & Navarro (1999) and Navarro & Steinmetz (1999) find a discrepancy between the observed and "theoretical" TF on the basis of CDM simulations of disk galaxy formation. It is hence possible that WDM helps out also on this point, but this has to be checked with more detailed simulations.

4. Magnetic fields in galactic disks and disk galaxy formation

Rögnvaldsson (1999) showed how the typical magnetic field strengths observed in galactic disks can be explained as a result of disk galaxy *formation*, as an alternative to the usually assumed dynamo amplification of an initially very weak magnetic field in the disk: Hot, virialized gas ($T \sim 2 \cdot 10^6$

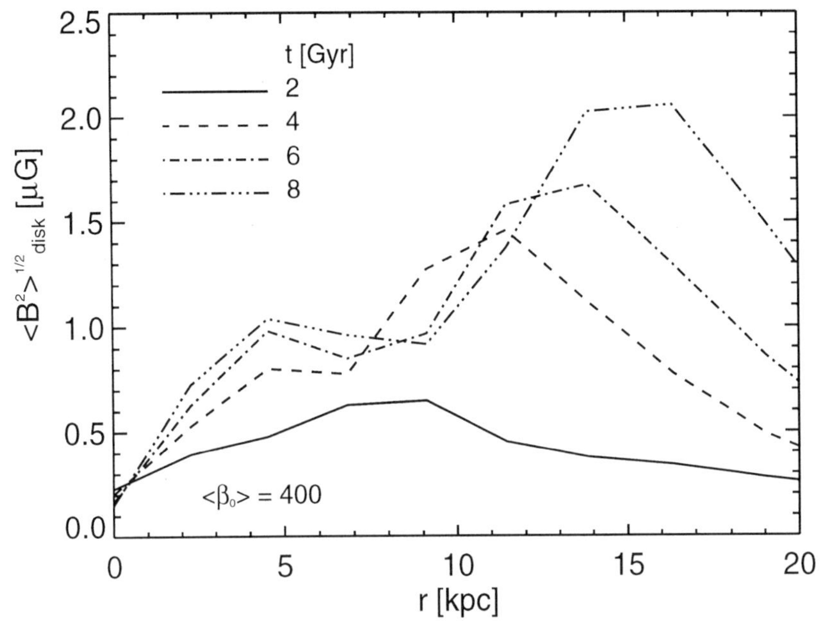

Figure 5. The azimuthally averaged field strength in the disk at various times for the $\langle\beta_0\rangle$=400 experiment.

K) in a dark matter halo is assumed to initially follow the dark matter distribution and to be rotating slowly, corresponding to a spin-parameter $\lambda \sim 0.05$, typical of galactic, dark matter halos. The hot gas is assumed to be threaded by a weak and random magnetic field. As the hot gas cools radiatively, gravity forces it to flow inwards and due to the spin it forms a growing, cold, galactic disk in the central parts of the dark matter halo. The magnetic field follows the cooling gas inwards and is strongly amplified by compression and shear in the forming disk. Rögnvaldsson (1999) carried out magnetohydrodynamical (MHD) simulations of this process using an eulerian mesh MHD code – the work is described in detail in Rögnvaldsson, Sommer-Larsen & Nordlund (2000). The simulations were run with various initial magnetic field strengths in the hot gas, parameterized by the initial ratio between the gas pressure and magnetic pressure $\langle\beta_0\rangle = \langle\frac{P_{gas}}{B^2/8\pi}\rangle_0$. The temporal evolution of the average magnetic field strength in the disk gas is shown in Figure 4. For weak initial fields (we have taken $\langle\beta_0\rangle$=100-400 as a starting point, since these values are typical values in the hot, intergalactic gas in clusters of galaxies) the average magnetic field strength grows gradually from about t=1 Gyr (after an initial relaxation phase). The average values of 1-2μG reached after about 5 Gyr are quite reasonable for typical disk galaxies, indicating a route to the explanation of the magnetic field strengths observed in galactic disks alternative to the usual dynamo one.

Another aspect of the growth of the field strength is reflected in the radial average in the disk, shown in Figure 5 at various times for a simulation with $\langle\beta_0\rangle$=400. The field strength is always highest in the outermost part of the growing disk, since the fieldlines brought in with the cooling flow are stacked on top of the already existing field there and the field is further amplified by the disk shear.

5. Acknowledgements

I am grateful to Örnólfur Rögnvaldsson for permitting me to present unpublished results and figures from his PhD thesis, and I have benefited from comments by him, Sasha Dolgov and Jens Schmalzing. Finally, I would like to thank the organizers for a magnificent conference.

References

1. Beers, T. C., & Sommer-Larsen, J. 1995, *Astrophys. J. Suppl.*, **96**, 175
2. van den Bosch, F. C., et al. 1999, *Astron. J.*, submitted (astro-ph/9911372)
3. Burkert, A. 1995, *Astrophys. J.*, **488**, L55
4. de Blok, W. J. G., & McGaugh, S. S. 1997, *MNRAS*, **290**, 533
5. Chiappini, C., Matteucci, F, & Gratton, R. 1997, *Astrophys. J.*, **477**, 765
6. Dubinski, J., & Carlberg, R. 1991, *Astrophys. J.*, **378**, 496
7. Fall, S. M., & Efstathiou, G. 1980, *MNRAS*, **193**, 189
8. Fukushige, T., & Makino, J. 1997, *Astrophys. J.*, **477**, L9
9. Gelato, S., & Sommer-Larsen, J. 1999, *MNRAS*, **303**, 321
10. Giovanelli, R., et al. 1999, *Astrophys. J.*, **477**, L1
11. Huss, A., Jain, B., & Steinmetz, M. 1999, *Astrophys. J.*, **517**, 64
12. Klypin, A., Kravtsov, A. V., Valenzuela, O., & Prada, F. 1999, *Astrophys. J.*, **523**, 32
13. Kravtsov, A. V., Klypin, A. A., Bullock, J. S., & Primack, J. R. 1998, *Astrophys. J.*, **502**, 48
14. Moore, B., Governato, F., Quinn, T., Stadel, J., & Lake, G. 1998, *Astrophys. J.*, **499**, L5
15. Moore, B., Ghinga, S., Governato, F., Lake, G., Quinn, T., Stadel, J., & Tozzi, P. 1999a, *Astrophys. J.*, **524**, L19
16. Moore, B., Quinn, T., Governato, F., Stadel, J., & Lake, G. 1999b, *MNRAS*, submitted (astro-ph/9903164)
17. Navarro, J. F., & Benz, W. 1991, *Astrophys. J.*, **380**, 320
18. Navarro, J. F., & White, S. D. M. 1994, *MNRAS*, **267**, 401
19. Navarro, J. F., Frenk, C. S., & White, S. D. M. 1995, *MNRAS*, **275**, 56
20. Navarro, J. F., Frenk, C. S., & White, S. D. M. 1996, *MNRAS*, **462**, 563
21. Navarro, J. F., & Steinmetz, M. 1999, *Astrophys. J.*, in press (astro-ph/9908114)

22. Rögnvaldsson, Ö. 1999, PhD thesis, University of Copenhagen
23. Rögnvaldsson, Ö., Sommer-Larsen, J. & Nordlund, Å. 2000, in preparation
24. Sommer-Larsen, J., Gelato, S., & Vedel. H. 1999, *Astrophys. J.*, **519**, 501
25. Sommer-Larsen, J., & Dolgov, A. 1999, *Astrophys. J.*, submitted (astro-ph/9912166, SD99)
26. Steinmetz, M., & Navarro, J. F. 1999, *Astrophys. J.*, **513**, 555

HAS THE IMF OF EG'S ALWAYS BEEN THE SAME?

C. CHIOSI[1]
Department of Astronomy, University of Padua
Vicolo Osservatorio 5, 35122 Padova, Italy

Abstract. Several properties of elliptical galaxies are examined in the light of the Larson galactic wind model and the classical recipes for the star formation rate and initial mass function. The suggestion is advanced that several points of disagreement between theory and observational data could be alleviated if not removed by adopting an initial mass function whose typical mass scale at early galactic epochs was higher than the present-day estimate for the Solar Vicinity. Galaxy models with this type of initial mass function are finally used to study the problem of the iron and gas content in the intra-cluster medium.

1. Introduction

Deciphering the past history of star formation in Elliptical Galaxies (EG's) is basic to set clues on the galaxy formation process itself. Two competing scenarios are proposed: the monolithic picture in which a single dominant episode of star formation occurred in the past, ever since followed by quiescence and passive evolution (secondary episodes are however always possible); the hierarchical scheme, in which a series of mergers of smaller sub-units has taken place over the billions of years, each of those likely accompanied by star forming activity (mergers of inert objects with no additional star formation cannot be excluded).

The key question to be addressed and hopefully answered is what kind of signatures the stellar populations of a galaxy have inherited from the forming mechanism. Prior to any other consideration, it is worth recalling that the diagnostic to our disposal is based on integrated spectra, magnitudes and colors, luminosity weighed line strength indices, and estimates of chemical abundances. Unfortunately, there is always a severe degeneracy between age and metallicity effects, i.e. the spectrum and hence the colors of an old and metal-poor stellar population may look like those of a

F. Matteucci and F. Giovannelli (eds.), The Evolution of the Milky Way, 599–610.

young and metal-rich one. This is intrinsic to stellar structure and cannot be eliminated. Long ago Larson (1974) suggested that the Color-Magnitude Relation (CMR) of EG's (see below) stems from a mass-mean metallicity relation, whereby the massive EG's are more metal-rich, and proposed the SN-driven galactic wind (SDGW) mechanism as the key process by which massive EG's, owing to their deeper gravitational potential, *retain gas and form stars for longer periods of time and hence get more metal-rich than the low-mass ones.* This model has long been applied to understand many (possibly all) of the properties of EG's. In what follows we examine a few of these to check whether they are fully compatible with the SDGW scheme.

2. G-Dwarf Analog, CMR and Enhancement in α-elements

G-Dwarf Analog. The near UV spectrum of E-galaxies shows that the *relative percentage of old, metal-poor stars is small,* thus suggesting that early evolution proceeded either via the so-called infall model (Bressan et al. 1994; Tantalo et al. 1996) or prompt enrichment (Greggio 1996).

CMR. On the average, EG's get redder at increasing luminosity, i.e follow a mean CMR, which is tight for cluster galaxies (Bower et al. 1992) and more dispersed for objects in small groups and field (Schweitzer & Seitzer 1992). As already mentioned the CMR is interpreted by means of the SDGW model of Larson (1974). The tightness of the cluster CMR seems to indicate that EG's are old (say 13 - 15 Gyr) and nearly coeval. The alternative that the CMR is an age sequence, with bluer galaxies younger than the red ones, has been proved not to be viable by Kodama & Arimoto (1997). The situation with field galaxies is more intrigued, perhaps compatible with star formation spreaded over longer periods of time.

Enhancement in α-elements. Gradients in the line strength indices Mg_2 and $\langle Fe \rangle$ with different slopes are observed across individual EG's and systematic differences in both quantities are detected among the nuclear regions of different galaxies. The current interpretation of the data is that they reflect a trend in the mean value of the ratio [α/Fe], where α stands for O, Mg, etc... (see Matteucci 1997 and Peletier 1999). Bright EG's seem to have $[Mg/Fe] \geq 0.3$, faint EG's $[Mg/Fe] \simeq 0$. It is commonly concluded that in a galaxy the ratio [α/Fe] increases toward the center or passing from galaxy to galaxy with the luminosity (mass). These are rather demanding constraints on the kind of stellar progenitors at work and on the overall duration of the star forming period. *This should get shorter at increasing galaxy mass. It is exactly the opposite of what implied by the CMR.* See Bressan et al. (1996) and Matteucci (1997) for more details.

3. What about ages and metallicities ?

Field Galaxies. Attempts aimed at reconstructing the past history of star formation (SF) and metal enrichment in field EG's from the line strength indices have been made by many authors (Gonzáles 1993, Worthey et al. 1994, Bressan et al. 1996, Greggio 1996; Tantalo et al. 1998a). Looking at the indices H_β, $[MgFe]$, Mg_2, $\langle Fe \rangle$ of the Gonzáles (1993) galaxies, and taking into account the information provided by the UV excess and velocity dispersion σ, when available, the following remarks can be made:

(i) EG's scatter more in H_β than $[MgFe]$, and therefore do not follow the relation expected on this diagram for old objects matching the CMR. Furthermore, the data cannot be accounted for by closed-box models. In contrast they better agree with the so-called infall models (Bressan et al. 1996) or prompt enrichment (Greggio 1996). All this suggests an intrinsic large dispersion in age and a narrow range in metallicities ($Z_\odot \leq Z \leq 3 \times Z_\odot$).

(ii) To reconcile the constraints set by the CMR and the H_β-$[MgFe]$ diagnostic, the suggestion is made that *each galaxy of the sample had a different SF history.* Some completed their stellar activity in the far past with no evidence of subsequent episodes. Others had a more prolonged star forming history, perhaps in recurrent episodes of short duration.

(iii) Using the galaxy to galaxy differences δH_β and δMg_2, $\delta\langle Fe \rangle$ in the nuclear values, transferring these into $\Delta log(t)$, $\Delta log Z/Z_\odot$, $\Delta[Mg/Fe]$, and finally examining the correlation between $\Delta log(t)$ and absolute magnitude M_V for a sample of galaxies, Tantalo et al. (1998a) suggested that a sort of *upper limit* to the age of the bulk stars exists, traced by those galaxies with no sign of re-juvenation. Indeed the brightest quiescent galaxies ($M_V \simeq -23$) are compatible with a very old age, whereas the fainter objects ($M_V \simeq -20$) are consistent with a time delay up to a few Gyr ago.

EG's in clusters and groups. Since the Gonzáles (1993) sample is essentially made by *field* galaxies, what about EG's in clusters and groups? Kuntschner & Davies (1998) measured the H_β and $[MgFe]$ indices for a number of E and S0 galaxies in the Fornax cluster. In this sample, EG's crowd toward low values of H_β and high values of $[MgFe]$ and seem compatible with old ages, whereas S0's scatter on a much broader region perhaps indicating recent stellar activity. In addition to this, the *cluster EG's* of Kuntschner's & Davies' (1998) are much less dispersed along the H_β axis than the *field EG's* of Gonzáles (1993). Finally, compiling literature data, Forbes & Terlevich (1999) derive the H_β vs. $[MgFe]$ plane for EG's in *loose* and *compact groups*. Despite the small number of objects in total, EG's in *loose groups* behave similarly to field EG's, whereas EG's in *compact groups* mimic *cluster* EG's. *Does all this imply that a galaxy's environment play an* *important role in determining its evolution? In the context of the Bressan et al. (1996) and Tantalo et al. (1998a) analyses, does it mean that secondary stellar activity is favored in physical conditions such as those found in the field and loose groups?*

Interacting versus normal galaxies. Longhetti et al. (1999a,b) posed two important questions, i.e. whether (i) interacting and normal galaxies differ in the properties of their spectral indices, (ii) whether there is any combination of spectral indices useful to dating the interaction and the accompanying stellar activity in turn. To this aim, they selected 51 early-type galaxies located in low density environments with clear signatures of interaction or fine structures. The sample contains 21 shell galaxies and 30 members of pairs. The reference *normal galaxies* are those of the Gonzáles (1993) sample. The results can be summarized as follows:

(i) Normal and pair-galaxies follow the universal σ vs. Mg_2 relation, whereas shell-galaxies lie above it. The Fe vs. Mg_2 relation of normal, pair- and shell-galaxies is flatter than the theoretical prediction. Once more this fact hints for enhancement of α-elements (Mg in particular).

(ii) In the σ vs. H_β plane normal galaxies seem to follow a nice relation, along which galaxies with shallower gravitational potential have stronger H_β, whereas shell- and pair-galaxies scatter all over the plane. A group of galaxies with deep gravitational potential and strong H_β is also found. *Is this a signature of recent star formation?*

(iii) Surprisingly, in the H_β vs. $[MgFe]$ plane shell, pair and normal galaxies share the same distribution. There is, however, a group of peculiar galaxies with much stronger H_β as compared to the normal ones. *Does it mean that the scatter seen in this diagram has a common origin, perhaps secondary episodes of star formation that can occur independently of whether or not a galaxy is interacting?*

(iv) The distribution of all galaxies in the H_β vs. $[MgFe]$ is confined within a narrow strip (narrow range of metallicity) whose slope is much steeper than the path followed by aging Single Stellar Populations. This was interpreted as a signature of both large age ranges for the bulk stars and substantial chemical enrichment (up to about two times solar).

Cluster dwarf EG's. An important piece of information comes from the colors of dwarf EG's in clusters at different red-shift. Abell 851 is one of the best known clusters at high red-shift (z=0.41). It contains a rich population of late-type spirals, irregulars, ellipticals, mergers, and many post-star-burst galaxies E+A. It has recently been observed by Bertola et al. (2000) in the ultraviolet (F300W) and I-band (F702W) of the HST-WFPC2. In the following we will not consider irregular and spiral galaxies, for which ongoing star formation is obvious, but concentrate on E-S0 objects, those of relatively faint luminosity, in particular. The key result is that faint ellipticals to be detected as they are, must have undergone recent stellar activity. Similar behavior is seen for the dwarf EG's in the Coma cluster (z=0.01) observed by Donas et al. (1995) in their balloon-borne UV/optical pass-bands (m_{UV}, m_b) which are centered at 2000 Å and 4500 Å, respectively. Thanks to the red-shift displacement, $m_{UV} \simeq m_{300}$ and $m_b \simeq m_{702}$ so that comparison is meaningful. The remarkable coincidence between the two sets of data (see Bertola et al. 2000) means that no significant evolution between the two epochs (from z=0.4 to the present) is detectable. Since there is no reason why Abell 851 and Coma faint EG's should coordinate their effort in triggering star formation, *the unavoidable conclusion is that faint (low mass) galaxies are able to undergo recent (likely current) stellar activity and only those in which the process is underway or just over are detected. In both cases massive EG's are fully compatible with old stellar activity followed by quiescence.*

4. The Intra-Cluster Medium

The intra-cluster medium (ICM) of clusters of galaxies is long known to be filled by hot, metal-rich gas revealed by the X-ray emission caused by thermal brems-strahlung at temperature of about 10^7 to 10^8 K. Characterizing parameters of the ICM gas are: (i) the total mass which is of the order of one-third of a given cluster's total gravitational mass (White et al. 1993). (ii) the abundance of iron which goes from about 1/2 solar (Rothenflug & Arnaud 1985) to 1/3 solar (Renzini 1997, Fukazawa et al. 1998). (iii) the abundances of α-elements, for which the ASCA observations (Mushotzky et al. 1996; Fukazawa et al. (1998) provide the following estimates $[\frac{\alpha}{Fe}] \simeq 0.2{\pm}0.3$, $[\frac{Si}{Fe}] \simeq 0.37^{+0.17}_{-0.35}$ $[\frac{O}{Fe}] \simeq 0.48^{+0.24}_{-0.55}$, Ishimaru & Arimoto (1997) and Arimoto et al. (1997) argue that considering the uncertainties all the data are consistent with the solar value. It is widely accepted that the ICM metals have been processed in galaxies and subsequently either ejected or stripped away. Indeed, the SDGW model of Larson (1974) is the favored mechanism for ejecting heavy elements (Matteucci & Vettolani 1988; Mihara & Takahara 1994; Matteucci & Gibson 1995; Elbaz et al. 1995; Gibson & Matteucci 1997). The ram pressure stripping of the enriched interstellar medium during the initial phase of star formation cannot, however, be excluded (Sarazin 1979), even if arguments can be given against this possibility (White 1991; Renzini et al. 1993). Summarizing the results of many studies, the interpretations of these data soon encounters two difficulties: (i) if the iron mass is recovered, too low a gas mass is predicted; (ii) iron is over-produced with respect of α-elements. The conclusion was that a large fraction of the gas must be primordial. *Does it reflects what really happened in galaxy clusters, or does it simply stem from inadequacies of current galaxy models in predicting the right ejecta of metals and gas?*

5. General remarks

Summarizing the key points of the above discussion, there are several aspects of the standard SDGW model to be revised in order to have: (i) The duration of the bulk of stellar activity increasing at decreasing galaxy mass. More precisely the hint is that star formation was short, perhaps confined in a single dominant episode, in massive galaxies and long, maybe in a series of recurrent episodes (bursts), in the low mass ones. (ii) The ejecta by galactic wind able to solve the iron-gas problem in the ICM. (iii) The observational trend for the enhancement of α-elements.

It is worth recalling that the current modeling of galaxy evolution stands on rather simple recipes concerning the efficiency of star formation: the Schmidt law for the SF rate, and the Salpeter (1955) law (or Salpeter-like) for the initial mass function (IMF). In the following we want to explore the possibility that in the past the IMF has been significantly different from the one we see today in the Solar Vicinity for which the Salpeter was originally established.

6. The IMF in early stages

Is there any plausible reason why the first generations of stars in primeval galaxies had an IMF different from the one we see today? The present-day data for the IMF of stars in the Solar Vicinity (Scalo 1998) can be described by a declining power law for masses above $1\,M_\odot$ (Salpeter 1955), flattening below $\simeq 0.5\,M_\odot$, and possibly even declining below $\simeq 0.25\,M_\odot$. Following Larson (1998) this type of IMF is suitably represented by

$$\frac{dN}{dlog\,M} \propto M^{-\alpha} exp(-\frac{M_s}{M}) \tag{1}$$

where $x = \alpha - M_s/M$ is the logarithmic slopes, $\alpha = 1.35$ is the Salpeter case, and M_s is the mass-scale. Therefore, the IMF approaches the Salpeter (1995) law at large masses, peaks at $M_p = M_s/\alpha$, and falls exponentially at lower masses.

In this context, the whole problem is reduced to understanding whether α and M_s have always been the same or they have changed with time and hence physical conditions of the medium.

The mass scale M_s might be related to the Jeans mass $M_J \propto T^2\,P^{-1/2}$, where T and P are the temperature and pressure of the gas (Larson 1995). The temperature is controlled by the balance between heating by external radiation fields and cooling by line emission from atoms and molecules and thermal emission from dust. More intense radiation fields and lower heavy elements abundances would imply higher temperatures and larger M_s unless there is a sufficiently large compensating increase in pressure. *How were temperature and pressure at the early times?* While there are no compelling arguments for the pressure in star-forming clouds has to be much higher at early times, there are several reasons for the temperature has indeed to be higher. First, the cosmic background radiation (CBR) temperature T_{CBR}, which is the minimum possible cloud temperature, scales with red-shift as $T_{CBR} = 2.73(1+z)$, and it becomes higher than the present day temperature in molecular clouds $\simeq 10$ K at red-shifts greater than about 2. Second, at early stages the abundance of heavy elements is lower which implies lower cooling efficiencies per given heating rate. Finally, the heating rate was certainly higher in the far past because of the intense field radiation and likely strong stellar activity (supernova explosions, stellar winds, cosmic rays, etc.)

The above arguments suggest that M_p at early stages was significantly skewed towards relatively high masses. It is worth clarifying here that strong effects would result even by changing M_s by a modest amount. If the present-day value in the Solar Vicinity is $M_p = 0.20\,M_\odot$, an increase up to $1\,M_\odot$ would be fully sufficient to change the IMF significantly. As far as the slope of the IMF is concerned very little can be said. At early epochs the mean slope above M_p was likely flatter than today.

Is there any IMF in literature having the kind of behavior we have discussed so far? Observations are not yet able to provide a hint, but there are theoretical studies on the subject. For instance Padoan et al. (1997) derived an IMF whose functional dependence is much similar to that of eq. (1) with

$$M_p \simeq 0.2 M_\odot \times (T/10)^2 \times (n/1000)^{-0.5} \times (\Sigma_g/2.5)^{-1} \qquad (2)$$

where T, n and Σ_g are the temperature, number density, and velocity dispersion of the gas. Furthermore, the logarithmic slope continuously change as function of T, n and Σ_g. The IMF gets flatter and M_p gets higher at increasing T or decreasing n and Σ_g.

7. Galaxy models with variable IMF

Chiosi et al. (1998) made use of the Padoan et al. (1997) IMF to calculate galaxy models in which the IMF continuously varied with time passing from the situation in which M_p was skewed toward the high mass end during the early epochs to that in which, thanks to the increased metallicity and hence cooling efficiency (lower temperatures), M_p gradually shifted to lower masses and the Salpeter IMF was recovered. Starting from this idea, they build a thermal model in which heating by SN explosion, stellar winds, UV radiation and mechanical sources, and cooling processes by various agents were taken into account (see Chiosi et al. 1998 for all details) All this was implemented in a simplified version of the chemo-spectro-photometric SDGW models of elliptical galaxies developed by Tantalo et al. (1998b) in which the spatial gradients in density and star formation were taken into account.

General properties All the models obey some general properties determined by the interplay between the physical conditions of the star forming gas and the IMF:

(i) In the very early stages of a galaxy, the temperature of the star forming gas is no cooler than the current T_{CBR}, and the mean gas density ρ_g and velocity dispersion Σ_g determine the starting value of M_p. Since they are lower and larger, respectively, in galaxies of high mass as compared to those of small mass, M_p is skewed towards more massive stars, and the slope of the IMF above M_p tends to flatten at increasing galaxy mass.

(ii) The energy from supernova explosions and stellar winds is injected into the interstellar gas making it hotter and hotter till a sort of equilibrium stage is reached, during which both the gas temperature and M_p remain nearly constant.

(iii) However, as star formation proceeds, the gas gets richer in metals thus increasing the cooling efficiency. The situation is eventually met in which cooling is so efficient that the gas temperature and peak mass of the IMF in turn start decreasing. This implies that less energy is injected into the interstellar medium with consequent catastrophic fall-off of both temperature and peak mass.

(iv) The duration of this initial phase is longer at increasing mass M_G of the proto-galaxy, but it is short always compared to the total galaxy lifetime. However, it bears very much on the final outcome, because it determines the kind of stellar populations one would expect in galaxies of different mass.

Star formation history. The overall duration of the star forming activity is found to increase at decreasing galaxy mass *as suggested by the abundance ratios* $[\alpha/Fe]$. Furthermore, in low mass galaxies, following the dominant initial activity of rather short duration, traces of star formation might occur over much longer periods of time (see Chiosi et al. 1998 for details), as perhaps hinted by the H_β, Mg_2, $\langle Fe\rangle$, and $\langle MgFe\rangle$ diagnostic, and the trend in the enhancement of α-elements. Finally, as the central regions of a galaxy are denser than the external ones, star formation is expected to last longer over there. This secures that the mean metallicity of a galaxy increase with its mass, in compliance with the requirements of the CMR, and that inside individual galaxies gradients in metallicity exist as indicated by the observed gradients in colors and line strength indices.

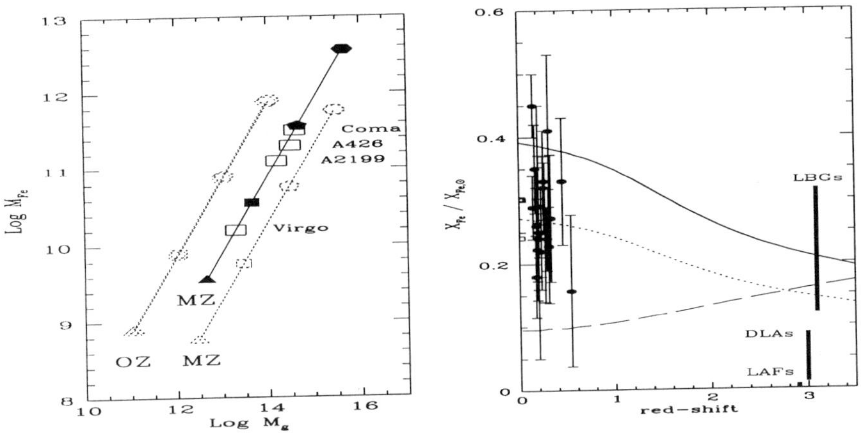

Figure 1. *Left Panel*: The iron-gas mass relationship for galaxies with the Salpeter IMF (dotted lines) and the Padoan et al. (1997) IMF (solid line). The labels OZ and MZ stand for one-zone and multi-zone galaxy models, respectively. The triangles, squares, pentagons and hexagons along the three lines are the results for normalization factors to the total baryonic mass in the cluster of 10^{13}, 10^{14}, 10^{15} and $10^{16}\,M_\odot$ respectively. The perturbation index is $n = -1.5$. The open squares are the data for four clusters taken from Matteucci & Vettolani (1988). *Right Panel*: The variation of the abundance $X_{Fe}/X_{Fe,\odot}$ as a function of the red-shift. The filled circles are the data by Mushotzky & Loewenstein (1997). The open squares are the four clusters of the left panel. The solid heavy line is the evolution of Case C models, the solid thin line is the same but with the addition of 30% of primordial gas. Finally, the dashed line is the prediction of Case B models. The vertical bars show the metallicity range of damped Ly-α systems (DLAs), Ly-break galaxies (LBGs) and Ly-α forest (LAFs). Data from Pettini (2000)

Remnants & living stars. Since M_p depends also on the mean gas density and hence galaxy mass, we expect the proportions of living stars and remnants to change from galaxy to galaxy. For instance, within the sphere with effective radius R_e of the $1 \times 10^9 M_\odot$ galaxy about 80% of the original mass is in visible stars, the rest is in remnants, and there are almost no traces of gas. Conversely, in the same region of a $3 \times 10^{12} M_\odot$ galaxy only 25% of the original mass is in living stars, about 20% is in collapsed remnants, and 55% has remained in form of gas, partly expelled at the stage of galactic winds and partly slowly ejected by dying stars. From the same arguments advocated to predict the gradient in the duration of the star forming activity across a galaxy, we expect the fraction of living stars to decrease with galacto-centric distance. It goes without saying that as a consequence of this, one expects the mass-luminosity ratio to increase with the galaxy mass (*gate to understanding the tilt of the FP*) and the radial distance.

Metallicity and enhancement of α-elements. How do the metallicity and enhancement of α-elements vary as a function of time and galactic mass under this continuously changing IMF ? Examining the Chiosi et al. (1998) models we notice that in the $3 \times 10^{12} M_{\odot}$ galaxy there are very few stars stored in low metallicity bins say up to $Z = 0.01$. *This implies that for this galaxy the so-called G-Dwarf analog does not occur.* In the opposite extreme, i.e. the $1 \times 10^{9} M_{\odot}$ object, about 50% of the stars are in the bins up to $Z = 0.01$. However, as $[Fe/H]$ rapidly increases to a tenth of solar, even in this case few stars of very low Z are found and the G-Dwarf analog is avoided. *Infall-models are somehow ruled out by the new IMF.* In addition to this, the totality of stars in the $3 \times 10^{12} M_{\odot}$ galaxy have $[O/Fe]$ well above solar. The opposite occurs with the $1 \times 10^{9} M_{\odot}$ galaxy in which almost the totality of stars have $[O/Fe]$ below solar. These results are particularly interesting because two of the most intriguing and demanding constraints imposed by the observational data, i.e. the G-Dwarf analog and enhancement in α-elements as a function of the galaxy luminosity (mass in turn), are ruled out by the same token.

The CMR. Despite the shorter duration of the star forming activity and earlier winds, the mean metallicity in massive galaxies is higher than in the low mass ones, because of the shift of the IMF toward higher stellar masses. *This means that the CMR is accounted for as in the standard SDGW models, but the old contradiction between CMR and and α-enhancement does no longer occur.*

8. The ICM of galaxy clusters

Galaxy models with variable IMF have been recently used by Chiosi (2000) to study the gas and metal content of the ICM. The analysis stands on the following key ingredients: (i) The adoption of multi-zone models of EG's in the framework of the SDGW scheme (see Tantalo et al. 1998b for details). They yield a more realistic description of the galactic ejecta in which the effects of gradients in star formation and chemical enrichment are taken into account. (ii) Since the stellar IMF of Padoan et al. (1997) is sensitive to the physical condition of the gas and no cooling process exists decreasing the temperature of a galaxy's gas below the limit set by the current value of the CBR, the immediate consequence follows that the stellar initial mass function of proto-galaxies whose stellar activity began at high red-shift (when the CBR temperature was higher than the present-day mean temperature of molecular clouds) is different from the one in galaxies which did the same but at lower red-shift. Therefore galaxies of given mass but whose bulk activity of star formation took place at different red-shift are expected to eject different quantities of gas and metals.

In order to exploit this possibility the following model is proposed: at any time (red-shift) the mass distribution of proto-galaxies is similar to the Press-Schechter (1974) law holding for Dark Matter

$$\frac{d\Phi}{dM} = \frac{1}{M}\Big(\frac{2}{\pi}\Big)^{1/2}\Big(\frac{n+3}{6}\Big)\Big(\frac{M}{M*}\Big)^{\frac{(n+3)}{6}} \times exp\Big[-\frac{1}{2}\Big(\frac{M}{M*}\Big)^{\frac{(n+3)}{3}}\Big] \qquad (3)$$

where $n = -1; -2$ is the power spectrum of the perturbations (Lacey & Cole 1994) and M^* is the high-end cut-off mass. This in turn varies with the red-shift

$$M^* = M_N \times (1+z)^{-\frac{6}{n+3}} \qquad (4)$$

where M_N is a suitable normalization mass. Since the above relations strictly hold for Dark Matter, they must be suitable scaled to the range of galaxy masses. This is obtained by choosing $M_N = 10^{13}\, M_\odot$ at $z = 0$. At any red-shift (time), the galaxy masses are confined between M^* and the minimum mass for an object prone to collapse (Tegmark et al. 1994). The galactic ejecta $M_i^{ej}(M, z)$ of gas and metals are then convolved with the re-scaled Press-Schechter function by integrating upon mass and red-shift. The results are shown in Fig.7. When the same stellar IMF and galactic ejecta of previous studies, for instance those by Matteucci & Vettolani (1988), Matteucci & Gibson (1995) and Gibson & Matteucci (1997), are adopted in which there is no dependence on the red-shift, we recover the old results. This means that we fall back into the well known difficulty common to all of those studies. However, when the above scheme is let work completely the total amount of iron and gas can be simultaneously matched (see the left panel of Fig. 7). The abundance of Fe of multi-zone models with variable IMF are fully compatible with the observational determinations by Mushotzky & Loewenstein (1997) shown in the right panel of Fig.7, whereas those from models with the Salpeter IMF are too low. The oxygen to iron ratio is $[O/Fe] = -0.11$ in the case of the variable IMF, and $[o/Fe] = -0.315$ in the case of the Salpeter IMF. Three final considerations: (i) Any IMF of the type given by eq. (1) would lead to similar results provided the mass scale M_s increases with the temperature of the star forming medium. (ii) The adoption of the multi-zone scheme, though much improving on the final result, is not enough to solve the problem if not accompanied by an IMF of the type we have adopted. (iii) Finally, even in the case of multi-zone models with variable IMF case some dilution (up to about 30%) by primordial gas (never cycled through galaxies) is required to match the mean Fe abundances observed in galaxy clusters. In any case this is less than the older estimates of 65% to 90%.

References

1. Arimoto N., Matsushita K., Yshimaru Y., Osaki T., & Renzini A., 1997, ApJ, 477, 128
2. Bertola F., Capellari M., Chiosi C., Buson L.M., Dressler A., Oemler A., 2000, ApJ, in press
3. Bower R.G., Lucey J.R., & Ellis R.S., 1992, MNRAS 254, 601
4. Bressan A., Chiosi C., & Fagotto F., 1994, ApJS 94, 63
5. Bressan A., Chiosi C., & Tantalo R., 1996, A&A 311, 425
6. Chiosi C., 2000, A&A submitted
7. Chiosi C., Bressan A., Portinari L., & Tantalo R., 1998, A&A 339, 355
8. Donas J., Milliard B., Laget M., 1995, A&A 303, 661
9. Elbaz D., Arnaud M., & Vangioni-Flam E., 1995 A&A 303, 345
10. Forbes D. A., Terlevich A. I., 1999, astro-ph/9908072
11. Fukazawa Y., Makishima K., Tamura T. et al., 1998, PASJ 50, 187
12. Gibson B.K., & Matteucci F., 1997, ApJ 475, 47
13. Gonzáles J.J., 1993, Ph.D. Thesis, Univ. California, Santa Cruz
14. Greggio L., 1996, MNRAS, 285, 151
15. Ishimaru & Arimoto N., 1997, PASJ 49, 11
16. Kodama T., & Arimoto N., 1997, A&A, 320, 41
17. Kuntschner H., Davies R.L., 1998, MNRAS, 295, L29
18. Lacey C., & Cole S., 1994, MNRAS 271, 676
19. Larson R.B., 1974, MNRAS 166, 585
20. Larson R.B., 1995, MNRAS 272, 213
21. Larson R.B., 1998, MNRAS 301, 569
22. Longhetti M., Bressan A., Chiosi C., Rampazzo R., 1999a, A&A, 345, 419
23. Longhetti M., Bressan A., Chiosi C., Rampazzo R., 1999b, A&A in press
24. Matteucci F., 1997, Fund. Cosmic Phys. 17, 283
25. Matteucci F., & Gibson B.K., 1995, A&A 304, 11
26. Matteucci F., Molaro P., Vladilo G., 1997, A&A, 321, 45
27. Matteucci F., & Vettolani G., 1988, A&A 202, 21
28. Mushotzky R.F., & Loewenstein M., 1997, ApJ 481, L63
29. Mushotzsky R.F., Loewenstein M., Arnaud M., Tamura T., Fukazawa Y., et al., 1996, ApJ 466, 686
30. Padoan P., Nordlund A.P. & Jones B.J.T., 1997, MNRAS 288, 145
31. Peletier R., 1999, in *The Evolution of Galaxies on Cosmological Time-scales*, ASP Conf. Series, Vol. 3, J.E. Beckman, and T.J. Mahoney, eds., astro-ph/9904280
32. Pettini M., 2000, astro-ph/0001075
33. Press W.H., & Schechter P., 1974, ApJ 187, 425
34. Renzini A., 1997, ApJ. 488, 35
35. Renzini A., Ciotti L., D'Ercole A., & Pellegrini S., 1993, ApJ 419, 52
36. Rothenflug R., & Arnaud M., 1985, A&A, 144, 431
37. Salpeter E.E., 1955, ApJ 121, 161
38. Sarazin C.L., 1979, Astrophys. Lett. 20, 93
39. Scalo J., 1998, in *The Initial Mass Function*, Gilmore G. & Howell D., eds., ASP Conf. Ser. Vol. 142, p. 201
40. Schweizer F., & Seitzer P., 1992, AJ 104, 1039
41. Tantalo R., Chiosi C., Bressan A., & Fagotto F., 1996, A&A 311, 361
42. Tantalo R., Chiosi C., Bressan A., 1998a, A&A, 333, 419
43. Tantalo R., Chiosi C., Bressan A., P. Marigo, & Portinari L., 1998b, A&A, 335, 823
44. Tegmark M., Silk J., Rees M.J., Blanchard A., Abel T., & Palla F., 1997, ApJ 474, 1
45. White R.E.III, 1991, ApJ 367, 69
46. White S.D.M., Navarro J.F., Evrard A.E., & Frenk C., 1993, Nature 366, 429
47. Worthey G., Faber S.M., Gonzáles J.J., Burstein D., 1994, ApJS, 94, 687

CONCLUSIONS

CONCLUDING REMARKS

B.E.J. PAGEL
Astronomy Centre
CPES, Sussex University, Brighton BN1 9QJ, UK

Abstract.
I comment on three of the topics discussed at this meeting: The O/Fe ratio and its implications; aspects of nucleosynthesis related to supernovae, the s-process and primary nitrogen; and the G-dwarf problem.

1. Introduction

The discussions at this interesting and enjoyable conference have ranged over abundances deduced from observation, star clusters, the relations between halo, bulge and disk(s), star formation, the IMF, stellar evolution and nucleosynthesis and the form and evolution of the Milky Way. My few random comments will cut across these categories.

2. O/α/Fe ratios as functions of metallicity and halo-bulge-disk relations

Garik Israelian presented us with the claim of a *fait accompli* regarding the increase in O/Fe with diminishing metallicity based on OH bands, dismissing the apparent conflict with results by Fulbright & Kraft (1999) based on [OI]. I would not be so sure, nor would I write off the forbidden line as a useful source of information. Better signal-to-noise is required, but can be got, and the effects of different atmospheric parameters and opacity sources in the UV still need to be sorted out. In the meantime, the issue remains undecided.

Independently of that, there are significant patterns in the behaviour of oxygen and α-elements relative to [Fe/H]. As was mentioned in Sean Ryan's review, Nissen & Schuster (1997) studied disk and halo stars with overlapping metallicity, [Fe/H], and found that while thick-disk and some

F. Matteucci and F. Giovannelli (eds.), The Evolution of the Milky Way, 613–618.

halo stars have the same characteristic excess of O/Fe and α/Fe, there is a group of so-called 'anomalous' halo stars with more solar-like abundance ratios, presumably having been formed on a long enough time scale for SNIa to contribute. As Raffaele Gratton mentioned, the young clusters Pal 12 and Ruprecht 106 are similar and Fe/Si can be different in different classical globular clusters with the same Fe/H. These results bring to my mind three issues:

- Is the 'anomaly' discrete or continuous (Ryan)?
- Does its existence constrain low-metallicity inhibition of SNIa (Chiaki Kobayashi)?
- Why do at least some have excess Eu as well as Fe (Jehin et al., mentioned by Pagel)?

More light is cast on the halo and thick and thin disks by the work of Fuhrmann (1998) on Mg; Gratton et al. (1996) have very similar results for oxygen that were shown by Cristina Chiappini. Halo and thick-disk stars within the range of metallicity [Fe/H] ≥ -2 have a uniform Mg and O overabundance also shown by the mildly deficient globular clusters in the bulge described by Sergio Ortolani and Beatriz Barbuy: [Mg/Fe] $\simeq 0.4$ up to Mg/H $>$ solar, which may also apply to a star with solar [Fe/H] in the bulge (McWilliam & Rich 1999). Thus all these stars bear the signature of core-collapse supernovae indicating star formation on short time-scales as discussed by Chiappini and Gerry Gilmore, followed by a period of low star formation where iron but no O or Mg was formed, before the initiation of star formation in the thin disk, as there is a gap in Fe/O and Fe/Mg. The thin disk itself follows evolution according to a longer time-scale, fitting well to the disk models by Pagel & Tautvaišienė (1995), Chiappini, Matteucci & Gratton (1997) and others, but there is also a substantial overlap in metallicity (larger in Mg/H or O/H than in Fe/H) between the thick and thin disks not explained in the current models; this suggests perhaps dilution by infall, following a merger that thickened the disk.

Publication of the Gratton et al. results is eagerly awaited, not least to see whether they throw any light on the O/Fe controversy.

3. Nucleosynthesis theory

There are clearly many uncertainties remaining in the calculation of theoretical yields from supernovae of both main types. I like the title of Friedel Thielemann's talk in the Preliminary Programme: 'SNIa yields (to be confirmed)'! Alessandro Chieffi gave us an excellent clear introduction to the relevant basic physics and sources of uncertainty, and John Danziger revealed the great variety and uncertainty of supernova properties from an observational point of view. It was interesting to hear two quite different interpretations of the manganese abundance in metal-poor stars, which Thielemann was making efforts to get out of SNIa whereas Nikos Prantzos — in agreement with an earlier discussion remark by Ryan — identified it as a consequence of metallicity-dependent yields in SNII.

While on the subject of supernovae, I was interested in Massimo Turatto's update on supernova rates. He had a marginal indication that the ratio of SNIa to core-collapse supernovae increases towards later-type galaxies, but said it was insignificant. This is relevant to the relatively high ratio of Ia/II type remnants found from X-ray observations in the LMC by Hughes et al. (1995), which I have blamed on a recent decline of the SFR in the latest burst appearing in some of the Holtzmann-type diagrams shown by Gilmore whereas others blame it on a steeper IMF (cf. Pagel & Tautvaišienė 1998; Tsujimoto et al. 1995).

We have, in fact, heard a lot about the IMF at this meeting, from Francesco Paresce, Manuela Zoccali, Rosie Wyse and others, and it seems pretty uniform in different systems apart from local fluctuations and pretty flat below $1 M_{\odot}$. Combined with a Salpeter slope at higher masses, this gives a pretty high yield which would be helpful in giving enough metals to intergalactic gas, but may raise problems in modelling chemical evolution in the Milky Way (cf. Pagel 1999).

Regarding the s-process, Oscar Straniero has abolished the elegant fit that used to be obtained for s-process abundances in the Solar System from a simple exponential distribution of neutron exposures, which I regard as a great pity having once described that as arguably the most elegant result in the whole of nucleosynthesis theory (Pagel 1997)! Roberto Gallino discussed the strange predominance of s-process elements in ω Cen; here I take issue with him slightly over the metallicity where s-process sets in among the field stars (see Fig 1): in my view it does so (for whatever reason) between [Fe/H] -2 and -2.5. Ryan reminded us of the differences in behaviour between strontium and barium which are not understood.

4. Mixing and primary nitrogen

André Maeder gave a most beautiful presentation of new work on the effect of rotation on massive stars and the possibility of their producing primary nitrogen. This point has been argued over to and fro ever since Edmunds & Pagel (1978) introduced the idea that variations in N/O between H II regions with a given low O/H might be due to varying effective ages of the underlying stellar population combined with a fixed time delay for primary N production by intermediate-mass stars. More sophisticated versions of this idea have been based on 'bursting' or 'gasping' models of the star formation rate with N/O varying either in course of the burst cycle (Garnett

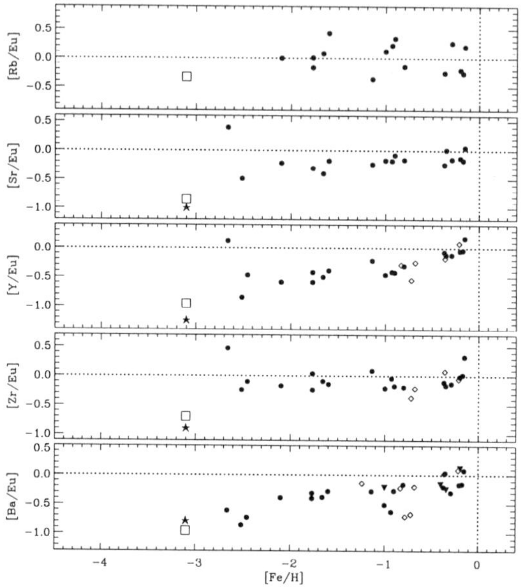

Figure 1. s-process element to europium ratios as a function of metallicity, after Pagel & Tautvaišienė (1997). The empty square gives an estimate of a pure r-process ratio based on solar-system data, whereas the black five-pointed star gives the ratio measured in CS 22892–052 by Sneden et al. (1996).

1990; Pilyugin 1992, 1993, 1999) or as a result of variations in the mass range of stars undergoing third dredge-up (Matteucci & Tosi 1985; Marconi, Matteucci & Tosi 1994).

Now Izotov & Thuan (1999) claim that, among blue compact galaxies with low metallicity, no significant scatter in N/O exists — a result which would favour the idea that their primary nitrogen comes from massive stars without a significant time delay — but there is evidence of real scatter in larger samples (e.g. Kobulnicky & Skillman 1996) and in damped Lyman-α systems (Lu, Sargent & Barlow 1998), so — as discussed by Pilyugin (1999) — the question as to whether primary nitrogen in low-metallicity systems comes mainly from massive or intermediate-mass stars still remains open.

5. The G-dwarf problem

In the solar neighbourhood, as noted by Monica Tosi and Chiappini, the metallicity distribution is nicely accounted for by inflow (see Fig 2), originally suggested on different grounds by Oort (1970) and Larson (1972), and now receiving welcome support from the excellent presentation on the compact high-velocity H I clouds by Butler Burton. In the Galactic bulge, the metallicity distribution function is somewhat broader than in the disk

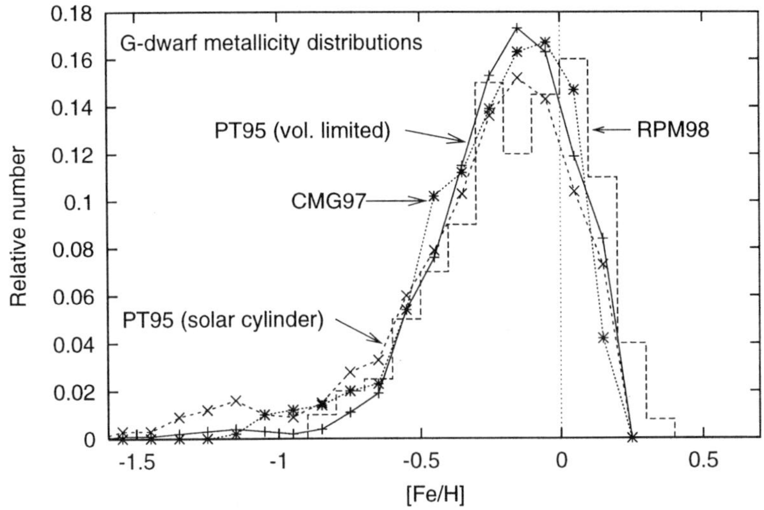

Figure 2. Metallicity distribution function in the solar neighbourhood after Rocha-Pinto & Maciel (1998) (histogram) compared with inflow models by Pagel & Tautvaišienė (1995) (continuous and dashed curves) and by Chiappini, Matteucci & Gratton (1997) (dotted curve).

(McWilliam & Rich 1994) and shifted to higher Z, if not Fe/H, and its relationship with closed-box or other models is not clear. Some sort of model resembling the old 'concentration model' in what Cesare Chiosi described as the obscure but thoughtful paper by Lynden-Bell (1975) may be appropriate here, but Walter Maciel has introduced some complications by deriving an oxygen abundance distribution function from planetary nebulae that fits the stellar one only if one assumes a solar O/Fe ratio, contrary to the results mentioned in Section 1 above. There could be complications in the interpretation of PN spectra, as pointed out by Danziger when referring to recombination lines, but in general the agreement between PN and adjoining H II regions is good.

Elsewhere, as Chiosi has just reminded us, the G-dwarf problem keeps turning up like a bad penny, notably in elliptical galaxies where in general single stellar population models assuming, e.g., solar metallicity with Mg enhanced give a better fit to the integrated spectrum in the UV than models including chemical evolution, at least in a simple way; the same applies to some dwarf systems like Ursa Minor, as reported by Wyse. These results are a humbling reminder of how little we still know about cosmic chemical evolution; indeed, about 90 per cent of baryonic matter in the universe is dark and its chemical composition unknown!

It remains to thank Franco Giovanelli and Francesca Matteucci and their assistants and colleagues for organizing this magnificent conference, not forgetting some great snorkelling!

References

1. Chiappini, C., Matteucci, F. & Gratton, R. 1997, *Ap. J*, **477**, 765
2. Edmunds, M.G. & Pagel, B.E.J. 1978, *MNRAS*, **185**, 77P
3. Fuhrmann, K. 1998, *A & A*, **338**, 161
4. Fulbright, J.P. & Kraft, R.P. 1999, *Astr. J.*, **118**, 527
5. Garnett, D.R. 1990, *Ap. J.*, **363**, 142
6. Gratton, R., Carretta, E., Matteucci, F. & Sneden, C. 1996, preprint
7. Hughes, J.P., Hayashi I., Helfand D., Hwang U., Itoh M., Kirshner R., Koyama K., Markert T., Tsunemi H., Woo J. 1995, *Ap. J. Lett.*, **444**, L81
8. Izotov, Y.I. & Thuan, T.X. 1999, *Ap. J.*, **511**, 639
9. Kobulnicky, H.A. & Skillman, E.D. 1996, *Ap. J.*, **471**, 211
10. Larson, R.B. 1972, *Nature Phys. Sci.*, **236**, 7
11. Lu, L., Sargent, W.L.W. & Barlow, T.A. 1998, *Ap. J.*, **115**, 55
12. Lynden-Bell, D. 1975, *Vistas in Astronomy*, **19**, 299
13. Marconi, G., Matteucci, F. & Tosi, M. 1994, *MNRAS*, **270**, 35
14. Matteucci, F. & Tosi, M. 1985, *MNRAS*, **217**, 391
15. McWilliam, A. & Rich, R.M. 1994, *Ap. J. Suppl,*, **91**, 749
16. McWilliam, A. & Rich, R.M. 1999, in *Chemical Evolution from Zero to High Redshift*, J. Walsh & M. Rosa (eds.), Springer-Verlag,, p. 73
17. Nissen, P.E. & Schuster, W.A. 1997, *A & A*, **326**, 751
18. Oort, J.H. 1970, *A & A*, **7**, 381
19. Pagel, B.E.J. 1997, *Nucleosynthesis and Chemical Evolution of Galaxies*
20. Pagel, B.E.J. 1999, in Ringberg Workshop: *Galaxies in the Young Universe II*, Hans Hippelein (ed.), Springer-Verlag, in press; astro-ph 9911204
21. Pagel, B.E.J. & Tautvaišienė, G. 1995, *MNRAS*, **276**, 505
22. Pagel, B.E.J. & Tautvaišienė, G. 1997, *MNRAS*, **276**, 505
23. Pagel, B.E.J. & Tautvaišienė, G. 1998, *MNRAS*, **288**, 108
24. Pilyugin, L.S. 1992, *A & A*, **260**, 58
25. Pilyugin, L.S. 1993, *A & A*, **277**, 42
26. Pilyugin, L.S. 1999, *A & A*, in press; astro-ph 9904157
27. Rocha-Pinto, H.J. & Maciel, W. 1998, *A & A*, **339**, 791
28. Sneden, C., McWilliam, A., Preston, G.W., Cowan, J.J., Burris, D.L. & Armosky, B.J. 1996, *Ap. J.*, **467**, 819
29. Tsujimoto T., Nomoto K., Yoshii Y., Hashimoto M., Yanagida S., Thielemann F.-K. 1995, *MNRAS*, **277**, 945

CONCLUDING ADDRESS

F. GIOVANNELLI[1]
Istituto di Astrofisica Spaziale, CNR - Area di Ricerca di Tor Vergata
Via del Fosso del Cavaliere 100 - I 00133 Roma, Italy

1. Some comments

I do not want to make some concluding remarks for two reasons: the first because we have listen to interesting concluding remarks presented by experts that Francesca and I invited for this purpose; the second because I am not directly involved in the specific field of the chemical evolution of the Galaxy and then I could add only marginal remarks.

However, before to conclude officially this workshop, I would like to express few personal comments coming out from a fruitful week of discussions on the chemical evolution of the Galaxy.

- I think that the necessity of such a workshop has been clearly demonstrated during the different sessions, full of interesting updated papers presented by most of participants. With their presentations and following discussions, all participants rendered actual and alive this workshop. Many thanks to everybody.
- I think that we have to be careful in accepting too many models, without a critic evaluation of them through experimental data. Figure 1 contains actually my synthetic conclusions, which I can translate, in the caption, for the benefit of almost the total number of participants.

Finally, I would like to express on behalf of all members of the Scientific Committee many thanks to the participants for the excellent level of their talks and comments. Many thanks also to the Local Organizing Committee that has been able to conduct and smooth the workhop over. On behalf of the Local Organizing Committee, I wish to thank once more the participants for their good attitude in attending the workshop and for facilitating the it not easy work of the organizers. Special thanks to Mr. Massimo Frutti (Istituto di Astrofisica Spaziale, CNR), Mr. Dario Chiummariello (Dipartimento di Astronomia, Universit/'a di Trieste), and to the Angel of the

F. Matteucci and F. Giovannelli (eds.), The Evolution of the Milky Way, 619–620.

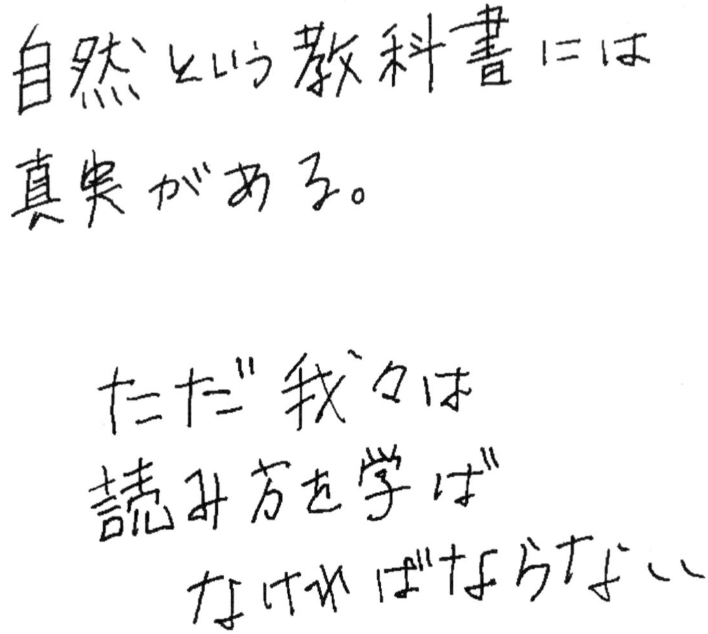

Figure 1. The truth is written in the book of the Nature. We must learn to read this book.

Question-Answer forms: Dr. Simonetta Fabrizio (Osservatorio Astronomico di Trieste).

I could say more words, but I sincerely think that the moment to close this workshop has actually arrived. Thanks a lot to everybody.

The workshop is closed, see you next time!

List of First Authors

Barbuy, B.: *Abundances of globular clusters in the Galactic bulge*

Bellazzini, M.: *HST-WFPC2 photometry of the globular cluster NGC288: bynary systems, blue stragglers and very blue stars*

Bono, G.: *Radial variables as tracers of stellar populations*

Bragaglia, A.: *Old open clusters as tracers of Galactic evolution*

Burton, W.B.: *Compact high-velocity clouds at high resolutions*

Carraro, G.: *The age of the Galactic disk*

Carretta, E.: *Mixing along the red giant branch in metal-poor field stars*

Cassisi, S.: *The effect of external pollution on the evolution of low-mass metal-free stars*

Chiappini, C.: *The effects of a variable IMF on the chemical evolution of the Galaxy*

Chieffi, A.: *The explosive yields of massive stars*

Chiosi, C.: *Has the IMF of EG's always been the same?*

Danziger, I.J.: *Abundances in supernovae*

Della Valle, M.: *The role of classical novae in the galactic nucleosynthesis*

Dominguez, I.: *The evolution of zero metal intermediate mass stars*

Feast, M.: *Can Mira variables tell us the chemical abundances in stellar systems?*

Ferraro, F.R.: *HST-UV observations of stellar clusters: looking for BSS*

Gallino, R.: *Evolution of heavy elements in* ω *Cen: the puzzle of s-enrichment*

Giovannelli, F.: *Concluding address*

Giovannelli, F.: *The origin of X-Ray pulsars in binary systems*

Girardi. L.: *Clump stars in the solar neighbourhood*

Gratton, R.G.: *Metal abundances in globular clusters*

Grenon, M.: *The origin and kinematical evolution of SMR stars*

Hernandez, X.: *Reconstructing the star formation history of the Galaxy*

Israelian, G.: *Early Galactic evolution of carbon, nitrogen and oxygen*

Kobayashi, C.: *Metallicity effects on type Ia supernovae and galactic and cosmic chemical evolution*

Limongi, M.: *Explosive nucleosynthesis of massive stars. The effect of the metallicity*

Maciel, W.J.: *Planetary nebulae: abundances and abundance gradients*

Maeder, A.: *Stellar evolution with rotation*

Malagnini, M.L.: *Observations of candidate SMR stars*

Maraston, C.: *Galactic globular clusters as calibrators of synthetic line indices*

Marigo, P.: *Chemical yields from low- and intermediate-mass stars*

Marconi, M.: *The metal abundance of Galactic cepheis as derived from their pulsational properties*

Matteucci, F.: *The chemical evolution of the Milky Way: some important facts*

Origlia, L.: *Infrared observations of stellar clusters*

Ortolani, S.: *The globular clusters and field stars in the Galactic bulge*

Pagel, B.E.J.: *Concluding remarks*

Pagel, B.E.J.: *S/α/Fe abundance ratios in halo field stars: is there a globular cluster connection?*

Panagia, N.: *Observational constraints to the evolution of massive stars*

Paresce, F.: *On the globular cluster IMF below 1 $M_\odot$*

Phelps, R.L.: *Abundances in open star clusters: what we have to work with*

Piotto, G.: *Galactic globular cluster relative ages: clues on the Milky Way early evolution*

Portinari, L.: *On the abundance gradients in the Galactic disk*

Prantzos, N.: *The evolution of the Milky Way disk*

Pulone, L.: *Infrared observations of globular clusters*

Ripepi, V.: *Absolute magnitude of RR Lyrae in globular clusters from pulsational properties*

Romano, D.: *The influence of nova nucleosynthesis on the chemical evolution of the Galaxy*

Rood, R.T.: *The effect of CNO and enhanced α-elements on globular cluster stars*

Ryan, S.G.: *Abundance ratios in Galactic metal-poor stars*

Smartt, S.J.: *Abundance gradients along the Galactic disk*

Sommer-Larsen, J.: *Formation of disk galaxies: on the angular momentum problem, the Tully-Fisher relation and magnetohydrodynamics*

Stanghellini, L.: *Planetary nebulae as probes of stellar evolution and populations*

Straniero, O.: *Asymptotic giant branch stars*

Thielemann. F.-K.: *Yields from type Ia supernovae*

Thomas, D.: *Stellar yields and chemical evolution*

Tosi, M.: *The chemical evolution of the Milky Way*

Tsujimoto, T.: *Inhomogeneous chemical evolution of the Galactic halo*

Turatto, M.: *The present rate of supernovae*

Vanbeveren, D.: *The evolution of binaries*

Van der Hucht, K.A.: *Wolf-Rayet stars, binaries, black holes*

Ventura, P.: *AGB stars new models: lithium production in the Galaxy and Magellanic Clouds*

Wyse, R.F.G.: *Faint stars in the Ursa Minor dwarf spheroidal galaxy: implications for the stellar IMF*

Zoccali, M.: *The initial mass function of the Galactic bulge down to ~ 0.15 $M_\odot$*

Author Index